Environmental Fluid Mechanics

IAHR Monograph

Series editor

Peter A. Davies
Department of Civil Engineering,
The University of Dundee,
Dundee,
United Kingdom

The International Association for Hydro-Environment Engineering and Research (IAHR), founded in 1935, is a worldwide independent organisation of engineers and water specialists working in fields related to hydraulics and its practical application. Activities range from river and maritime hydraulics to water resources development and eco-hydraulics, through to ice engineering, hydroinformatics and continuing education and training. IAHR stimulates and promotes both research and its application, and, by doing so, strives to contribute to sustainable development, the optimisation of world water resources management and industrial flow processes. IAHR accomplishes its goals by a wide variety of member activities including: the establishment of working groups, congresses, specialty conferences, workshops, short courses; the commissioning and publication of journals, monographs and edited conference proceedings; involvement in international programmes such as UNESCO, WMO, IDNDR, GWP, ICSU, The World Water Forum; and by co-operation with other water-related (inter)national organisations. www.iahr.org

IAHR AIRH

Supported by

CEDEX

Environmental Fluid Mechanics

Memorial Volume in honour of Prof. Gerhard H. Jirka

Editors

Wolfgang Rodi & Markus Uhlmann

Institute for Hydromechanics, Karlsruhe Institute of Technology (KIT), Karlsruhe Germany

CRC Press
Taylor & Francis Group
Boca Raton London New York Leiden

CRC Press is an imprint of the
Taylor & Francis Group, an **informa** business
A BALKEMA BOOK

Cover information:
"Vortical flow structures induced by a jet at Reynolds number 6500 discharging into a closed container. Laboratory model visualization by planar laser-induced fluorescence. Image by M. Vaas and M. Lenz."

CRC Press/Balkema is an imprint of the Taylor & Francis Group, an informa business

Typeset by V Publishing Solutions Pvt Ltd, Chennai, India
Printed and Bound by CPI Group (UK) Ltd, Croydon, CR0 4YY

Published by: CRC Press/Balkema
P.O. Box 447, 2300 AK Leiden, The Netherlands
e-mail: Pub.NL@taylorandfrancis.com
www.crcpress.com – www.taylorandfrancis.com

Library of Congress Cataloging-in-Publication Data

Applied for

ISBN: 978-0-415-67045-6 (Hbk)
ISBN: 978-0-203-80396-7 (eBook)

About the IAHR Book Series

An important function of any large international organisation representing the research, educational and practical components of its wide and varied membership is to disseminate the best elements of its discipline through learned works, specialised research publications and timely reviews. IAHR is particularly well-served in this regard by its flagship journals and by the extensive and wide body of substantive historical and reflective books that have been published through its auspices over the years. The IAHR Book Series is an initiative of IAHR, in partnership with CRC Press/Balkema – Taylor & Francis Group, aimed at presenting the state-of-the-art in themes relating to all areas of hydro-environment engineering and research.

The Book Series will assist researchers and professionals working in research and practice by bridging the knowledge gap and by improving knowledge transfer among groups involved in research, education and development. This Book Series includes Design Manuals and Monographs. The Design Manuals contain practical works, theory applied to practice based on multi-authors' work; the Monographs cover reference works, theoretical and state of the art works.

The first and one of the most successful IAHR publications was the influential book "*Turbulence Models and their Application in Hydraulics*" byW. Rodi, first published in 1984 by Balkema. I. Nezu's book "*Turbulence in Open Channel Flows*", also published by Balkema (in 1993), had an important impact on the field and, during the period 2000–2010, further authoritative texts (published directly by IAHR) included *Fluvial Hydraulics* by S. Yalin and A. Da Silva and *Hydraulicians in Europe* by W. Hager. All of these publications continue to strengthen the reach of IAHR and to serve as important intellectual reference points for the Association.

Since 2011, the Book Series is once again a partnership between CRC Press/Balkema – Taylor & Francis Group and the Technical Committees of IAHR and I look forward to helping bring to the global hydro-environment engineering and research community an exciting set of reference books that showcase the expertise within IAHR.

Peter A. Davies
University of Dundee, UK
(Series Editor)

Table of contents

Preface

This book contains written versions of papers presented at the Gerhard H. Jirka Memorial Colloquium on Environmental Fluid Mechanics held June 3–4, 2011, at the Karlsruhe Institute of Technology (KIT) in Karlsruhe, Germany. The colloquium was organized to honour and commemorate Prof. Gerhard H. Jirka who was one of the most outstanding scientists of our days in the area of Environmental Fluid Mechanics. 23 world leading scientists in the field accepted the invitation to present keynote lectures at the colloquium and paid tribute to Prof. Jirka by reporting on their recent work.

Environmental Fluid Mechanics represents the scientific and applied study of fluid motion and its implications on the transport and distribution of mass and heat in natural and man-affected systems in the environment. Due to the growing awareness of and concern about the state of our environment and the strong influence of flow processes on this, the discipline is of prime topicality and importance. The contributions in this book are restricted to the hydrosphere as environment because this was the working area of Prof. Jirka and, naturally, also the colleagues paying tribute to him as authors work in this area. However, even this subarea is very rich in different complex flow-related phenomena, which is reflected by the wide range of contributions to this book. These provide information on recent developments in the following key areas of Environmental Fluid Mechanics: Shallow Flows (5), Fluvial Hydraulics (5), Small-Scale Phenomena (3), Jets and Stratified Flows (3), Gravity Currents (4), and Mass Transfer (3) and include experimental, theoretical and numerical studies. The numbers in parentheses indicate the number of contributions in each area. Heading off the book is a paper by former co-workers of Prof. Jirka, summarizing his scientific work and achievements, and this paper forms an essential element of the memorial volume. The CD-ROM accompanying this book contains digital versions of all contributions, most in full colour.

All contributions to this book are invited papers and full responsibility for the contents rests with the authors. We should like to thank all of them for their efforts in preparing their contributions which commemorate in an excellent way the life and work of Gerhard H. Jirka.

We are also grateful to the Deutsche Forschungsgemeinschaft (DFG) and the Karlsruhe Institute of Technology (KIT) for sponsoring the Memorial Colloquium without which this book would not have been produced. Finally, we thank Alistair Bright and Lukas Goosen of CRC Press/Balkema for the good cooperation in the preparation of this book.

Wolfgang Rodi
Markus Uhlmann

Gerhard H. Jirka 1944–2010

Professor Gerhard H. Jirka was one of the leading scientists in hydraulic and environmental engineering of our days. Born on September 14, 1944, in Kasten, Austria, he studied at the Agricultural University of Vienna, from where he received his Diploma in 1969. He then moved to the US becoming a research assistant at the Massachusetts Institute of Technology (MIT), where he pursued studies in water resource systems and hydrodynamics and earned his Ph.D. in 1973. He stayed at MIT as research engineer and lecturer until 1977 when taking up a faculty position at Cornell University, and was first assistant, then associate and from 1987 full professor. In 1984 he founded at Cornell the *DeFrees* Hydraulics Laboratory specializing in environmental fluid mechanics research and became its first director. In 1995 he accepted an offer to become a chair holder at and director of the Institute for Hydromechanics at the University of Karlsruhe, now Karlsruhe Institute of Technology (KIT), Germany. This position he held until his retirement in September 2009. He there provided excellent leadership to the laboratory he directed and had considerable impact on modernizing the education of civil engineers. From 2008 he was also Associate Director of the Centre for Climate and Environment at KIT and, also beyond his retirement, provided vision and guidance to KIT in forming an Excellence Centre for water research.

For more than 30 years Gerhard Jirka made consistently important and lasting contributions to many areas of hydraulic and environmental engineering, some of which will be reviewed and summarized in the first contribution to this Memorial Volume. His research always covered a wide range from fundamental studies to the development of engineering methods and hence had both great scientific and practical impact. Gerhard Jirka was a prolific writer, disseminating the results of his research and his knowledge through some 250 publications, many of them in the most renowned journals, and also many providing the primary source of information on a number of important topics, which will remain standard references on these. Through this he earned himself worldwide recognition as one of the top experts in the field of hydraulic and environmental engineering as is manifested by prestigious awards he received, among them the Freeman Hydraulics Prize and the Walter L. Huber

Civil Engineering Research Prize as well as the Hunter Rouse Hydraulic Engineering Lecture Award, all from ASCE, the Arthur T. Ippen Award of IAHR and a membership of the Academy of Sciences of Argentina, to name only a few. He was often invited as lecturer, as visiting scientist and as a member of international expert panels such as the one for the storm flood barrier of the Venice Lagoon.

During all his professional life, Gerhard Jirka provided extensive and valuable service to the Hydraulic and Environmental Engineering Community. He served on various committees of ASCE and was the chairman of the Hydraulics Division of ASCE from 1989 to 1990. In Germany, he was an influential member of various hydraulics committees and advised the German Research Foundation (DFG) on their strategies on water research. He was closely associated with and particularly active in IAHR, where he served on various committees and was the chairman of the Fluid Mechanics Committee from 1990 to 1996. He helped to set up the European Graduate School – Environment Water (EGW) and coordinated a number of Summer Schools in the field of Environmental Fluid Mechanics. From 2001 to 2009 he was a member of the IAHR Council and from 2005 to 2009 a Vice-President of the IAHR. In this function he chaired its Structure Change Task Force and was instrumental in introducing a new, modern structure in IAHR and a new name that represents better IAHR's environmental activities.

Gerhard H. Jirka passed away unexpectedly and far too early on February 14, 2010. He will be remembered not only for his great scientific work and professional engagement, but also as a fine and interesting human being who managed to transfer his vitality, enthusiasm and optimism to all who were around him. He is sadly missed by the scientific community and by his friends and family.

Chapter 1

Mixing and transport processes in environmental fluid systems: Gerhard Jirka's scientific achievements

Tobias Bleninger[1], H. Herlina[2], Volker Weitbrecht[3], and Scott Socolofsky[4]

[1] *Department of Environmental Engineering (DEA), Federal University of Paraná (UFPR), Curitiba, Brazil*

[2] *Institute for Hydromechanics (IfH), Karlsruhe Institute of Technology (KIT), Karlsruhe, Germany*

[3] *Laboratory of Hydraulics, Hydrology and Glaciology (VAW), Swiss Federal Institute of Technology (ETH Zurich), Zurich, Switzerland*

[4] *Coastal and Ocean Engineering Division, Zachry Department of Civil Engineering, Texas A&M University, College Station, USA*

ABSTRACT: This paper highlights some of the scientific and engineering contributions of Prof. Gerhard H. Jirka. In broad terms, his life's work related to contributions to and the formation of the field of Environmental Fluid Mechanics. We review three major areas of his achievements here, beginning with his work on jets and plumes and their related integral methods, resulting in the development of the CORMIX model for evaluating the fluid mechanical performance and dilution characteristics of pollutant and waste heat discharges. Second, his studies on gas transfer at the air-water interface are discussed, demonstrating his ability to combine findings from experiments with theory and to cover a wide range of topics, from fundamental studies of the detailed phenomena to the development of methods for solving practical engineering problems. A third area strongly influenced by his recent work is shallow flows, wherefrom significant results on shallow turbulent wake flows, shallow vortex dynamics, and groyne field flows will be presented here as a few examples.

1 INTRODUCTION

Since the beginning of his career in Vienna, Austria, 1969, where Gerhard Jirka received his undergraduate diploma, he was interested in combining hydraulic engineering with transport phenomena, oriented to environmental problems. He made many important and lasting contributions in that regard, considerably forming the field of Environmental Fluid Mechanics. Gerhard Jirka advanced this field as researcher, lecturer, and as associate editor of the journal *Environmental Fluid Mechanics* using the definition: "Environmental fluid mechanics is concerned with the fluid motions and associated mass and heat transport processes that occur in the earth's hydrosphere and atmosphere on local or regional scales. A particular emphasis within these scales - and in contrast to the yet larger domain of Geophysical Fluid Dynamics - is the influence of these flows on and their interaction with man-made facilities and structures and their response to anthropogenic releases of mass and heat" (Socolofsky and Jirka, 2005). Based on that definition, Gerhard Jirka initiated, already at Cornell University (beginning 1984) and continuing (since 1995) at the Institute for Hydromechanics (IfH) at the Karlsruhe Institute of Technology (KIT, former University of Karlsruhe), the shift from research on technical flows and the related mainly empirical descriptions of mixing and transport processes towards mixing and transport in environmental flows described by process-based parameterizations and flow classifications. He established the research division "Environmental Fluid Mechanics" at the IfH that was not only responsible for a variety of research projects but also provided three regular classes on Environmental Fluid Mechanics I – III (mixing and transport, stratified flows, and modeling applications). In the framework of the European Engineering Graduate School-Environment Water he initiated a very successful series of summer schools under the title "Environmental Fluid Mechanics: Theory, Experiments, and Applications." He further introduced the German translation "Umweltfluidmechanik," and proposed a working group with that name within the German Association for Water, Wastewater and Waste (DWA). Besides the formalization, and integration of that research field in Europe and the further scientific development on the international level, he contributed significantly to modify academic curricula accordingly, combining engineering education with a solid base in chemistry and biology, oriented to environmental problems, which now can be found in many Civil and Environmental Engineering departments.

The objective of this paper is to present key contributions of his work and their subsequent application in engineering and benefit to society, and to provide a framework for his publications and activities, allowing the reader to understand and experience Gerhard Jirka's passion and enthusiasm for Environmental Fluid Mechanics problems in science and engineering.

He is not only known for his scientific papers, but also for his summary and review articles that present deep interpretation and context within the field. He was intelligently interpreting and integrating different results, using flow classifications schemes, stability criteria, and mathematical models resulting in simple but effective tools for evaluating hydraulic performance or mixing characteristics of a wide range of environmental fluid systems. His major research activities were in the following fields:

Buoyant Jets and Plumes, distinguishing between pollutant and heat transport and classifying the related flow phenomena, resulting in the development of integral models and the CORMIX expert system as well as applied studies related to the mechanics of submerged multiport diffusers (see Section 2).

Gas Transfer at the Water-Air Interface, providing a basis for an in-depth understanding of the gas transfer process associated with near-surface turbulence phenomena through detailed experimental studies as well as proposing empirical relations for solving practical problems (see Section 3).

Stratified flows, with early fundamental works on the selective withdrawal from stratified fluid layers (Jirka and Katavolab, 1997) and density currents and wedges (Jirka and Arita, 1987, Arita and Jirka, 1987) and recent findings related to the interfacial mixing phenomena of two-layer exchange flows (Negretti *et al.*, 2007, 2008a), and instability criteria (Negretti *et al.*, 2008b), mainly applying experimental and analytical methods.

Dynamics of cooling lakes and reservoirs, resulting in a predictive classification of cooling lakes (Jirka and Harleman, 1979a) based on a fluid mechanical analysis of the stratifying and destratifying mechanism (Jirka and Watanabe, 1980).

Transport phenomena at the sediment-water interface, providing fundamental information of the interaction of pressure and velocity fluctuations within gravel beds (Detert *et al.*, 2007), and developing a novel experimental design for a differential turbulence column to study the particle interaction under open channel turbulence conditions without advective flow (Kühn, 2008).

Shallow flows, contributing an understanding of their stability, the transport characteristics of large coherent structures within the flow, and the significance of the vertical confinement, which both modulates the turbulence and results in bottom friction playing a dominant role on their behavior, and summarizing this insight through correlation with the friction parameter (see Section 4).

It was Gerhard Jirka's childlike enthusiasm for fluid mechanical phenomena in the environment, starting from internal waves in a Latte Macchiato up to how to ship through vortex formations at river confluences, combined with his persistence and curiosity, which enabled him and motivated his students and co-workers to realize break-through experiments, and to analyze, explain and apply the results with optimal simplicity, always open and demanding for criticism and further improvements.

As already indicated in the above list, the remaining sections will cover three main topics, starting with his early works on jets and plumes, followed by Gerhard Jirka's achievements on gas transfer at the water-air interface, and closing with his contributions to mixing and transport in shallow flows. The authors, all former and recent Ph.D. and postdoctoral students of Gerhard Jirka, were closely working with him on these topics in the last 10 years, and are honored to be invited to provide this review.

2 JETS, PLUMES, AND OUTFALLS

Buoyant jet phenomena exist whenever fluid is discharged in environmental, engineering or industrial processes, generally with combinations of excess (or deficit) in momentum and buoyancy. Typical examples are pollutant spills, wastewater outfalls or smokestacks, and injection devices in treatment facilities or reactors. There are hundreds of papers published on experiments and analysis of jets and plumes; however, most of them consider discharges into unlimited ambients, directly at boundaries (wall jets), or in strongly confined ambients (reactors). One of Gerhard Jirka's greatest achievements was his contribution in bringing these studies together, associating the different flow regimes and discharge configurations to specific classes, and providing additional experimental studies on stability and interaction of jets in a weakly confined ambient, such as in shallow environmental flows.

Gerhard Jirka started his research career in 1969 studying jets and plumes at the Massachusetts Institute of Technology (MIT), working under Professor Harleman. At that time, numerous power plants were to be built in the United States (US), and new environmental laws stemming from the U.S. Clean Air and Clean Water Acts required more predictive methods than previously used for the permits required for contaminant and waste heat discharges. Though individual jet and plume phenomena had been studied intensively already by that time, there was no combined or integrated model available to incorporate those different phenomena. In his thesis work, Gerhard Jirka established the fundamental differences in the type of mixing characteristics between high-buoyancy diffusers (e.g., sewage) and low-buoyancy ones (e.g., cooling water), which results in completely different design strategies. He presented this work in Jirka (1982a), which was awarded the Freeman Hydraulics Prize of the American Society of Civil Engineers and forms the basis of his approach to flow classification. This practical approach not only compared initial flow characteristics between each other and for different discharge types, but also the subsequent dominant flow features. Furthermore, he incorporated boundary interaction processes by defining quantitative criteria for the stability of near-field mixing processes in constrained water bodies, distinguishing between a stable and unstable near-field. In the stable near-field, the jet entrains mainly "clean" ambient water; thus, the near-field jet entrainment processes have a negligible effect on the collapsed, established waste field in the intermediate and far-field region, where source induced effects disappear. In contrast, the near-field is considered as unstable when the jet entrains polluted water that is when the jet mixes with its own fluid, thereby creating a pollutant or heat build-up.

A first breakthrough summary paper in this area was his contribution on "Turbulent Buoyant Jets in Shallow Fluid Layers" (Jirka, 1982b) in the book *Turbulent Buoyant Jets and Plumes*, where he showed his ability to fully exploit and synthesize existing individual paper contributions to obtain new, fundamental insights. A second methodology review was provided in 1986 (Chu and Jirka, 1986) on "Buoyant Surface Jets and Plumes in Environmental Fluid Mechanics" in the *Encyclopedia of Fluid Mechanics*.

Applications of his approaches using flow classification and stability criteria in applied predictive models for cooling water diffusers demonstrated them to be intrinsically different from the jet mixing regimes which had been studied in the sewage discharge field. This was largely a result of the differing density and flow rate of the discharges. Thus, the design of most thermal power plant diffusers constructed in the US and elsewhere in the 1970s made use of these criteria and predictive techniques. One particular diffuser design of Gerhard Jirka's which has a particularly high mixing efficiency was, in fact, patented (Jirka and Harleman, 1973, 1979b).

After his Ph.D. study and the subsequent work on waste heat disposal and hydraulic model studies for multiport diffuser design carried out at the Ralph M. Parsons Laboratory for Water Resources and Hydrodynamics at MIT, he widened the applications of his models, covering the full range of phenomena occurring in different discharge situations. These were conventional single-port discharges, efficient multiport diffusers, and free surface channel discharges, with either positively or negatively buoyant effluents discharging in stagnant or cross-flowing shallow flows with uniform or stratified density distribution. Gerhard Jirka hereby initiated, coordinated, and executed several experimental studies on these topics, the results of which led to his well known flow classification schemes, the modeling framework CORMIX, and applications for outfall systems as described in the following subsections.

2.1 Experiments and flow classification

Gerhard Jirka's analysis was initially focused on the near-field and transitional region, where source-induced characteristics dominate and ambient turbulence or variability is negligible. In these regions, steady-flow and spatial averages are often sufficient to describe the effect of the ambient on the discharge flow field. Thus, mean depths, and widths, together with (cross-sectional) mean flow velocities, and temporal mean density profiles are the only parameters needed in such studies to describe the receiving water body. The discharge is usually described by its coordinates and geometry in the receiving waters, and its flux variables in kinematic form, namely, the volume flux $Q_o = U_o a_o$, with the cross-sectional (a_o) average velocity U_o, the momentum flux, $M_o = U^2_o a_o = U_o Q_o$, and the buoyancy flux $J_o = U_o g'_o a_o = Q_o g'_o$, with the reduced gravity $g'_o = g(\rho_a - \rho_o)/\rho_a$, and the ambient and discharge density, ρ_a, and ρ_o respectively. The pollutant mass flux can be described using the initial pollutant concentration C_o through $\dot{M}_{C,O} = U_O C_O a_O = Q_O C_O$.

Gerhard Jirka's idea behind flow classification uses these variables to apply two principles. The first one is to start with an order-of-magnitude analysis in order to distinguish between typical, realistic discharge types and to establish the limitations of his approximations. As example, in Jirka (1982a) the differences between wastewater and cooling water discharges, and the effect and limitations of modeling approaches for each were determined. Second, he employs a length-scale analysis as described in Fischer *et al.* (1979), comparing discharge length scales with receiving water length scales (see Figure 1), such as the average water depth H, width B, and average ambient velocity u_a. The length scales used were the discharge length scale $L_Q = Q_0/\sqrt{M_0}$, the momentum length scale $L_M = M_0^{3/4}/\sqrt{J_0}$, which denotes a scaling for the transition from jet to plume behavior in a stagnant ambient (note that the product of the discharge pipe diameter D with the densimetric Froude number $F_o = U_0/\sqrt{g'D}$ is proportional to L_M), $L_m = M^{1/2}/u_a$ the jet to crossflow length scale, which denotes a scaling for the distance of transverse jet penetration beyond which strong deflection by the crossflow u_a occurs, and $L_b = J_0/u_a^3$ the plume to crossflow length scale, which denotes a scaling for the distance of plume penetration beyond which strong deflection by the crossflow occurs. This approach allows for application of several different models designed for different flow classes to be applied to a large set of processes, from the jet exit to the regions after the jets or plumes impinged with usually one or two boundaries (surface, bottom, river bank, or coastline), thus beyond the near-field. This has an important practical applicability, as most water quality regulations require information throughout the near field and about these boundary interactions.

Gerhard Jirka's contributions to such flow descriptions and classifications can be divided into the following areas:

- *Experiments and resulting Flow Classifications* based on non-dimensional or length scale analysis for positively buoyant (Jirka and Akar, 1991) or negatively buoyant single jets (Bleninger *et al.*, 2010a) and multiport diffusers (Jirka and Doneker, 1991)
- *Processes at Boundaries or for Asymptotic Regimes*, such as buoyant surface or interface spreading, upstream spreading, and plume trapping for positively buoyant jets (Akar and Jirka, 1994, 1995, Jirka and Fong, 1981), density currents

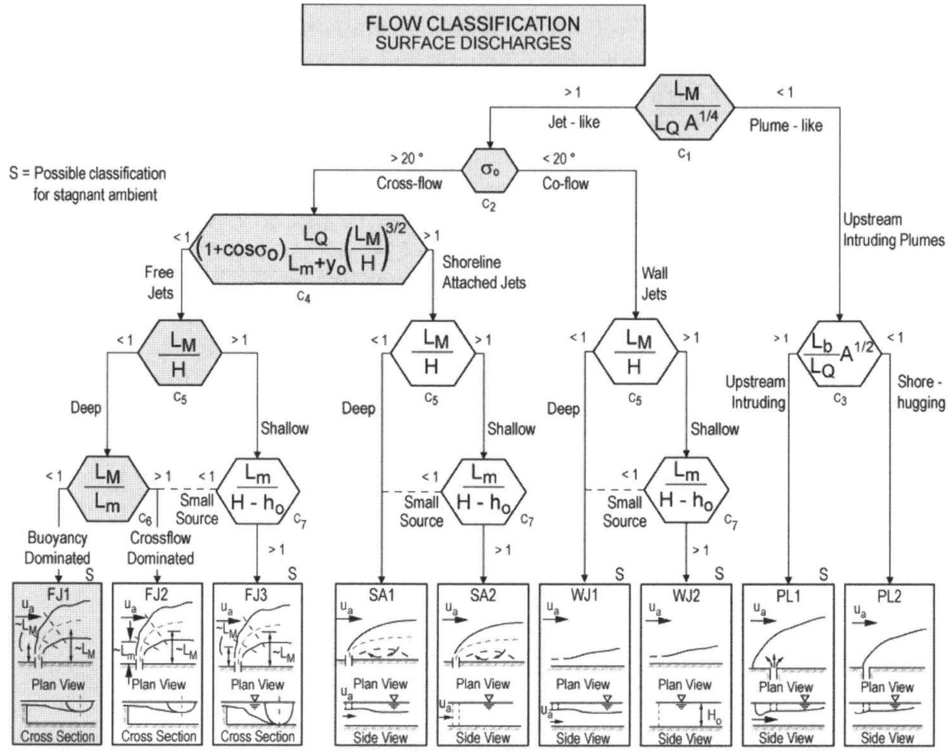

Figure 1 Flow classification tree for positively buoyant surface discharge into uniform density layer (Source: www.cormix.info, modified from Jones *et al.*, 2007).

for negatively buoyant jet discharges (Jirka and Arita, 1987), jets in shallow (Jirka, 1982a, 1994, Lee and Jirka, 1981) and jets in confined flows (Jirka and Harleman, 1973, 1979b)

- *Modeling approaches for these transitional regimes*, such as boundary interaction models for dense discharges (Doneker *et al.*, 2004) or near-field, far-field transition models (Bleninger and Jirka, 2004, Bleninger, 2006) or simplified far-field water quality related models (Holley and Jirka, 1986, Jirka and Lee, 1994).

For many of these studies Gerhard Jirka conducted his own experiments, and always made use of numerous experimental results published by others to improve and validate the parameterization of his classification schemes.

The resulting classification schemes are not only the fundamental base of Gerhard Jirka's modeling approach implemented in the CORMIX modeling system (see below), but they indeed have a value of their own which unfortunately is still underused. Figure 1 illustrates one out of more than 10 classification trees developed by Gerhard Jirka and colleagues. The application of these classification schemes requires only very few and general input data related to the discharge and ambient characteristics. The simple quantitative comparison of the mixing length scales with the geometrical scales of the discharge and receiving waters is possible by following the guidance given in the classification trees, and the quantitative criteria defined by Gerhard Jirka and his co-workers. Thus, a discharge screening analysis can be done with these classification trees alone, without running a model at all.

For example, sensitive coastlines or river banks may require that no plume contact with the bank occurs in the near-field, downstream vicinity. For a surface discharge with given buoyancy and momentum flux, only flow classes of type FJ (free jet, see Figure 1) are in compliance with such a water quality protection criterion. The other flow classes will have downstream bank contact in the near-field region. Figure 1 shows that for the given example (grey shaded boxes) the discharge seems to be in compliance, predicting an FJ1 flow class. Furthermore it can be seen that buoyancy will keep the plume stratified in the surface layer (plan view in lowest shaded box), with no bed contact in the near-field region, in contrast to the situation in class FJ3 with bed contact. Flow classification was Gerhard Jirka's passion, and provided several characteristic non-dimensional flow type indicators, leading to flow classification trees used in several modeling approaches.

2.2 CORMIX discharge model

Gerhard Jirka's practical design work and scientific contributions showed that the implementation of models in environmental impact assessments requires a framework for the conglomerate of all existing, often competing and sometimes misleading mixing equations and predictive methods. The US Environmental Protection Agency (USEPA) contracted Gerhard Jirka in 1988 to develop an expert system fulfilling the objective to provide a scientifically correct, but practical and efficient discharge modeling system, especially considering boundary interactions, which were lacking in existing models at that time. This work was accomplished by Gerhard Jirka during his period at Cornell University, resulting in the Cornell Mixing Zone Expert System, CORMIX (www.cormix.info), which had been modified, updated, and maintained by Gerhard Jirka from that time until his passing, and is still a powerful and popular tool for analyzing practical problems associated with thermal and pollutant discharges, continuing its life though maintenance and development at MixZon, Inc.

The principle idea behind the CORMIX expert system approach is a rulebase describing the flow classification trees (Figure 1), with altogether more than 100 flow classes and related flow class descriptions representing more than 100 individual flow simulation codes, each with variable complexity. Moreover, the length-scales based approach applied within CORMIX helped to improve the parameterization of the individual flow processes in general. For example, the normalization of the coordinates (x, y, z) of buoyant plume trajectories using the length scale L_M resulted in one single curve instead of the multiple curves for trajectories that result from normalizations with the jet diameter D and different Froude numbers. This simple modification allowed for much more accurate predictions and representations of combined phenomena over a large flow spectrum. In CORMIX, each flow class is associated with a sequence of mixing and transport submodules, which are triggered by CORMIX according to the classification, and transitions indicated by the related length-scales. In the initial part of the near-field region, where no boundary interactions yet occur, a jet integral model (CorJet for single port and multiport discharges, and CorSurf for Surface Discharges) is usually run until detecting boundaries in the vicinity of the plume, and handing over the results for example to buoyant spreading modules or passive advection-diffusion models, either based on simplifications of 3D equations for jets or based on semi-empirical equations. Figure 2 illustrates typical CORMIX results, for a heated ($\Delta T = 10°C$) effluent discharge ($Q_o = 1$ m³/s) through a single port pipe ($D = 0.9$ m) at 20 m depth, 50 m offshore into a receiving water with crossflow

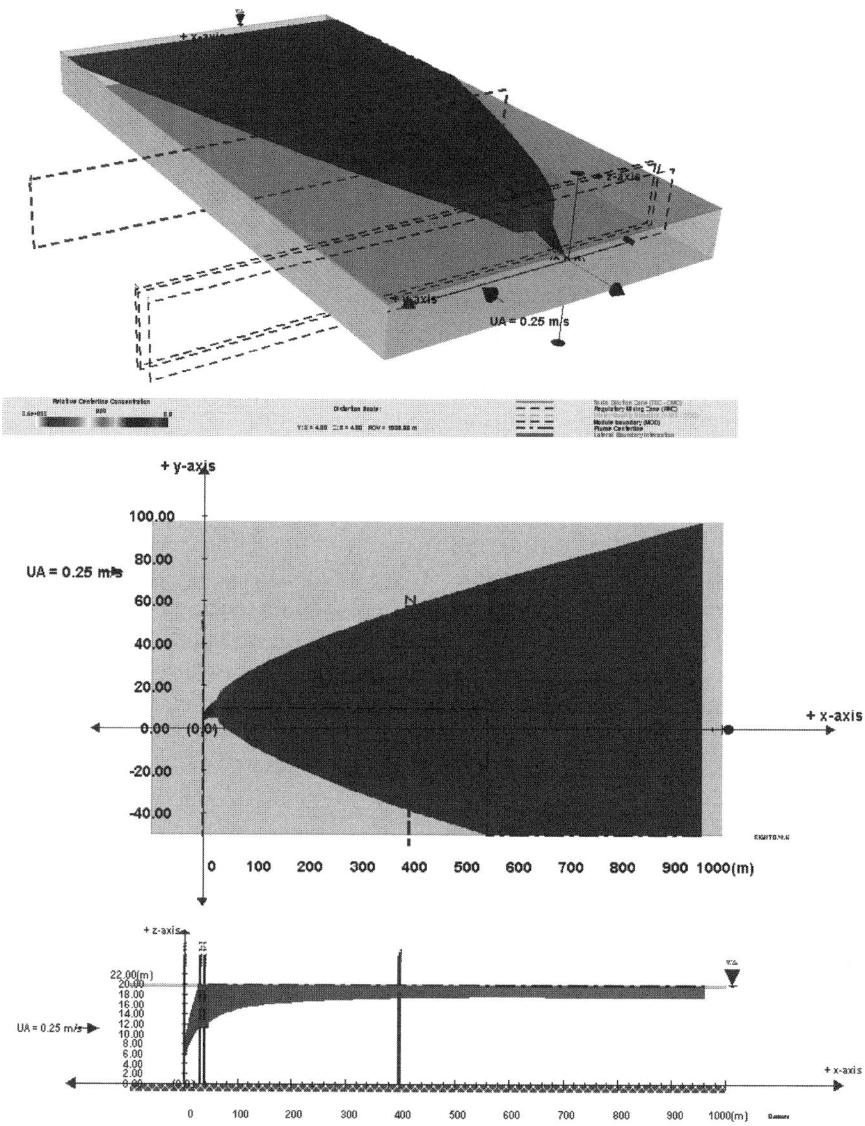

Figure 2 Flow visualization of CORMIX. Top: 3D image with dashed lines indicating submodule regions. Middle: Plan view indicating bank interaction at the right bank. Down: Side view showing the thin surface spreading layer developing soon after surface impingement.

of 0.25 m/s and uniform temperature distribution. The results are also showing the regions where CORMIX switches between the submodules (dashed lines).

The initial CORMIX version was capable of modeling single port and multiport discharges into stagnant or flowing, uniform or stratified environments, such as lakes, rivers, and coastal waters. Since then the model has been significantly extended, initially by adding surface discharge configurations, then by adding dense discharges for produced water from oil-drilling platforms and brine discharges from seawater desalination plants. Furthermore, he added modifications of the steady integral methods to simulate simple unsteady effects, such as simple tidal reversals with related pollutant build-up effects, and near-field stability issues. A further effort was undertaken to describe yet other flow classes beyond the near-field region, such as density currents establishing along different slopes with different roughnesses, resulting from dense discharges after jet impingement at the bed. Gerhard Jirka showed that the integral model approach can be modified for these flow types, and implemented them in the CORMIX submodules (Jirka *et al.*, 1996, Doneker and Jirka, 1990). This allows CORMIX to compute not only concentration distributions for positively or negatively buoyant discharges, but also to predict particle distributions for sediment-laden jet discharges, such as dredging effluents or slurries (Doneker and Jirka, 1998, Doneker *et al.*, 2004).

The CORMIX model has been widely validated by laboratory and field studies (Jirka, 2004, 2006, 2007), many of them being independent studies by other authors: for example, remote sensing field studies for single port and surface cooling water discharges into coastal waters (Davies *et al.*, 1997), field studies of multiport diffuser discharges into the Great Lakes (Tsanis *et al.*, 1994) or a shallow river (Zhang and Zhu, 2011), and laboratory studies on buoyant surface discharges (Summer *et al.*, 1994). Figure 3, for example, shows results from laboratory studies of a surface discharge configuration with crossflow for a positively buoyant effluent. The solid lines indicate the measured results, showing lines of constant normalized concentrations. The dash-dotted line shows the result of the corresponding CORMIX simulation, illustrating the trajectory of the predicted centerline, where maximum normalized concentrations C_c were computed. The dashed lines show the predicted plume outline, here plotted for a value of $\sqrt{2}$ times the predicted half-width b_h, defined as the location where the concentration is $1/e$ times lower than the centerline concentration ($C(b_h) = C_c/e$ seen from a local cross-section). The comparison shows that the dominant flow features are reproduced, even though CORMIX uses only very few general parameters, such as one mean, uniform ambient velocity and top hat profile descriptions within the plume. The location of downstream bank interaction, the downstream layering effect without bed contact, and the mixing characteristics shown by the concentration reductions are close to the measured results. These features are important for water quality protection measures.

Gerhard Jirka's anticipation of future developments and trends also resulted in studies which have now become standard procedures or software tools for outfall systems, for example, providing the Environmental Hydraulics Framework for the design of discharges from desalination plants (Bleninger and Jirka, 2007, 2008, Bleninger *et al.*, 2010b), the procedure CorTime which can run hundreds of CORMIX simulations in time-series mode for diffuser performance analysis, procedures to couple CORMIX with far-field models (Bleninger, 2006), and the Internal Diffuser

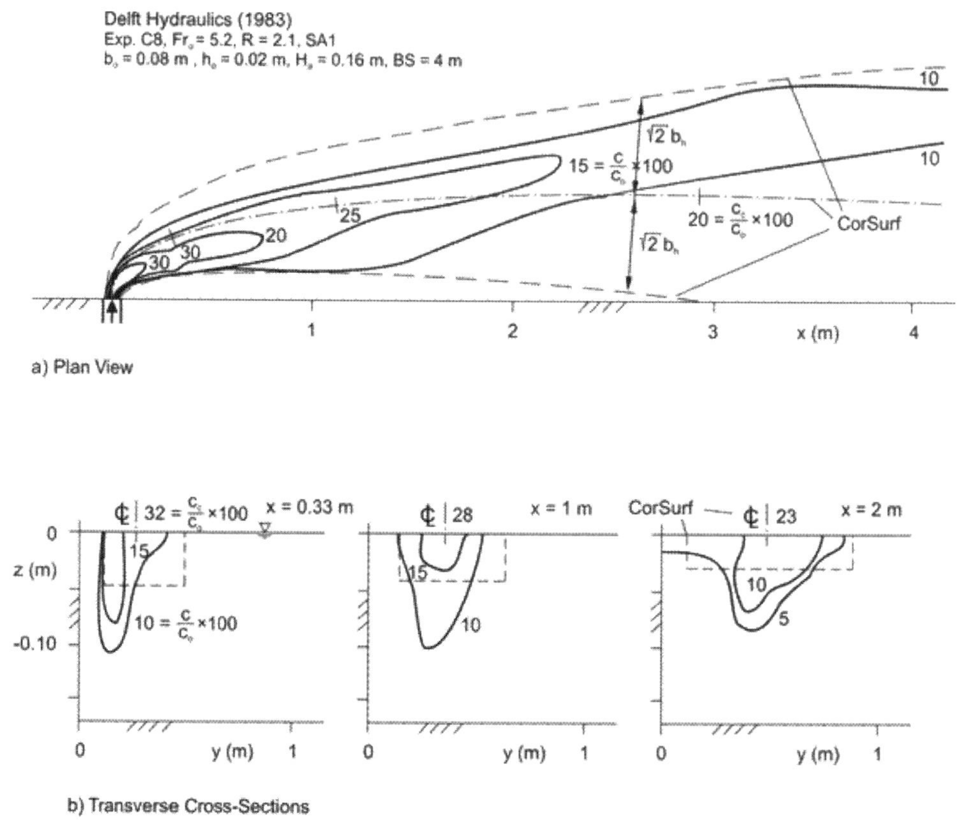

Figure 3 Independent CORMIX validation for a surface discharge studied in laboratory experiments (Source: Summer *et al.*, 1994).

Hydraulics Program CorHyd within the CORMIX system (Opila *et al.*, 2009) which computes the flow distribution along a diffuser line. Furthermore he integrated these individual contributions also beyond fluid mechanical topics, as described in the next subsection for environmental impact mitigation measures.

2.3 Applications for outfall systems

It was also Gerhard Jirka's special analytical treatment of scientific problems which allowed not only the successes highlighted above, but also established a more holistic approach, integrating individual processes into an effective analysis tool and overall methodology. A historical perspective from the first studies on buoyant jet phenomena up to the most recent investigations with optical measurement techniques can be found in Gerhard Jirka's (2004) paper outlining his view of the topic, including all his previous efforts in the field. This paper was the first in a series of four papers summarizing and integrating recent works by him and others on mixing and flow phenomena, flow classification, and modeling. The first paper details the integral model development of a turbulent buoyant jet discharging into unbounded stratified ambient flows (Jirka, 2004). This integral model, called CorJet, is a very good example of Gerhard Jirka's capabilities and philosophy, in which he describes an entrainment and drag model based on the relevant physical phenomena using a combination of 4 entrainment coefficients to span the five asymptotic flow regimes (jet, plume, wake, advected line puff, and advected line thermal), instead of using only one coefficient for all as was done in previous approaches. Flow classification optimizes which flow regime is dominant, resulting in the model using the correct physics throughout the simulation. Thus, he strongly influenced and supported the change in environmental hydraulic research from empirical models to more deterministic descriptions and a better understanding of the underlying phenomena. Gerhard Jirka extended this description to plane jets resulting from multiport diffusers (CorJet, Jirka, 2006), and to free surface discharges (CorSurf, Jirka, 2007), using amplifications of the integral method. An initial classification for surface discharges to improve the parameterization and prediction methodology was published in the same year (Jones *et al.*, 2007). These four papers clearly show Gerhard Jirka's achievements on implementing his understanding of fluid mechanics phenomena into practical tools.

The study of flow phenomena, the integration of these modeling approaches, and the development of practical model applications alone deserve respect and honor, but Gerhard Jirka was always interested to go beyond science and engineering to provide further service to society, especially as regards improving environmental engineering measures. An example of his legacy in this regard includes his successful lobbying to include mixing regulations within the European Framework Directive (EC, 2000, Jirka *et al.*, 2004) which led to amendments within those regulations (Bleninger and Jirka, 2011) which will improve the implementation of a combined approach (end-of-the-pipe and receiving water standards) to effluent regulation, enhancing water quality for the benefit of society. This all together summarizes the active role he played in the field, and all related disciplines from jets and plumes to outfall hydraulics and including discharge regulations.

3 GAS TRANSFER

By 1980, Gerhard Jirka's research interests extended also to the problem of gas transfer. With his students, numerous works on this subject were performed at Cornell University and later at IfH, Karlsruhe Institute of Technology. The most important achievements are briefly discussed in the following. In his early stage of interest in this subject, Gerhard Jirka and colleagues recognized the need of an interdisciplinary exchange among those who are working on gas transfer. In 1983, together with Brutsaert, Gerhard Jirka was co-convenor of the 1st International Symposium on Gas Transfer at Water Surfaces held at Cornell University, the motivation being reflected in their foreword of the proceedings: "This interfacial mass transfer is, by its nature, highly complex. ... the transfer involves a wide variety of physical phenomena occurring over a wide range of scales. As a consequence, scientists and engineers from diverse disciplines and problem areas, have approached the problem, often with greatly differing analytical and experimental techniques and methodologies" (Brutsaert and Jirka, 1984). The initiation of the symposium (now having had its 6th convening in

2010) was a great contribution, achieving a summary of the state-of-the-art as well as prompting new ideas and approaches in the effort to understand the interfacial gas transfer problem.

3.1 Detailed measurement of near surface turbulence

Until the 1980's, the common approach in estimating the interfacial gas transfer velocity K_L, defined as $K_L = j/\Delta C$, w where j = the surface gas flux and ΔC dissolved gas concentration difference between the interface and the bulk in the water, was through empirical relations and conceptual models. At that time, the conceptual models were not supported by detailed near-surface hydrodynamic data. The reason was obviously due to the small thickness of the boundary layer on the liquid side in which the transfer process is concentrated and the state of technology available to interrogate this boundary layer. For most environmentally important gases (e.g., CO_2, O_2, NO, CH_4), this thickness is only on the order of tens to hundreds of micrometers.

It was the late 1980's when Brumley and Jirka (1987) made a great contribution by taking up this challenge. They performed detailed near-surface velocity measurements in a far-field homogeneous isotropic turbulence water environment by employing a grid-stirred tank system. Experiments in such a system provide a convenient analogy to the near-surface turbulence generated by bottom-shear in flowing streams. Their measurement, shown in Figure 4 confirmed the picture of the spatial structure proposed by the Hunt-Graham (HG) theory (Hunt and Graham, 1978). Further, on the basis of certain features in Chan and Scriven's idealized stagnation flow (Chan and Scriven, 1970) and Hariott's random model (Harriot 1962), Brumley and Jirka proposed a surface-divergence-type model, which "... demonstrates how fluctuating surface divergence can enhance the gas transfer rate, a mechanism that is sensitive to

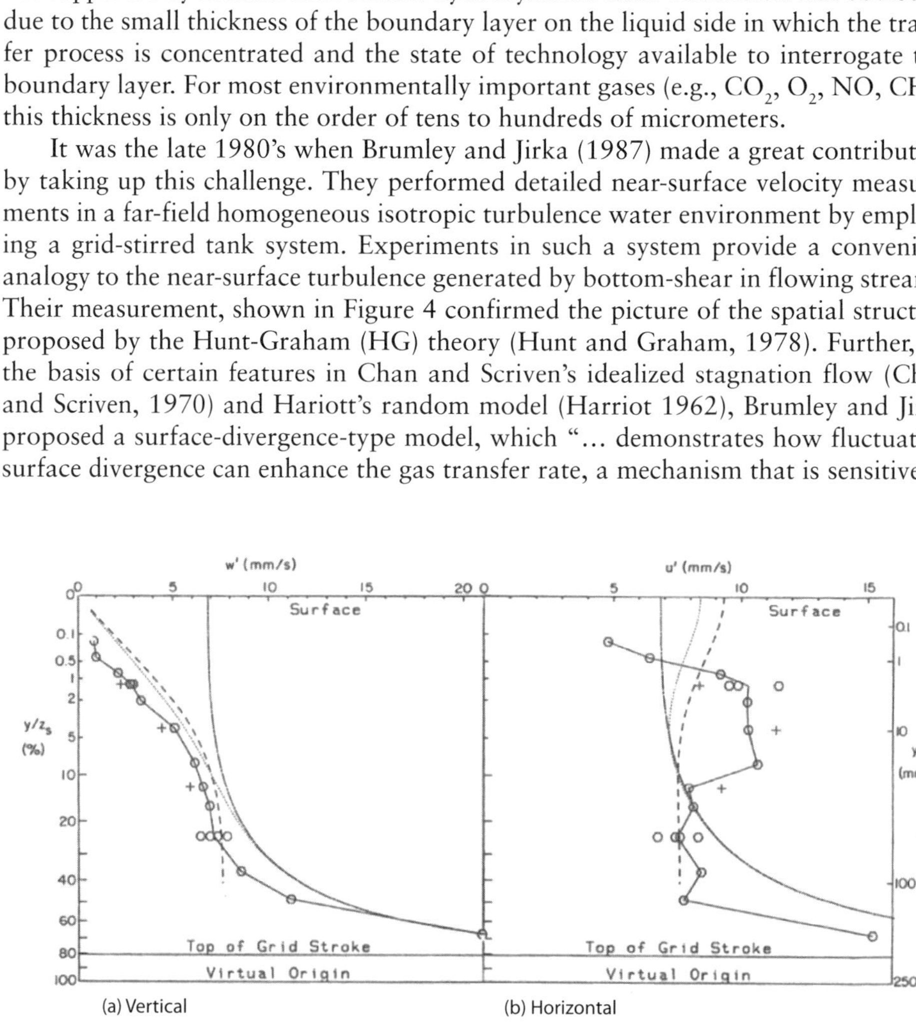

Figure 4 Detailed near-surface turbulent velocity profiles measured in a grid-stirred tank. (a) Vertical fluctuations; (b) horizontal fluctuations, -.- Hopfinger-Toly relation; ----Hunt-Graham profile; ... combined profile; o and + indicate two data sets taken on different days. (Source: Brumley and Jirka 1988).

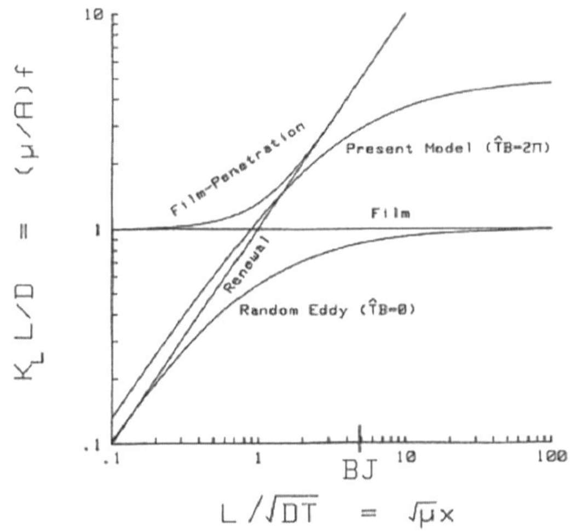

Figure 5 The dependence of the gas transfer velocity (KL) upon the parameter $L/\sqrt{DT}$ Brumley and Jirka compared their surface-divergence-type model with existing conceptual models. (Source: Brumley and Jirka 1988).

damping of surface divergence fluctuations by slight surface contamination" (Brumley and Jirka 1988). Using the HG theory, they generated theoretical surface divergence spectra and used this information to compare their model with some of the conceptual models (see Figure 5). To date, the surface-divergence-type models are in wide use and have been developed further (e.g., Turney and Banerjee, 2008). They are favored as a tool for estimating the mean gas transfer velocity (K_L) because of their robust applicability over wide-ranging flow conditions.

3.2 Direct measurement of gas transfer

A drawback of conceptual models is that they do not provide detailed insight into the actual mechanisms by which dissolved gas in the boundary layer is mixed into the bulk. Direct measurements of the gas concentration and velocity were needed. Around the 1990's, researchers had reported such measurements. However, none of them performed simultaneous measurements of velocity and concentration. Derived from the advective-diffusive mass transport equation, the total mean gas flux $\overline{J}$ can be written as

$$j = -D\frac{\partial \overline{c}}{\partial z} + \overline{c'w'}, \tag{1}$$

where D is the molecular diffusivity, $\overline{c}$ the mean concentration, c' and w' are the concentration and vertical velocity turbulent fluctuations, respectively and z denotes the vertical direction. The above equation clearly suggests the necessity of quantifying the turbulent mass flux term $\overline{c'w'}$.

In 1992, Chu and Gerhard Jirka reported first simultaneous measurements of turbulent velocity and gas concentration near the surface by coupling hot-film probes (Brumley and Jirka, 1987) with a polarographic oxygen probe (Jirka and Ho, 1990). They were able to show that the integral length scale of the concentration fluctuation was of the same order as that of the vertical velocity fluctuations. Although the mean of the direct measured flux $\overline{c'w'}$ data looked rather scattered, the values were of the same order of magnitude as the total mean flux $\overline{J}$, which was determined from independent reaeration measurements in the bulk. This gave experimental evidence that a significant portion of the total gas flux is due to turbulent transport.

3.3 Predictive reaeration equations

Gerhard Jirka's research was not confined only to the fundamental studies of the detailed phenomena associated with gas transfer; he and his co-workers also developed empirical relations for use in practical engineering to assess the gas transfer rate K_2 that represents the surface transfer velocity K_L divided by the water depth H (i.e., $K_L = K_2 /H$). Many empirical models had been proposed by different researchers, mostly relating the reaeration coefficient K_2 to wind velocity or global stream parameters, such as the flow velocity, slope and water depth. The representation of complex stream systems based only on these parameters often leads to errors. Motivated by this, Gerhard Jirka and his group developed empirical relations including additional important factors and considerations, such as bed morphology or high and low wind regimes, to obtain more accurate predictions (e.g., Moog and Jirka 1995, 1999, Chu and Jirka 2003).

As mentioned already, numerous predictive reaeration equations employing depth, slope and velocity had been proposed over the last few decades. Examples of some commonly used models are summarized briefly in Jirka and Herlina 2008. It was known that such empirical equations frequently over-predict the K_2 value, in some cases by up to five times. In 1998, Moog and Jirka proposed a new metric called mean multiplicative error (MME). In their paper, they defined MME as being equal to the geometric mean of the factors, greater than unity, by which the estimates have to be multiplied or divided to match the corresponding measurements (Moog and Jirka 1998). By examining a number of reaeration equations, they showed that the MME method offers several advantages such as i) not being biased toward under-prediction as is the case when the differential errors method is used, ii) yielding identical results for both reaeration coefficient K_2 and gas transfer velocity K_L and iii) less sensitivity to extreme errors. MME has, thus, been found to be useful for recalibrating reaeration equations.

3.4 Measurements with different turbulence forcing mechanisms

Beginning in 2000, Gerhard Jirka and his students at IfH continued the effort, previously initiated at Cornell University, to obtain detailed insight into the actual mechanisms in which dissolved gas in the boundary layer is mixed into the bulk through simultaneous full-field measurements of velocity and concentration. Around this time, optical measurement techniques, such as Particle Image Velocimetry (PIV) and Laser

Induced Fluorescence (LIF), were preferred in laboratories to avoid disturbance of the thin concentration boundary layer. By coupling the PIV and LIF techniques, Gerhard Jirka and his students were able to obtain for the first time 2D synoptic information of velocity and concentration near the surface during the gas transfer process under grid-agitated turbulent forcing (Herlina and Jirka 2007, 2008). A snapshot of such a synoptic measurement is depicted in Figure 6a, showing eddy structures impinging on the surface from below. The downwelling motions of such eddies initiate the peeling process related to surface renewal events, sweeping part of the concentration boundary layer and transporting this oxygen-rich fluid down into the bulk. The upwelling motions, on the other hand, bring oxygen-poor fluid from the bulk up to the surface and at the same time cause a thinning of the boundary layer leading to a higher gas transfer rate.

Whilst the first attempt in measuring the turbulent mass fluxes by Chu & Jirka (1992) were inconclusive, probably due to the invasive and spatially non-coincident measurement techniques, the normalized turbulent flux $\overline{c'w'}$ profiles measured by Herlina and Jirka (2008) showed a typical increase from around 0 at the interface to about 1 within a depth of approximately $2\delta_e$, where δ_e is the mean concentration boundary-layer thickness (Figure 6). The observed trend is inline with the known fact that near the interface molecular diffusion dominates the mass transport and that turbulent transport takes over within a very short distance. Such experimental evidence was at that time novel and should provide a useful database for the development and validation of highly resolving numerical models. Further, rather than showing a preference for the large-eddy or small-eddy structures as being dominant for the transport mechanisms, Herlina and Gerhard Jirka's interrelated results based on visual observations, statistical analysis of the concentration profiles and spectral analysis of the flux term $\overline{c'w'}$ suggest that the gas transfer process is controlled by a spectrum of varying eddy sizes. Their results also indicated that the small-scale eddies are predominant for the gas transfer at high turbulence levels, while the large-scale turbulent structures are predominant at lower turbulence levels. Thus, their data supports the two-regime model proposed by Theofanous *et al.* (1976).

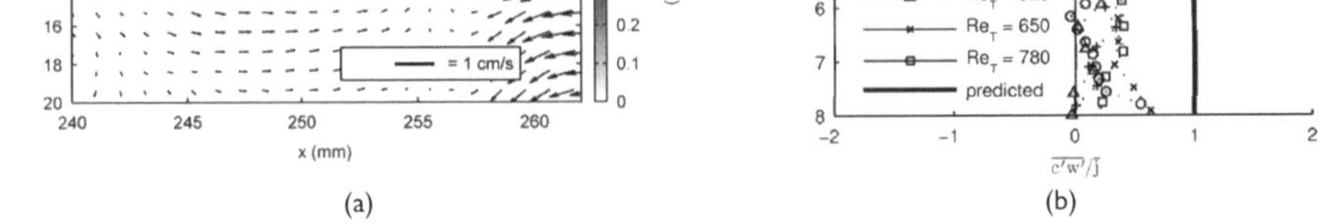

Figure 6 Gas transfer under grid-generated turbulence. (a) Visualisation of synoptic measurements of velocity (arrows) and oxygen concentration (LIF, dark colors for high saturated and light colors for low background concentration) providing structural details on the near-interface turbulent transfer (resolution 7 μm); (b) Normalized mean turbulent mass flux.

Over the years, Gerhard Jirka and his students had performed significant fundamental studies of the detailed phenomena of gas transfer related mostly to bottom-shear generated turbulence. Bottom-shear is one of the typical turbulence generation sources in nature, but generation due to wind-shear and buoyant-convective instability are also of importance. Around 2005, gas transfer experiments induced by the latter turbulence forcing mechanism were also performed at IfH employing a similar synoptic PIV-LIF technique as in the gas transfer experiments with grid-agitated turbulence. Typical instantaneous concentration images under grid-stirred (Herlina and Jirka 2004) and buoyant-convective instability (Herlina, Jirka, Murniati 2007) are presented in Figure 7a and b, respectively. The process in the buoyant-convective case is initiated by introducing cold air above the water surface which in turn generates a cold thermal boundary layer on the water side. The visualization gives direct insight into the transfer mechanisms in which oxygen-rich fluid is transported from the boundary layer to the bulk and the differences between the two cases. While, as described above, the continuous motion of eddy structures approaching the interface from below is the main mechanism controlling the gas transfer under grid-agitated turbulence (bottom-shear turbulence), the transfer mechanism with buoyant-convective instability is dominated by the continuous movement of sinking and rising plumes. These plumes form shorly after initiation of cooling as colder plumes in the form of mushroom-like structures start to plunge into the ambient fluid as visualized in Figure 7b. Further discussion can be found in Jirka, Herlina, Niepelt (2010).

In addition to these laboratory methods, the use of direct numerical simulation (DNS) methods promises future opportunities to reveal many details of this complex gas transfer problem. Details that, in laboratory experiments, might be affected or biased by inevitable disturbances (e.g., surface contamination) and limitation of measurement accuracy. Gerhard Jirka saw this great potential and initiated the present numerical simulation work on gas transfer at IfH (Herlina, Wissink, Jirka, 2008, 2010). The experimental results, which Gerhard Jirka and his co-workers obtained

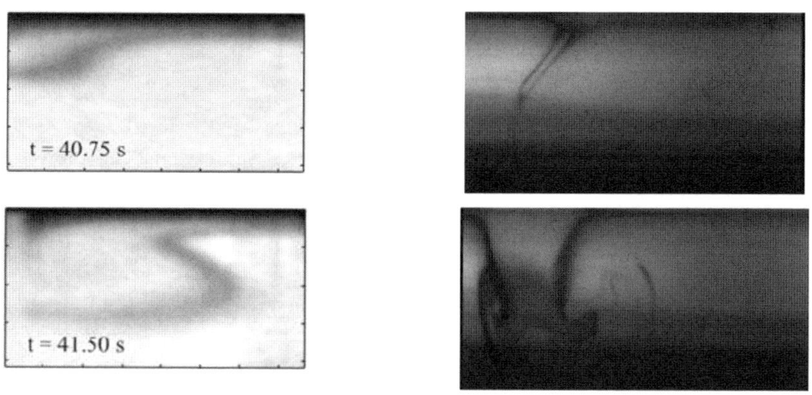

(a) Grid-stirred case, size ≈ 8 mm x 4 mm

(b) Buoyant-convective case, size ≈ 65 mm x 35 mm

Figure 7 Typical instantaneous concentration images obtained during the gas transfer experiments with (a) grid-generated turbulent forcing and (b) buoyant-convecitve instability forcing. The dark and light colors indicate oxygen-rich and oxygen-poor regions, respectively.

over the years, provide the observational framework for the theory and are of course important for intercomparison with the results from highly resolved numerical simulations.

4 MIXING AND TRANSPORT IN SHALLOW FLOWS

Shallow flows are ubiquitous in nature and were an important aspect of Gerhard Jirka's research since his sabbatical at the Swiss Federal Institute of Technology in Zurich in 1993–1994, when he began investigating the stability and entrainment processes for shallow jets (Giger *et al.*, 1991). His work on this topic continued at Cornell University, using both laboratory and theoretical methods applied to shallow jets, wakes and mixing layers, and later expanded at the IfH to also include groyne fields and the behavior of generalized shallow water vortices. Shallow flows occur whenever the region of interest has a lateral extent much greater than the water depth, as in, "lowland rivers, lakes, estuaries, bays and coastal areas, but also in density-stratified atmospheric and oceanic flows," (Jirka and Uijttewaal 2004a). These vertically-confined flows exhibit unique flow structures and resulting mixing processes, which are summarized nicely by Gerhard Jirka and Wim Uijttewaal in their foreword to *Shallow Flows*, which contains selected papers from the 1st International Symposium on Shallow Flows, initiated and hosted by them in 2003. They write, "Shallow flows are – by virtue of their large size – intrinsically turbulent flows. Their most fascinating aspect is the separation into and interaction of highly disparate scales of turbulence: on one hand, there are the strongly three-dimensional small scale turbulent motions generated by the vertical shear at the confining boundaries, and on the other hand, the large scale quasi-two-dimensional turbulence due to various kinds of horizontal instabilities in the flow" (Jirka and Uijttewaal 2004b). Gerhard Jirka loved the opportunity to apply stability analysis (using laboratory, analytical and numerical means) to environmentally relevant flows, and his work focused on the large-scale vortical coherent structures of the flow (2DCS), and in particular, their generation, evolution, and mixing properties. Here, we present a few of his more important results on wakes, vortex dynamics, and groyne field flows.

4.1 Background

As in his analysis of the different flow classifications for buoyant jet flows, Gerhard Jirka had a keen eye to distinguish different types of stability regimes in shallow flows and to categorize them using key non-dimensional parameters of the flow. His early laboratory work on shallow wake flows at high depth Reynolds number ($Re_h = uh/\nu > 10^4$) identified three wake types: VS, the vortex street type with an oscillating vortex shedding mechanism; UB, an unsteady bubble in which vortices grow and detach downstream of a recirculating region behind the wake body; and, SB, a steady bubble wake in which no large-scale vortical structures form (Chen and Jirka 1995). Figure 8 shows an example of these three wake classes behind a shallow circular cylinder. Chen and Jirka (1995) mapped the flow behavior as a function of the depth Reynolds number $Re_D = uD/\nu$ and the shallow flow stability parameter $S_f = c_f D/h$, introduced by Ingram and Chu (1987), where D is the wake body

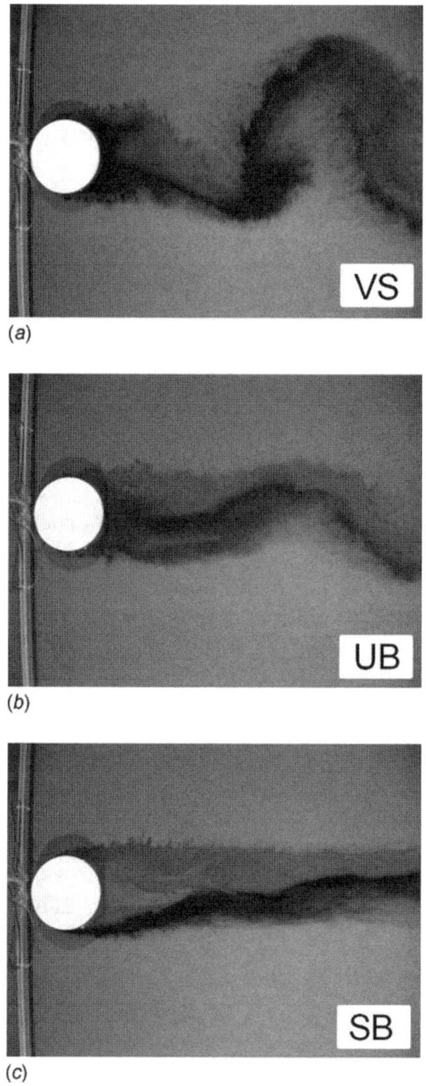

Figure 8 Plan view of shallow turbulent wake flows behind a circular cylinder showing the three wake types identified in Chen and Jirka (1995) (from von Carmer *et al.*, 2009).

diameter, u is the ambient velocity, h is the water depth, ν is the kinematic viscosity, and c_f is a quadratic-law friction factor for the bed roughness. For wake flows with Re_D above 1500, the stability was uniquely dependent on S_f, with the transition values dependent on the shape of the wake body (cylinder, plate, or porous plate). Gerhard Jirka describes the stability parameter as a ratio of the competing dynamic effect of bottom friction to the kinematic instability mechanisms of the aspect ratio; thus, higher values of S_f promote stability while lower values indicate growth of two-dimensional coherent structures. This parameter is effective in a wide array of shallow flows to describe these dynamics and further classify the flow regimes.

In their quest to discern and quantify the mechanisms of stability in shallow wake flows, Chen and Jirka published a seminal paper in the *Journal of Fluid Mechanics* (Chen and Jirka 1997) that incorporated the effect of bottom friction and viscosity on shallow wake stability, extending the work on shallow flow stability initiated by Chu *et al.* (1991) and comparing to their own 1995study. The paper lays the foundation of many future applications of linear stability analysis in shallow shear flows to study absolute and convective instabilities of the flow (see e.g., Chen and Jirka 1998; Socolofsky and Jirka 2004; White and Nepf 2007). The analytical stability analysis agreed well with experiments, identifying the VS wake flow as a predominantly absolute instability and the UB wake as convectively unstable. They extended their analysis to shallow jets and mixing layers in Chen and Jirka (1995), including the effect of lateral confinement. The results explain the onset of meandering in shallow jets and predict that stabilization of the jet wake would occur after the jet reaches a critical lateral width and the effect of bottom friction overwhelms the instability of the shear layer of the jet. They also determined that the effect of viscosity, including turbulent eddy viscosity due to three-dimensional fine-scale turbulent motion, is relatively minor in comparison to the bed friction effect; hence, the shallow stability parameter S_f (generalized using the lateral length scale of the 2DCS to replace D) explains the dominant physics of the evolution of 2DCS in shallow flows.

Through this work on shallow flow stability, Gerhard Jirka developed a classification system for the generation of 2DCS in shallow flows. In his summary paper in the *Journal of Hydraulic Research* (Jirka, 2001) and again in his paper appearing in *Shallow Flows* (Jirka and Uijttewaal, 2004b), he identified three mechanisms that generate shallow flow instability and result in 2DCS. Type A mechanisms result from topographic forcing, as in the flow around an island, headland, or other large-scale structure in the flow that creates a velocity deficit and wake-like instability. Type A mechanisms are the most energetic, with S_f typically less than 0.2 (Jirka and Uijttewaal 2004b). The Type B mechanism results from internal transverse shear instabilities. These are the instabilities resulting in meandering in shallow jets and in 2DCS generation in general lateral shear flow instability; they occur for S_f between 0.2 and 0.5. The weakest generation mechanism is Type C, resulting from secondary instabilities of the base flow. There is limited experimental evidence for Type C mechanisms, but they are expected especially in unsteady and reversing base flow, as in tidal flow in a wide channel or unbounded domain. This classification system is useful to help understand the underlying physics of 2DCS generation and provides a simple, predictive parameterization to assess the importance of 2DCS to a given shallow base flow.

While linear stability analysis and this classification of the generation mechanism describe the onset of 2DCS formation in the flow, laboratory experiments shed light on their growth and decay. As stated previously, shallow flows in nature are fully-turbulent flows, containing a broad spectrum of three-dimensional turbulent motions. However, the vertical confinement also exerts a strong two-dimensionality to the flow when the domain is much wider than the depth. Dimensional analysis of turbulent motion for a theoretically two-dimensional flow field yields a -3 slope to the energy spectrum, indicating an inverse turbulent cascade from smaller- to larger-scale motion. This is observed in shallow flows as eddies on the order of the water depth and slightly larger readily combine to form larger scale structures. While the turbulence remains a three-dimensional characteristic of the flow, Gerhard Jirka was adamant that 2DCS experience the inverse cascade in their growth and retain many features of two-dimensionality. Experimental observations of the inverse cascade are shown for the concentration fluctuations in a shallow jet in Jirka and Uijttewaal (2004b). Near the jet source, the cascade follows the classical –5/3 slope, and the dominant length-scale is on the order of the water depth; several water depths downstream, the spectral slope is clearly –3, and the peak length scale of the flow has increased by about a factor of 5.

The inverse cascade also helps to extend the life of 2DCS in the flow. The stability analyses and laboratory experiments demonstrate that the growth stage of the 2DCS is quite short, usually terminating when the structure is 2 to 4 times larger than the water depth. At that point, bottom friction arrests the growth and begins to stabilize the structure. Decay by bottom friction, while the dominant decay mechanism, is rather slow, and structures continue to combine with other structures with nearly no loss in net momentum (hence the inverse cascade).

Gerhard Jirka's passion for shallow flows, their stability, and wide application in environmental fluid mechanics is witnessed by the fact that he spent a significant portion of his start-up funds when moving to Karlsruhe in 1995 to build the shallow water table at the IfH (see Figure 9). The basin measures 5.5 m in width, 13.5 m in length and is equipped with a reversing flow capability. In the following, we highlight some of the work Gerhard Jirka and his students conducted in this and similar facilities at the IfH, with specific focus on the shallow wake, dynamics of generalized shallow vortices, and the flow in groyne fields. This work significantly advanced the field and likewise gave Gerhard Jirka joy as he could apply his love for physics and mathematics to study the wide array of flow types in the natural environment.

Figure 9 The shallow water basin at the Institute for Hydromechanics, Karlsruhe Institute of Technology, showing the basin, typical experimental lighting, and the three-dimensional instrument carriage (from von Carmer 2005).

4.2 Shallow wakes

Work on shallow wakes at IfH continued the analysis of their stability, and significantly advanced understanding of their mixing and transport properties. This was facilitated by the new shallow basin and by intensive development of new measurement techniques, notably applications of particle image velocimetry (PIV) to shallow flows and planar concentration analysis (PCA) methods. These methods were applied to single shallow cylinders to study individual 2DCS and their interaction with neighboring structures, to multiple cylinders and the mixing caused by shallow grid turbulence, and to engineering methods to stabilize wake flows by strategic placement of enhanced bottom roughness in the flow.

Gerhard Jirka had a great ability to apply the latest measurement methods and to see innovative ways to extend them to new, sometimes complicated, situations. Much of the measurement capability of the shallow basin was developed by his student Carl von Carmer through his Ph.D. dissertation work (von Carmer 2005), and focused on the ability to make time-resolved, synoptic measurements of instantaneous concentration and velocity to educe the turbulent transport properties in shallow wake flows. Initial steps yielded a coupled laser Doppler anemometer (LDA) and laser induced fluorescence (LIF) point measurement to yield data highly resolved in time, capable of uncovering the turbulent dynamics of the flow (von Carmer 2000). Coupled with an ultrasonic depth profiler, these instruments also validated the quality of the shallow basin: repeatable, uniform shallow flows with fully-developed log-velocity profiles are possible in depths down to 1 cm, with accuracy of 0.5 mm in the water depth, uniformity of the bottom within 1.5 mm, and depth-averaged velocity in excess of 25 cm/s. Hence, this world-class facility greatly extends the experimental domain available for shallow flow research. Following these initial measurements that also validated aspects of the –3 inverse turbulence cascade, full-field measurements using shallow PIV and PCA were developed. Here, Gerhard Jirka and his students pioneered new measurement techniques at scales not previously possible. Floating tracer particles were used for the PIV measurement, eliminating the need for a laser light sheet to identify the measurement plane, and allowing for a much simpler set-up than traditional PIV (Weitbrecht *et al.*, 2002). Measurement fields were on the order of square meters (as compared to a few 100 square centimeters for traditional PIV), capturing data at up to 30 frames per second. The planar concentration method was also novel, taking advantage of the fact that the highly turbulent flow is very quickly well mixed in the vertical. Thus, a food-coloring dye is used, and the method measures the absorption of ambient light by the dye, which is related by calibration to the depth-averaged concentration (von Carmer *et al.*, 2009). Together, these techniques allow quantification of the evolution of velocity and concentration within 2DCS in shallow wake flows.

Gerhard Jirka realized that analysis of this full-field information required robust vortex identification methods and that the results would yield a strong understanding of the transport properties of 2DCS. Figure 10 shows an example result of the combined PIV/PCA measurement of a VS-type wake behind a shallow circular cylinder. The result in the figure is a phase average over several shedding cycles, indicating the repeatability of the vortex shedding in the flow and allowing for combined analysis of the PIV and PCA fields, which were obtained independently. Vortex identification

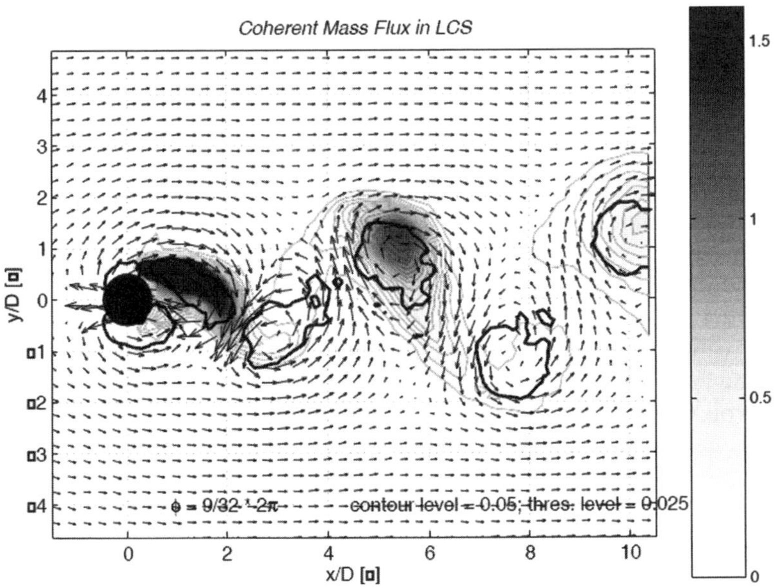

Figure 10 Phase-averaged distribution of absolute values of the coherent specific mass flux (gray-scale color) in g/(m^2s) with the phase-averaged velocity field. Bold lines denote Q-values of the velocity field to identify 2DCS (from von Carmer 2005).

in the velocity field can be achieved using various methods that calculate the local rotation independent of the local advection and strain rate of the flow (see e.g., Adrian *et al.*, 2000); here, the Q-value is used which is based on the Weiss function that relates rate-of-strain with vorticity to obtain local rotation (von Carmer *et al.*, 2009). Thus, the solid black lines in the figure outline the boundaries of the 2DCS in the wake. Similar phase averaging of the concentration measurements coupled with the velocity information allows for the identification of the mass fluxes associated with the two-dimensional coherent motion in the wake. The shaded values in the figure quantify this coherent specific mass flux so that the mass transport in the wake can be visualized. As seen in the figure, tracer mass is significantly redistributed and concentrated within the 2DCS within a distance of about 6 cylinder diameters downstream. The stability analysis of this flow class also indicates that the 2DCS are unstable and largely growing in this domain of the wake. As the wake stabilizes, tracer mass flux becomes uniformly distributed, dominated by the lateral turbulent diffusion of the base flow, indicated in the figure by the light color (low coherent mass flux) in the last two vortices in the field of view. Hence, actively growing 2DCS rapidly accumulate mass from the surrounding flow, and this process dissipates in the stabilized wake. As a result of their initial rapid mass accumulation, 2DCS in nature are observed to contain highly concentrated, uniform pockets of mass that transport passively for long times as the 2DCS are slowly dissipated by bottom friction and lateral turbulent diffusion.

These methods were also applied to study turbulence behind a row of shallow circular cylinders, which Gerhard Jirka referred to as shallow grid turbulence. Uijttewaal and Jirka (2003) applied LDA and PIV to the flow field behind a row of shallow cylinders to understand the formation of 2DCS in the flow and the decay of this turbulence downstream of the grid. Their work confirmed the –3 slope in the energy density spectrum and observed the merging of smaller structures to form larger 2DCS downstream, as predicted by this inverse cascade. They also determined that the decay of the vortices that shed from the grid is determined by the cylinder diameter, independent of the grid spacing. Rummel *et al.* (2005) further applied the PCA methods to evaluate the lateral turbulent diffusion of a constant, point-source tracer injection occurring at various points in the flow. For experiments in the absence of the grid, measurements agreed with the literature for lateral turbulent diffusion in open-channel flow (e.g., Fischer *et al.*, 1979). For experiment with the shallow grid turbulent, eddy diffusivities were an order of magnitude higher. In both cases, the spreading agrees with Taylor's theorem, and the eddy diffusivity correlates with the size and turbulent intensity of the largest-scale coherent structures in the flow.

From these insights on the stability mechanisms in shallow wake flows, Gerhard Jirka had the idea to explore stabilization of wake flows by localized enhanced bottom roughness. This work was inspired by the many studies in unbounded cylinder wakes that study the effect on wake stability of inserting small control cylinders upstream and within the wake flow. Here, the focus was on bottom friction, with the hypothesis that stability could be controlled by manipulating (enhancing) bottom friction at strategic points in the flow. The enhanced bottom roughness was achieved using a course wire mesh having characteristic roughness height 2 mm, which should be compared to the hydraulically smooth basin bottom. The experiments were applied to an UB-type cylinder wake, and roughness meshes were place in either the lateral shear layer directly downstream of the cylinder or at the downstream edge of the recirculation bubble. In both cases, the previously unstable wake appeared identical in nature to the SB-type wake, and the formation of a recirculation bubble behind the wake was suppressed (Negretti *et al.*, 2005). This work highlighted the engineering aspects of shallow flows and underscored the importance of bottom friction for determining the stability and dynamics of 2DCS in shallow flows.

4.3 Shallow vortex dynamics

In order to explore the dynamics of a shallow water vortex without the effects of background turbulence or lateral shear, Gerhard Jirka conceived of a set of experiments to generate and track a single shallow vortex in a quiescent, laterally-unbounded domain. These experiments had a very strong analytical component, of which he was very proud and very excited to undertake. The experiments were initiated in a new, smaller shallow water tank with a transparent, smooth bottom. The vortex is generated as a type of solid body motion. This is accomplished using a segmented cylinder with an open bottom that is rotated in the center of the tank and then rapidly removed so that the subsequent vortex dynamics can be followed. The experimental setup, characteristic of Gerhard Jirka's approach, was highly sophisticated, yet minimalistic and elegant. In order to have good repeatability and visual access to the flow field, the cylinder is precisely computer controlled to spin up and then automatically rise out of the flow and move completely out of the measurement domain along a traverse mounted to the top of the tank. Both PIV and PCA were applied to these experiments, utilizing a smaller floating glitter tracer in the PIV measurements than in the shallow basin, and the results of this work were reported posthumously in the *Journal of Fluid Mechanics* with his post-doctoral student, Dong-Guan Seol (Seol and Jirka 2010) and in a recent review article by Weitbrecht *et al.* (2011).

Figure 11 shows a schematic of the experimental apparatus and locations of the imaging system. The experimental matrix included variation of the water depth, initial angular velocity of the vortex, and cylinder diameter. The spin-up time in each case was 4 s. During the vortex formation stage, four principle disturbances occur (Seol and Jirka 2010): the water surface deflects downward at the cylinder wall (resulting in the generation of a wave upon removal of the cylinder), internal circulation develops within each sector of the cylinder, there is small leakage of fluid from inside the cylinder through the gap between the cylinder and tank floor, and removal of the cylinder also induces mixing and surface waves due to the previously displaced volume of the cylinder wall. Each of these effects were minimized in the experiments, and observations of the velocity field shortly after cylinder removal showed that they dissipated quickly relative to the much more rigorous and long-lasting dynamics of the shallow vortex.

In addition to these processes, the no-slip condition at the cylinder wall results in a ring of opposite-rotating vorticity outside the shallow vortex within the cylinder. The outer vorticity is observed to organize by inverse cascade of small vorticity patches and, in shallow flow, always forms two large patches of opposite rotation to the central vortex, thus, forming a tripole system. In each case, the resulting vortices have lateral extent much greater than the water depth, yielding a shallow vortex system. Three typical cases for different water depths are shown in Figure 10 following the development of the tripole system.

After this formation stage, the experiments tracked the evolution and decay of the vortex system. For shallowness ratio (cylinder diameter to water depth) above 4, the tripolar system appears to be the system's natural state, and the formation dynamics are Reynolds number independent for $\mathrm{Re} = V_0 R_0 / \nu$ above 14,000, where V_0 is the initial azimuthal velocity of the cylinder wall and R_0 is the cylinder radius. As the vortex

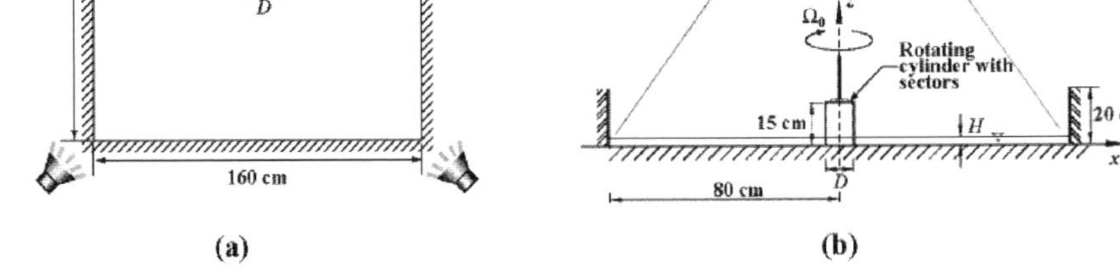

Figure 11 Schematic of the experimental setup for a shallow water vortex a) plan view and b) side view (from Weitbrecht *et al.*, 2011).

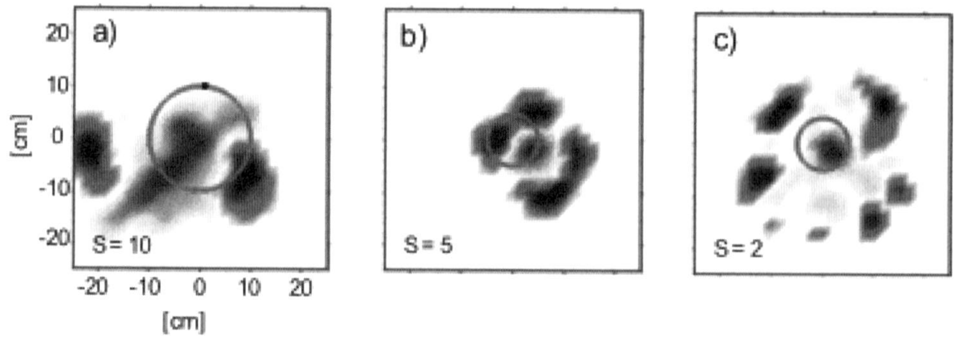

Figure 12 Vorticity maps normalized by the observed maximum vorticity. Gray circle indicates the original cylinder location. S is the shallowness, defined as the ratio of the cylinder diameter to the flow depth. Adapted from Weitbrecht *et al.* (2011).

system decays, there is a transition from the initial high turbulence regime to a laminar flow in later stages. The vortex energy is observed to decay at t^{-1} in the turbulent regime and e^{-t} for the laminar case (Seol and Jirka 2010). The results further show that turbulent energy is transferred from the central vortex to the two satellite vortices under strong shallowness conditions, a common feature of the inverse turbulence cascade. Seol and Jirka were also able to show from their data that the fastest growing mode in the flow is the perturbation wave number mode 2, consistent with previous stability analysis results, and that increasing the water depth suppresses this mechanism, hence, showing that the vortex instability mechanism in shallow flows is indeed different from that in unbounded flows, and results in the quasi two-dimensional nature of 2DCS in the flow (Seol and Jirka 2010).

4.4 Exchange flows in groyne fields

Gerhard Jirka applied these studies on shallow flow dynamics to another important environmental flow in order to understand the mechanisms for mass exchange between a groyne field and the main stem in rivers. Transport models for river management, as for example the Rhine River Alarm Model (Spreafico and van Mazajk 1993), require knowledge of the one-dimensional dispersion coefficient in a wide range of flow conditions. Dead zones, especially those created by river groynes, have a significant impact on the one-dimensional dispersion coefficient, and predictive models for dispersion in managed rivers were lacking. Gerhard Jirka and his students applied similar methods as developed for shallow wakes (PIV and PCA) to measure mass exchange in groyne fields and combined this information with various numerical models to elucidate the exchange mechanics and predict the dispersion coefficient needed by one-dimensional river transport models. Much of this work was guided through the Ph.D. dissertation work of Volker Weitbrecht (Weitbrecht 2004), and the results are summarized in Weitbrecht *et al.* (2008), which was honored by receiving the American Society of Civil Engineers (ASCE) Karl Emil Hilgard Hydraulics Prize for the best paper appearing in the Journal of Hydraulic Engineering in the selection period 2007/2008.

Figure 13 shows a typical result of the laboratory measurements. The experiments were performed in a laboratory flume with adjustable bottom slope so that uniform normal flow conditions existed in the main channel. Groynes of different lengths were placed on one side of the flume at various spacing and orientation to reflect natural groyne fields, with the opposite flume wall assumed to be the channel midpoint. Panel a) of the figure shows a typical instantaneous velocity field. Eddies shed from the tip of the groyne grow in the shear layer between the groyne pocket and the main channel. Depending on the spacing of the groynes, various recirculation cells develop in the groyne pocket, visualized here by the gray-scale vorticity in the flow. Panel b) shows a PCA concentration field result. As with the shallow vortex experiment, Gerhard Jirka and students devised a clever means to achieve uniform instantaneous dye injection throughout the groyne pocket without disturbing the steady-state flow in the flume. The injection device consisted of a 3 cm by 3 cm array of small needle injectors attached to a box containing the dye reservoir. A uniform pressure was applied to the dye reservoir for a short period, injecting the dye equally through all needles in the array, and then the box was lifted and removed from the visualization plane so that the evolution of dye tracer in the groyne pocket could be measured by PCA over time. The PIV data provide the information necessary to understand the fluid mechanical properties controlling exchange and the PCA data yield an independent measure of the tracer exchange coefficient.

Gerhard Jirka followed Valentine and Wood (1979) and modeled the exchange process using an entrainment hypothesis. From the PIV data, an average entrainment velocity could be obtained, and from the PCA data, a direct measurement of the exchange coefficient yielded an independent measurement of the entrainment rate of the exchange velocity. Previous studies had not produced a relationship between the groyne field geometry and the entrainment coefficient k. The experiments at IfH identify two mechanisms that control the exchange: 2DCS in the mixing layer between the dead zone and the main channel and the recirculating gyres present within the dead zone (Weitbrecht *et al.*, 2008). A length-scale that characterizes the gyres is a type of hydraulic radius $R_h = WL/(W + L)$, where W is the length of the groin and L is the spacing between successive groynes. R_h is also a characteristic length of the shear layer that forms between two successive groynes. The formation of 2DCS within this

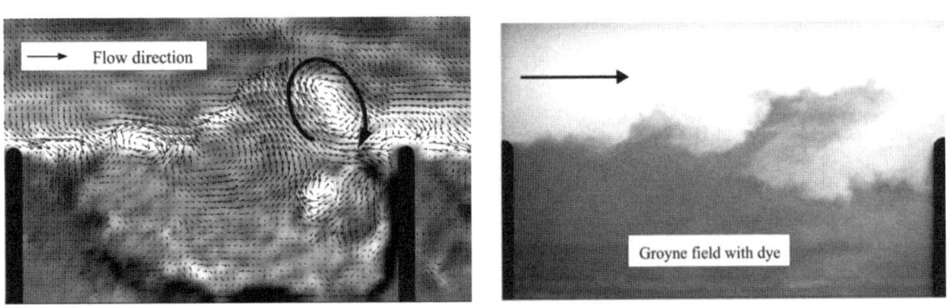

a) b)

Figure 13 Plan view of an instantaneous a) velocity and vorticity field (light color indicates clockwise rotation and dark color anti-clockwise rotation) and b) dye concentration measurement in a groyne field. Water depth is 5 cm (equal to the width of each groyne).

layer is dependent on the stability of the shear layer, which depends on S_f. Gerhard Jirka proposed an analogy to S_f in groyne fields that is termed the nondimensional morphometric parameter

$$R_D = \frac{WL}{h(W + L)}, \tag{2}$$

in which h is the mean water depth in the main channel. Correlation of the entrainment coefficient with R_D yields $k = 0.003R_D + 0.015$, which gives good agreement over all the experiments conducted at IfH, those by Uijttewaal *et al.* (2001), and by large eddy simulations conducted by Hinterberger (2004) and reported in Hinterberger *et al.* (2007).

Once the exchange rate between the groyne field and main channel is known, the dispersion characteristics of the river can be determined. Weitbrecht *et al.* (2004) present a novel method using two-dimensional particle tracking in a simple numerical model of the river. The flow in the numerical model is prescribed, and is taken equal to the mean depth-integrated channel flow, allowed to vary across the channel cross section. A cloud of tracer particles is released at the upstream section of the model and tracked as it propagates downstream. Particles advect deterministically with the mean flow and have a stochastic random component related to the two-dimensional dispersion coefficient in depth-average open channel flow (see e.g., Fischer *et al.*, 1979). The dead zones are modeled by a sticking factor that retards particles that touch the boundary between the channel and the groyne field, chosen to match the measured exchange coefficient k. By modeling the tracer cloud until it achieves a fully-developed Taylor dispersion, the effective dispersion coefficient can be evaluated from the tracer spreading. This has proven to be an effective tool to estimate dispersion for one-dimensional river models where field tracer data are lacking and directly helps to improve the reliability of early warning systems that exist for many managed rivers.

5 CONCLUDING REMARKS

This review of Gerhard Jirka's achievements has shown the great width and depth of his contributions, but only a small part of his publications could be cited. Altogether, he has authored or co-authored more than 230 journal and conference papers, most of them in the most renowned journals, such as the *Journal of Hydraulic Engineering*, the *Journal of Hydraulic Research* and the *Journal of Fluid Mechanics*, and not only the quantity of publications and citations, but also the quality, is exceptional. Each of his more than 30 Ph.D. students maintains his devotion and excellent teaching efforts in memory. This article is intended as a dedication to Gerhard Jirka not only from the authors, but from all his Ph.D. students, his colleagues, friends and family, and to give insight in his working ability and extraordinary engineering and research ethics, and to encourage the reader to look for new ideas and hidden secrets, whenever reading one of his many publications. His spirit, example, and enthusiasm will remain among us and will continue to live not only in the water related research and academic community, but also among his friends, and family.

ACKNOWLEDGEMENTS

The authors deeply acknowledge the efforts of the editors Wolfgang Rodi and Markus Uhlmann to plan and realize the Memorial Symposium, and to publish this book, and providing their full support for this article. We acknowledge furthermore the help com Dr. Cornelia Lang, Dr. Rob Doneker, and all former Ph.D. students of Gerhard Jirka, who provided their help.

REFERENCES

Adrian, R.J., Christensen, K.T. and Liu, Z.C. (2000). Analysis and interpretation of instantaneous turbulent velocity fields. *Exp Fluids*, 29 (3), 275–290.

Akar, P.J. and Jirka, G.H. (1994). Buoyant Spreading Processes in Pollutant Transport and Mixing. Part I: Lateral Spreading in Strong Ambient Current. *J. Hydraulic Research*, 32, 815–831.

Akar, P.J. and Jirka, G.H. (1995). Buoyant Spreading Processes in Pollutant Transport and Mixing. Part II: upstream spreading in weak ambient current. *J. Hydraulic Research*, 33 (1), 87–100.

Arita, M. and Jirka, G.H. (1987). Two layer model of saline wedge. I. Entrainment and interfacial friction. *J. Hydraul. Eng.* 113, 1229.

Bleninger, T. (2006). *Coupled 3D hydrodynamic models for submarine outfalls: Environmental hydraulic design and control of multiport diffusers.* Dissertation series of the Institute for Hydromechanics, Karlsruhe Institute of Technology, University Press, Karlsruhe, Germany.

Bleninger, T. and Jirka, G.H. (2004). Near- and far-field model coupling methodology for wastewater discharges. *Proc. 4th Intl. Symposium on Environmental Hydraulics and 14th Congress of Asia and Pacific Division, International Association of Hydraulic Engineering and Research*, Hong Kong, China, 15–18.12., Lee J.H.W., Lam K.L., Eds.

Bleninger, T. and Jirka, G.H. (2007). First steps in modeling and design of coastal brine discharges. *Desalination and Water Reuse*, Aug/Sept - 2007.

Bleninger, T. and Jirka, G.H. (2008). Modeling and environmentally sound management of brine discharges from desalination plants. *Desalination*, Vol. 221/1–3 pp. 585–597.

Bleninger, T. and Jirka, G.H. (2011). Mixing zone regulation for effluent discharges into EU waters. *Proceedings of the ICE - Water Management*, 165.

Bleninger, T., Niepelt, A. and Jirka, G.H. (2010a). Desalination plant discharge calculator. *Desalination and Water Treatment*, 13, 2010, 156–173.

Bleninger, T., Niepelt, A., Jirka, G.H., Lattemann, S., Purnama, A., Al-Barwani, H.H. and Doneker, R.L. (2010b). Environmental hydraulics framework of the design of discharges from desalination plants. *Proc. Intl. Symposium on Environmental Hydraulics*, 23.-25.06.10, Editors: Christodoulou G.C. and Stamou A.I., Athens, Greece.

Brumley, B.H. and Jirka, G.H. (1987). Near-Surface Turbulence in a Grid-Stirred Tank. *J. Fluid Mechanics*, Vol. 183, 235–263.

Brumley, B.H. and Jirka, G.H. (1988). Air-Water Transfer of Slightly Soluble Gases: Turbulence, Interfacial Processes and Conceptual Models. *J. Physico-Chemical Hydrodynamics*, Vol. 10, No. 3, 295–319.

Brutsaert, W.H. and Jirka, G.H. (1984). *Gas Transfer at Water Surfaces.* D. Reidel Publishing Co.

Chan, W.C. and Scriven, L.E. (1970). Absorption into irrotational stagnation flow. A case study in convective diffusion theory. *Industrial & Engineering Chemistry Fundamentals*, volume: 9 issue: 1 pages: 114.

Chen, D. and Jirka, G.H. (1995). Experimental-study of plane turbulent wakes in a shallowwater layer. *Fluid Dyn. Res.*, 16 (1), 11–41.
Chen, D. and Jirka, G.H. (1997). Absolute and convective instabilities of plane turbulent wakes in a shallow water layer. *Journal of Fluid Mechanics*, 338, 157–172.
Chen, D. and Jirka, G.H. (1998). Linear stability analysis of turbulent mixing layers and jets in shallow water layers. *Journal of Hydraulic Research*, 36:5, 815–830.
Chu, C.-R. and Jirka, G.H. (1992). Turbulent Velocity and Oxygen Concentration Fluctuations in the Surface Layer of a Grid-Stirred Tank. *Intl. Journal of Heat and Mass Transfer*, 35, No. 8, 1957–68.
Chu, C.-R. and Jirka, G.H. (2003). Wind and stream flow induced reaeration. *J. Env. Eng.* 129 No. 12, 1129–1136.
Chu, V.H. and Jirka, G.H. (1986). Buoyant surface jets and plumes in environmental fluid mechanics. *Encyclopaedia of fluid mechanics*, N. Cheremisinoff, ed., Chap. 27, Gulf, Houston.
Chu, V.H., Wu, J.H. and Khayat, R.E. (1991). Stability of transverse shear flows in shallow open channels. *J. Hydraul. Eng.-ASCE*, 117 (10), 1370–1388.
Davies, P.A., Mofor, L.A. and Neves, M.J. (1997). Comparisons of Remotely Sensed Observations with Modeling Predictions for the Behaviour of Wastewater Plumes from Coastal Discharges. *International Journal of Remote Sensing*, Vol. 18, No. 9, pp. 1987–2019.
Detert, M., Klar, M., Wenka, T. and Jirka, G.H. (2007). Pressure- and velocity-measurements above and within a porous gravel bed at the threshold of stability. *Developments in Earth Surface Processes*, Volume 11, 2007, pp. 85–105.
Doneker, R.L. and Jirka, G.H. (1990). CORMIX1: An Expert System for Hydrodynamic Mixing Zone Analysis of Conventional and Toxic Submerged Single Port Discharges. Tech. Rep., *DeFrees Hydraulics Laboratory, School of Civil and Env. Eng., Cornell University* (also published by U.S. Environmental Pro-tection Agency, Tech. Rep. EPA/600/3–90/012, Environmental Research Lab, Athens, Georgia, 1990).
Doneker, R.L. and Jirka, G.H. (1998). D-CORMIX: A decision support system for hydrodynamic mixing zone analysis of continuous dredge disposal sediment plumes. *Proc. of the 25th Annual Conference on Water Resources Planning and Management, ASCE, Jun. 1998.*
Doneker, R.L., Nash, J.D. and Jirka, G.H. (2004). Pollutant transport and mixing zone simulation of sediment density currents. *J. Hydraul Eng-Asce*, 130 (4), 349–359.
EC - European Community (2000). *Water Framework Directive.* Official Publication of the European Community, L327, Brussels.
Fischer, H.B., List, E.J., Koh, R.C.Y., Imberger, J. and Brooks, N.H. (1979). *Mixing in Inland and Coastal Waters.* Academic Press, New York.
Giger, M., Dracos, T. and Jirka, G.H. (1991). Entrainment and mixing in plane turbulent jets in shallow water. *J. Hydraul Res*, 29 (5), 615–642.
Harriott, P. (1962). A random eddy modification of the penetration theory. *Chemical engineering science*, 17, 149–154.
Herlina and Jirka, G.H. (2004). Application of LIF to investigate gas transfer near the air-water interface in a grid-stirred tank. *Experiments in Fluids* 37 (3), 341–349.
Herlina and Jirka, G.H. (2007). Turbulent gas flux measurements near the air-water interface in a grid-stirred tank. C.S. Garbe, R.A. Handler, and B. Jähne, editors, *Transport at the Air Sea Interface - Measurements, Models and Parameterizations.* Springer Verlag.
Herlina and Jirka, G.H. (2008). Experiments on gas transfer at the air-water interface induced by oscillating grid turbulence. *J. Fluid Mechanics*, Vol. 594, pp. 183–208.
Herlina, H., Jirka, G.H. and Murniati, E. (2007). Oxygen transfer dominated by buoyant convective instability. *Proc. 5th International Symposium on Environmental Hydraulics.*
Herlina, H., Wissink, J.G. and Jirka, G.H. (2008). Gas transfer through the air-water interface in a turbulent flow environment. *7th Euromech Fluid Mechanics Conference*, Manchester.

Herlina, H., Wissink, J.G. and Jirka, G.H. (2010). Direct numerical simulation of interfacial mass-transfer with far-field homogeneous turbulent flow. *6th International Symposium on Gas Transfer at Water Surfaces*, Kyoto.

Hinterberger, C. (2004). Dreidimensionale und tiefegemittelte large eddy simulation von flachwasserströmungen. Dissertation series of the Institute for Hydromechanics, Karlsruhe Institute of Technology, University Press, Karlsruhe, Germany.

Hinterberger, C., Fröhlich, J. and Rodi, W. (2007). Three-dimensional and depth-averaged large-eddy simulations of some shallow water flows. *Journal of Hydraulic Engineering* 133 (8), 857–872.

Holley, E.R. and Jirka, G.H. (1986). *Mixing and Solute Transport in Rivers*. Field Manual, U.S. Army Corps of Engineers, Waterways Experiment Station, Tech. Report E 86 11.

Hunt, J.C.R. and Graham, J.M.R. (1978). Free stream turbulence near plane boundaries. *J. Fluid Mech.* 84, 209–235.

Ingram, R.G. and Chu, V.H. (1987). Flow around islands in Rupert Bay: An investigation of the bottom friction effect. *Journal of Geophysical Research*, 92(C13), 14521–14533.

Jirka, G.H. (1982a). Multiport Diffusers for Heat Disposal: A Summary. *J. Hydraulics Division, ASCE*, (108), HY12, pp. 1423–68.

Jirka, G.H. (1982b). *Turbulent Buoyant Jets in Shallow Fluid Layers*. In: Turbulent Jets and Plumes, Rodi, W. (Ed.), Pergamon Press.

Jirka, G.H. (1994). Shallow Jets. In: *Recent Advances in the Fluid Mechanics of Turbulent Jets and Plumes*, P.A. Davies and M.J. Valente Neves (Eds), Kluwer Academic Publishers, Dordrecht.

Jirka, G.H. (2001). Large scale flow structures and mixing processes in shallow flows. *J. Hydraul Res*, 39(6), 567–573.

Jirka, G.H. (2004). Integral model for turbulent buoyant jets in unbounded stratified flows. Part I: Single round jet. *Environ Fluid Mech*, 4(1), 1–56.

Jirka, G.H. (2006). Integral model for turbulent buoyant jets in unbounded stratified flows, Part 2: Plane jet dynamics resulting from multiport diffuser jets. Environ Fluid Mech, 6(1), 43–100.

Jirka, G.H. (2007). Buoyant surface discharges into water bodies. II: Jet integral model. *J. Hydraul Eng-Asce*, 133(9), 1021–1036.

Jirka, G.H. and Akar, P.J. (1991). Hydrodynamic Classification of Submerged Single-Port Discharges. *J. Hydraulic Engineering*, ASCE, (117), 1095–1111, HY9.

Jirka, G.H. and Arita, M. (1987). Density currents or density wedges: boundary layer influence and control methods. *J. Fluid Mechanics*, Vol. 177, 186–206.

Jirka, G.H., Bleninger, T., Burrows, R. and Larsen, T. (2004). Management of point source discharges into rivers: where do environmental quality standards in the new EC-water framework directive apply? *Journal of River Basin Management*, Vol. 2, Issue 1, 2004, www.jrbm.net

Jirka, G.H. and Doneker, R.L. (1991). Hydrodynamic Classification of Submerged Multiport Diffuser Discharges. *J. Hydraulic Engineering*, ASCE, (117), 1113–1128, HY9.

Jirka, G.H., Doneker, R.L. and Hinton, S.W. (1996). User's Manual for CORMIX: A Hydrodynamic Mixing Zone Model and Decision Support System for Pollutant Discharges into Surface Waters. Tech. Rep., DeFrees Hydraulics Laboratory, Cornell University (also published by U.S. *Environmental Protection Agency*, Tech. Rep., Environmental Research Lab, Athens, Georgia).

Jirka, G.H. and Fong, H.L.M. (1981). Vortex dynamics and bifurcation of buoyant jets in crossflow. *J. of the Engineering Mechanics Div.*, ASCE, Vol. 107, EM 6.

Jirka, G.H. and Harleman, D.R.F. (1973). The mechanics of submerged multiport diffusers for buoyant discharges in shallow water. R.M. Parsons Lab. Water Res. Hydrodyn., Tech. Rep. 169, *Mass. Inst. Tech., Cambridge*, Mass.

Jirka, G.H. and Harleman, D.R.F. (1979a). Cooling impoundment: Classification and analysis. *J. Energy Div. ASCE* 1 05:29 1–309.

Jirka, G.H. and Harleman, D.R.F. (1979b). Stability and mixing of vertical plane buoyant jet in confined depth. *J. Fluid Mechanics*, Vol. 94, pp. 275–304.

Jirka, G.H. and Herlina, H. (2008). Ecological Processes: Reaeration. In Sven Erik Jorgensen and Brian D. Fath, *Encyclopedia of Ecology*, 1st Edition, Elsevier B.V., Oxford, pp. 2975–2981.

Jirka, G.H., Herlina and Niepelt, A. (2010). Gas transfer at the air-water interface: experiments with different turbulence forcing mechanisms. *Exps. Fluids*, Vol. 49, Nr. 1, pp. 319–327.

Jirka, G.H. and Ho, A.H.-W. (1990). Gas Transfer at the Water Surface: Measurements of Gas Concentration Fluctuations. *Journal of Hydraulic Engineering*, 116, No. 6, 835–847.

Jirka, G.H. and Katavolab, D.S. (1979). Supercritical withdrawal from two-layered fluid systems. *J. Hydraulic Research*, Vol. 17, Issue 1, pp. 53–62.

Jirka, G.H. and Lee, J.H.W. (1994). Waste disposal in the ocean. *Water Quality and its Control*, Vol. 5 of Hydraulic Structures Design Manual, M. Hino (Ed.), A.A.Balkema Publishers, Rotterdam.

Jirka, G.H. and Uijttewaal, W.S.J. (2004a). Shallow Flows: A Definition. *Proc. Shallow Flows*, Editors: Jirka and Uijttewaal, Taylor and Francis Group, London.

Jirka, G.H. and Uijttewaal, W.S.J. (2004b). Shallow flows: A definition. In: *Shallow Flows*, G.H. Jirka and W.S.J. Uijttewaal, eds., Selected papers of the International Symposium on Shallow Flows, 16–18 June 2003, Delft, The Netherlands, A.A. Balkema Publishers, Leiden, The Netherlands.

Jirka, G.H. and Watanabe, M. (1980). Thermal structure of cooling ponds. *J. Hydraul. Div. ASCE*. 106:70 1–15.

Jones, G.R., Nash, J.D., Doneker, R.L. and Jirka, G.H. (2007). Buoyant surface discharges into water bodies. I: Flow classification and prediction methodology. *J. Hydraul Eng-Asce*, 133(9), 1010–1020.

Kühn, G. (2008). *Untersuchungen zur Feinsedimentdynamik unter Turbulenzeinfluss*. Dissertation series of the Institute for Hydromechanics, Karlsruhe Institute of Technology, University Press, Karlsruhe, Germany, 2007/2, ISBN: 978-3-86644-205-4

Lee, J.H.W. and Jirka, G.H. (1981). Vertical round buoyant jet in shallow water. *Journal of the Hydraulics Division*, ASCE, 107(HY12), 1651–1675.

Moog, D.B. and Jirka, G.H. (1995). Tilting Wind-Water Tunnel. *Air-Water Gas Transfer*, Jähne, B. and Monahan, E. Ed.s, pp. 89–99, AEON Verlag, Hanau.

Moog, D.B. and Jirka, G.H. (1998). Analysis of Reaeration Equations Using Mean Multiplicative Error. *Journal Environmental Engineering*, 124, 104–110.

Moog, D.B. and Jirka, G.H. (1999). Stream Reaeration in Non-Uniform Channel Flow: Macro Roughness Enhancement. *Journal Hydraulic Engineering*, 125, 1, 11–16.

Negretti, E.M., Socolofsky, S.A. and Jirka, G.H. (2008a). Linear stability analysis of inclined two-layer stratified flows. *Phys. Fluids* 20, 094104.

Negretti, M.E., Socolofsky, S.A., Rummel, A.C. and Jirka, G.H. (2005). Stabilization of cylinder wakes in shallow water flows by means of roughness elements: an experimental study. *Exp Fluids*, 38 (4), 403–414.

Negretti, M.E., Zhu, D.Z. and Jirka, G.H. (2007). Barotropically induced interfacial waves in two-layer exchange flows over a sill. *J. Fluid Mech*. 592, 135.

Negretti, E.M., Zhu, D.Z. and Jirka, G.H. (2008b). The effect of bottom roughness in twolayer flows down a slope. *Dynamics of Atmospheres and Oceans*, Volume 45, Issues 1–2, May 2008, pp. 46–68.

Opila, F., Bleninger, T., Ramachandran, A. and Doneker, R.L. (2009). CorHyd User Manual, Internal Diffuser Hydraulics Model, MixZon Inc., www.mixzon.com

Rummel, A.C., Socolofsky, S.A., von Carmer, C.F. and Jirka, G.H. (2005). Enhanced diffusion from a continuous point source in shallow free-surface flow with grid turbulence. *Phys Fluids*, 17 (7), paper 075105.
Seol, D.G. and Jirka, G.H. (2010). Quasi-two-dimensional properties of a single shallow-water vortex with high initial Reynolds numbers. *Journal of Fluid Mechanics*, 665, 274–299.
Socolofsky, S.A. and Jirka, G.H. (2004). Large-scale flow structures and stability in shallow flows. *J. Environ Eng Sci*, 3 (5), 451–462.
Socolofsky, S.A. and Jirka, G.H. (2005). Special Topics in Mixing and Transport Processes in the Environment. Course Script Environmental Fluid Mechanics, *IfH, Karlsruhe Institute of Technology, and Coastal and Ocean Engineering Division, Texas A&M University*.
Spreafico, M. and van Mazajk, A. (1993). Alarmmodell Rhein. Ein Modell für die operationelle Vorhersage des Transportes von Schadstoffen im Rhein. Technical Report I-12, *Internationale Kommission zur Hydrologie des Rheingebiets*, Lelystad.
Summer, W., Schmidt, O. and Zhang, W. (1994). *Buoyant Surface Discharges Analysed by the Expert System CORMIX and Compared with Delft Hydraulics Laboratory Data Under Varying Flow Conditions*. Hydroinformatics '94, Proceedings of the International Association for Hydraulic Research, Rotterdam; Balkema, Vol. 1.
Theofanous, T.G., Houze, R.N. and Brumfield, L.K. (1976). Turbulent mass transfer at free, gas liquid interfaces with applications to open channel, bubble and jet flows. *Intl J. Heat Mass Transfer*, 19, pp. 613–624.
Tsanis, I.K., Valeo, C. and Diao, Y. (1994). Comparison of Near-Field Mixing Models for Multiport Diffusers in the Great Lakes. *Canadian Journal of Civil Engineering*, Vol. 21. Feb.
Turney, D.E. and Banerjee, S. (2008). Transport phenomena at interfaces between turbulent fluids. *Aiche Journal*, Vol. 54, Issue 2, pp. 344–349.
Uijttewaal, W.S.J. and Jirka, G.H. (2003). Grid turbulence in shallow flows. *Journal of Fluid Mechanics*, 489, 325–344.
Uijttewaal, W.S.J., Lehmann, D. and van Mazajk, A. (2001). Exchange processes between a river and its groyn fields: Model experiments. *J. Hydraul. Eng.-ASCE*, 127 (11), 928–936.
Valentine, E.M. and Wood, I.R. (1979). Experiments in longitudinal dispersion with dead zones. *Journal of the Hydraulics Division-ASCE*, 105 (HY8), 999–1016.
von Carmer, C.F. (2000). LDA-LIF System zur Untersuchung großräumiger kohärenter Strukturen in flacher turbulenter Strömung. Delgado A., Werner F., Ruck b., Leder A., Dopheide D., eds, Lasermethoden in der Strömungsmesstechnik, *8 GALA-Fachtagung*, pp. 18.1–18.9 Shaker Verlag, Aachen.
von Carmer, C.F. (2005). Shallow turbulent wake flows: Momentum and mass transfer due to large-scale coherent vortical structures. Dissertation series of the Institute for Hydromechanics, Karlsruhe Institute of Technology, University Press, Karlsruhe, Germany.
von Carmer, C.F., Rummel, A.C. and Jirka, G.H. (2009). Mass Transport in Shallow Turbulent Wake Flow by Planar Concentration Analysis Technique. *J. Hydraul Eng-Asce*, 135 (4), 257–270.
Weitbrecht, V. (2004). Influence of Dead-Water Zones on the Dispersive Mass Transport in Rivers. Dissertation series of the Institute for Hydromechanics, *Karlsruhe Institute of Technology*, University Press, Karlsruhe, Germany, www.uvka.de/univerlag/volltexte/2004/11/
Weitbrecht, V., Kuhn, G. and Jirka, G.H. (2002). Large scale PIV-measurements at the surface of shallow water flows. *Flow Measurement and Instrumentation*, Vol. 13 (5–6), pp. 237–245.
Weitbrecht, V., Seol, D.-G., Negretti, E., Detert, M., Kühn, G. and Jirka, G.H. (2011). PIV measurements in environmental flows: Recent experiences at the Institute for Hydromechanics in Karlsruhe. *J. Hydro-environ. Res.*, 5 (4), 231–245.
Weitbrecht, V., Socolofsky, S.A. and Jirka, G.H. (2008). Experiments on mass exchange between groin fields and main stream in rivers. *J. Hydraul. Eng.-ASCE*, 134 (2), 173–183.

Weitbrecht, V., Uijttewaal, W.S.J. and Jirka, G.H. (2004). 2-D particle tracking to determine transport characteristics in rivers with dead zones. In: *Shallow Flows*, G.H. Jirka and W.S.J. Uijttewaal, eds., Selected papers of the International Symposium on Shallow Flows, 16–18 June 2003, Delft, The Netherlands, A.A. Balkema Publishers, Leiden, The Netherlands.

White, B.L. and Nepf, H.M. (2007). Shear instability and coherent structures in shallow flow adjacent to a porous layer. *Journal of Fluid Mechanics*, 593, 1–32.

Zhang, W. and Zhu, D.Z. (2011). Near-Field Mixing Downstream of a Multiport Diffuser in a Shallow River. *J. Environ. Eng.* 137, 230.

Part I

Shallow flows

Chapter 2

Horizontal mixing in shallow flows

Wim S.J. Uijttewaal
Delft University of Technology, Delft, The Netherlands

ABSTRACT: Many environmental flows can be considered as shallow. Clear examples are found in low-land rivers, lakes and coastal areas. The large width to depth ratio of shallow flows can give rise to flow structures that have two-dimensional characteristics. Those eddies with vertical axes contribute to the transverse exchange of mass and momentum and are therefore important for mixing processes. In most cases large eddies are generated in the wake of an obstacle or in the unstable shear layer further downstream. With a strong vertical confinement and little dissipation, large structures could be formed by the merging of vortices of equal sign (upcascading). In this paper various experiments will be addressed that were aimed at revealing the mechanisms of the generation and evolution of large eddies in shallow flows. The flow configurations comprise the shallow grid turbulence and shallow mixing layers. The experiments reveal that all large eddy structures carry the signature of the vertical confinement. The growth in length scale can be due to large-scale instability of the shear layer, selective dissipation of small-scale turbulence, and merging of vortices. The latter is difficult to observe in a dissipative flow containing 3D-turbulence because of the disturbing effects of the fluctuations and the limited lifetime of an eddy. The observations made with the experiments provide us with information regarding modeling approaches. When a full 3D-LES modeling is not feasible, the 3D-turbulence can be parameterized whereas the flow structures can be resolved. However, care should be taken of the effect of the 3D-fluctuations on the evolution of the large eddies as well as the disturbing effects of secondary circulation.

1 INTRODUCTION

Predicting the flow in rivers accurately is important for river engineering and flood control. As rivers are rather variable regarding discharge, as well as bathymetry the flow is continuously adapting to its boundary conditions. The associated velocity gradients lead to the production of large eddies that, due to their dimensions, can contribute substantially to the transverse exchange of mass and momentum. It is therefore important to understand the generation mechanisms of coherent structures as well as their development. As current computational resources do not allow solving for the flow in river reaches with a resolution high enough to resolve the large eddy structures, these phenomena have to be accounted for in a simplified and parameterized model. A proper understanding of the phenomena helps in formulating these parameterizations. Many environmental flows such as rivers are bounded in the vertical by a free surface and a bed. The vertical confinement puts constraints to the length scale by which momentum is exchanged between the bed and the water column. The properties of the bed are therefore of paramount importance for the overall flow pattern. Due to the space available, mixing lengths in the horizontal directions can however be much bigger thus allowing for more effective mixing and momentum exchange. For these large mixing lengths to develop in the presence of high dissipation the shear layer should be hydrodynamically unstable giving rise to the accumulation of kinetic energy in large eddy structures. The shallowness causes the motion of these structures to lie pre-dominantly in the horizontal plane resulting in quasi two-dimensional features.

Though the notion of shallow flows dates far back, its important attraction has been the simplified modelling possibilities with two-dimensions as the ultimate limit. This has been done for the case of idealised 2D-flows in a stratified environment on one hand and the simplified depth and Reynolds averaged evironmental flows on the other.

It has been one of the merits of Gerhard Jirka to arouse interest in the two-dimensional features of an otherwise high-Reynolds number turbulent flow. The presence of a turbulent boundary layer gives rise to vertical mixing and substantial energy dissipation against bed friction. This strong mixing literally blurs the image of vortex dynamics and makes it difficult to distinguish between 2D and 3D behavior. Nevertheless, its importance makes it worthwhile trying to sharpen the picture.

This paper addresses the various mechanisms that govern the horizontal exchange of mass and momentum in shallow flows in general and mixing layers in particular. It builds on the important contributions to this field of research by Gerhard Jirka. From the classical examples of mixing layers, wakes and jets the first one has our main interest. This is mainly because the mixing layer is found in many applications like river confluences, compound channels, groyne fields and harbor entrances. It is also the archetype of a simple unstable shear flow which is well studied for a large variety of conditions.

Figure 1 View on a shallow low-land river with groyne fields and floodplains (source: https://www.beeldbank.rws.nl).

2 SHALLOW FLOWS

All turbulent flows are essentially three-dimensional and for each dimension characteristic length and velocity scales can be identified that apply to the largest flow structures. At smaller scales the turbulence properties are not so much linked to the flow geometry resulting in isotropy towards the smallest dissipative scales. With much space available in the horizontal directions the limitations in the size of the flow structures is governed by local geometrical length scales and the dynamic response of the flow to that. A separation of length scales could result in a flow with little interaction between 3D boundary layer turbulence and the horizontal motion, other than acting as homogeneous mixing over a small length. In that case the flow could be considered as 2D with a rather uniform eddy viscosity, thereby effectively reducing the Reynolds number of the large-scale motion.

In applications in rivers and in laboratory settings, the length scales are not always so disparate. The neglect of the interactions between the different scales is therefore not justified beforehand.

In environmental flows coherent quasi two-dimensional flow structures generally have their velocity components in the horizontal plane and extend from the bed to the free-surface with a strong deformation close to bed as a result of the no-slip boundary condition. They can be generated in various ways. Jirka (2001) distinguishes three mechanisms that transfer kinetic energy to the large-scale eddies: topographical forcing, internal transverse shear instability and secondary instability of the base flow. The strongest generation mechanism is formed by topographical forcing where an obstacle gives rise to flow separation and a transverse shear layer in its wake. Typically, vortices are shed from the obstacle and their downstream evolution is supported by the shear in the wake.

As with all vortices, the velocity gradients in the base flow are important for the energy balance. Particularly with an inflexion point present, like in a classical mixing layer, the flow is unstable and shear instabilities will result in the formation of coherent structures. The length scale and strength of the structures will depend on the local velocity gradient and on the past growth/decay of the long-living structures. With shallow flows also the energy dissipation by the bed friction is important as it covers a major part of the flow domain.

The secondary instabilities of the base flow are more difficult to perceive as they are hardly reported in literature. Only under well controlled laboratory conditions arbitrary motion can be converted into organized coherent structures e.g., when vertical fluctuations are suppressed due to stratification or with the interaction with the free surface (see e.g., Maassen, 2002).

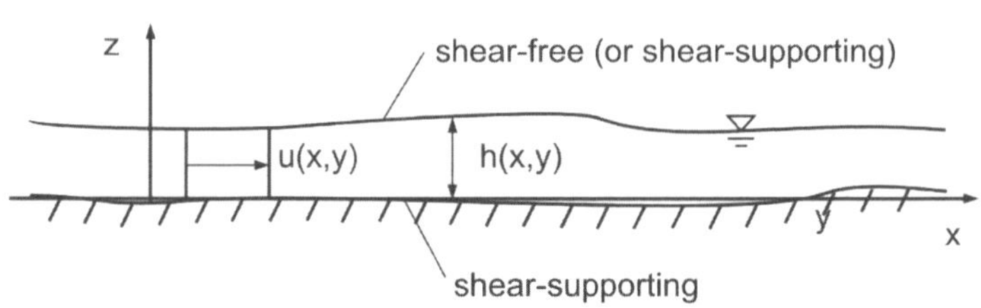

Figure 2 Side view of a free-surface shallow flow (Jirka & Uijttewaal, 2004).

In the following a number of shallow flow experiments will be addressed in which large scale coherent structures were generated in different geometries and under different conditions: grid turbulence, jets, wakes and mixing layers. The experiments are highly schematized and simplified in order to shed some light on the physical mechanisms.

2.1 Jet, wake and grid turbulence

Studies by Dracos, Giger & Jirka (1992) on shallow jets, and Chen & Jirka (1995, 1997) on shallow wakes revealed the characteristic dynamics caused by the vertical confinement. These characteristics are identified by a clear -3-slope in the turbulence energy density spectra and pronounced quasi-periodic oscillations in the velocity signal.

In jet and wake flows the mean flow structure consists of two interacting mixing layers containing opposite vorticity, prohibiting a strong interaction between the alternating vortices. The turbulence length scales are primarily governed by shear instabilities and their growth is established by the entrainment of ambient fluid in the shear layers and the absorption by the large vortices of vorticity present at much smaller scales. The merging of vortices is hardly possible because vortices of equal sign are separated by vortices of opposite sign.

In order to get to a better understanding of the dynamics of interacting vortices, Uijttewaal & Jirka (2003) created grid turbulence in a shallow basin. A uniform approach flow disturbed by an array of cylinders equidistantly placed across the flow produces large-scale vortical structures as a result of the interaction of the wakes behind and the jets between the cylinders (see Figure 5). Since the mean flow

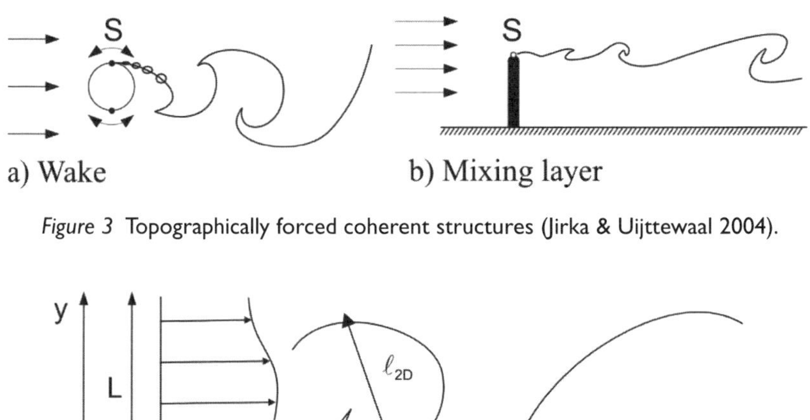

Figure 3 Topographically forced coherent structures (Jirka & Uijttewaal 2004).

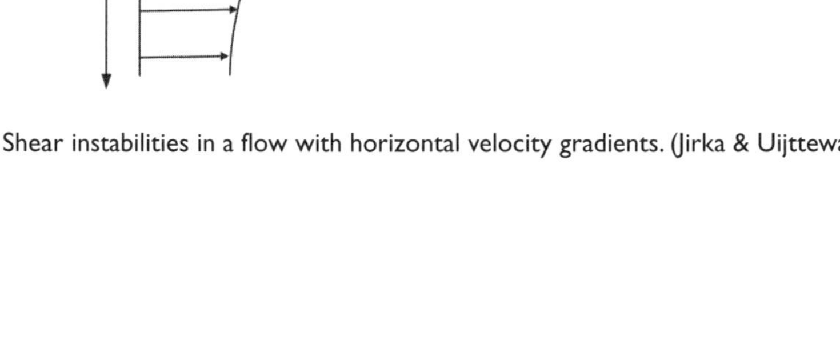

Figure 4 Shear instabilities in a flow with horizontal velocity gradients. (Jirka & Uijttewaal, 2004).

Figure 5 Visualisation (top-view, 2.7 m × 2.2 m) of large scale coherent structures in a flow downstream of a 2D-grid. The flow is from left to right and the blockage by the grid is 80%. (Uijttewaal & Jirka, 2003)

downstream of the grid is uniform again, all vertical vorticity is produced around the cylinders and the dynamics further downstream is fully governed by vortex interaction and bed friction. In idealized 2D condition without a strong dissipation, like stratified flows or soap films, such experiments give rise to a growth in length scales due to the merging of vortices (e.g., Vorobieff, 1999; Sommeria, 1986).

The shallow grid turbulence experiments showed a rapid decay of the turbulent kinetic energy due to the 3D-boundary layer. The spectral distribution of turbulent kinetic energy indicated the downstream development of large eddy structures as recognized from the clear peak in energy density and the -3-slope at the high-frequency side of the peak, Figure 6. One might argue that the peak is not so much formed by the dynamics of the vortices but more by the selective removal of energy at the other frequencies. Nevertheless, the same characteristics are were found in experiments by others (Dracos *et al.*, 1992) and even vortex merging has occasionally been observed. The resulting structures found at 24 cylinder diameters downstream had a small velocity but length scales more than 10 times the water depth. In contrast, experiments that were performed at deeper water did not show such behavior. The experimental difficulty here is to create a turbulent flow with little dissipation despite the presence of 3D-turbulence and at the same time having a strong 2D-dominated vortex interaction. It will be shown below that the energy level can be kept high through the presence of a velocity gradient in the mean flow field.

2.2 Mixing layers

The shallow mixing layers in this paper are defined as open channel flows with a transverse velocity gradient. It is shallow because the mixing layer width and the structures therein are generally larger than the water depth. The lateral velocity

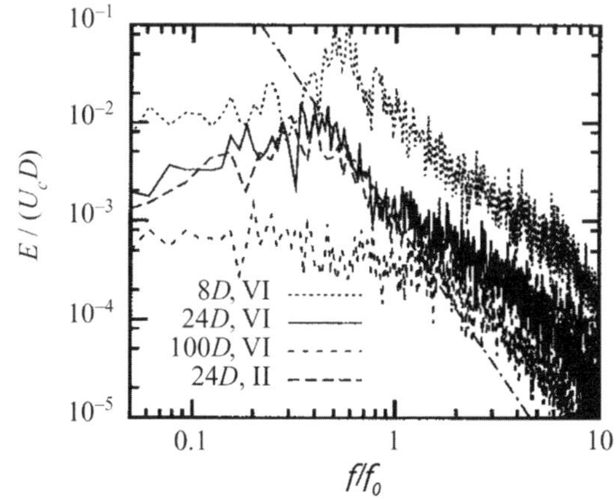

Figure 6 Energy density spectra of transverse velocity components at various downstream distances, expressed in cylinder diameters *D*, from a 2D-grid with 0.05 m water depth, $D = 63$ mm, f_0 is the natural vortex shedding frequency of a cylinder with Strouhal number 0.2. The straight line indicates a -3 slope (Uijttewaal & Jirka, 2003).

difference giving rise to the shear layer in the mean flow can have various causes depending on the specific boundary conditions: e.g., velocity difference at the inflow, transverse differences in depth or roughness.

2.2.1 Differences in inflow velocities as it is found with river confluences, lateral expansions and side cavities

Two regions with different streamwise velocities form a mixing layer at the interface between them (Uijttewaal & Booij, 2000). With a uniform horizontal bed and a transversely uniform free-surface slope the velocity difference is disappearing gradually with downstream distance because the high-velocity side is decelerated by bed friction whereas the low velocity side is accelerated. The eventual transversely uniform flow is established through friction and gravity rather than through a horizontal momentum exchange. Nevertheless, the instabilities in the shear layer develop into eddy structures that give rise to an increase in the horizontal mixing length and consequently a growth of the mixing layer width. The figure shows that the mixing layer width can grow to more than 10 times the water depth over a distance of approximately 100 times the water depth.

The energy density spectra of the transverse velocity fluctuations are depicted in Figure 8. Here the peaks associated with the large eddies are clearly visible as well as their -3-slopes. The peak energy density remains high or even increases going downstream. This is caused by the instable shear layer that supplies the necessary energy for the eddies.

Considering the shallow mixing layer in its depth averaged form, the shear induced Kelvin-Helmholz instabilities result in vortex structures with vertical axes of rotation. From a linear stability analysis that includes bed shear stress and the effective eddy viscosity due to small scale 3D turbulence, a range of wave lengths can be

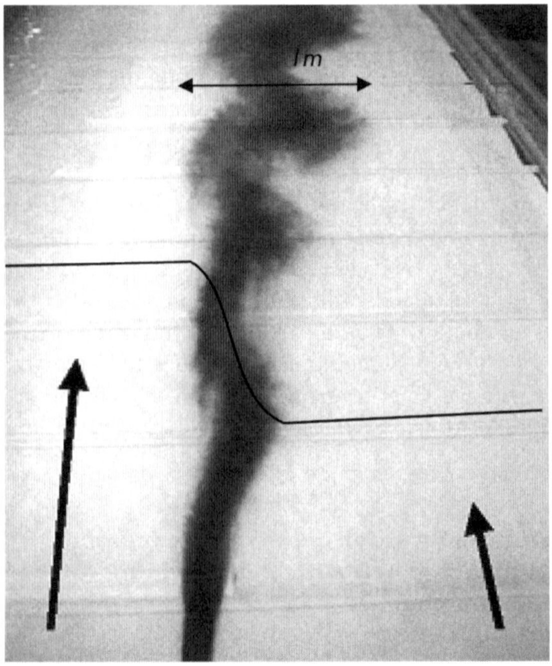

Figure 7 Perspective view on the visualisation of a shallow mixing layer in a horizontal laboratory flume, 3 m wide, 67 mm deep, showing large eddies up to 1 m diameter as well as the small-scale diffusive bottom turbulence (van Prooijen & Uijttewaal, 2002). The arrows indicate the mean streamwise velocities determined by inflow conditions.

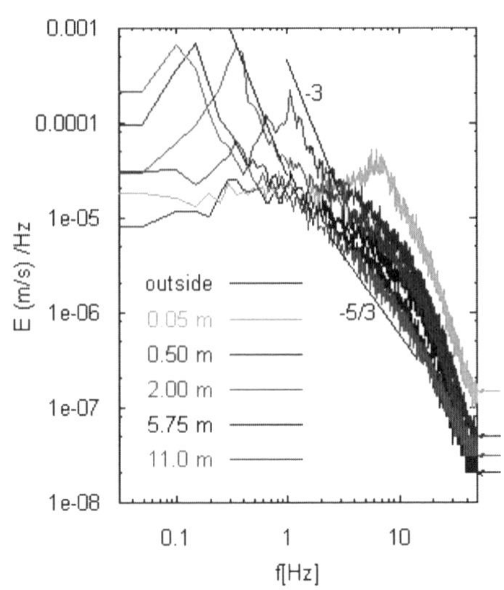

Figure 8 Spectral density of turbulent kinetic energy for various downstream distances of a shallow mixing layer, 67 mm deep. The peaks indicate the presence of large eddies and their -3-slope indicate 2D-dynamics.

identified that have a positive growth rate (Chen & Jirka, 1998). A structure that is advected downstream will grow in accordance with is size and the local mean velocity profile. The energy density distribution for a certain downstream position can be obtained by calculating the accumulated growth along the mixing layer for each wave length. An example of such an analysis is given in figure 9 together with experimental data. It shows that the spectral distribution for the large scale motion is reasonably well predicted and that the energy level from which the structures start to grow is that of the background turbulence (indicated by the horizontal line). The latter conclusion indicates the importance of a correct representation of the disturbances that form the seedlings of the eddy structures. It was demonstrated by Van Prooijen and Uijttewaal (2009), that for a depth averaged simulation of a mixing layer the technique of kinematic simulation is an effective tool. By artificially replacing the small scale dynamics that is filtered out by the depth averaging procedure the resolved shear instabilities find the proper disturbance level to grow from. Note that the growth of the length scales is completely determined by the length scales of the shear instabilities, rather than by vortex merging.

2.2.2 *Lateral variation in water depth as it is found with compound channels with shallow floodplains and a deep main channel*

A schematized version of a river with a high water stage, a so called compound channel flow, is depicted in figure 10. In the main channel and the flood plain two parallel streams are formed with different velocities in accordance with the depth. The flow in the shallower part experiences higher friction resulting in a lower mean velocity. The interfacial shear layer leads to the formation of eddy structures that contribute to the momentum transfer from the main channel to the floodplain.

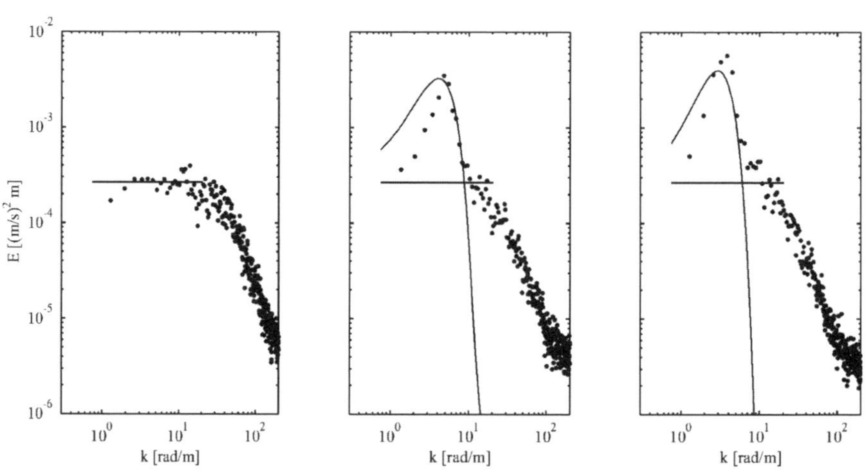

Figure 9 Spectral density of turbulent kinetic energy at 0, 4.5 and 10 m distance downstream of a shallow mixing layer, 67 mm deep. The horizontal line indicates the energy level of the boundary layer flow, the curved line is the cumulative net growth for each wavenumber as determined using a linear stability analysis. (van Prooijen & Uijttewaal, 2002).

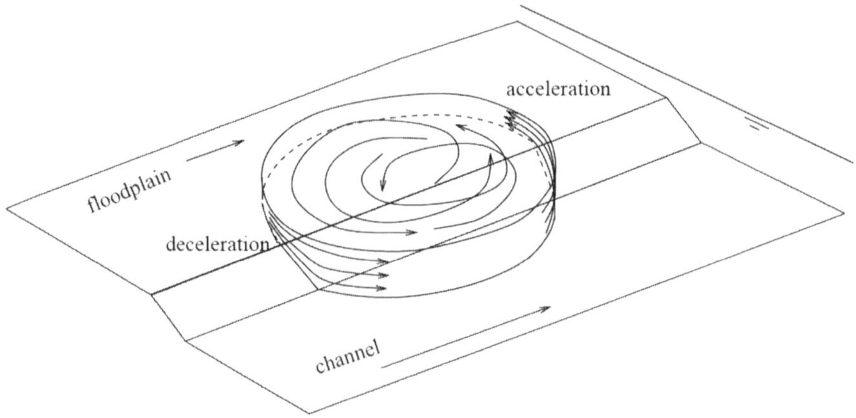

Figure 10 Sketch of a mixing layer generated by a transverse change in water depth. (van Prooijen *et al.*, 2005).

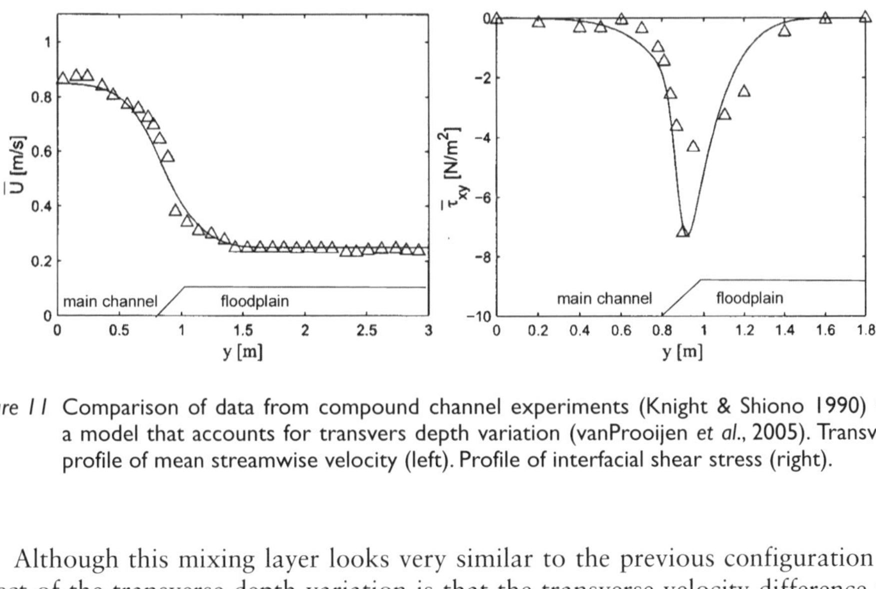

Figure 11 Comparison of data from compound channel experiments (Knight & Shiono 1990) with a model that accounts for transvers depth variation (vanProoijen *et al.*, 2005). Transverse profile of mean streamwise velocity (left). Profile of interfacial shear stress (right).

Although this mixing layer looks very similar to the previous configuration the effect of the transverse depth variation is that the transverse velocity difference will not disappear with downstream distance. Furthermore, any transverse motion in the mixing layer will sense the variation in depth. The vertical compression will accelerate the flow towards the flood plain and decelerate the reverse flow leading to a deformation of the eddy structures. This effect is supposed to enhance mixing proportional to the relative change in water depth (vanProoijen *et al.*, 2005).

Implementing this idea in a simple momentum balance produces good agreement with experimental data. Figure 11 shows that the mixing layer shape and stress distribution is well captured by the model. It should be noted that the good agreement is only an indirect justification of the model assumptions. There are no direct observations of the eddy structures in those experiments.

2.2.3 *Lateral variation in bed friction due to variation in bed material, bed forms or vegetation*

In natural systems the bed is seldom smooth and roughness distributions can be heterogeneous on various scales. In order to study what happens at the transitions from hydraulically smooth to rough beds in streamwise and transverse directions a geometry was arranged as depicted in Figure 12, since in comparison with a smooth bed the flow above a rough bed attains a lower mean velocity. Above the transition between the smooth and the rough bed a mixing layer develops. As with the compound channel case, at a certain downstream distance an equilibrium situation establishes for the transverse distribution of the streamwise velocity. For the case shown it is of the order of 50 times the depth. In contrast with the mixing layer of Figure 7, the width of the mixing layer remains of the order of the water depth indicating that another mechanism is governing the momentum transfer (Vermaas *et al.*, 2011).

With the two previous cases in mind one would think to find similar observations for the mixing layer caused by roughness variation. Surprisingly, no large eddy structures are found in this case. This explains why the mixing layer in Figure 12 remains narrow despite the large downstream distance. Apparently the water depth determines the dominant length scale of mixing. For this configuration the transverse roughness change gives rise to a circulation cell in the plane perpendicular to the main stream as shown in Figure 13. Though the magnitude of the transverse velocity is small, it is large enough to prevent mixing layer eddies to be formed. It is clear from Figure 13 that near the bed at the transition the flow is pushed away from the rough towards the smooth side. The formation of streamwise vorticity is known to occur

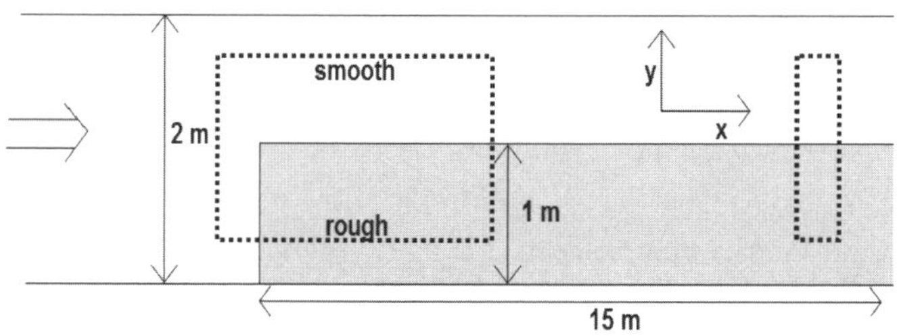

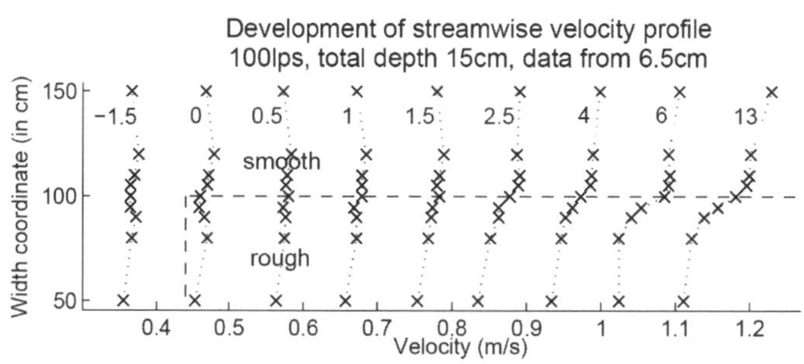

Figure 12 Sketch of an experiment on mixing layer formation due to roughness difference. Top view of the experimental configuration (upper panel). Measured transverse profiles of streamwise velocity are labeled with downstream distance (m) and the velocity scale is shifted 0.1 m/s for each curve (lower panel) (Vermaas *et al.*, 2011).

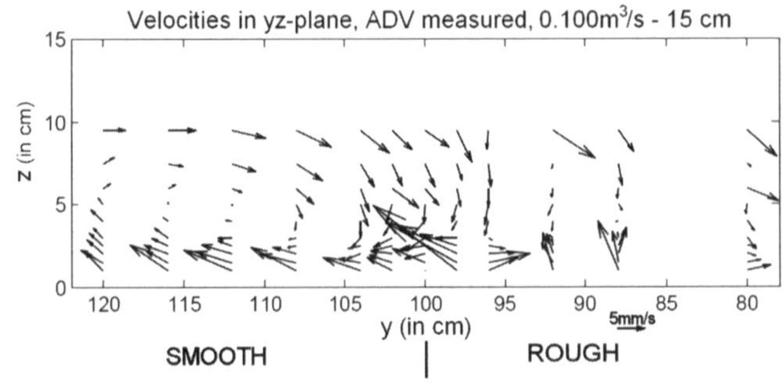

Figure 13 Time averaged velocities in a cross-sectional plane in a developed flow over parallel lanes that differ in roughness. ADV measurements (Vermaas *et al.*, 2011).

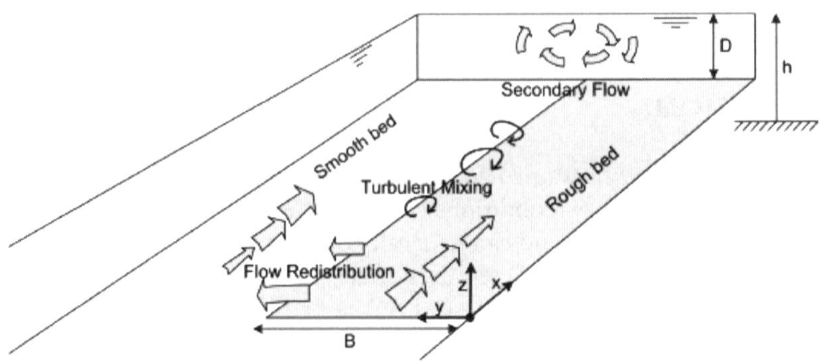

Figure 14 Different mechanisms of momentum transfer over a developing shear layer: flow redistribution, turbulent mixing and secondary flow. (Vermaas *et al.*, 2011).

where the anisotropy in the turbulence is strong (Perkins, 1970). The corner eddies in a straight open channel are examples of the same phenomenon. The strength of the cell is influenced by the abruptness of the lateral change in bed roughness. As the circulation cell is bounded by the vertical dimensions of the flow the mixing layer width is also restricted to this size. Any large-scale structure that would develop in the mixing layer is advected and deformed by the circulation cell before it can attain a significant amplitude. These phenomena can also be encountered in compound channel flows when the transition between the main channel and flood plane is rather abrupt (Tominaga & Nezu, 1991).

3 DISCUSSION

The three examples as addressed in Section 2.2 all lead to the formation of a shallow shear layer. However, the development of the shear layer and the coherent structures therein will be different for each case, despite the fact that the profiles of mean velocity can look very similar.

The turbulence in the shear layer is characterised by three types of flow structures: large-scale quasi-2D eddies with vertical vorticity, small-scale 3D turbulence, and secondary circulation with streamwise vorticity. The dominance of one over the other with respect to the transfer of momentum will depend on the specific flow configuration and the turbulence generation mechanism.

It is important to notice that for all cases mentioned boundary layer turbulence is present with an essentially 3D nature and a length scale typically much smaller than the water depth. This 'background' 3D-turbulence is found throughout the whole flow domain and is not restricted to the mixing layer. However it has an effect on the generation of eddy structures because it acts as a disturbance on the main flow. At the same time its dissipative character drains energy from the large eddies.

Another complicating factor is that in natural systems often a combination of the three causes is found. For example, the shallow flood plains are usually covered with vegetation whereas the deeper main channel is not. Moreover, with movable beds the roughness in the form of ripples and dunes can develop in mutual interaction with the flow thereby affecting the flow resistance.

4 CONCLUSIONS

The examples of shallow flows provided in this paper reveal that coherent structures can be formed under various conditions. Despite the fact that the mean streamwise velocity distribution is simple and can generally be considers as two-dimensional (i.e., uniform over the depth), subtle 3D features affect the horizontal mixing substantially. These are mainly related to the properties of the bed and the small-scale turbulence generated in the boundary layer including turbulence anisotropy. In order to properly predict the horizontal mixing the effects of bottom boundary layer should be represented in the modeling approach. This requires either a (large eddy) simulation with a resolution sufficiently high so that the energy containing part of the 3D turbulence spectrum is resolved, or a proper parameterization of its effects on the large-scale flow. The former has become feasible for schematized laboratory experiments at moderate Reynolds numbers. The latter will still be necessary during the coming decades when it concerns the simulation of high Reynolds-number river flows on a prototype scale (vanProoijen & Uijttewaal, 2009).

ACKNOWLEDGEMENTS

The author like to thank Bram van Prooijen, David Vermaas and Ton Hoitink for their valuable contributions to the various parts of the research described in this paper.

REFERENCES

Chen, D. and Jirka, G.H. (1995). Experimental study of plane turbulent wakes in a shallow water layer. *Fluid Dynamics Res.* 16, 11–41.

Chen, D. and Jirka, G.H. (1997). Absolute and convective instabilities of plane turbulent wakes in a shallow water layer, *J. Fluid Mech.* 338, 157–172.

Chen, D. and Jirka, G.H. (1998). Linear stability analysis of turbulent mixing layers and jets in shallow water layers, *J. Hydraulic Research*, 36 (5), 815–830.

Dracos, T., Giger, M. and Jirka, G.H. (1992). Plane turbulent jets in a bounded fluid layer. *J. Fluid Mech.* 241, 587–614.

Jirka, G.H. (2001). Large scale flowstructures and mixing processes in shallowflows, *J. Hydraulic Research* 39 (6), 567–573.

Jirka, G.H. and Uijttewaal, W.S.J. (2004). Shallow flows: a definition. *in Shallow Flows*, Jirka & Uijttewaal eds. Balkema, The Netherlands. pp. 3–11.

Knight, D.W. and Shiono, K. (1990). Turbulence measurements in a shear layer region of a compound channel. *J. Hydraul. Res.* 28 (2), 175–196.

Maassen, S.R., Clercx, H.J.H. and van Heijst, G.J.F. Self-organization of quasi-two-dimensional turbulence in stratified fluidsin square and circular containers, *Phys. Fluids* 14 (7), 2150–2169.

Perkins, H.J. (1970). The formation of streamwise vorticity in turbulent flow. *J. Fluid Mech.* 44 (4), 721–740.

Prooijen, B.C. van and Uijttewaal, W.S.J. (2002). A linearized model for the evolution of large-scale turbulence structures in shallow mixing layers. *Physics of Fluids* 14 (12), 4105–4114.

Prooijen, B.C. van and Uijttewaal, W.S.J. (2009). The relevance of a back-scatter model for depth-averaged flow simulation, *Flow Turbulence and Combustion* 82 (1), 73–91.

Prooijen, B.C. van, Battjes, J.A. and Uijttewaal, W.S.J. (2005). Momentum exchange in uniform compound channel flow. *J. Hydraulic Engineering,* Vol. 131 (3), pp. 175–183.

Sommeria, J. (1986). Experimental study of the two-dimensional inverse energy cascade in a square box. *J. Fluid Mech.* 170, 139–168.

Tominaga, A. and Nezu, I. (1991). Turbulent structure in compound open-channel flows, *J. Hydraulic Engineering*, 117 (1).

Uijttewaal, W.S.J. and Booij, R. (2000). Effects of shallowness on the development of free-surface mixing layers. *Physics of Fluids* 12 (2), 392–402.

Uijttewaal, W.S.J. and Jirka, G.H. (2003). Grid turbulence in shallow flows, *J. Fluid Mech.* 489, 325–344.

Vermaas, D.A., Uijttewaal, W.S.J. and Hoitink, A.J.F. (2011). Lateral transfer of streamwise momentum caused by a roughness transition across a shallow channel, *Water Resources Research*, 47, W02530.

Vorobieff, P., Rivera, M. and Ecke, R.E. (1999). Soap film flows: Statistics of two-dimensional turbulence. *Phys. Fluids* 11, 2167–2177.

Chapter 3

Onset and development of instabilities in shallow shear flows

Mohamed S. Ghidaoui[1], *Man Y. Lam*[1] *and Jun H. Liang*[2]

[1] *Department of Civil & Environmental Engineering, The Hong Kong University of Science and Technology (HKUST), Kowloon, Hong Kong*

[2] *Formerly: Department of Civil & Environmental Engineering, HKUST; Currently: Department of Atmospheric and Oceanic Sciences, University of California, Los Angeles, California, USA*

ABSTRACT: Large scale coherent structures are prominent in free surface flows including estuaries, oceans, lakes and rivers. The structures are in the form of vortices with vertical axis which extend from the bed to the water surface and possess diameters that are far larger than the water depth. Understanding of such coherent structures is important for the mixing and transport of mass (e.g., pollutants and sediments), momentum and energy in surface water flows. The life of the structures involves birth, growth with downstream distance for part of the flow domain followed by decay and ultimately full disappearance. The mechanisms leading to birth and growth are believed to involve flow instabilities which, because of the near two-dimensionality of the flow, evolve under the constraint of enstrophy and energy cascade. The energy and enstrophy constraint – a result of the suppression of vortex stretching due to the vertical confinement by the bed and free surface – promotes growth via vortex merging. On the other hand, the bottom friction, which represents the effect of the background three-dimensional turbulence on the large scale quasi-two dimensional turbulence, suppresses the large scale instabilities, limits their growth and causes them to eventually disappear with distance downstream. In this paper, the role of linear, weakly nonlinear and nonlinear hydrodynamic stability theories in illuminating the mechanisms of formation, growth and then decay of large scale structures in free shear flows is explained. For illustration purpose, the shallow mixing layer is used.

1 INTRODUCTION

I, the lead author, thank the organizing committee for the invitation to take part in the memorial of the late Professor Gerhard Jirka. The first time Gerhard and I met was in the mid-nineties when he delivered a lecture in Hong Kong and we became good friends ever since. His great and inspirational personality and leadership as well as his open, friendly, upbeat and helpful demeanor will be greatly missed.

Current and future researchers in shallow flows are lucky that Gerhard devoted a good part of the past 20 years to this field. He placed the field on strong fundamental footing, stimulated interest in the field amongst researchers, and founded and organized the first shallow flows conference in Delft in 2003 – an event that was instrumental in bringing researchers in this field together for the first time. He provided unconditional and continued support to the lead author of this paper throughout the organization of the second shallow flows conference in Hong Kong, and extended the same support to the organizers of the third shallow flows in the US until his untimely death on February 14, 2010.

Jirka and Uijttewaal (2004) defined shallow flows as being "largely unidirectional, turbulent shear flows ... occurring in a confined layer of depth scale H. This confinement leads to a separation of turbulent motions between small scale three-dimensional turbulence, 3D ≤ H, and large scale two-dimensional turbulent motions, 2D ≥ H, with mutual interaction" and classified their mode of generation into three classes: type A – topographical forcing, type B – internal transverse shear instabilities and type C – secondary instabilities of the base flow. Such turbulent shallow shear and the large scale vortical structures associated with them are a common occurrence in free surface water flows (rivers, estuaries, lakes and oceans) as well as in atmospheric flows. The dynamics of large scale coherent structures in shallow shear flows is an interesting fundamental fluid mechanics problem and plays a key role for mixing and transport of mass, momentum and energy in shallows flows. For example, the understanding gained in this field is needed to explain and model (i) the trapping of contaminants and sediments in the lee side of islands and headlands, (ii) the larger than usual concentration of pollutants within the core of the structure (5 to 10 times larger than the mean), and (iii) the excessive decrease in discharge capacity of composite and compound channels and at river confluences.

Figure 1 sketches the role of the various processes (hydrodynamic instability, bed friction, the flow of energy from mean flow to small and large scale turbulence) on the formation of large scale coherent structures in shallow flows. Consider a free surface water flow with a depth H and where the mean flow velocity has a transversal shear as well as a vertical shear. The transverse shear maybe generated by type A (e.g., wake flow), type B (e.g., compound and composite channel, channel junction, tidal jet) or type C (e.g., boils). The vertical shear is generated by bed friction. At scales smaller than the water depth, instabilities are generated once the Reynolds number exceeds a threshold value and results in three-dimensional turbulence subject to a single constraint: energy. The result is that the energy extracted from the mean flow by the instabilities is fed to the turbulent structures whose size is of the order of the water depth and then it cascades down to the small scales. This cascading is made possible by the action of vorticity stretching.

At scales larger than the water depth, the transverse shear promotes the development of Kelvin-Helmholtz (KH) while the bed friction, which is a gross representation of the effect of sub-depth three-dimensional turbulence on turbulent motions with size much greater than the water H, suppresses it. This suggests that the proper control parameter for turbulent motions with scales larger than water depth is the ratio of the resistance force (bed friction) to the driving force (transverse shear). In fact, Wolanski *et al.* (1984) developed a dimensionless parameter, P, by considering the ratio of the generation of vorticity by transverse shear to the dissipation of vorticity by the bottom friction. Ingram and Chu (1987) proposed a stability parameter, S, on the basis of

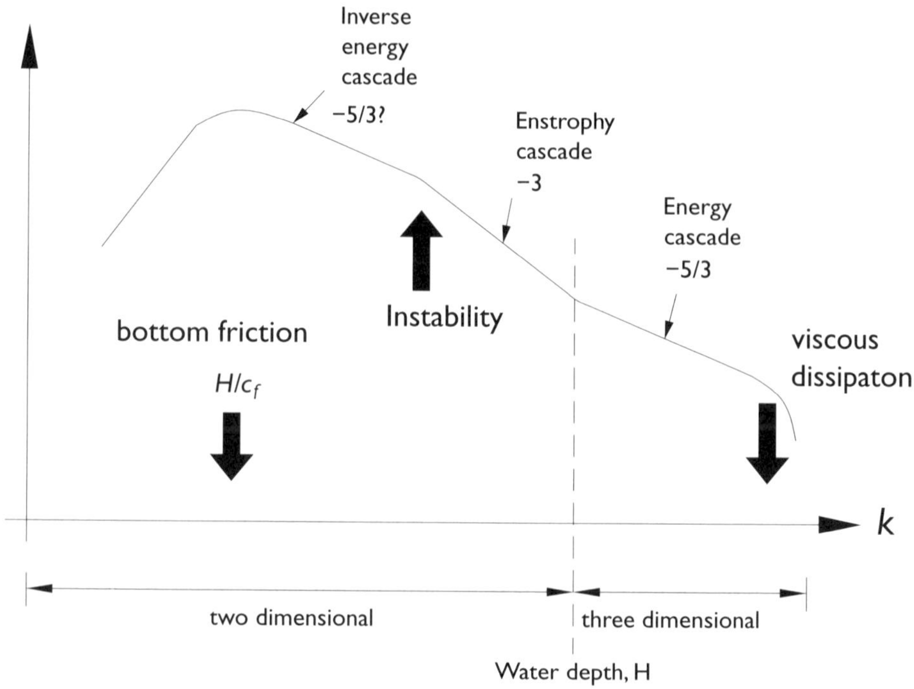

Figure 1 Formation of large scale turbulent structures.

the ratio of the rate of energy loss from the large scale turbulent structures due to the bottom friction to the rate of energy gained by these structures due to the work done by the transverse shear force. While both parameters are one and the same, S is predominantly used in the study of shallow flows perhaps because it is easier to estimate and also emerges naturally in linear stability models derived from the shallow water equations.

When S is smaller than one (i.e., the resistance force (bed friction) is smaller than the driving force (transverse shear)), large scale instabilities develop. The development of such instabilities is subject to two constraints: energy and enstrophy. The enstrophy constraint arises due to the vertical confinement of the flow between the bed and free surface and acts to suppress the stretching of large structure vortices along the vertical direction. The result is a double cascade of energy where a part of the energy that is extracted by the instability from the mean flow and injected into the turbulent field cascades upscale according to the enstrophy constraint and the remainder cascades downscale according to the energy constraint. The energy upscale is evidenced by the spectral analysis in Uijttewaal and Booij (2000) and Uijttewaal and Jirka (2003) and the flow visualization in Uijttewaal and Jirka (2003).

The onset, formation and spatial development of coherent structures in shallow shear flows are the product of hydrodynamic instabilities. Hydrodynamic instabilities can be broadly classified into linear, weakly nonlinear and nonlinear theories. Several studies examined the linkage between the spatial development of coherent structures by performing linear stability studies of the local time mean velocity profiles at a several positions along the streamwise direction (e.g., Chen and Jirka 1997, van Prooijen and Uijttewaal 2002, Socolofsky *et al.*, 2003, Ghidaoui *et al.*, 2006). In these studies, the nonlinear turbulent interactions among modes are invariably reflected in the time mean velocity profiles being used as base flows; thus, while the evolution equations of the instability are linear the base flows implicitly accounts for nonlinearities. In general, it appears that there is a reasonable connection between linear stability and experiments in the early stages of development of the mixing layer, where the structure is of plane wave type. This connection appears to deteriorate once vortical structures begin to form. In fact, van Prooijen and Uijttewaal (2002) concluded that most linearly unstable mode do not represent the spatial development of the shallow mixing layer. Whatever agreement there is between linear stability of the time mean velocity profile and the vortical structures is likely due to the fact that the perturbation of the time mean velocity profile is known to rediscover some of properties of the flow dynamics of the data from which the time mean profile was derived (Ghidaoui *et al.* 2006). In this article, the onset and development of instabilities in turbulent shallow shear free surface flows is examined with the view of mapping the appropriate stability theory to the appropriate stage of development of the large scale structures. Due to the space limitation, the focus of the study is on shallow mixing layers.

2 STABILITY AND THE DEVELOPMENT OF LARGE SCALE STRUCTURES

The mixing layer, jet and wake velocity profiles form the basic building blocks of shallow turbulent shear flows (Figure 2). These profiles contain inflection points and are thus susceptible to the Kelvin-Helmholtz (KH) instability. As stated in the introduction, the force that promotes the growth of a perturbation and, thus, the formation of the KH instability is the mean transverse shear. The force that suppresses the growth of a perturbation and, thus, the formation of the KH instability is the bed friction. Let ΔV be the scale of the velocity difference across the shear layer, V_c the scale of the velocity at the inflection point (the free stream velocity could equally serve and often used for wakes while the centerline velocity is often used for mixing layers), L the scale of the width of the shear layer as well as the horizontal turbulent mixing length and C_f the bed friction coefficient. Therefore, LV_c is the scale of the horizontal turbulent viscosity; $LV_c\Delta V/L$ is the scale of the destabilizing transverse shear stress and $V_c\Delta VLH$ is the scale of the force associated with this force; and $0.5C_fV_c^2\,L^2$ is the stabilizing bed shear force (note that the 0.5 in the bed friction force comes from the often used definition of the Chezy shear stress in terms of the kinetic energy head). As a result, the flow is expected to be susceptible to the KH instability when $0.5\,C_fV_c^2\,L^2 < V_c\Delta VLH$ (i.e., $S = 0.5\,C_fV_cL/H\Delta V < 1$).

A number of laboratory studies have examined the spatial development of instabilities in shallow wakes (e.g., Chen anf Jirka 1995, Uijttewaal and Jirka 2003), in shallow mixing layer (Chu and Babarutsi 1988, van Prooijen and Uijttewaal 2002) and in shallow jets (e.g., Dracos *et al.*, 1992, Biggs *et al.*, 2010). Field observations of the spatial development of instabilities in shallow shear flows are reported in

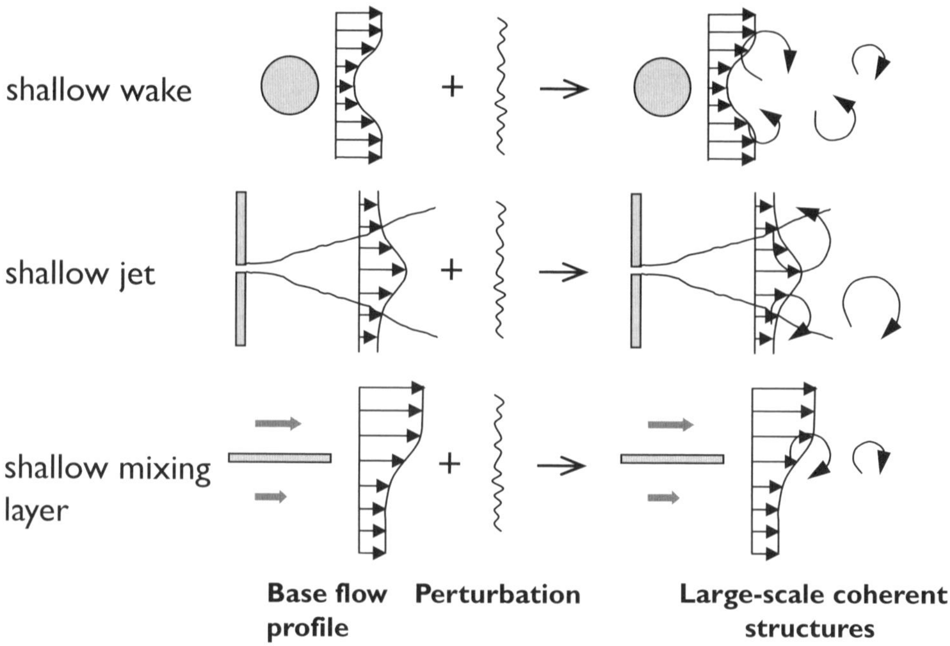

Figure 2 Stability basic building blocks of shallow shear flows.

(Ingram and Chu 1987, Wolanski *et al.*, 1984, Rhoads and Sukhodolov 2008). Numerical modeling results for the spatial development of instabilities in shallow shear flows can be found in Lloyd and Stansby (1997-a & b), Nadaoka and Yagi (1998), Ghidaoui *et al.* (2006), Hinterberger *et al.* (2007), Ghidaoui and Liang (2008), Chu (2010) and others.

Field, laboratory and numerical studies resulted in a consistent picture of the spatial development of instabilities: instabilities arise near the point of generation of the shear layer, they then grow in scale with distance downstream over a certain region and then commence to decay and eventually disappear. These stages of development are due to the increase in S with distance downstream. For a given flow setting, the water depth, the free stream velocity and the velocity difference are fixed. Therefore, $S = C_f LV_c/H\Delta V$ is minimum (i.e., most unstable) at the location where the width of the shear layer L is minimum, which occurs at or near the location where the shear layer is generated. The growth of L with downstream distance is accompanied by the increase in S and leads to the reversion of the flow to stability at point somewhere downstream where S exceeds a critical value. From this point onwards, all large scale instabilities are damped and the coherent structures commence to loose energy with downstream distance and eventually disappear. It is important to note that most shallow water tables available to researchers around the world are not long enough to allow the observation of the full life – onset, growth, decay and demise – of instabilities in shallow shear flows.

Consider the numerical test rig for a shallow mixing layer (Figure 3) chosen to correspond to the experimental test rig of Uijttewaal and Booij (2000) and van Prooijen and Uijttewaal (2002a). In order to minimize artificial reflection from the outflow boundary, a radiation boundary condition is imposed at $x = 15$ m. At the lateral boundaries (i.e., $y = 1.5$ m and $y = -1.5$ m), a free-slip condition is prescribed. For the test case considered here, the fast stream velocity is 0.23 m/s, the slow stream velocity is 0.11 m/s, the water depth is 42 mm and the friction coefficient is 0.0064. Therefore, $S = 0.5C_f LV_c/H\Delta V = 0.1L$. The prescribed small amplitude perturbation at the inflow is white noise. Details of the numerical code with which the current study is conducted can be found in Ghidaoui *et al.* (2006) and the references given in it. A Cartesian mesh with 600 cells in the streamwise direction and 90 cells in the lateral direction is adopted. The cells are of uniform size in the x direction. In order to ensure high flow resolution in the shear layer region, the density of cell distribution along the transverse direction is highest within the shear layer region and becomes stretched exponentially outside this region. The smallest cell width in the transverse direction is the same as the horizontal cell width, i.e., $\Delta y_{min} = \Delta x$. The Courant number for the computation is 0.8, i.e., $\Delta t = 0.8\Delta x_{min}/\left(V + \sqrt{gh}\right)_{max}$ where V is the local flow speed. Note that it is important to ensure that the grid resolution is fine enough to resolve all flow instabilities and structures that are larger than the water depth.

Figures 4.1 and 4.2 display the vorticity field as well as the passive scalar field and illustrate the spatial development of the KH instability. The flow domain can be divided into a plane wave region, a transition region from plane wave to large scale vortical (eddy) structure, a vorticity merging region and decay region. The plane

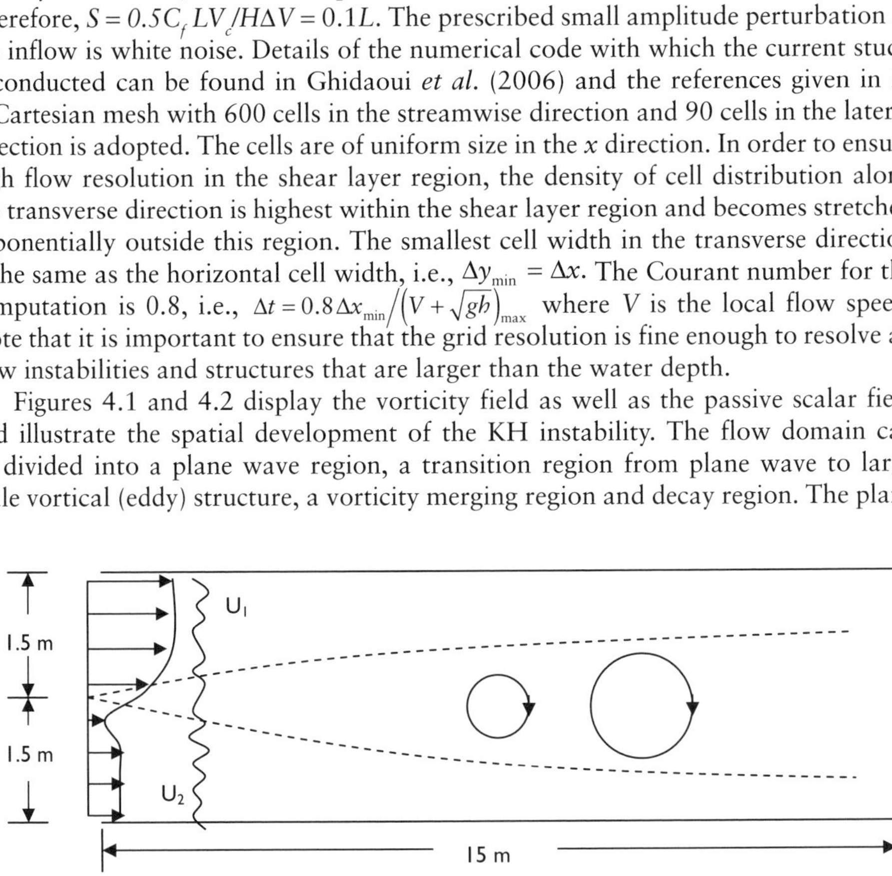

Figure 3 Formation of large scale turbulent structures.

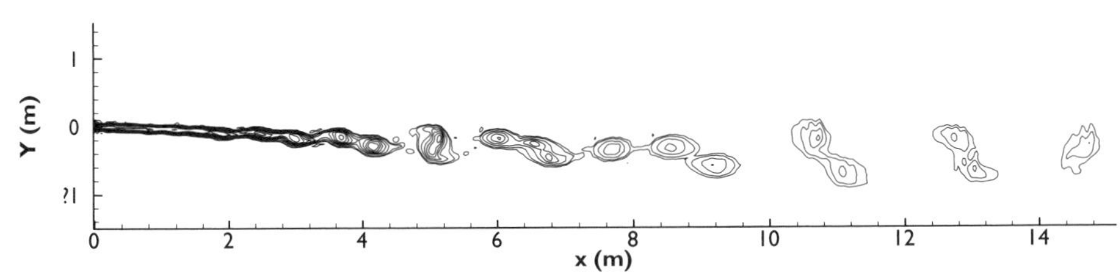

Figure 4.1 Vorticity contour.

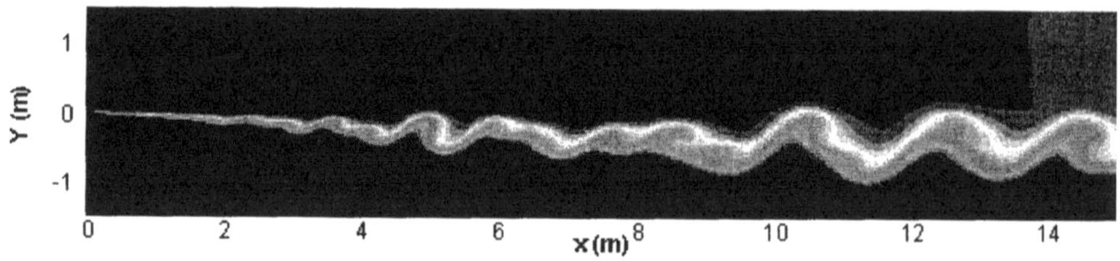

Figure 4.2 Scalar contours.

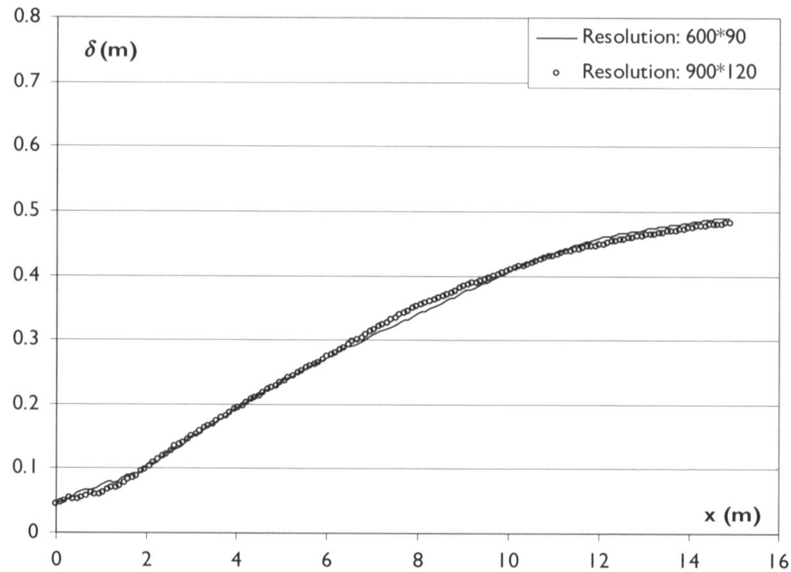

Figure 4.3 Mixing layer width versus downstream distance.

wave region extends from the inflow boundary and to about $x = 2.0$ m and is characterized by a small, but spatially growing, amplitude wave. The transition region is extends from about $x = 2.0$ m to about 3.0 m. The first fully developed large scale vortex is formed around $x = 3.0$ m. The vortex merging region extends from about $x = 3.0$ m to about 12.0 m. Within this region, from $x = 3.0$ m to 5.0 m, the vortices experience slight lateral displacement and the mechanism of vortex induction, to be discussed later in the paper, commences. The process of first vortex merging appears to begin around $x = 5.0$ m and is complete around $x = 9.0$ m. The second vortex merging takes place in the region $x = 9.0$ m to about $x = 12.0$ m. Beyond $x = 12.0$ m, the mechanism of vortex merging appears to become suppressed. In fact, a reduction in strength of the large scale vortex is clearly seen in Figure 4.1 just after $x = 14$ m.

The development of the time-mean of the mixing layer width, δ, associated with the spatial evolution of the mixing layer is given in Figure 4.3 and exhibits a relatively slow growth in the plane wave range $x < 2.0$ m, followed by a relatively large growth associated with vortex formation and merging in the range $2.0 \text{ m} < x < 12.0$ m and

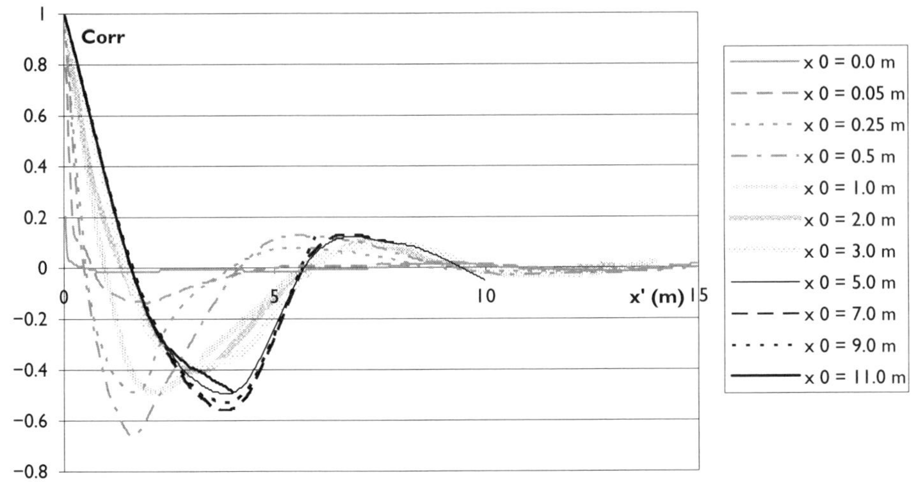

Figure 4.4 Correlation coefficients for a frame of reference moving at 0.9 of the centerline velocity of the shallow mixing layer.

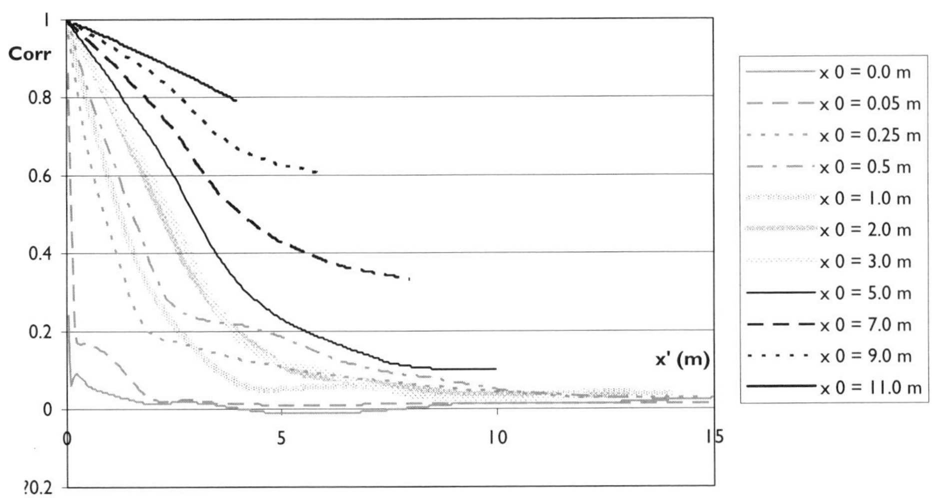

Figure 4.5 Correlation coefficients for a frame of reference moving at the centerline velocity of the shallow mixing layer.

finally a relatively slow growth with δ having a horizontal asymptote in the vortex suppression region. Given that δ is a natural measure of the horizontal length scale of the flow structure L, then $S = 0.1L = 0.1\delta$ and illustrates that the behavior of the stability parameter is analogous to the mixing layer width. Thus, S ranges increases from 0.005 at the inflow to the asymptotic value of 0.05 at the downstream end.

A linkage between the flow development in Figures 4.1 and 4.2 and hydrodynamic instabilities can be gleaned from the study of the spatial evolution of frequencies

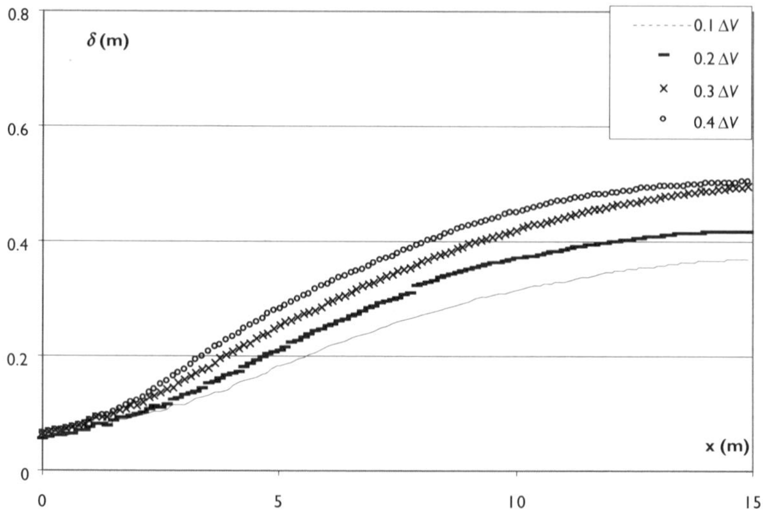

Figure 4.6 Sensitivity of shallow mixing layers to inflow forcing amplitude.

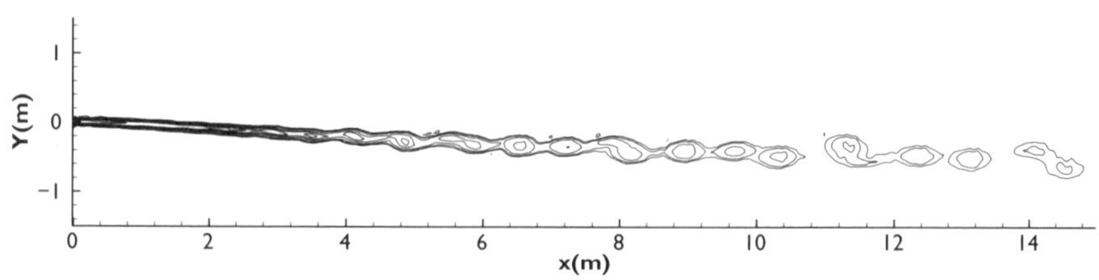

Figure 4.7 Vorticity contour for inflow perturbation of amplitude of 0.1ΔV (the case with amplitude of 0.3ΔV is shown in Figure 4.1).

shown in Figures 5.1 through 5.8. At the inflow, the amplitude of the transverse velocity perturbation is wide band and nearly evenly distributed amongst the different frequencies as is expected from a white noise type function. The amplitude of the spectrum of the transverse velocity at the inflow is chosen to be of the order of 0.1% of the fast free stream velocity. The amplitude of the streamwise velocity perturbation at the inflow is chosen to be of the order of $0.3\Delta V$. A glance at Figures 5.1 through 5.8 reveals that (i) frequencies are selectively amplified and damped with distance (i.e., the flow chooses to amplify some frequencies and damp others); (ii) the large frequencies (small scale) motion are damped; (iii) the dominant (most amplified) frequency switches in a discrete manner towards smaller frequencies (larger scale) with x. These processes are discussed in more details in the remainder of this section.

2.1 Plane wave region

Consider the spectra at $x = 0.5$ m and $x = 1.0$ m. It is clear that the modes with frequencies ranging from about 0.4 to about 3 rad/s are growing with downstream distance

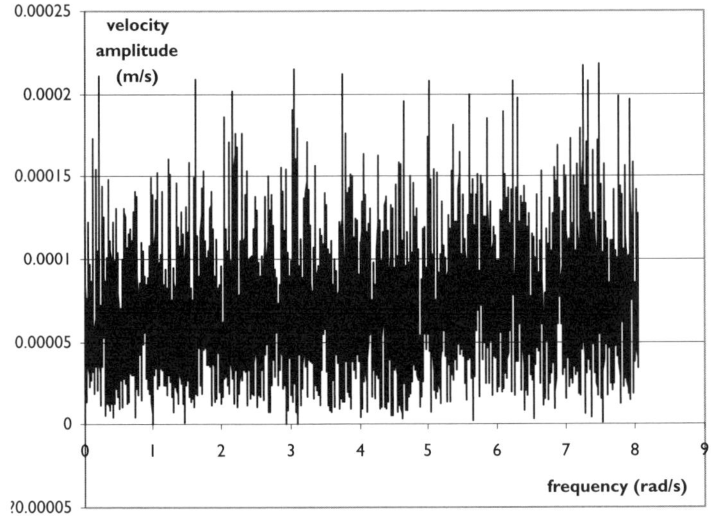

Figure 5.1 Spectrum of the transverse velocity at the inflow ($x = 0$).

while all other modes are decaying. The frequency of the mode that grows the fastest (i.e., most unstable) is around 1.5 rad/s. Its amplitude is about 0.00023 m/s at $x = 0.5$ m and 0.00047 m/s at $x = 1.0$ m, which is very small compared to the base velocity; thus, linear stability theory is expected to be valid in this range. Linear stability analysis for shallow mixing show that the most unstable wavenumber (fundamental) is around $k_1L = 0.45$ and that the range of unstable modes and growth rates vary strongly with the bed friction parameter (e.g., Chu *et al.*, 1991) but only weakly with the Froude number, F_r, (Ghidaoui and Kolyshkin 1999) provided that F_r is below about 0.7. Given that S = 0.0025 at the inflow boundary and referring to the linear stability results of Chu *et al.* (1991), Ghidaoui and Kolyshkin (1999) and others reveals that (i) the range of unstable dimensionless wave number, kL, is 0.1 to 0.9, (ii) the most unstable mode is 0.45 and (iii) the dimensionless growth rate of the instabilities, $\sigma L\Delta V$, ranges from 0 for $kL = 0.1$ and $kL = 0.9$ to 0.1 for the most unstable mode, where σ is the dimensional temporal growth rate of instabilities. Using the relation $f = kV_c = 0.17k$ to translate wave numbers to frequencies gives the following linearly unstable frequency range at the inflow (L = 0.05 m): 0.34 rad/s $\leq f \leq$ 3.06 rad/s and that the most linearly unstable frequency is $f \approx 1.53$ rad/s. In addition, the amplitude of the most unstable mode grows from about 0.00012 at the inflow boundary (see Figure 5.1) to 0.00012 exp $(\sigma x/V_c) = 0.00012$ exp $(0.24{*}0.5/0.17) = 0.00024$ m/s at $x = 0.5$ m and 0.00012 exp $(0.24{*}1.0/0.17) = 0.00048$ m/s at $x = 1.0$ m. Therefore, the unstable frequency range, the most amplified mode and the spatial growth of amplitude obtained from linear analysis are in good agreement with those in Figures 5.2 and 5.3. This implies that the early stage of the spatial development of the mixing layer is governed by linear hydrodynamic stability theory of the steady base flow at the inflow boundary. Where possible ambiguity may arise, the instability that controls the plane wave region will be referred to as the primary instability.

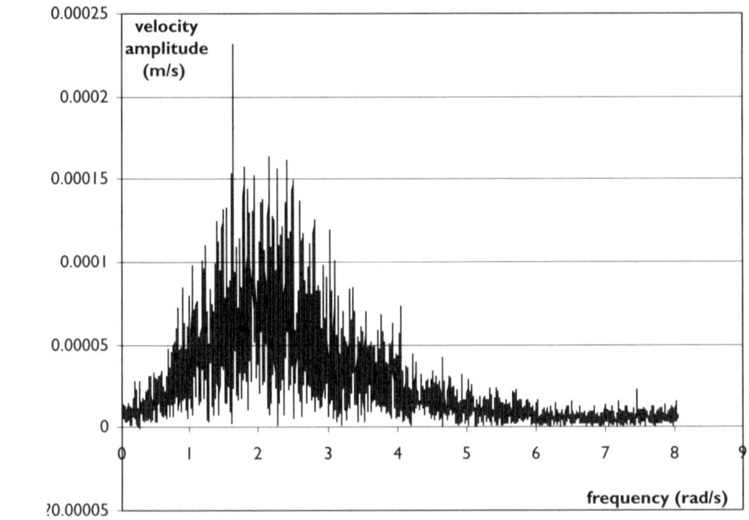

Figure 5.2 Spectrum of transverse velocity at x = 0.5 m.

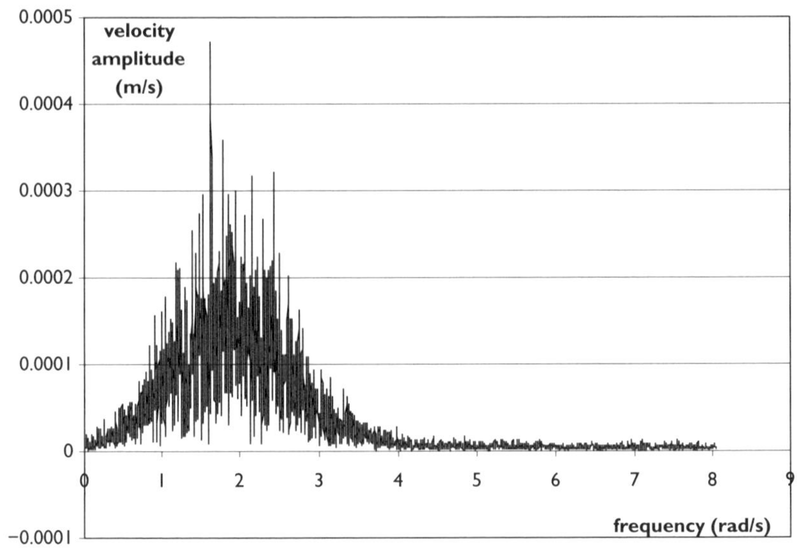

Figure 5.3 Spectrum of transverse velocity at x = 1.0 m.

Chen and Jirka (1997) show that instabilities of shallow shear flows are either of convective or absolute type. In the context of shallow wakes, Chen and Jirka (1997) connected absolute instability to vortex street flow and convective instability to unsteady bubble flow and provided a convincing link between linear stability theory and experimental data. Briefly, a flow is absolutely unstable if a small perturbation

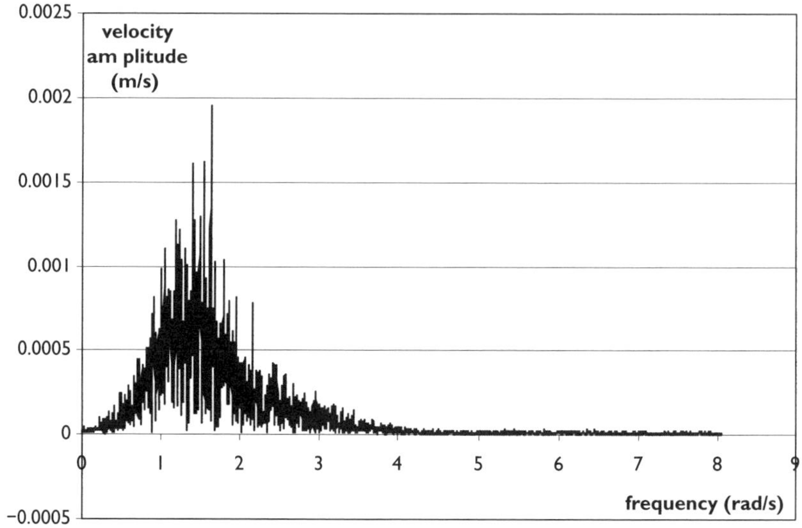

Figure 5.4 Spectrum of transverse velocity at $x = 3.0$ m.

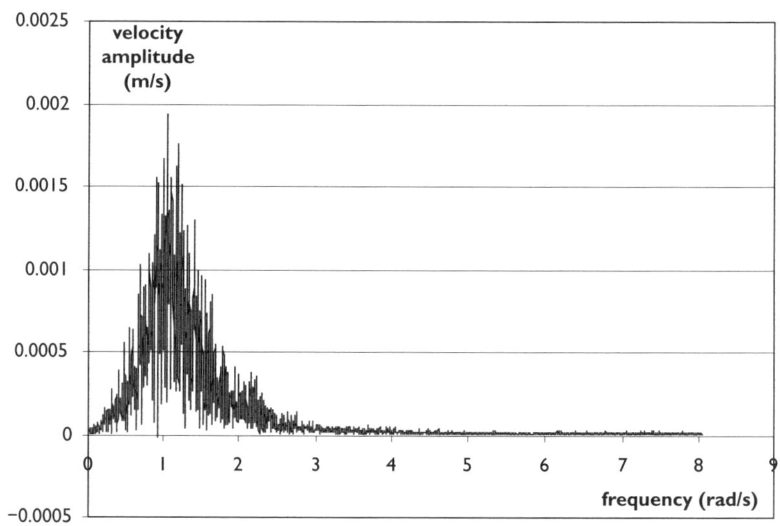

Figure 5.5 Spectrum of transverse velocity at $x = 5.0$ m.

grows with time at a fixed point and propagates in all directions; it is convectively unstable if a perturbation at any fixed point decays with time but grows in a frame of reference that moves with the perturbation. Therefore, perturbations in a convectively unstable flow are carried downstream; thus, require continued forcing at the inflow boundary to maintain the flow structures associated with them. To ascertain the type of primary instability controls the plane wave region, the simulation was repeated

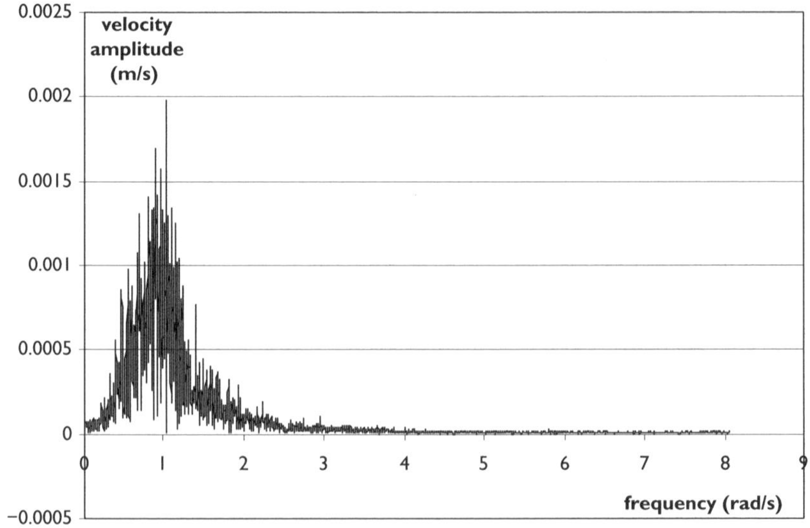

Figure 5.6 Spectrum of transverse velocity at $x = 7.0$ m.

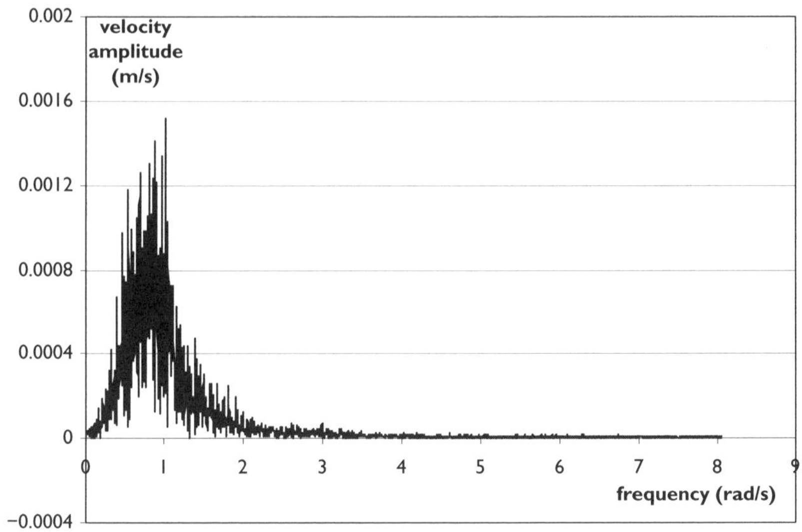

Figure 5.7 Spectrum of transverse velocity at $x = 9.0$ m.

but with no perturbation at the inflow boundary. The results (not shown here) reveal that, in the absence the perturbation, the flow remains stable. The fact that the existence of instability is contingent on the imposition of perturbations at the upstream boundary indicates that the instability is convective and that the flow development is sensitive to the type and amplitude of perturbation.

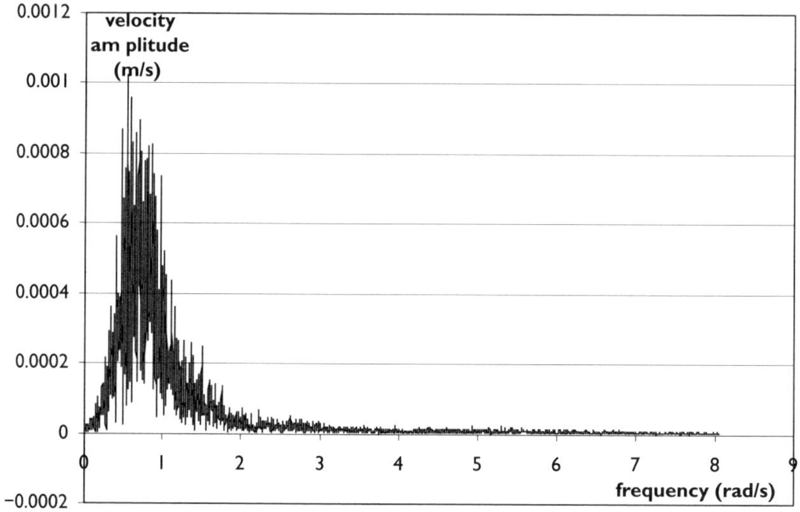

Figure 5.8 Spectrum of transverse velocity at $x = 13.0$ m.

Further evidence into the convective nature of the instability can be gleaned from examining the behavior of the spatial-temporal correlation coefficient for different frame of reference speed:

$$Corr(u(x_0, t_0), u(x_0 + x', t_0 + t)) = \frac{Cov(u(x_0, t_0), u(x_0 + x', t_0 + t))}{SD(u(x_0, t_0))SD(u(x_0 + x', t_0 + t))} \quad (1)$$

where *Corr* stands for correlation, *Cov* denotes covariance, *SD* indicates standard deviation, x_0 and $x_0 + x'$are two spatial sampling locations, t_0 is the start time of sampling, and t is the time lag between the two sampling location such that $t = x'/U$, where U is the velocity of the frame of reference. Figures 4.4 and 4.5 compare the correlation coefficients under different convective velocities. It is shown that the downstream perturbations are positively correlated to the upstream perturbations when the speed of the frame of reference is equal to the convective velocity of the perturbations. When the frame of reference is moved at speed different from the convective velocity (e.g., 0.9 U), negative correlation coefficients are prominent. The conclusion is that the instability is convective and is being advected at the mean speed of the two streams in shallow mixing layers.

The fact that the existence of instability is contingent on the imposition of perturbations at the upstream boundary indicates that the instability is convective and that the flow development is sensitive to the type and amplitude of perturbation. This fact is illustrated in Figure 4.6, where the mixing layer is forced with different amplitudes at the inflow boundary. Near the inflow ($x < 1.5$ m), the growth of the mixing layer width is almost the same for different inflow forcing amplitudes. This occurs in the plane wave region where the large vortical turbulent structure that would responsible for much of the spread of the mixing layer has not yet formed. From $x \approx 2$ m onwards, the larger is the perturbation amplitude the faster is the growth of the width of the mixing layer. For the mixing layer forced by the perturbation with amplitude 0.3ΔV, vortices begin to roll up between $x \approx 2.0$ m and $x \approx 3.0$ m, while vortices begin to roll up between $x \approx 3.0$ m and $x \approx 4.0$ m for the mixing layer forced by the perturbation with amplitude 0.1ΔV. The sensitivity to inflow perturbations implies that caution is needed when comparing results of different experiments, repeating sets of experiments and comparing models with data.

2.2 Transition region from plane wave to vortical structure

The transition from the linear to nonlinear regimes causes the flow structure to change from a plane wave type near the inflow boundary to a vortex type near $x = 3$ m (see Figures 4.1 and 4.2). While the dominant frequency of the vortical structure when it first forms agrees well with linear stability theory, the amplitude does not. In fact, the actual variation in amplitude is ¼ that of the amplitude predicted by the primary instability. This slow growth in amplitude of the instability as one follows its progress from its onset at $x \approx 3.0$ m to its development at $x \approx 3.0$ m suggests that weakly nonlinear analysis be invoked. Starting from the shallow water equation and using multiple scale and singular perturbation theory, Kolyshkin and Ghidaoui (2003) derived the following governing equation for the amplitude of the instability in the weakly nonlinear region:

$$\frac{\partial A}{\partial \tau} = \sigma A + \gamma \frac{\partial^2 A}{\partial \xi^2} - \mu |A|^2 A, \tag{2}$$

which is a Ginzburg-Landau (GL)-type equation. In it, A is the amplitude, $\xi = \varepsilon(x - c_g t)$ is a slow streamwise variable which moves with a group velocity c_g, and $\tau = \varepsilon^2 t$ is a slow time variable, ε is the small perturbation parameter and reflects that weakly-nonlinear analysis is performed within the unstable domain but very close to the neutral stability conditions, and $\sigma = \sigma_r + i\sigma_i$, $\gamma = \gamma_r + i\gamma_i$ and $\mu = \mu_r + i\mu_i$ are complex coefficients. The slow time variables define the variation of the amplitude of the instability in transition range, while the fast variables define its oscillatory part.

The right hand side of the GL contains terms representing linear amplification (1st term), diffusion (2nd term) and nonlinear saturation (3rd term). If $\sigma_r < 0$ then the amplitude will decay for sufficiently large time. On the other hand, if $\sigma_r > 0$ then A grows. If $\mu_r > 0$ the dissipation increases with the growth of A and the nonlinearities tend to saturate the instability. This is the case here since the amplitude of the fundamental mode with frequency 1.5 rad/s grows initially, saturates in the region $x \approx 3.0$ m to $x \approx 7.0$ m where the saturation amplitude is about 0.002 m/s and then decays afterwards (see sequence of spectral plots in Figures 5.1 to 5.8). Such instability is usually referred to as supercritical instability in the hydrodynamic stability literature. At saturation, a balance between the 1st and 3rd term in the right hand side of the GL and leads to a saturation amplitude relation $A_s = \sqrt{\sigma_r / \mu_r}$.

The saturation state is associated with the appearance of the vortical structures at the end of the transition region between $x \approx 2.0$ to 3.0 m. This implies that that vortex structure is the result of the instability of the plane wave structure that existed from the inflow boundary up to $x \approx 2.0$ (see Figure 4.1 and 4.2). Stuart and Diprima (1978) considered the stability of a perturbed plane wave solution that is governed by the GL equation. They show that indeed the plane wave solution is unstable provided $1+\mu_i\gamma_i/\mu_r\gamma_r < 0$. Ghidaoui *et al.* (2006) solved for the coefficients of the GL equation for the case of shallow wakes and found that plane wave solution are unstable. The computed results indicate that $1+\mu_i\gamma_i/\mu_r\gamma_r < 0$. Further confirmation of this condition by direct calculation of the GL coefficient is being pursued and will be communicated in future research papers.

2.3 Vortex merging region

From $x = 3$ m to $x = 5$ m, the vortices are fully developed and their dominant frequency is shifted to about 1.2 rad/s (see Figure 5.5 and as well as Figures 4.1 and 4.2). There are two sources for this shift: nonlinearities and the spreading of the mixing layer with downstream distance (i.e., the basic state, assumed parallel in linear stability studies, is in fact non-parallel). The array of fully developed vortices with the dominant frequency of 1.2 rad/s formed around $x = 5$ m are perturbed by a band of frequencies (see Figure 5.5). As a result, the vortices are displaced laterally in an irregular manner (see Figure 4.1). This lateral displacement combined with the vortex induction mechanism promotes secondary instability, discussed below, of the array of vortices to perturbation with frequencies smaller than of 1.2 rad/s. In fact, from $x = 5$ m to around $x = 7$ m, the amplitude of the array of fully developed vortices with the dominant frequency of 1.2 rad/s grows little, the amplitude of the motion with frequencies higher than 1.2 rad/s gets damped and the amplitude of the motion with frequencies smaller than 1.2 rad/s grows. From $x = 7$ m to around $x = 9$ m, the amplitude of the vortices with frequency of 1.2 rad/s are damped, the amplitude of the motion with frequencies higher than 1.2 rad/s continue to get damped and the amplitude of the motion with frequencies smaller than 1.2 rad/s continue to grow. The shift towards smaller frequencies (large scale motion) in the region $x = 5$ m to $x = 9$ m is associated with the process of vortex pairing. The complete shift of the dominant frequency from 1.2 rad/s to its sub-harmonic frequency of 0.6 rad/s occurs between $x = 9$ m to $x = 13$ m and signals the full merging of pairs of vortices as can be gleaned from Figures 5.7 and 5.8 and visually seen in Figures 4.1 and 4.2.

The connection between secondary instability of the array of vortices and the shift towards lower frequencies and vortex merging, proposed as possible mechanism for the doubling in scale of the dominant structures in the above paragraph, is examined. Temporal stability analysis with the following longitudinal velocity perturbation (u'), lateral velocity perturbation (v') and water depth perturbation h' is performed:

$$u'(X,Y,0) = [2A\sin(2\pi k_1 X/L_X) + A\sin(2\pi k_2 X/L_X)]\exp(-Y^2/\delta_0^2) \quad (3)$$

$$v'(X,Y,0) = [-2A\sin(2\pi k_1 X/L_X) - A\sin(2\pi k_2 X/L_X)]\exp(-Y^2/\delta_0^2) \quad (4)$$

$$h'(X,Y,0) = \left[2A\frac{H}{\sqrt{gH}}\sin(2\pi k_1 X/L_X) + A\frac{H}{\sqrt{gH}}\sin(2\pi k_2 X/L_X)\right]\exp(-Y^2/\delta_0^2)$$

(5)

in which k_1 is the fundamental (most linearly unstable mode); $k_2 = k_1/2$ is its first sub-harmonic; $A = 0.134\Delta V$ is the amplitude of the sub-harmonic velocity perturbation; and $2A$ is amplitude of the velocity perturbation of the fundamental. The amplitude of the water height fluctuation is set according to the relationship $h' = u' \, H/\sqrt{gH}$.

The mapping between the temporal and the spatial instability is made using $x = V_c \, t$. The size of the moving frame (window) used to follow the development of the mixing layer is 5.6 m × 5.6 m. The window is divided into 512 × 512 cells. In the temporal simulation, periodic boundary conditions are enforced in the streamwise edges of the moving frame and the free slip boundary condition is imposed on its transverse edges. Linear stability analysis as well as the simulation results presented in Figures 5.2, 5.3 and 5.4 show that most unstable (fundamental mode) is $k_1 L_0 = 0.45$, where $k_1 = 2\,\pi/\lambda_1$, where λ_1 is the wavelength of the most unstable mode and $L_0 = 0.025$ m is the scale of the initial width of the mixing layer. Therefore, $\lambda_1 = 6.28/0.45/0.025 = 0.35$ m; thus, the number of unstable structures in a window is $5.6/0.6977 = 16$ (i.e., $k_1 = 16$).

The development of the primary and secondary instability is illustrated in Figures 6.1 through 6.6. At $t = 0$, the vorticity at the center of the mixing layer is perturbed and exhibits a plane wave-like structure (Figure 6.1). At $t = 7.14$ s (i.e., $x_{frame} = 7.14{*}0.17 = 1.21$ m ≡ distance by which the frame moves downstream from its initial position), a roll of 16 vortices is formed as a result of the KH instability (Figure 6.2). At $T = 15.00$ s (i.e., $x_{frame} = 2.55$ m) the vortices are displaced laterally in an alternate manner with respect to the centerline of the mixing layer due to the sub-harmonic perturbation (Figure 6.3). Vortex induction and velocity difference across the mixing layer cause any vortex displaced in the positive y to acquire a faster advective speed while the vortex immediately downstream of it, displaced in the negative y direction, to acquire a slower advective speed. As a result, the vortices become arranged in pairs where the vortices in each pair twist around one another for sometime (Figures 6.3 and 6.4), then merge (Figure 6.5) and acquire a new structure in which the number of vortices is half what it was (Figure 6.6). The merger causes the scale to double and hence the dominant mode to shift from k_1 to $k_1/2$. Analysis whose results are not shown here due to space limitation indicate that the vortical structure in Figure 6.6 is unstable to the second sub-harmonic perturbation (i.e., mode with $k_1/4$) and leads to another merger, doubling in scale and shift of the dominant mode from $k_1/2$ to $k_1/4$ further downstream. The process of vortex merging continues downstream up the point where the scale of the vortices is large enough so that the bed friction force dominates the mean transverse shear force and prevents further sub-harmonic instabilities from developing.

The energy of the fundamental and the sub-harmonic wavenumbers is plotted with time in Figure 6.7. The fundamental (most unstable) mode $k_1 = 16$ grows linearly for the first to 2 to 3 seconds ($x = 0$ to $x \approx 0.5$ m) where the corresponding structure is plane wave type. From about 3 s to about 7.14 s ($x \approx 0.5$ m to $x \approx 1.21$ m) energy continues to grow but at a diminished rate and saturates at $t = 7.14$ s. This phase of energy growth corresponds to the region where the transition from plane wave structure to a vortex structure occurs. The maximum energy $t = 7.14$ s of the fundamental mode corresponds to the full development of the vortical structure shown in Figure 6.2. In both the plane wave region and the transition region from plane wave to vortex structure, the energy of the sub-harmonic is negligible. In fact, the energy of mode $k_1/2 = 8$ commences to experience significant growth only around

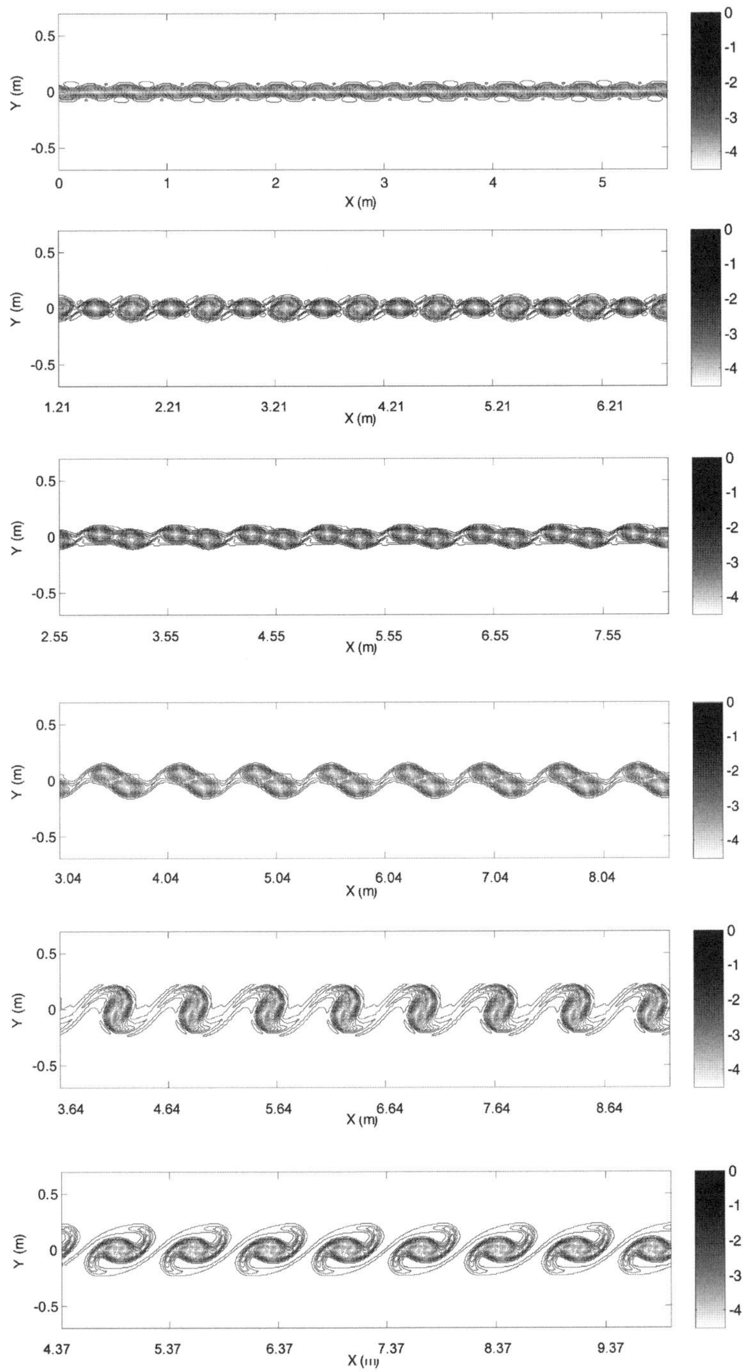

Figures 6.1 to 6.6 (From top to bottom): Vorticity contours at T = 0 s (6.1), 7.14 s (6.2), 15.00 s (6.3), 17.86 s (6.4), 21.43 s (6.5) and 25.71 s (6.6).

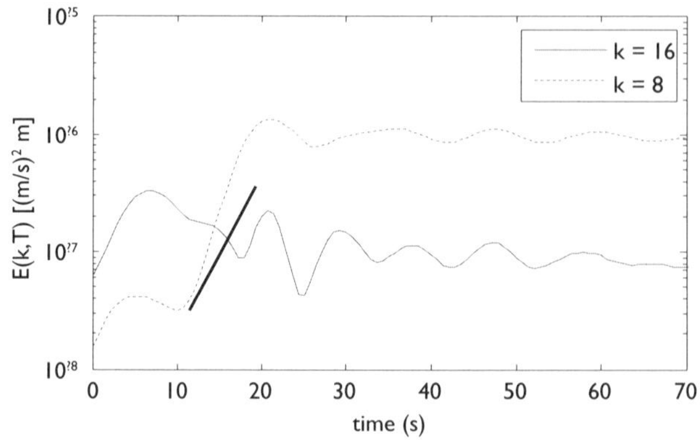

Figure 6.7 Modal energy content against time. The dark line represents the initial growth rate of the sub-harmonic.

$t = 12$ s ($x \approx 2$ m) and is associated with the lateral displacement of the vortices by the sub-harmonic mode and results in zigzag pattern visible at $t = 15$ s (see Figure 6.3). The rate of energy of growth of the sub-harmonic instability is initially exponential, suggesting secondary linear instability of the string of 16 vortices to mode $k_1/2 = 8$, followed by followed by reduced rate of growth and saturation at about t 22 s. This saturation marks the merging of vortices as can be seen in Figure 6.5. It is noted that at this stage, the energy is mostly contained in $k_1/2 = 8$. The flow of energy from base flow to mode $k_1 = 16$ in the early phase of the development then to mode $k_1/2 = 8$ and so on is a clear evidence of the upscale of energy in shallow mixing layer flows.

3 CONCLUSION

Large scale coherent structures are prominent in free surface flows including estuaries, oceans, lakes and rivers. The structures are in the form of vortices with vertical axis which extend from the bed to the water surface and possess diameters that are far larger than the water depth. Understanding of such coherent structures is important for the mixing and transport of mass (e.g., pollutants and sediments), momentum and energy in surface water flows.

The life of the structures involves birth, growth with downstream distance for part of the flow domain followed by decay and ultimately full disappearance. These stages have been investigated in this work for the case of a shallow mixing later. It is found that the early stage of development, the structure is plane wave type and is well represented by linear stability theory under the parallel base flow assumption. The next stage of development involves the transition from plane wave to vortex type structure. In this region, the frequency of the instability is well approximated by the primary linear instability, but the amplitude grows at a significantly slower rate than the primary instability and reaches a saturation state. The slow growth in amplitude and the appearance of a saturation state suggests that the transition is due to supercritical (secondary) instability and that its dynamics is governed by weakly nonlinear stability theory. The next stage of development is dominated by vortex merging process. Each merger in the sequence of mergers is governed by a secondary instability of the already established vortical structure to its sub-harmonic. The result is an upscale of energy. The downscale of energy is linked to the large strains in the braid region connecting the vortices. The last stage defines the suppression of merging, decay and final demise of the large scale vortical structures. In this region, all large scale instabilities are suppressed due to the fact that the force that promotes their growth (mean transverse shear) is smaller than the force that promotes their suppression. It is this rich physics of shallow shear flows combined with their highly important practical implications and relevance that fascinated and attracted Gerhard to this field.

ACKNOWLEDGEMENT

The financial support by the Research Grant Council of Hong Kong, Project No. 612908, is acknowledged.

REFERENCES

Biggs, C., Nokes, R. and Vennell, R. (2010). The dynamics of a steady shallow turbulent jet. *Proceedings of the 17th Congress of the Asia and Pacific Division of the International Association of Hydraulic Engineering and Research*, 21–24 Feb. 2010.

Chen, D. and Jirka, G.H. (1995). Experimental study of plane turbulent wake in a shallow water layer. *Fluid Dyn. Res.* 16(1), 11–41.

Chen, D. and Jirka, G.H. (1997). Absolute and convective instabilities of plane turbulent wakes in a shallow water layer. *J. Fluid Mech.*, 338, 157–172.

Chu, V.H. (2010). Shear instability, wave and turbulence simulations using the shallow-water equations. *J. of Hyd. Env. Res*, Special Issue I on Shallow Flows, Edited by M.S. Ghidaoui and Y.K. Tung, Volume 3, Issue 4, 173–178.

Chu, V.H. and Babarutsi, S. (1988). Confinement and bed-friction effects in shallow turbulent mixing layers. *J. Hydr. Engrg., ASCE*, 114(10), 1257–1274.

Chu, V.H., Wu, J.H. and Khayat, R.E. (1991). Stability of transverse shear flows in shallow open channels. *J. Hydr Engrg, ASCE*, 117(10), 1370–1388.

Dracos, T., Giger, M. and Jirka, G.H. (1992). Plane turbulent jets in a bounded fluid layer. *J. Fluid Mech*, 241, 587–614.

Ghidaoui M.S. and Kolyshkin, A.A. (1999). Linear stability analysis of lateral motions in compound channels with free surface. *J. Hydr Engrg, ASCE*, 125(8), 871–880.

Ghidaoui, M.S. and Liang, J.H. (2008). Investigation of Shallow Mixing Layers by BGK Finite Volume Model. *Int. J. of Com. Fluid Dyn.*, 22(7), 523–537.

Ghidaoui, M.S., Kolyshkin, A.A, Chan, F.C., Liang, J.H. and Xu, K. (2006). Linear and Nonlinear Analysis of Shallow Wakes. *J. Fluid Mech.*, 548, 309–340.

Hinterberger, C., Fröhlich, J. and Rodi, W. (2007). Three-Dimensional and Depth-Averaged Large-Eddy Simulations of Some Shallow Water Flows. *J. Hydr. Engrg, ASCE*, 133(8), 857–872.

Ingram, R.G. and Chu, V.H. (1987). Flow around islands in Rupert Bay: An investigation of the bottom friction effect. *J. Geophys. Res.*, 92, 14521–14533.

Jirka, G.H. and Uijttewaal, W.S.J. (2004). Shallow flows: a definition, in *Jirka, G.H. and Uijttewaal, W.S.J. (Eds) Shallow flows: selected papers of the International Symposium on Shallow Flows*, 16–18 June 2003. Taylor and Francis, 3–11.

Kolyshkin, A.A and Ghidaoui, M.S. (2003). Stability Analysis of Shallow Wake Flows. *J. Fluid Mech.*, 494, 355–377.

Liang, J.H. (2006). Linear and nonlinear analysis of shallow mixing layers. MPhil thesis. Hong Kong University of Science and Technology.

Lloyd, P.M. and Stansby, P.K. (1997-a). Shallow-water flow around model conical islands of small side slope. I: Surface piercing. J. Hydr. Engrg., ASCE. 123(12), 1057–1067.

Lloyd, P.M. and Stansby, P.K. (1997-b). Shallow-water flow around model conical islands of small side slope. II: Submerged. *J. Hydr. Engrg., ASCE.* 123(12), 1068–1077.

Nadaoka, K. and Yagi, H. (1998). Shallow-water turbulence modeling and horizontal large-eddy computation of river flow. *J. Hydr. Engrg., ASCE*, 124(5), 493–500.

Rhoads, B.L. and Sukhodolov, A.N. (2008). Lateral momentum flux and the spatial evolution of flow within a confluence mixing interface. *Water Resources Research*, 44, W08440.

Socolofsky, S.A., von Carmer, C. and Jirka, G.H. (2003) Shallow turbulent wakes: Linear stability analysis compared to experimental data. In *Proc. Intl Symposium on Shallow Flows, Delft*, vol. 1, 133–140.

Stuart, J.T. and Diprima, R.C. (1978). The Eckhaus and Benjamin-Feir resonance mechanisms, *Proc. R. Soc. Lond. A*. 362, 27–41.

van Prooijen, B.C. and Uijttewaal, W.S.J. (2002). A linear approach for the evolution of coherent structures in shallow mixing layers. *Phys. Fluids*, 14(12), 4105–4114.

Uijttewaal, W.S.J. and Booij, R. (2000). Effect of shallowness on the development of free-surface mixing layers. *Phys. Fluids*, 12(2), 392–402.

Uijttewaal, W.S.J. and Jirka, G.H. (2003). Grid turbulence in shallow flows, *J. Fluid Mech*, 489, 325–344

Wolanski, E.J., Imberger, J. and Heron, M.L. (1984). Island wakes in shallow coastal waters, *J. Geophys. Res.*, 89, 10553–10569.

Chapter 4

Shallow flows with bottom topography

G.J.F. van Heijst, L.P.J. Kamp and R. Theunissen
J.M. Burgers Centre and Fluid Dynamics Laboratory,
Department of Physics, Eindhoven University of Technology,
Eindhoven, The Netherlands

ABSTRACT: This paper discusses laboratory experiments and numerical simulations of dipolar vortex flows in a shallow fluid layer with bottom topography. Two cases are considered: a step topography and a linearly sloping bottom. It is found that viscous effects – i.e., no-slip conditions at the non-horizontal parts of the bottom – play an important role in shallow flows with bottom topography. The combination of these viscous effects and the 3D nature of the bottom topography gives rise to generation of vorticity (with a vertical component), which may significantly influence the flow evolution. Obviously, such shallow flows cannot be described simply by the two-dimensional Navier-Stokes equation with an additional 'bottom friction' term.

1 INTRODUCTION

Shallow flows are encountered in many environmental situations, like rivers, harbours, lakes, and coastal regions. Also in industrial configurations shallow flows occur in a rich variety. Because the vertical scale of motion is significantly smaller than the horizontal scales of the large-scale motion, such 'shallow flows' exhibit very interesting dynamics. Due to the pioneering work of Gerhard Jirka and others, the topic of 'shallow flows' has become a well-recognised field of research, and it was Gerhard Jirka (together with Wim Uijttewaal) who took the initiative of organizing the first conference 'Shallow Flows' (see Jirka & Uijttewaal, 2004). By now, laboratory experiments and numerical simulations have revealed a rich phenomenology of shallow flows, in the form of anisotropy of turbulence, meandering of currents, and the behaviour of coherent vortex structures (see e.g., Jirka, 2001; Uijttewaal & Jirka, 2003).

The concept of shallow flows has also been used in an attempt to study quasi-two-dimensional (quasi-2D) turbulence in the laboratory, by creating turbulent flow (driven by electromagnetic forcing) in a shallow layer of electrolyte (see e.g., Tabeling *et al.*, 1991; Xia *et al.*, 2009). In these studies it is assumed that the shallowness of the flow domain ensures a quasi-2D motion, i.e., a planar flow with a Poiseuille-like velocity distribution in the vertical, caused by the no-slip condition at the bottom. Recent experimental and numerical studies by Akkermans *et al.* (2008a, b) and Cieślik *et al.* (2009a, b) have demonstrated that this assumption is generally not correct, however, and that significant vertical motion may be present in such shallow flows.

In a recent paper, Duran-Matute *et al.* (2010) have demonstrated that the simple scaling argument based on the continuity equation commonly adopted for proving that shallow flows are (quasi-) 2D is in general not correct, and that a more subtle scaling argument should be applied.

The present paper focuses on the effect of bottom topography on vortex structures in a shallow fluid layer. A central question is: what happens when a dipolar vortex structure moves over a sloping bottom? Such a situation is observed in the case of an inhomogeneous obliquely incident breaking wave train on a beach, where two vortices of opposite circulation are generated (see e.g., Bühler & Jacobson, 2001). As a first step to analyse this problem, a single vortex over a linearly sloping bottom may be considered as being a segment of a larger 3D vortex ring. In this inviscid approach the motion of the vortex is identical to the self-propulsion speed of the ring, and – in the case of a vortex ring with a uniform vorticity distribution and a circular cross-section – it will thus move without change of shape in a direction parallel to the coast (see Bühler & Jacobson, 2001). As a slightly better model, one could take into account the stretching/squeezing of vortex tubes in the shallow vortex, according to Kelvin's circulation theory leading to changes in its vorticity distribution. This 'modulation' of the vorticity structure implies a shape change and hence a more complicated motion of the vortex (see also Peregrine, 1998).

For the case of a vortex pair moving over a sloping bottom towards the coast, one could argue that each vortex behaves according to the 'vortex-ring model', and that the oppositely signed vortices separate when approaching the coast, subsequently moving away from each other in opposite directions along the coast. This behaviour is similar to that of two oppositely signed point vortices approaching a wall (see Lamb (1932), §155.3), the motion of which is governed by the 'image principle'. Inclusion of viscous effects drastically changes this behaviour, as shown in the numerical simulations by Orlandi (1990) of a 2D dipole structure colliding against a no-slip wall. The boundary layer formed at this wall contains vorticity with a sign opposite to that of the neighbouring vortex core. When the dipole moves closer to the wall, the oppositely signed vorticity is removed by advection induced by the dipole and moves around the dipole cores, which leads to a widening of the primary structure, a splitting of the dipole, and eventually a rebound from the wall.

In their modelling of vortex structures over a sloping bottom, Bühler & Jacobson (2001) have introduced a parameterized 'bottom friction' in order to take into account the drag induced by the no-slip bottom boundary. However, in essence this model is still 2D.

In the present paper we will examine to what extent and how 3D effects play a role when shallow vortices encounter topographic features. Laboratory experiments have been carried out in a shallow layer of electrolyte, with a dipolar vortex forced electromagnetically. Particle Image Velocimetry has been used in order to obtain quantitative information about the flow. In parallel with the experiments, numerical flow simulations have been performed with a finite-element method. The experimental and numerical techniques are described in Sections 2 and 3, respectively. Results on a step topography are discussed in Section 4, while Section 5 presents the experimental and numerical results obtained for the slope configuration. Finally, some general conclusions a formulated in Section 6.

2 LABORATORY ARRANGEMENT

The experimental set-up consists of a horizontal square tank of dimensions 52 cm × 52 cm, which is filled with a layer of sodium chloride solution of thickness H. Beneath the bottom plate of the tank a disk-shaped permanent magnet is placed, with a diameter of 25 mm and a thickness of 5 mm, producing a magnetic field strength of approximately 1 tesla. Two electrode plates are mounted in the fluid along two facing sidewalls of the tank and connected to a power supply. Application of an electric current I through the fluid implies a Lorentz force acting on the fluid above the magnet, by which a motion is induced locally. As the dominant part of the magnetic field has a vertical orientation above the magnet, the combination with the horizontal electrical field lines results in a localized horizontal Lorentz force, which generates a jet-like flow. A pulsed forcing then results in a dipolar flow structure, which propels itself through the fluid. By mounting a step or a sloping bottom on the tank bottom, one may thus examine the behaviour of the dipole when encountering these topographic features.

It is convenient to introduce a Cartesian (x, y, z) – coordinate system, with x, y the horizontal and z the vertical coordinates. The origin of this coordinate system is located above the centre of the magnet on the bottom of the tank. When the electrical current I is made to run in the x – direction, while the principal magnetic field lines are oriented vertically (in the z – direction), the resulting jet/dipole structure will move in the y – direction. The flow field is given by $\mathbf{v} = (x, y, z)$, with u, v, and w the velocity components in the coordinate directions (x, y, z). A schematic drawing of the experimental configuration is presented in Figure 1.

In the experiments the flow was visualized either by sprinkling small dust-like particles on the free surface of the fluid or by injecting a neutrally-buoyant dye in the neighbourhood of the magnet. Quantitative information about the flow was obtained by applying Particle Image Velocimetry (PIV). For this purpose small, passive, almost neutrally buoyant particles were introduced in the fluid, the motion of which near the free surface was visualized by illumination with a horizontal laser light sheet, and recorded from above by a high-resolution digital camera. In this way the horizontal flow field (u, v) at the free surface could be reconstructed in detail, and derived quantities such as the vertical vorticity

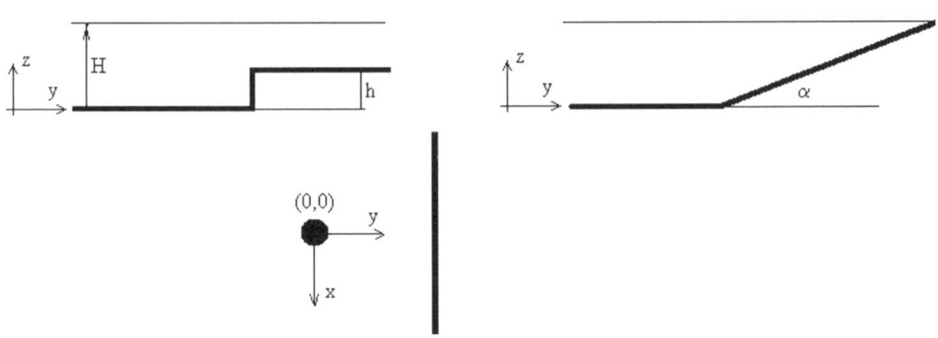

Figure 1 Definition sketches of the coordinate system, the step and slope topographies, and the magnet configuration.

$$\omega_z = \frac{\partial v}{\partial x} - \frac{\partial u}{\partial y}, \tag{1}$$

could be determined. For a detailed description of this laboratory set-up and the PIV method applied, see Akkermans *et al.* (2008a).

Three sets of laboratory experiments were carried out: experiments with a flat bottom, with a step topography, and with a linearly sloping bottom. The flat-bottom experiments served as a reference set, enabling to see the evolution of the undisturbed vortex dipole. The step topography was simply created by mounting a solid plate of thickness h on the tank bottom, whereas the slope topography was created by mounting a thin plate in the tank, inclined under a certain angle α. The forcing magnet was mounted at some distance from the topography, so that the dipole was generated in the flat-bottom part of the tank; after travelling over some distance and being fully developed at that stage, it would then encounter the topography. The step topography and the bottom edge of the slope topography were mounted at $y = y_0 = 35$ mm. In all experiments reported here, the depth of the undisturbed fluid layer measured H = 8 mm. The flow was forced by applying a constant electrical current I of amplitude 2 or 3 A during some time Δt. In the experiments reported here the forcing duration was taken to be $\Delta t = 1$ s, which yielded rather compact dipolar vortices. The Reynolds number Re = UD/v based on the maximum velocity U at the end of the forcing period and the initial diameter D of the dipole measured approximately 1400.

3 NUMERICAL SIMULATIONS

Three-dimensional numerical flow simulations were carried out with a finite-element code, which essentially solves the Navier-Stokes equation (extended with the Lorentz forcing term)

$$\frac{\partial \mathbf{v}}{\partial t} + (\mathbf{v} \cdot \nabla)\mathbf{v} = -\frac{1}{\rho}\nabla p + \nu\nabla^2 \mathbf{v} + \frac{1}{\rho} J_0 \mathbf{e}_x \times \mathbf{B} \tag{2}$$

numerically. Here v is the 3D velocity vector, p is the pressure, ρ the mass density, ν the kinematic fluid viscosity, $J_0 e_x$ the uniform electrical current density in the x – direction, and **B** is the magnetic field produced by the magnet. This magnetic field is known in closed analytical form in terms of elliptical integrals (see e.g., Jackson, 1998). The computational domain was chosen slightly smaller than the experimental flow domain. For a typical run, this domain was meshed with approximately 50,000 mesh elements, with a fine mesh near the bottom and near the step. Proper convergence of the computations was checked by additional mesh refinements in a number of runs. No-slip conditions were applied at the solid boundaries (bottom, step surfaces, sloping bottom), while a stress-free condition was applied at the flat free surface and the lateral side wall (in order to reduce their influence on the evolution of the dipole). Free surface deformations were excluded.

4 STEP TOPOGRAPHY

Laboratory experiments have been performed for different step heights h, although only a few cases will be discussed here: $h = 0$ (no step), $h = 0.5\ H$, and $h = 0.625\ H$. The two latter cases correspond with step heights 4 mm and 5 mm, respectively, with the reference layer depth being $H = 8$ mm. The vortex dipole was created at an initial position $(x, y) = (0, 0)$ and launched in the y – direction, i.e., perpendicular to the step (positioned at $y = y_0 = 45$ mm). This distance was chosen such that the vortex dipole was allowed to be fully developed when arriving at the step topography. Figure 2 presents snapshots of the flow evolution, in terms of the horizontal velocity field (black arrows) and the vertical vorticity ω_z (grey shades) as measured at the free surface. The left column shows the evolving dipolar vortex for $t = 3$ s, 5 s, 7 s, and 9 s for the no-topography case ($h = 0$), while the middle and right columns show the corresponding moments in the flow evolution for the cases $h/H = 0.5$ and 0.625, respectively. The field of view is limited and does not show the full extent of the flow region; in order to show more relevant details, we have zoomed in on one half of the dipolar flow in particular, viz. on the part with negative vorticity ($\omega_z < 0$, visible as lighter grey shading in Figure 2). The horizontal black line in each plot represents the position $y = y_0$ of the step (at least, in the middle and right columns).

As can be seen in the left column, in the no-step case ($h = 0$) the dipole quickly forms and moves in the y – direction. Although initially some small-scale irregularities are visible (due to the forcing, see also Akkermans *et al.*, 2008a, b) the dipole soon acquires a smooth structure, while moving steadily along its axis. The tail behind the dipole is probably due to a slightly too long forcing period. The presence of the step topography (see middle and right columns) changes this behaviour drastically: although part of the dipole structure passes the step, it shows a relatively fast decay, while the longest surviving part of the dipole is arrested in front of the step. Moreover, a patch of oppositely signed vorticity is observed to form alongside the main dipole structure, at the deep side of the step. This remarkable feature is visible even in the earlier stages ($t = 3$ s), and is subsequently advected towards the rear of the dipole. In the later stages it takes the form of a shield around the primary dipole core. As the snapshots in the right column clearly show, these features are even more pronounced when the step height h is increased. In this case, the last stage of the flow consists of a shielded dipole arrested at the deep side of the step; the part that has crossed the step decays quite rapidly and is virtually absent after the initial stage.

The patch of opposite vorticity flanking the primary dipole halves and located at the deeper side of the step is entirely due to the no-slip condition at the vertical step wall. This conjecture has been confirmed by 3D numerical simulations, which allow examining in detail the 3D flow structure, i.e., the sequence of events at different levels in the fluid layers. These flow simulations have clearly revealed that the vertical sidewall of the step acts in the same way as in the case of a dipole colliding against a sidewall. As shown by the 2D simulations by Orlandi (1990), when the dipole approaches the wall, a boundary layer is formed, containing vorticity of opposite sign compared to the approaching vortex. In the case of the dipole, these two patches of oppositely signed vorticity would be wrapped around the primary dipole cores, resulting in a widening of the dipole, and subsequently a rebound from

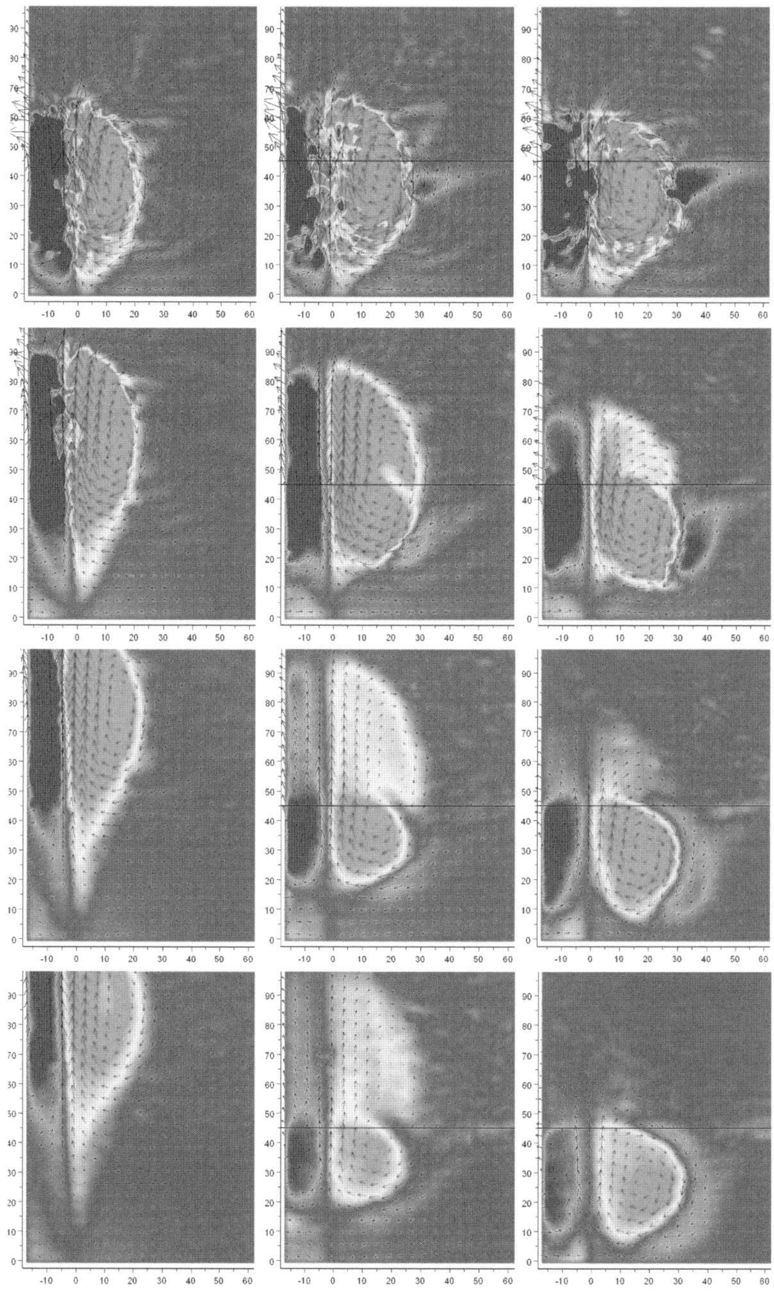

Figure 2 Experimentally measured flow evolution for the dipole encountering a step topography for $h/H = 0$ (left), 0.5 (centre) and 0.625 (right). The snapshots are taken at intervals of 2 s, starting from t = 3 s (top) to t = 9 s (bottom). The graphs show the horizontal flow field (vectors) and the vertical vorticity component ω_z (grey shades) measured at the free surface. The black lines indicate the position $y = y_0$ of the step.

the wall. Our 3D simulations have revealed that the vertical sidewall of the step acts in a similar way, namely as a source of oppositely signed vorticity ω_z. Apparently, the signature of this locally produced vorticity is observed at the free surface of the fluid layer, see Figure 2, middle and right columns. Figure 3 shows snapshots of the flow field and the vorticity at the free surface at $t = 4$ s according to the numerical simulations for the three cases $h/H = 0.625$ (left), 0.875 (middle), and 1.0 (right), this latter case corresponding with a vertical wall extending over the full depth of the fluid layer. In the latter case one clearly observes how oppositely signed vorticity is gradually advected away from the wall, leading to a rebound of the dipole cores, as in the purely 2D simulation of Orlandi (1990).

In the case $h/H = 0.875$ (middle column) one observes the same effects of the wall-induced vorticity, and similarly – although more weakly – for the case $h/H = 0.625$ (left column). Note that the tail of the dipole containing doubly-signed vorticity is due to the forcing with a realistic 3D magnetic field. The main conclusion of the laboratory observations and numerical simulations is that (i) a portion of the dipole structure may pass the step, but decays quickly, and (ii) the sidewall of the step induces oppositely

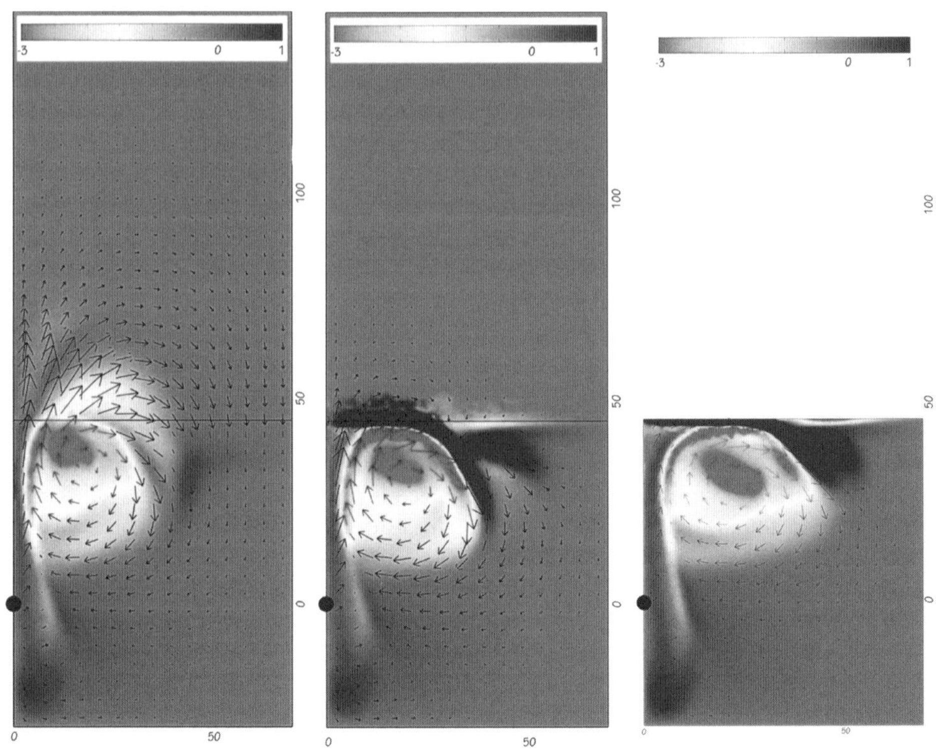

Figure 3 Snapshots of the numerically simulated flow due to a dipolar vortex encountering the step topography at $t = 7$ s for three different step heights: $h/H = 0.625$ (left), 0.875 (centre) and 1.0 (right). The graphs show the horizontal flow field (vectors) and the vertical vorticity component ω_z (grey shades) at the free surface. The black lines indicate the position $y = y_0$ of the step, while the black dot denotes the centre of the magnet. Field of view: $x > 0$.

signed vorticity, which leads to a shield around the remaining primary dipole halves, by which the dipole remains arrested at the deeper side of the step.

5 SLOPING BOTTOM

A second set of bottom-topography experiments has been carried out, now with the dipolar vortex climbing a sloping bottom. In order to allow the dipole to be fully developed before encountering the topographic slope, it was generated over the horizontal part of the tank bottom, in $(x, y) = (0, 0)$, at some distance from the edge of the slope (which is positioned at $y = y_0 = 35$ mm). Experiments have been performed for a range of slope angles α, but only a few characteristic cases will be discussed here.

Figure 4 shows snapshots of the evolution of the dipolar vortex moving over a weak slope with angle $\alpha = 3.4°$ (left column) and over a steeper slope with inclination $\alpha = 6.8°$ (right column). For the weakest slope, the dipole seems not to be affected too much by the sloping bottom, at least not for the duration of the experiment shown in the figure. The dipole behaviour changes when the slope angle α is increased, as visible for $\alpha = 6.8°$ (right column). Once the dipole moves over the slope, a band of oppositely signed vorticity is generated over the sloping bottom on either side of the dipole (note that in the limited field of view of the graphs in Figure 4 only the band on the right-hand side of the dipole is visible). The effect of this oppositely signed vorticity shield is twofold: It leads to a widening of the dipole structure and a decrease of its upslope translation speed, as can be seen by comparing with the case of the weaker topographic slope ($\alpha = 3.4°$, left column).

Figure 5 shows numerical simulation results for $\alpha = 3.4°$ (left column) and $\alpha = 6.5°$ (right column): snapshots of the horizontal flow field (vectors) and vorticity ω_z (grey shades) at the free surface for $t = 4$ s (upper panels) and 7 s (lower panels). For the steeper slope $\alpha = 6.5°$) the simulation results reveal the same features as observed in the experiment (Figure 4), namely a pronounced band of oppositely signed vorticity around the primary dipole, which tends to widen the dipole and to decrease its propagation speed. The formation of this band of oppositely signed vorticity is due to the no-slip boundary condition imposed by the sloping bottom. In order to explain this feature, we focus the attention to the right-hand side of the dipole (the half that is only visible in Figure 5). Although the vorticity in this core vortex is predominantly in the z – direction, with $\omega_z < 0$, in the boundary layer at the no-slip bottom the vorticity vector is directed parallel to the bottom and radially outwards, i.e., directed away from the vortex centre. Because of continuity, the swirl velocity above the shallower part will be larger than in the deeper part, implying a more intense boundary vorticity over the shallower part of the slope. In front of the dipole the boundary-layer vorticity is directed along the bottom, in the y, z – plane: this implies a z – component $\omega_z > 0$, which is visible also at the free surface in the shallower part of the water column (see Figure 4, right column, and also Figure 5, right column). Because of advection by the dipole flow, this vorticity $\omega_z > 0$ becomes distributed in a band around the frontal part of the dipole structure.

Figure 6 shows the distribution of the vertical component ω_z of the vorticity in vertical cross-sections (y, z – plane) through the negative dipole core at $t = 1.5$ s, 2.5 s, and 5.0 s, according to the numerical simulations for a slope $\alpha = 6.5°$: these ω_z– plots clearly show the presence of oppositely signed vorticity ($\omega_z > 0$) at the bottom near the

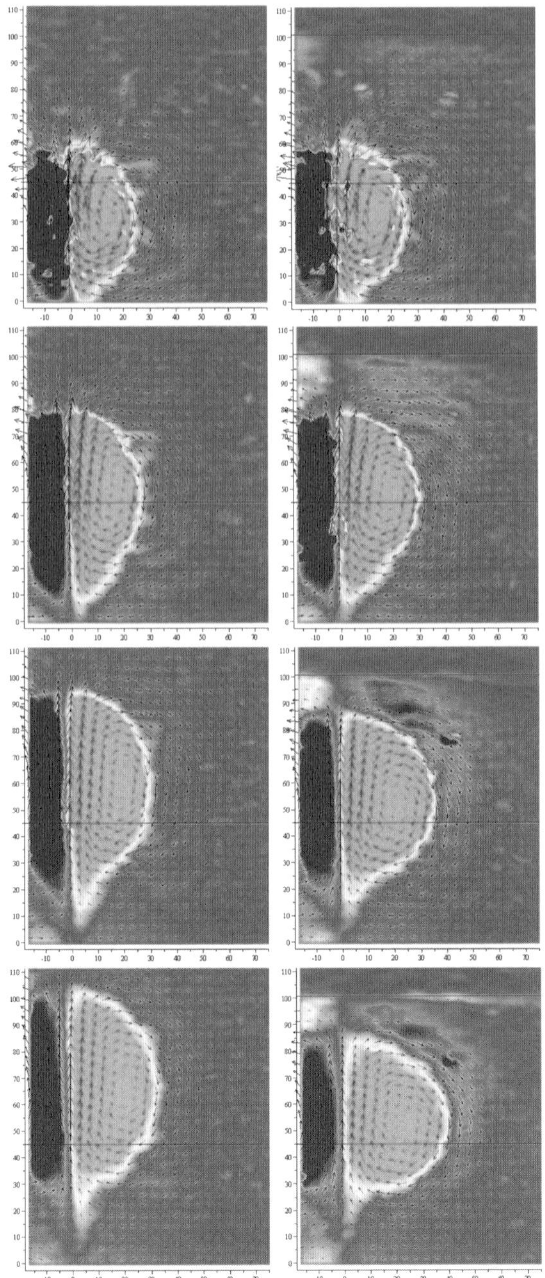

Figure 4 Experimentally measured flow evolution for the dipole encountering a slope topography for $\alpha = 3.4^{\circ}$ (left) and $\alpha = 6.5^{\circ}$ (right). The snapshots are taken at intervals of 2 s, starting from $t = 3$ s (top) to $t = 9$ s (bottom). The graphs show the horizontal flow field (vectors) and the vertical vorticity component ω_z (grey shades) measured at the free surface. The black lines indicate the position $y = y_0$ of the edge of the slope.

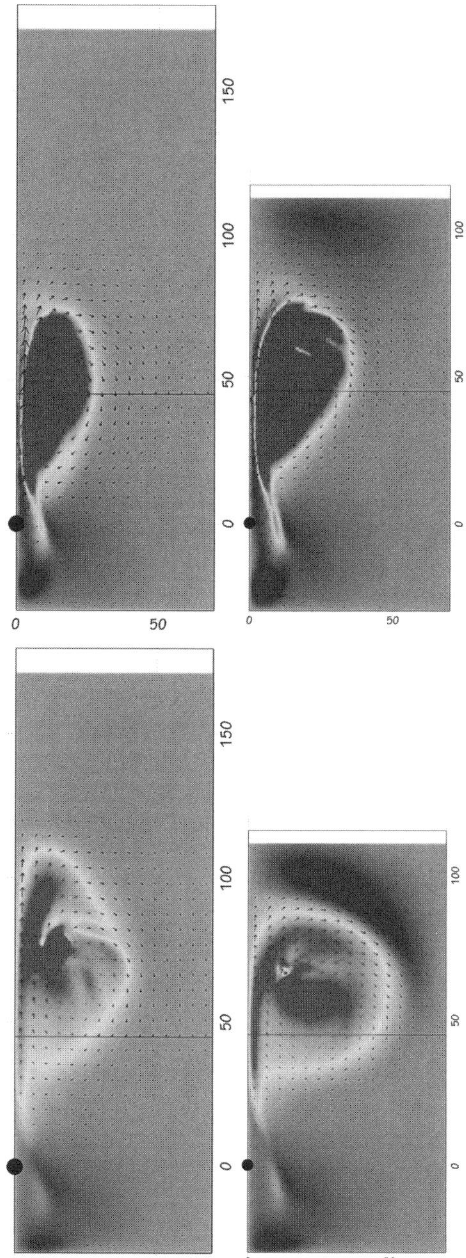

Figure 5 Snapshots of the numerically simulated flow due to a dipolar vortex encountering the slope topography at $t = 4$ s (top) and 7 s (bottom) for two different inclination angles: $\alpha = 3.4^{\circ}$ (left) and $\alpha = 6.5^{\circ}$ (right). The graphs show the horizontal flow field (vectors) and the vertical vorticity component ω_z (grey shades) at the free surface. The black lines indicate the position $y = y_0$ of the edge of the slope, while the black dot denotes the centre of the magnet. Field of view: $x > 0$.

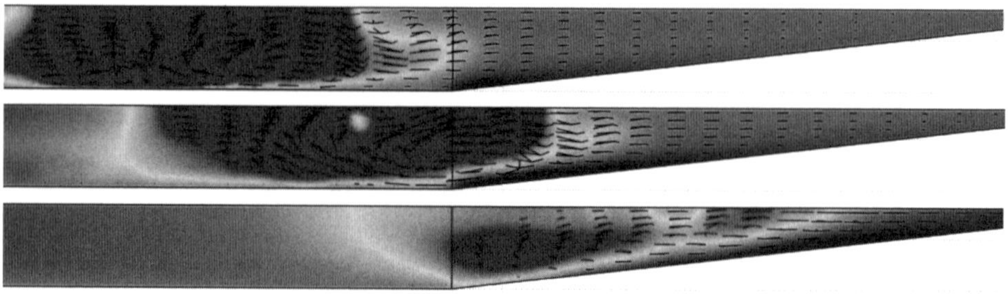

Figure 6 Numerical simulation results of the slope topography with $\alpha = 6.5^{\circ}$: Vertical cross-sections (y, z – plane) showing the distribution of the vertical vorticity component ω_z while the dipolar vortex is moving uphill over the sloping bottom for $t =$ 1.5 s (top), 2.5 s (middle), and 5.0 s (bottom). The very dark grey shading at the bottom indicates high positive values of ω_z.

front of the dipole, with the bottom vorticity becoming more intense when the dipole advances further into shallower parts.

Apparently, the non-horizontal no-slip bottom leads to vorticity with a non-zero vertical component ω_z, which eventually extends all the way up to the free surface, and which significantly influences the behaviour of the dipole vortex: it results in its widening and a decrease of its propagation speed. In this respect the flow evolution has similarities with the dipole-wall collision (see Orlandi, 1990), although rebound of the vortices is not observed here because of the significant decay owing to the shallowness of the fluid layer higher up the slope.

6 CONCLUSION

The laboratory experiments discussed here have revealed that 3D effects may play a significant role in shallow flows with bottom topography, at least in the cases considered: a dipolar vortex approaching a step and a dipolar vortex moving over a linearly sloping bottom. In the case of the step topography, the no-slip sidewall of the step implies generation of vertical vorticity ω_z, the signature of which is also clearly observed in the upper parts of the fluid column. As a result, a shield of oppositely signed vorticity is formed around the dipole, causing its arrest at the deeper side of the step. Before this takes place, a portion of the dipole in the upper part of the fluid column has crossed the step, although it decays quickly because of the shallowness of the fluid layer on that side. The dipole moving uphill over the sloping bottom experiences a similar effect. The boundary layer at the no-slip bottom implies the generation of vorticity, which is strongest in the shallower part of the fluid layer. The vorticity vector pointing locally parallel to the bottom implies a vertical component ω_z which is strongest near the front of the dipole. This bottom-induced vorticity ω_z has a sign opposite to that of the primary vortex, and its signature is clearly observed even at the free surface. Like in the case of step topography, the bottom-induced vorticity component ω_z becomes concentrated in a band around the vortex dipole, which causes its widening and its slow-down.

As a general conclusion of this exploratory study, it may be stated that viscous effects – i.e., no-slip conditions at the non-horizontal parts of the bottom – play an important role in shallow flows with bottom topography. The combination of these viscous effects and the 3D nature of the bottom topography gives rise to generation of vorticity (including a vertical component ω_z), which significantly influences the flow evolution. Clearly, such flows cannot be described simply by the two-dimensional Navier-Stokes equation with an additional 'bottom friction' term.

A final remark should be made about the 3D flow structure of the dipolar vortex, before and during its encounter with the topography. As discussed in Akkermans *et al.* (2008a, b) and Cieślik *et al.* (2009a, b), dipolar flow structures and other flows in a shallow fluid layer may exhibit significant vertical velocities, in spite of the shallowness of the fluid layer. Although these vertical motions are to some extent rather local, they may disrupt the 2D nature of the flow that one might have assumed intuitively. Applying the same techniques as in the studies of Akkermans *et al.* and Cieślik *et al.* we have monitored and quantified the amount of vertical motion in the present topography experiments by determining the ratio of the kinetic energy associated with the vertical velocity w and that residing in the horizontal velocity (u, v). A detailed discussion of these aspects of shallow flow over topography, however, is beyond the scope of the present paper.

REFERENCES

Akkermans, R.A.D., Cieślik, A.R., Kamp, L.P.J., Trieling, R.R., Clercx, H.J.H. and van Heijst, G.J.F. (2008a). The three-dimensional structure of an electromagnetically generated dipolar vortex in a shallow fluid layer. *Phys. Fluids,* 20, 116601.

Akkermans, R.A.D., Kamp, L.P.J., Clercx, H.J.H. and van Heijst, G.J.F. (2008b). Intrinsic three-dimensionality in electromagnetically driven shallow flows. *Europhys. Lett.*, 83, 24001.

Bühler, O. and Jacobson, T.E. (2001). Wave-driven currents and vortex dynamics on barred beaches. *J. Fluid Mech.,* 449, 313–339.

Cieślik, A.R., Akkermans, R.A.D., Kamp, L.P.J., Clercx, H.J.H. and van Heijst, G.J.F. (2009a). Dipole-wall collision in a shallow fluid. *Eur. J. Mech.-B/Fluids*, 28, 397–404.

Cieślik, A.R., Kamp, L.P.J., Clercx, H.J.H. and van Heijst, G.J.F. (2009b) Meandering streams in a shallow fluid layer. *Europhys. Lett.*, 85, 54001.

Duran-Matute, M., Kamp, L.P.J., Trieling, R.R. and van Heijst, G.J.F. (2010). Scaling of decaying shallow axisymmetric swirl flows. *J. Fluid Mech.,* 648, 471–484.

Jackson, J.D. (1998) *Classical Electrodynamics*. Wiley, New York, p. 181.

Jirka, G.H. (2001). Large scale flow structures and mixing processes in shallow flows. *J. Hydraulic Res.*, 39, 567–573.

Jirka, G.H. and Uijttewaal, W.S.J. (eds) (2004). *Shallow Flows*. Taylor and Francis Group, London.

Lamb, H. (1932). *Hydrodynamics*. 6th edition. Cambridge University Press.

Orlandi, P. (1990). Vortex dipole rebound from a wall. *Phys. Fluids,* A2, 1429–1436.

Peregrine, D.H. (1998). Surf zone currents. *Theor. Comput. Fluid Dyn.* 10, 295–310.

Tabeling, P., Burkhart, S., Cardoso, O. and Willaime, H. (1991). Experimental study of freely decaying two-dimensional turbulence. *Phys. Rev. Lett.*, 67, 3772–3775.

Uijttewaal, W.S.J. and Jirka, G.H. (2003). Grid turbulence in shallow flows. *J. Fluid Mech.*, 489, 325–344.

Xia, H., Shats, M. and Falkovich, G. (2009). Spectrally condensed turbulence in thin layers. *Phys. Fluids,* 21, 125101.

Chapter 5

Characteristic scales and consequences of large-scale horizontal coherent structures in shallow open-channel flows

A.M. Ferreira da Silva[1], *H. Ahmari*[2] *and A. Kanani*[1]

[1] *Department of Civil Engineering, Queen's University, Kingston, Ontario, Canada*

[2] *Manitoba Hydro, Winnipeg, Manitoba, Canada (Formerly Queen's University, Kingston, Ontario, Canada)*

ABSTRACT: The nature and consequences of the largest horizontal coherent structures (HCS's) of turbulence in a shallow open-channel flow are investigated on the basis of three series of velocity measurements. These were conducted in a 21 m long, 1 m wide channel, conveying a turbulent, sub-critical and uniform flow, with a flow depth of 4 cm. The bed surface was flat, and the bed shear stress was substantially below the threshold for initiation of motion. In two of the measurement series, small 'discontinuities' were introduced at the channel walls to regularize the shedding of turbulence structures. The filtered oscillograms of fluctuating velocity exhibited regular (nearly periodic) cycles of variation consistent with the presence of HCS's affecting (occupying) the entire body of fluid. Moreover, these were consistent with the presence of persistent horizontal burst-sequences issued from (or promoted by) the discontinuities. A simple decomposition of the velocity records yielded an average horizontal burst length of approximately five times the flow width. This result was independently confirmed with the aid of different advanced techniques for the detection of coherent structures, including energy spectrum analysis, and analysis of velocity signals by continuous wavelet transform. A slight internal meandering of the flow caused by the superimposition of burst sequences on the mean flow was detectable. The resulting convective flow patterns were found to be capable of inducing bed deformation consistent with the trace of alternate bars.

1 INTRODUCTION

Following Hussain (1983), in this paper the term 'large-scale coherent structure' or simply 'coherent structure' refers to the largest conglomeration of turbulent eddies that has a prevailing sense of rotation. These structures are 'born', grow to a size where they eventually occupy the body of flow, and disintegrate (or 'die'), the term 'burst' in the following designating the evolution of a large-scale coherent structure during its life-span. In open-channel flows, the bursts can be vertical or horizontal, with the coherent structures of the former rotating in the vertical $(x; z)$-planes and scaling with the flow depth, those of the latter rotating in the horizontal (x; y)-planes and scaling with the flow width.

As follows from the reviews by Yalin (1992), Nezu and Nakagawa (1993) and Roy *et al.* (2004), substantial research has been carried out to date on large-scale vertical coherent structures (VCS's). On the basis of Blackwelder (1978), Cantwell (1981), Hussain (1983), Gad-el-Hak and Hussain (1986), Rashidi and Banerjee (1988), among others, da Silva (2006) briefly synthesized the sequence of events associated with the life-cycle of large-scale vertical coherent structures (see Figure 5 and the related text in the just mentioned reference). Although relevant to this paper, this synthesis will not be repeated in this text and the reader here is merely referred to Figure 5 in da Silva (2006). In this figure, the distance O_iO_{i+1} between the birth places of two consecutive bursts of a longitudinal sequence of vertical burst is the vertical burst length λ_V, the life span of a burst (i.e., the vertical burst period) being $T_V = \lambda_V/u_{av}$, with u_{av} as average flow velocity.

There seems to be general agreement that the vertical burst length is proportional to the flow depth h, i.e., $\lambda_V \approx \alpha_V h$ (Jackson, 1976, Nezu & Nakagawa, 1993, Roy *et al.*, 2004, Franca & Lemmin, 2008, etc.), the value of the proportionality factor α_V being comparable to the value (namely 6) of the proportionality factor in the relation of average dune length $\Lambda_d \approx 6\,h$ due to Yalin (1977). Considering this, Yalin (1992) advanced a detailed turbulence-based explanation for the occurrence of dunes, viewed by this author as 'imprints' on the mobile bed of periodic sequences of vertical bursts. da Silva (1991), Yalin (1992) and Yalin and da Silva (2001) considered earlier river and open-channel flow measurements and observations, such as those by Yokosi (1967a), Dementiev (1962) and Grishanin (1979), in light of the present understanding of turbulence and coherent structures, together with available field and laboratory data on alternate bars. These authors noted the striking similarity between the expressions of average dune length and average alternate bar length ($\Lambda_a \approx 6B$), which suggests that if dunes are caused by a certain mechanism in vertical turbulence, then alternate bars should be due to analogous mechanism inherent in horizontal turbulence (that is, large-scale horizontal coherent structures). The possibility that alternate bars are but the 'horizontal version' of dunes had already been suggested by Jaeggi (1984) and Kishi (1980). However, no bursts were mentioned by these authors. The possible role played by large-scale horizontal coherent structures in the origin of meandering was also explored by Yalin (1992), Yalin and da Silva (2001) and da Silva (2006). Yet, in the absence of a systematic study of HCS's, it is impossible to ascertain the exact role they play in determining river morphological features, or to that matter, also patterns of sediment or contaminant transport.

Considering this, an extensive study of HCS's is presently being carried out at Queen's University, with the goal of contributing to a better understanding of the characteristic scales, dynamics and morphological consequences of HCS's in the context of river flows. The results of the first phase of this study were presented by da Silva and Ahmari (2009). In this paper, the analysis in the just mentioned work is extended, with the aim of gaining further insight into the nature of HCS's and their consequences. For the present purposes, use is made of the flow data collected by da Silva and Ahmari (2009) in a shallow, alternate bar inducing flow, as well as data resulting from three new series of measurements of the same flow recently carried out by Ahmari and da Silva (In preparation[b]) with the purpose of extending the measurement region further towards the banks than in the original measurements reported by da Silva and Ahmari (2009).

2 HORIZONTAL COHERENT STRUCTURES AND BURSTS: FUNDAMENTALS AND EXISTING HYPOTHESES

The design of the experiments reported in this paper was guided by some fundamentals of HCS's and existing hypotheses regarding their life cycle, which are extensively presented in da Silva and Ahmari (2009). For the sake of clarity and completeness of this paper, these are also reviewed below.

1 Even though HCS's have not yet been the focus of directed research, there are reasons to believe that in their life cycle they follow a sequence of events similar to that of VCS's (Figure 5 in da Silva, 2006) with the difference that they occur in a horizontal 'flow ribbon' (Yalin, 1992). This means that HCS's are likely to originate at the 'points' P near the banks and free surface (Figure 1), where horizontal shear stresses τ_{xy} are the largest, and from there be conveyed by the mean flow away from the bank and downstream while growing in size. Once their lateral extent becomes as large as B, they must be expected to interact with the opposite bank and disintegrate, the neutralized fluid mass returning to its original bank so as to arrive there at a time $t = T_H$, the horizontal burst length in this case being $\lambda_H = \alpha_B B$. Here T_H is horizontal burst period. There seems to be agreement that the coherent structures forming the horizontal bursts of a wide open channel have the shape of horizontally positioned disks (Figure 1b) eventually extending (along z) throughout the flow thickness h (Grishanin, 1979, Jirka & Uijttewaal, 2004, Yalin, 2006).

2 Since bursts are randomly distributed in space and time, under completely uniform conditions of flow there is an equal probability (or frequency) of occurrence of bursts for any region Δx and time interval Δt. This applies to both vertical and horizontal bursts. Such a homogeneous or uniform distribution of bursts along

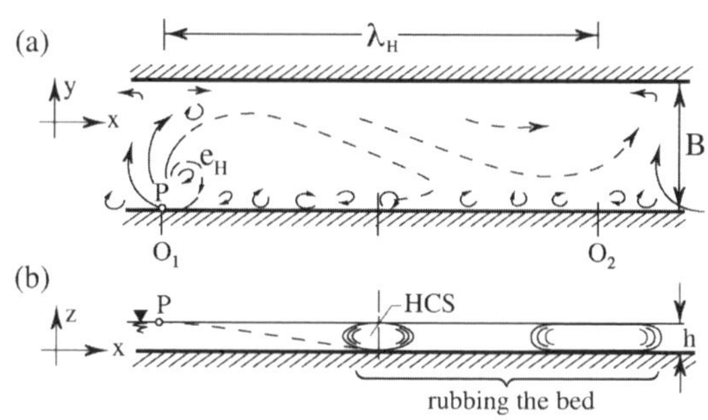

Figure 1 Conceptual representation of a horizontal burst cycle. (a) Plan view; (b) Side view.

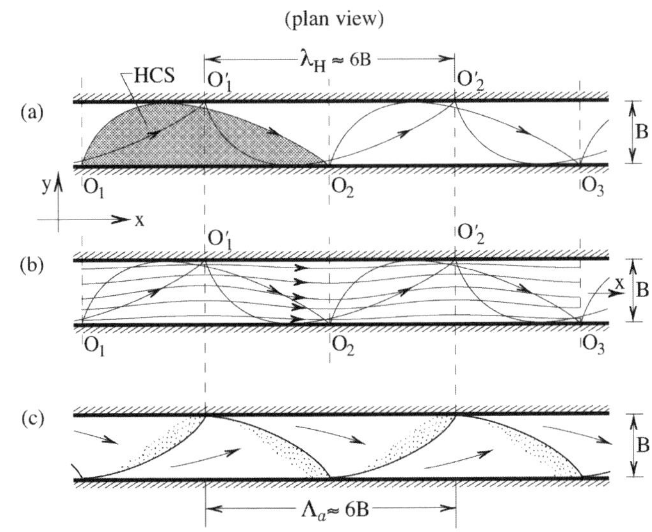

Figure 2 Plan view of open-channel flow: (a) Likely antisymmetrical arrangement of HCS's in flow plan; (b) Related internal meandering of flow; (c) Alternate bars on movable bed.

the flow direction x cannot lead to an internal periodic deformation of the flow that would 'imprint' itself as a periodic deformation of the bed surface (Yalin, 1992). There must thus be in the flow a 'location of preference' leading to the increment of frequency of bursts at that location – and since the break-up of one coherent structure (CS) triggers the 'birth' of the next CS, leading also to the more frequent generation of sequences of bursts initiating from it and, ultimately, to the periodic 'deformation' of the flow. According to Yalin (1992), an increased frequency of bursts at a location can only be realized by means of a local discontinuity (the section containing it thus becoming the preferential section, $x = 0$ say). In the case of laboratory conditions, the discontinuity can be the beginning of mobile bed or banks, an accidental ridge on the sand surface, etc.

3 If alternate bars are the 'imprints' of HCS's on the bed, and since alternate bars are antisymmetrical with respect to the x axis (Figure 2c), then in all likelihood so must be the sequences of horizontal bursts issued from the right and left banks, respectively. The antisymmetrical arrangement of the sequences of bursts is shown in Figure 2a.

3 EXPERIMENTAL SETUP AND DESCRIPTION OF MEASUREMENTS

The velocity measurements used in this paper were carried out in a 1 m wide, 21 m long and 0.4 m deep straight channel (Figure 3), with vertical side walls made of aluminium. The channel was installed in the 21 m long, 7 m wide river basin of the Civil Engineering Department at Queen's University. The complete details of the basin and present channel are given elsewhere (da Silva and El-Tahawy, 2008, Ahmari, 2010). The upstream end of the river basin consists of a 1.85 m wide and 8.8 m long stilling tank. The water entered the present experimental channel through a 1 m wide opening on the 0.60 m tall wall separating the stilling tank from the river basin. The channel bed was formed by a well-sorted silica sand, with an average grain size D (= D_{50}) of 2 mm and specific density of 2.65. To ensure that the bed material would not be removed from the bed at the channel entrance, a 1 m long stretch of gravel of diameter ≈ 2.5 cm was installed at the upstream end of the channel. The surface of the sand was scraped so as to produce the desired stream slope, and a flat bed surface. The flow free surface slope was adjusted by means of a tailgate at the end of channel.

The hydraulic conditions of the turbulent, sub-critical and uniform flow under investigation are summarized in Table 1. Here Q is flow rate, S is bed slope, $u_{av} = Q/(Bh)$ is average flow velocity, $\mathrm{R} = u_{av} h/\nu$ is flow Reynolds number (with ν as fluid kinematic viscosity), $\mathrm{F} = u_{av}/(gh)^{1/2}$ is Froude number (with g as acceleration due to gravity), $\mathrm{R}* = v* k_s/\nu$ is roughness Reynolds number (with $v* = (\tau_0/\rho)^{1/2} = (gSh)^{1/2}$ as shear velocity, and k_s as granular skin roughness, ρ being fluid density), and Y/Y_{cr} is relative flow intensity (with $Y = \tau_0/(\gamma_s D)$ as mobility number and Y_{cr} as the value of Y at the critical stage of initiation of sediment transport, γ_s being submerged specific weight). For the present sand, Y_{cr} was identified with 0.045. The following values were adopted for other pertinent fluid and sand properties: ρ = 1,000 kg/m^3, $\nu = 10^{-6}$ m^2/s, γ_s = 16,186.5 N/m^3, and $k_s = 2D_{50}$ (Kamphuis, 1974, Yalin, 1977). The bed slope was selected so that the resulting bed shear stress was substantially below the threshold for initiation of motion, ensuring that the bed remained flat throughout the measurements.

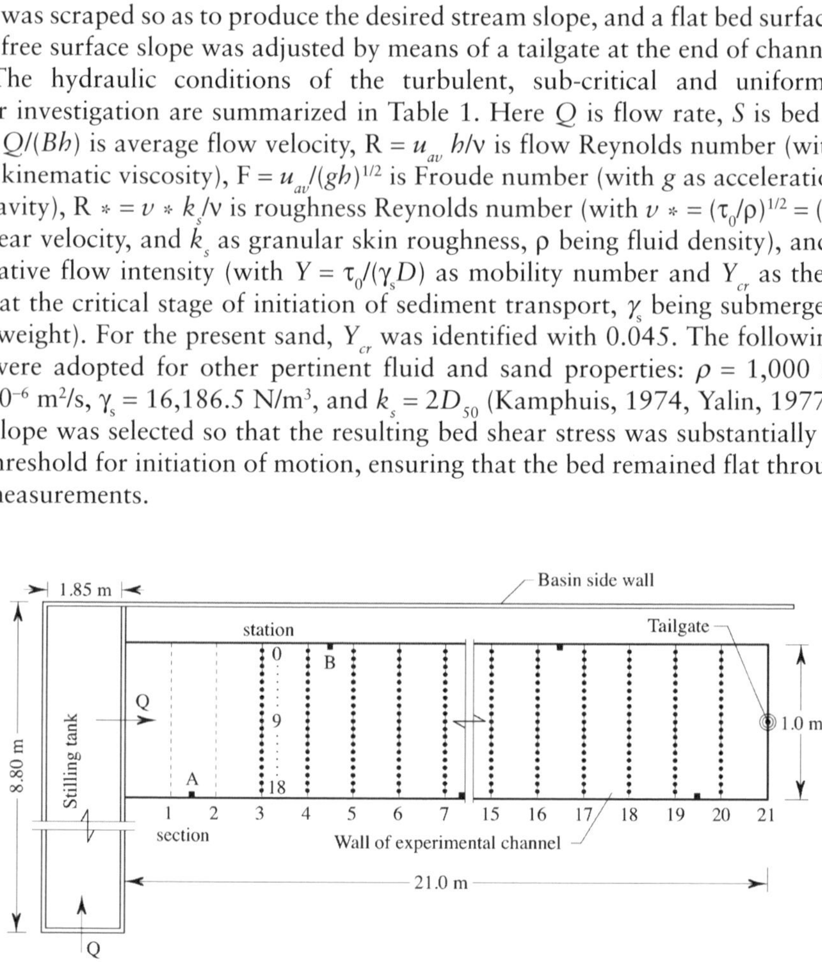

Figure 3 Schematic of experimental channel, with location of velocity measurements indicated by dots.

Table 1 Hydraulic conditions of the flow under investigation (B = 1 m, D_{50} = 2 mm).

Q (*L/s*)	*h* (*cm*)	*S*	u_{av} (*cm/s*)	R	F	R*	Y/Y_{cr}	*B/h*	*h/D*
9.0	4.0	0.0015	22.5	9000	0.36	97	0.4	25	20

The particular combination of values of B, h and D was selected so that the flow would lead to the occurrence of alternate bars if the bed material would be able to move. The flow thus purposely plots in the midst of the alternate bar region in the (B/h; h/D) existence region plan of bars in Figure 4.

As mentioned in the Introduction, in this work (Sections 4 and 5) use is made of the data resulting from the measurements by da Silva and Ahmari (2009). These consist of three series of flow velocity measurements, MS-2B, AS1-0B and AS2-7B, where the series MS-2B is to be viewed as the main series. In this test, and to establish 'locations of preference' (see Section 2, Point 2) for the generation of bursts and burst sequences on both walls, two 10 cm long blocks with 2 cm × 2 cm square bases were attached to the walls, with their longest sides standing vertically and their square bases lying ≈ 2 cm from the bed surface. The series of measurement AS1-0B and AS2–7B were carried out for the purposes of comparison and discussion. In series AS1–0B, no blocks were attached to the walls; in series AS2-7B, four blocks were attached to the right wall and three to the left, in an antisymmetrical arrangement as shown in Figure 3, the distance between consecutive blocks along each wall being 6 m (= $6B$). Note that 0B, 2B, and 7B in the test designation indicates the number of blocks used in the channel. The rationale for the plan arrangement of blocks is justified in view of Section 2, Point 3. In these series of measurements, 2 min long records of instantaneous flow velocity were collected at Sections 3 to 20, and in each section at 17 different points (henceforth referred to as stations). These were equally spaced along each cross-section, with stations 1 to 17 located 10 cm from the channel walls. Additionally, 20 min long records of instantaneous flow velocity were collected at the channel centerline (Section 9) of cross-Sections 6, 12, and 18. All velocity measurements were carried out at 1 cm below the free surface, with the aid of a 2D 16 MHz SonTeK™ Micro ADV™, mounted in an upstream-facing configuration, and operating at a sampling frequency of 20 Hz.

The considerations in Section 6 rest on three new series of measurements of the same flow, carried out in order to extend the measurement region further towards the banks, where the HCS's are believed to originate. These measurements were carried out with the aid of two synchronized 2D 16 MHz SonTek™ Micro ADV's,

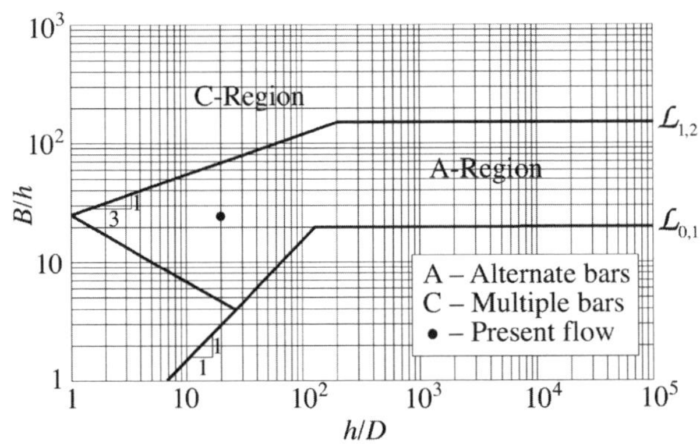

Figure 4 Existence region of alternate bars (A-region) in the (B/h; h/D)-plan of da Silva (1991), after modification by Ahmari and da Silva (2011), and plot of present flow in this region.

mounted in side-looking (bank-facing) configurations. In this case, 2 min long records of instantaneous longitudinal and transversal flow velocity were collected at Sections 3 to 20, and in each section simultaneously at two different stations at the same distance from the right and left walls of the channel (station 0 and 18, 1 and 17, 2 and 16, ..., and 6 and 12). Stations 0 and 18 were located 5 cm from the channel walls, stations 1 to 6 and 12 to 17 being the same as in the series of measurements described in the previous paragraph. In this case, no measurements were carried out at stations 7 to 11. The measurement level was the same as before, namely 1 cm below the free surface. In the following, these series of measurements will be referred to as MS-2B/2P, AS1-0B/2P and AS2-7B/2P.

4 DETECTION OF HCS'S; BURST PERIOD AND BURST LENGTH

4.1 Simple decomposition of 2 min long oscillograms of flow velocity

As is well-known (see e.g., Yokosi, 1967a, Yalin, 1992), after averaging the oscillogram of instantaneous flow velocity over consecutive time intervals Δt, the resulting 'smoothened' (or filtered) oscillograms contain only those velocity fluctuations whose period is larger than Δt. Thus, by selecting a sufficiently large Δt, it is possible to reveal those longest periods (or lowest frequencies) of velocity fluctuations that are due to the largest structures in the flow, and thus to reveal the burst period (Yalin, 1992). This procedure is illustrated in Figure 5, where it is applied to the 2 min long oscillogram of fluctuating component of longitudinal flow velocity u' collected at Section 10, station 17, series MS-2B. In this example, the longest periods of the

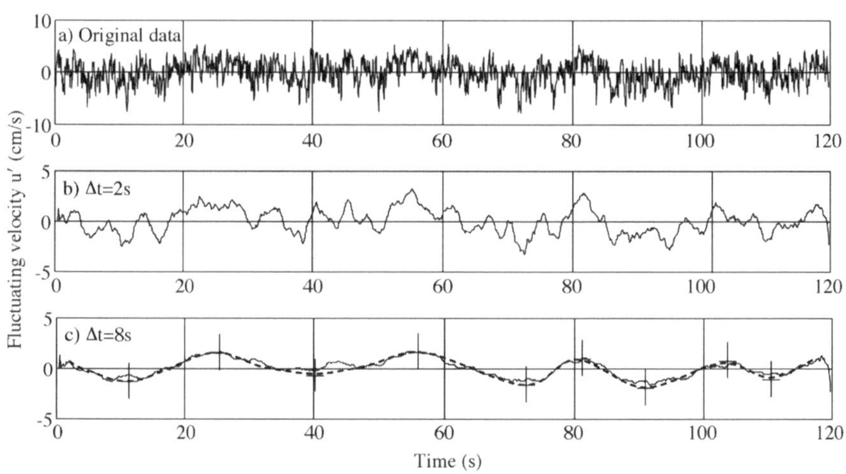

Figure 5 (a) Original 2 min long oscillogram of the fluctuating component of longitudinal flow velocity u' collected at Section 10, station 17, series MS-2B; (b) and (c) Resulting fluctuating velocity diagrams after averaging with the time-interval $\Delta t = 2$ s and $\Delta t = 8$ s, respectively. The dashed line in (c) is drawn so as to closely follow the trend of the smoothened oscillogram.

velocity fluctuations became evident when $\Delta t \approx 8$ s (see Figure 5c, where the solid line is the smoothened oscillogram, and the dashed line is drawn so as to closely follow the trend of the oscillogram and highlight its peaks and troughs). Note from Figure 5c that the smoothened oscillogram in this example exhibits a rather regular cyclic pattern of variation, with four cycles between the troughs at $t \approx 11$ s and $t \approx 111$ s, yielding an average period (averaged over the sampling time of 2 min) of ≈ 25 s = (111–11)/4 for the velocity fluctuations due to the largest structures in the flow. Some further examples of smoothened oscillograms resulting from the series of measurements MS-2B, AS2-7B and AS1-0B are shown in Figure 6. The averaging time intervals used in these examples are indicated in the figures themselves.

For each of the series of measurements MS-2B, AS2–7B and AS2–0B, the smoothening procedure described above was applied to the entire set of 306 (= 18 Sections × 17 stations) 2 min long oscillograms of u'. It was found that, as a rule, averaging time intervals of 8 s in series MS-2B and AS1-7B, and 4 s in series AS1–0B, were adequate to reduce the velocity oscillograms to the point where the periods of fluctuations due to the largest structures were clearly evident.

Through visual inspection of the entire set of oscillograms, it was found that in all three series of measurements, and irrespective of location in flow domain, the smoothened oscillograms, as a rule, exhibited long cycles of increasing-decreasing-increasing u'. These were by multiple times larger than those that can be expected

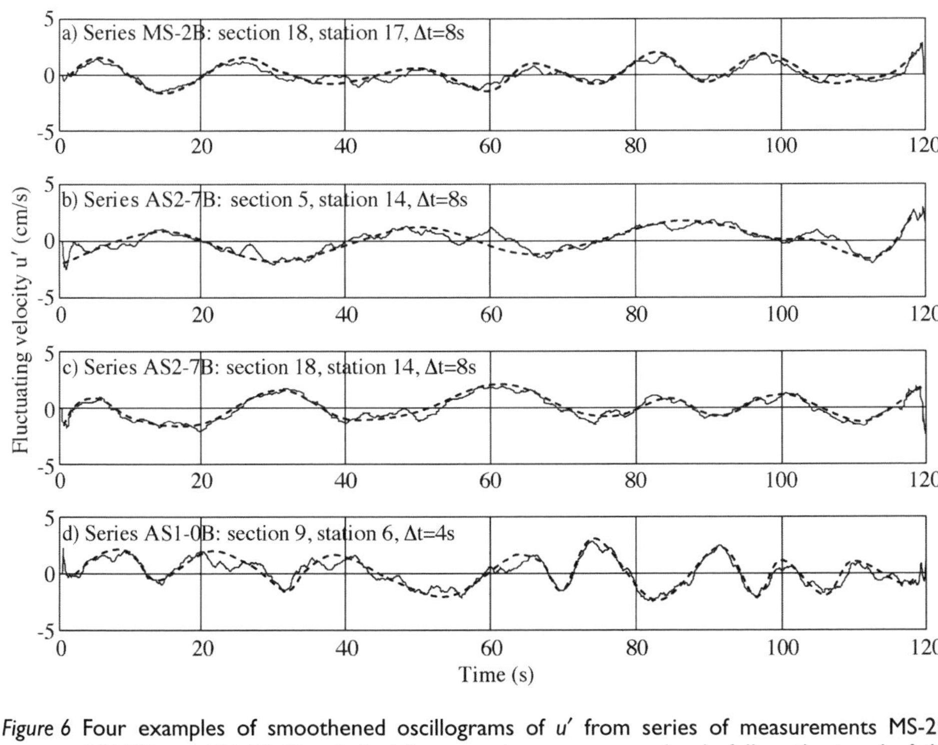

Figure 6 Four examples of smoothened oscillograms of u' from series of measurements MS-2B, AS2-7B and AS1-0B. The dashed lines are drawn so as to closely follow the trend of the smoothened oscillograms.

because of large-scale vertical turbulence. Indeed, identifying the burst length λ_V of VCS's with $\approx 6\ h$ (Jackson, 1976, Yalin, 1992, Roy *et al.*, 2004, del Álamo & Jiménez, 2003), it follows that in the present flow they are associated with periods of ≈ 1 s ($T_V = \lambda_V/u_{av} = (6 \times 0.04)/0.225 \approx 1$ s). Yet, as illustrated by the examples in Figure 6, the duration of each cycle of fluctuation of u' in present smoothened oscillograms was at its minimum of the order of ≈ 5 to 10 s, say, and as a rule much larger than 10 s. The presence of such cycles of fluctuation in the smoothened oscillograms, and the fact that they are present in all of the oscillograms, is a clear indication of the presence of HCS's, affecting (occupying) the entire body of fluid.

However, in their nature, the smoothened oscillograms were found to vary depending on the series of measurements, and within each series to vary depending also on location (along the flow direction x) of the velocity measurements. This aspect is amply discussed by da Silva and Ahmari (2009), who found that:

1. In series MS-2B, the smoothened oscillograms of u' in the upstream region of flow (from Section 3 to Section 7 or 8, say) tended to consistently exhibit rather long cycles of variation of fluctuating velocities, with a periodic nature marked by similar cycles of increase-decrease-increase of fluctuating velocity repeating themselves with the passage of time. This is typified by Figure 5c. In the downstream region of flow (from Section 14 or 15, say, to Section 20), the duration of the individual cycles of increase-decrease-increase of fluctuating velocity tended to vary more from cycle to cycle, with long duration cycles comparable to those frequently occurring in the upstream region becoming somewhat less frequent. This is typified by Figure 6a, where the duration of individual cycles of fluctuation of u' vary from ≈ 15 s for the cycle starting at $t \approx 60$ s, to ≈ 25 s for the cycle ending at $t \approx 40$ s. The middle region appeared as a transitional region, where the nature of the smoothened oscillograms tended to be of a mixed type between those more typical of the upstream and downstream regions.
2. In series AS2-7B, the smoothened oscillograms tended to consistently exhibit long and rather regular cycles of variation, throughout the channel. That is, the characteristics described above for the smoothened oscilograms in the upstream region of series MS-2B tended to predominate throughout the channel.
3. In series AS1-0B, the smoothened ascillograms throughout the channel tended to be of the same nature as those in the downstream region of series MS-2B, but with an even wider variation of cycle duration from cycle to cycle, and an increased frequency of shorter cycles in comparison to very large duration cycles.

The regularity ('periodicity') in the cycle of variation of fluctuating velocity of the smoothened oscillograms, as consistently observed in the upstream reach of series MS-2B and throughout the channel in series AS2-7B is consistent with the presence of large-scale periodic events in the flow, and more specifically with the presence of persistent longitudinal sequences of horizontal bursts originating at (or promoted by) the discontinuities. The variation along the flow direction x of the cyclic pattern of the smoothened oscillograms in series MS-2B, and the lack of such a noticeable variation in series AS2-7B suggest that, as hinted by Yalin (1992), the 'activity' started by the discontinuities, i.e., the generation of longitudinal sequences of bursts originating from the same location(s), is sustained only within a confined region downstream of the discontinuities. In the absence of further discontinuities along x, this activity attenuates, i.e., there is a deterioration in the degree of organization of the internal fluid motion, with the conditions sufficiently far from the discontinuities eventually reverting to those that would be observed in the absence of any discontinuities. After determining the average value of the periods of the velocity fluctuations for each of the smoothened oscillograms corresponding to the upstream region of series MS-2B and the entire channel in series AS2-7B, da Silva and Ahmari (2009) arrived at an estimate of 22.6 s for the value of the average burst period T_H of the bursts in the flow regions 'regulated' by discontinuities (i.e., $T_H \approx 22.6$ s). This corresponds to an average burst length $\lambda_H = u_{av} \times T_H \approx 5$ m (where $u_{av} = 0.225$ m/s), and thus $\lambda_H/B \approx 5$. The average duration of the velocity fluctuation cycles throughout the channel in series AS1-0B, namely 12.1 s, is substantially smaller than in series MS-2B and AS2-7B. This is a result of the greater degree of variability in the nature of the oscillograms, and a decreased frequency of very long duration cycles of fluctuation of flow velocity as described in point 3 above, both reflecting the greater degree of randomness in the internal structure of turbulence in the absence of discontinuities.

A variety of advanced tools are available for the detection of coherent structures and determination of their time and length scales (see e.g., Nezu and Nakagawa, 1993, Pope, 2000). In the following, and as a means to complement the considerations in this section, energy spectrum analysis and continuous wavelet transform are applied to the present velocity measurements.

4.2 Energy spectrum analysis

Compared with the technique based on smoothening of the oscillograms of fluctuating flow velocity presented in the previous section, the determination of the energy spectrum, a method closely related to the temporal autocorrelation function (see e.g., Ahmari 2010), is substantially more demanding in terms of the required temporal coverage of the velocity data.

For this reason, in the context of this work, one-dimensional energy spectra of instantaneous longitudinal velocity were determined for only three locations in the flow domain, namely at station 9 (centerline) of Sections 6, 12, and 18, where 20 min long records of velocity were collected. As an example, the energy spectrum corresponding to series MS-2B, Section 12, station 9 is shown in Figure 7. It is noted that, because of the coexistence of vertical and horizontal structures, the spectrum is multi-structural, in the sense of Ozmidov (1965), Monin and Ozmidov 1985, Yokosi (1967b) and Imamoto (1973). This aspect, however, is discussed elsewhere (Ahmari & da Silva, In preparation[a]), the energy spectra being used here exclusively to determine the frequency of the large scale horizontal coherent structures. This is associated with the left end of the '−1/2' power law range (dashed line in Figure 7) in the energy spectra. The results are summarized in Table. 2, where frequencies are converted to periods. For series MS-2B and AS2-7B, the values in Table. 2 vary from 20 s to 33 s, which is consistent with the range of variation of the average periods of velocity fluctuations of individual smoothened oscillograms in the previous section (see da Silva & Ahmari, 2009). The same applies to series AS1-0B.

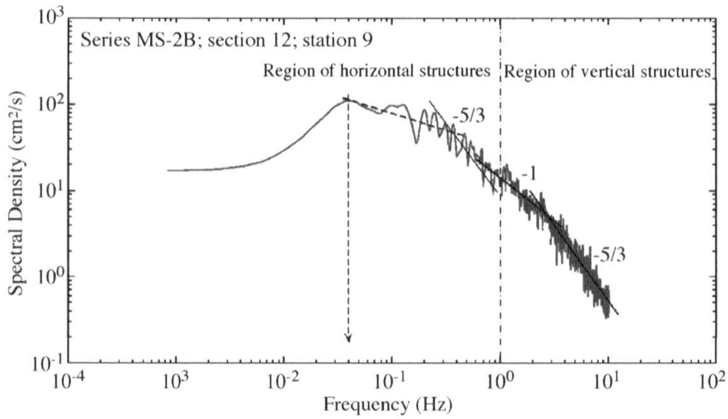

Figure 7 Example of plot of spectral density of longitudinal flow velocity.

Table 2 Sample values of characteristic time-scale (periods) of HCS's resulting from energy spectrum analysis for three stations on the channel centerline.

Series	*Section 6*	*Section 12*	*Section 18*
MS-2B	20 s	25 s	25 s
AS2-7B	20 s	33 s	29 s
AS1-0B	12 s	15 s	12 s

4.3 Continuous wavelet transform

As indicated by Polikar (1999) and Gurley and Kareem (1999), the continuous wavelet transform is a powerful tool for analyzing non-stationary signals. This makes it particularly suitable for the study of the velocity signal of turbulent flows. The technique involves the use of analyzing functions, called wavelets, which are localized in time (or space). The signals are analyzed at varying resolutions by decomposition of their frequency bands (Polikar, 1999, Miyamoto & Kanda, 2004). Wavelet transform has been used in different fields dealing with non-stationary signals, including wind, hydrological, and ocean engineering studies (see e.g., Kitagawa & Nomura, 2003, Mwale *et al.*, 2011, Massel, 2001). Wavelet transform was used for the detection of coherent structures in open-channel flows by Camussi (2002), Miyamoto and Kanda (2004), and Franca and Lemmin (2006).

In this work, continuous wavelet transform using (the complex-valued) Morlet wavelet mother function was applied to each of the 2 min long records of fluctuating longitudinal flow velocity of series AS2-7B. As an example the resulting contour plot of wavelet power in the time-frequency domain for Section 13, station 3 is shown in Figure 8. Note that this figure corresponds to the frequency bands of horizontal coherent structures. Indeed, as shown in Section 4.1, for the present flow $T_V \approx 1$ s (i.e., frequency = 1 Hz), while Figure 8 covers frequencies of 0.02 to 0.11 Hz, i.e., time-scales of 9.1 to 50 s (>> 1 s). The frequency components of velocity signal were then found through normalized integration over the measurement period of wavelet power for each different value of frequency. This yielded the global wavelet spectrum (GWS) shown in the graph on the right-hand side of Figure 8. The peaks in this graph represent the frequency components of the signal (Torrence & Compo, 1998). Accordingly, for this example, the representative frequency of HCS's is found to be 0.04 Hz, which corresponds to a period of 25 s. After applying the procedure above to the entire set of 2 min long records of velocity of series AS2-7B, the resulting values of characteristic periods were plotted in the form of probability density plots as shown in Figure 9, where the continuous curves are the plots of fitted log-normal distributions. Owing to paper length limitations, only the plots corresponding to the

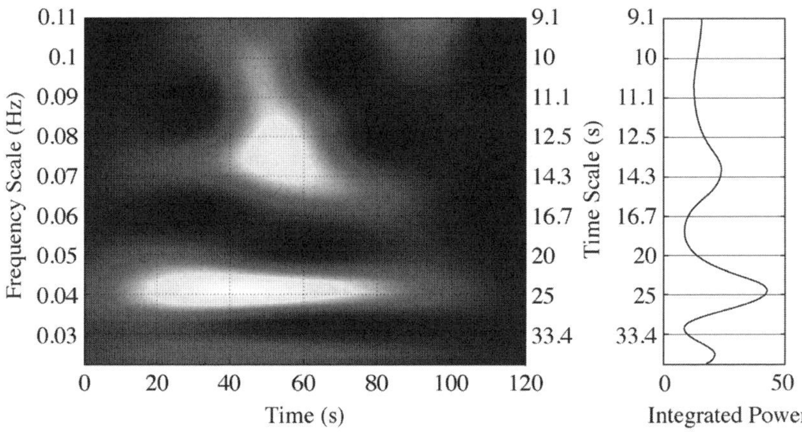

Figure 8 Contour plot of wavelet power in the time-frequency domain and corresponding global wavelet spectrum corresponding to Series AS2-7B, Section 13, station 3.

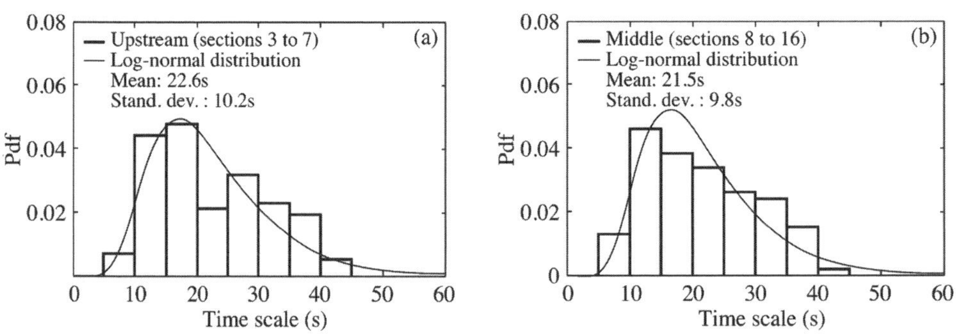

Figure 9 Probability density plots of characteristic time-scale (periods) of smallest frequency component in velocity records for series AS2-7B (upstream and middle channel regions). The values of mean and standard deviation shown are those of the fitted log-normal distributions.

upstream and middle regions of the channel are included in Figure 9. However, the plot for the downstream region is of the same nature as the plots shown, its mean value of characteristic period of smallest frequency component of the velocity signal being 19.2 s. Identifying the horizontal burst period with the mean value of the characteristic periods in Figures 9a and b, as well as the probability density plot for the downstream region, yields $T_H = 21.1$ s (= (22.6 s + 21.5 s + 19.2)/3). This value is quite comparable to that determined in Section 4.1.

5 EFFECT OF HCS'S ON THE MEAN FLOW; FORMATION OF ALTERNATE BARS

From the content of this paper, it follows that in the presence of a discontinuity, the straight time-averaged initial flow is subjected to a perpetual action of bursts 'fired' from the (ideally speaking) same location (the discontinuity at $x = 0$). Yalin (1992) argued that their action must inevitably render the flow to acquire a sequence of periodic (along x and t) non-uniformities. Consequently, the time-averaged streamlines – averaged over a multitude of burst periods – must vary along x with a period equal to the burst length. This is illustrated in the schematic Figure 2b.

To investigate this hypothesis, cross-sectional plots of local time averaged longitudinal flow velocity $\bar{u}$ (at the measurement level) were produced for all cross sections and all three series of measurements MS-2B, AS2-7B and AS1-0B. As examples, the resulting plots for cross Sections 8 to 15 corresponding to series AS2–7B are shown in Figure 10. Here, for the sake of facilitating the drawing, $\bar{u} - \hat{v}$ (where $\hat{v}$ is average flow velocity at the measurement level) is plotted instead of $\bar{u}$; each dashed horizontal line representing a cross section is, at the same time, the 'zero' of $\bar{u} - \hat{v}$. Additionally, the plots of average flow velocity (at the measurement level) for both left and right halves of the channel versus distance along the channel were also produced for all three series of measurements. Here, for space limitations, only the plots corresponding to series MS-2B and AS2-7B are shown in Figure 11, respectively, as these are the most pertinent to the present discussion.

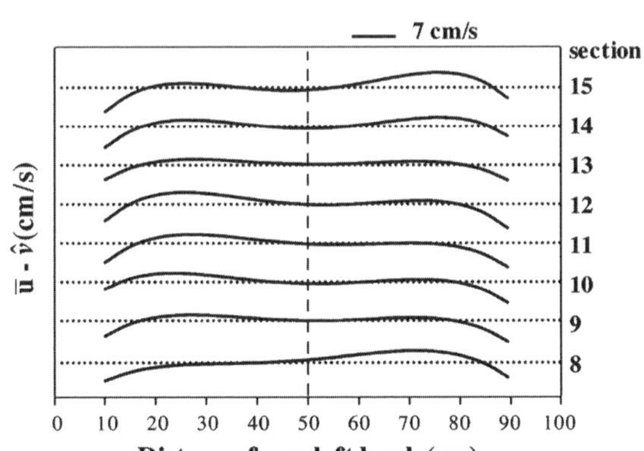

Figure 10 Plots of $\bar{u} - \hat{v}$ at Sections 8 to 15 for series of measurements AS2-7B.

Observe from the cross-sectional plots of $\bar{u} - \hat{v}$ in Figure 10 how the local time averaged flow velocity in series AS2-7B goes from being consistently slightly larger towards the left bank from Sections 9 to 12, to a nearly uniform distribution at cross Section 13, and then to becoming slightly larger towards the right bank at Sections 14 and 15. The shift of the locus of larger time-average flow velocity from one side of the channel to the other was observed throughout the entire length of the channel, as can be inferred from the plot of average flow velocity in the left and right halves of the channel in Figure 11b. The pattern of plan distribution of $\bar{u}$ implied by the plot in this figure is consistent with the wavelike deformation of the streamlines (the internal meandering) in the schematic Figure 2b. Moreover, the period of this wave-like deformation is generally comparable with the burst length $\lambda_H \approx 5B$, as estimated in previous sections. A periodic shift of the locus of larger time-average flow velocity from one side of the channel to the other is still quite evident in Figure 11a, corresponding to series MS-2B. The trend however, is less well defined than in Figure 11b, which is consistent with a flow structure less "regularized" throughout the channel.

Consider now the sediment transport continuity equation where, as shown for example in Yalin and da Silva (2001) and da Silva *et al.* (2006), the bed-load rate is a strongly increasing function of the flow velocity. This equation indicates that wave-like streamlines must, in turn, necessarily deform the bed so as to produce a sequence of periodic bed forms whose wavelength is the same as that of the wavelike deformation of the streamlines. To illustrate this point, the bed deformation patterns

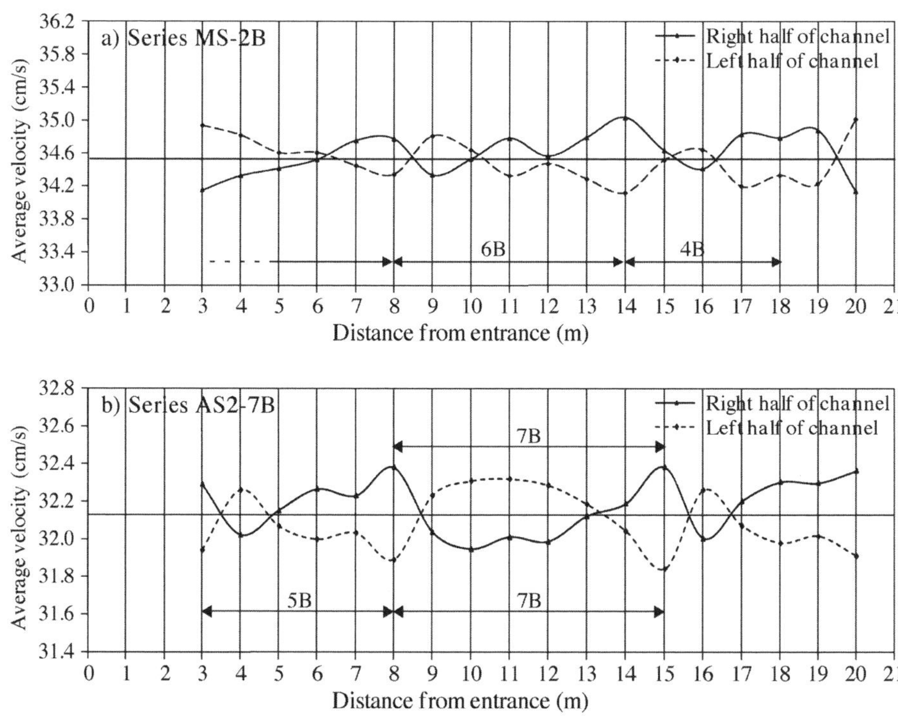

Figure 11 Plot of average flow velocity (at the measurement level) for the left and right halves of the channel. (a) Series MS-2B; (b) Series AS2-7B.

associated with the measured time averaged flow in series AS2-7B were calculated with the aid of the sediment transport continuity equation. It was assumed that the bed was formed by material with the same grain size (D = 2 mm) as the bed in the laboratory test, but having submerged specific weight γ_s = 9,712 N/m^3 (lighter material). The specific volumetric sediment transport rate was evaluated with the aid of Bagnold's bed-load equation. The details of these calculations are given elsewhere (Ahmari, 2010). The computed deformed bed is shown in Figure 12. This bed elevation contour plot is based on values of Δz, that is, it shows the changes in bed elevation in comparison to the flat bed over a time Δt (= 120 s) due to the convective patterns of time-averaged flow: $\Delta z > 0$ implies deposition, and $\Delta z < 0$ implies erosion. The computed deformed bed is consistent with the trace of alternate bars (see Figure 2c). The aforementioned lends support to the considerations by Yalin (1992), who suggested that bed and plan forms are generated by the turbulent bursts as the result of the convective patterns these superimpose on the mean flow (and perhaps to some extent, also by the direct action (the 'rubbing') of the coherent structures on the movable boundaries).

From the content of this section, it should not be concluded that a series of discontinuities along the walls (banks) of a channel is necessary to initiate alternate bars, or that these first appear in their full size. On the contrary, as illustrated by, e.g., the experiments of Ikeda (1983), the formation of alternate bars, just like that of dunes, is an 'activity' that progresses from upstream to downstream. As long as the frequency of eddy shedding will be increased at some near-bank location, then this will regularize the flow immediately downstream, generating the conditions for the appearance of the first alternate bar. This itself and the associated bed scour near the banks, will serve to direct the flow and regularize the eddy shedding further downstream. As a result, a second bar will start to appear as the first continues to grow, and so on, until the entire length of the channel will be covered by bed forms. It should be also noted that alternate bars are preceded by the appearance of smaller bar-like bed forms, which then 'grow' to a full size and shape consistent with the definition of alternate bars. This process too, is similar to the case of dunes, which do not originate so as to possess their full length in the first place. Where this aspect is concerned, the explanation given for the case of dunes by Yalin (1992), p. 72, is likely to be applicable to the present case of alternate bars. That is, the bursts dealt with in this paper are merely the largest horizontal bursts. In reality, at any given flow region the flow contains an "hierarchy" of superimposed horizontal burst sequences (along the flow direction x)

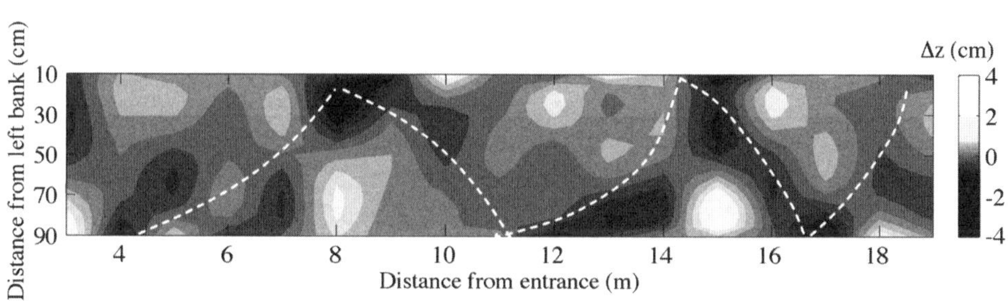

Figure 12 Bed deformation patterns associated with the internally meandering time averaged flow in series AS2-7B.

of different sizes, and time-scales, analogous to that in Figure 3.4 in Yalin (1992) for the case of vertical turbulence. The smaller the period of a sequence, the earlier it can imprint itself on the bed surface. Hence, the bed first becomes covered by smaller scale bars. Larger bursts then imprint themselves, and then even larger, and so on, with the larger bars at every step eliminating the previous bed forms. This process, which gives the impression that bars grow in length, terminates when the largest (alternate) bars are produced by the largest bursts in the flow.

6 PREFERENTIAL LOCATION OF EJECTIONS AND SWEEPS ALONG CHANNEL WALLS

On the basis of the data from the series of measurements MS-2B/2P, and AS2-0B/2P, and taking into account that HCS's rotate in the horizontal (x; y)-planes, quadrant analysis was used to study the joint behaviour of the fluctuating components of flow velocity in longitudinal and lateral directions (u' and v'). This analysis is extensively presented by Ahmari and da Silva (In preparation[b]). It seems, however, appropriate to end this paper with some results that follow from it, and in particular, those deriving from the determination of the probability of occurrence of each event (given by the ratio of number of data points in each quadrant to the total number of velocity measurements). In the central region of the channel, the data-points were found to be almost evenly distributed over the four quadrants, reflecting a large influence of flow structures associated with the opposite shear layer. However, towards the walls, the probability of occurrence of Q2 and Q4 events was different depending on location in the flow field. At some locations there was an increased probability of occurrence of ejection events, while at others, of sweep events. Considering this, the preferential locations of ejection and sweep events were estimated for series MS-2B/2P and AS2-7B/2P (at 5 cm from the channel walls) as shown in Figure 13. Observe that in both tests the distance between two consecutive ejection areas at either side of channel

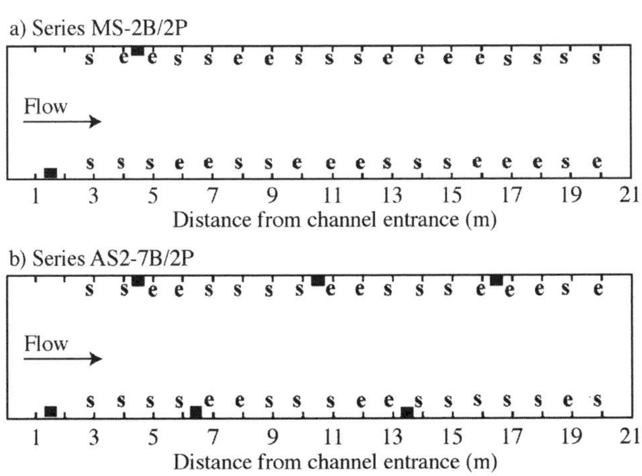

Figure 13 Preferential locations of ejections and sweeps along channel walls for series MS-2B/2P and AS2-7B/2P.

is between 4 to 7 m. Moreover, when an ejection region was observed at one bank, sweep events tended to predominate at the opposite bank. These results are consistent with a burst length of ≈ $5B$ as determined in Section 4. They also support the hypothesis in Figure 2a, regarding the arrangement in flow plan of HCS's.

7 CONCLUSIONS

The main findings of this paper can be summarized as follows:

1 In all series of measurements, and irrespective of location in flow domain, the smoothened oscillograms of u' were found to invariably exhibit cycles of fluctuation consistent with the presence of HCS's, affecting (or occupying) the entire body of fluid.
2 The smoothened oscillograms in the upstream reach of series MS-2B and throughout the channel in series AS2-7B consistently exhibited rather regular (nearly periodic) cycles of velocity fluctuation, consistent with the presence of persistent horizontal burst sequences promoted by the discontinuities.
3 The horizontal burst length in the regions of flow with an internal turbulence structure 'regularized' by the discontinuities was found to be $\lambda_H \approx 5B$, a value that is comparable to the values of average bar length and meander wavelength.
4 A wave-like deformation (internal meandering) of the time averaged flow was detectable. This was comparable to the burst length, which lends support to the hypothesis that the presence of periodic sequences of horizontal bursts 'fired' from both left and right walls as shown in Figure 2a, induces the internal meandering of flow, and in turn, the deformation of the bed into alternate bars (and possibly the meandering of the channel through direct action on the banks).
5 Preferential locations of ejections and sweeps at a spacing comparable with the burst length were found along the channel walls, with the location of ejections near one wall approximately coinciding with the locations of sweeps at the opposite wall. This further supports the hypothesis mentioned in point 4 above.

ACKNOWLEDGEMENTS

The first author is deeply grateful to Prof. Wolfgang Rodi and Prof. Cornelia Lang, for the invitation to write this paper honouring the memory of Prof. Gerhard Jirka, and orally present it at the Gerhard H. Jirka Memorial Colloquium on Environmental Fluid Mechanics, held at the Karlsruhe Institute of Technology (KIT), Karlsruhe, Germany, June 3–4, 2011. A debt of gratitude to Prof. Gerhard Jirka is acknowledged, for his generous friendship and the support he provided to the first author at various points of her professional career. The research in this paper was supported by funds from the Natural Sciences and Engineering Research Council of Canada, the Ontario Research and Development Challenge Fund, and Queen's University, Kingston, Canada.

NOTATION

B	flow width
D	typical grain size (usually D_{50})
F	Froude number ($Fr = u_{av}/(g\,h)^{1/2}$)
g	acceleration due to gravity
h	flow depth
k_s	granular skin roughness
O_i, O_i'	mark the flow cross section where a coherent structure originates ($i = 1, 2, \ldots$)
P	point marking the approximate 'birth' location of a coherent structure
Q	flow rate
R	flow Reynolds number ($R = u_{av}\,h/\nu$)
$R*$	roughness Reynolds number ($R* = v*\,k_s/\nu$)
S	bed slope
t	time
T	duration of life cycle of a coherent structure (burst period)
T_H	horizontal burst period
T_V	vertical burst period
u	component of local flow velocity vector in the longitudinal direction
v	component of local flow velocity vector in the transverse direction
u'	fluctuating component of u
$\bar{u}$	local time-averaged longitudinal flow velocity at the measurement level
u_{av}	average flow velocity ($u_{av} = Q/(Bh)$)
v'	fluctuating component of v
$\hat{v}$	average flow velocity at the measurement level (i.e., at 1 cm below the free surface)
$v*$	shear velocity ($v* = (\tau_0/\rho)^{1/2} = (gSh)^{1/2}$)
x	flow direction
y	direction horizontally perpendicular to the flow direction x
Y	mobility number ($Y = \tau_0/(\gamma_s\,D) = \rho v\,*^2/(\gamma_s\,D)$)
Y_{cr}	value of Y at the critical stage of initiation of sediment transport
z	vertical direction (direction vertically perpendicular to x and y)
α_V; α_B	coefficients of proportionality
γ_s	specific weight of grains in fluid
λ_H	horizontal burst length
λ_V	vertical burst length
Λ_a	average length of alternate bars
ν	fluid kinematic viscosity
ρ	fluid density
τ_0	bed shear stress

REFERENCES

Ahmari, H. (2010). Size, dynamics and consequences of large-scale horizontal coherent structures in open-channel flows. Ph.D. Thesis, Queen's University. Kingston, Canada.

Ahmari, H. and da Silva, A.M.F. (2011). Regions of bars, meandering and braiding in da Silva and Yalin's plan. *Journal of Hydraulic Research*, 49(6), 718–727.

Ahmari, H. and da Silva, A.M.F. (In preparation[a]) Energy spectrum of a multi-structural turbulent open-channel flow.

Ahmari, H. and da Silva, A.M.F. (In preparation[b]) Large-scale horizontal coherent structures in a shallow open-channel flow with upstream discontinuities: an experimental study.

Blackwelder, R.F. (1978). The bursting process in turbulent boundary layers. In: Smith, C.R. & Abbott, D.E. (eds.) *Coherent Structure of Turbulent Boundary Layers*. AFOSR/Lehigh University, Bethlehem, Penn., USA.

Camussi, R. (2002). Coherent structure identification from wavelet analysis of particle image velocimetry data. *Experiments in Fluids*, 32 (1), 76–86.

Cantwell, B.J. (1981). Organised motion in turbulent flow. *Annual Review of Fluid Mechanics*, 13 (1), 457–515.

da Silva, A.M.F. (1991). Alternate bars and related alluvial processes. M.Sc. Thesis, Queen's University. Kingston, Canada.

da Silva, A.M.F. (2006). On why and how do rivers meander: 14th IAHR Arthur Thomas Ippen Award Lecture. *Journal of Hydraulic Research*, 44 (5), 579–590.

da Silva, A.M.F. and Ahmari, H. (2009). Size and effect on the mean flow of large-scale horizontal coherent structures in open-channel flows: an experimental study. Special Issue in honour of Professor M. Selim Yalin (1925–2007), *Canadian Journal of Civil Engineering*, 36 (10), 1643–1655.

da Silva, A.M.F. and El-Tahawy, T. (2008). On the location in flow plan of erosion-deposition zones in sine-generated meandering streams. *Journal of Hydraulic Research*, 46 (1), Extra Issue, 49–60.

da Silva, A.M.F., El-Tahawy, T. and Tape, W. (2006). Variation of flow pattern with sinuosity in sine-generated meandering streams. *Journal of Hydraulic Engineering*, 132 (10), 1003–1014.

del Álamo, J.C. and Jiménez, J. (2003). Spectra of the very large anisotropic scales in turbulent channels. *Physics of Fluids*, 15 (6), L41–L44.

Dementiev, M.A. (1962). *Investigation of flow velocity fluctuations and their influences on the flow rate of mountainous rivers*. In: Technical Report of the State Hydro-Geological Institute (GGI). Report number: 98, 56–98 [In Russian].

Franca, M.J. and Lemmin, U. (2006). Detection and reconstruction of coherent structures based on wavelet multiresolution analysis. In: *Proceedings of River Flow2006, 3th International Conference on Fluvial Hydraulics, Lisbon, Portugal, Sept. 6–8, 2006*. Ferreira, Alves, Leal & Cardoso (eds.). Taylor & Francis Group, London, 181–190.

Franca, M.J. and Lemmin, U. (2008). Using empirical mode decomposition to detect large-scale coherent structures in river flows. In: *Proceedings of River Flow* 2008, *4th International Conference on Fluvial Hydraulics, Cesme-Izmir, Turkey, Sept. 3–5, 2008*. Altinakar, M.S., Kokpinar, M.A., Aydin, I., Cokgor, S. and Kirkgoz, S. (eds.). Kubaba Congress Department and Travel Services, 67–74.

Gad-el-Hak, M. and Hussain, A.K.M.F. (1986). Coherent structures in a turbulent boundary layer. Part 1: Generation of "artificial" bursts. *Physics of Fluids*, 29 (7), 2124–2139.

Grishanin, K.V. (1979). Dynamics of alluvial streams. *Gidrometeoizdat*, Leningrad [In Russian.].

Gurley, K. and Kareem, A. (1999). Applications of wavelet transforms in earthquake, wind and ocean engineering. *Engineering Structures*, 21 (2), 149–167.

Hussain, A.K.M.F. (1983). Coherent structures – reality and myth. *Physics of Fluids*, 26 (10), 2816–2838.
Ikeda, H. (1983). Experiments on bed-load transport, bed forms and sedimentary structures using fine gravel in the 4 m wide flume. *Env. Res. Center Papers* 2, p. 78., University of Tsukuba, Ibaraki.
Imamoto, H. (1973). Turbulence characteristics in free surface shear flows. *Bulletin Disaster Prevention Research Institute*, Kyoto University, 22, Part 3 (205), 153–186.
Jackson, G. (1976). Sedimentological and fluid-dynamic implications of the turbulent bursting phenomenon in geophysical flows. *Journal of Fluid Mechanics*, 77 (3), 531–560.
Jaeggi, M.N.R. (1984). Formation and effects of alternate bars. *Journal of Hydraulic Engineering*, 110 (2), 142–156.
Jirka, G.H. and Uijttewaal, W.S.J. (eds.) (2004). *Shallow Flows: Selected Papers of the International Symposium on Shallow Flows, 16–18 June 2003*. Delft, The Netherlands. A.A. Balkema, Rotterdam, The Netherlands.
Kamphuis, J.W. (1974). Determination of sand roughness for fixed beds. *Journal of Hydraulic Research*, 12 (2), 193–203.
Kishi, T. (1980). Bed forms and hydraulics relations for alluvial Streams. In: *Application of Stochastic Processes in Sediment Transport*. Shen, H.W. and Kikkawa, H. (eds.). Chapter 5. Water Resources Publications, Littleton, Colorado, USA.
Kitagawa, T. and Nomura, T. (2003). A wavelet-based method to generate artificial wind fluctuation data. *Journal of Wind Engineering and Industrial Aerodynamics*, 91 (7), 943–964.
Massel, S.R. (2001) Wavelet analysis for processing of ocean surface wave records. *Ocean Engineering*, 28 (8), 957–987.
Miyamoto, H. and Kanda, T. (2004). Extraction of Coherent Structure from PIV Data Using Wavelet Transform. In: *Proceedings of Building Partnerships; Joint Conference on Water Resource Engineering and Water Resources Planning & Management 2000*. ASCE, doi:10.1061/40517(2000)326.
Monin, A.S. and Ozmidov, R.V. (1985). Turbulence in the ocean. *D. Reidel Publishing Co.* Dovdrecht, The Netherlands.
Mwale, D., Gan, T.Y., Devito, K.J., Silins, U., Mendoza, C. and Petrone, R. (2011). Regionalization of Runoff Variability of Alberta, Canada, by Wavelet, Independent Component, Empirical Orthogonal Function, and Geographical Information System Analyses. *Journal of Hydrologic Engineering*, 16 (2), 93–107.
Nezu, I. and Nakagawa, H. (1993). *Turbulence in open-channel flows*. IAHR Monograph. A.A. Balkema, Rotterdam, The Netherlands.
Ozmidov, R.V. (1965) Energy distribution between oceanic motions of different scale. Izv. *Atmospheric and Oceanic Physics*, 1, 257–261.
Polikar, R. (1999). The story of wavelets. In: Mastorakis, N. (ed.), *Physics and Modern Topics in Mechanical and Electrical Engineering*. World Scientific and Eng. Society Press, 192–197. Available on-line at http://users.rowan.edu/~polikar/WAVELETS/WTtutorial.html [Accessed 20th August 2011].
Pope, S.P. (2000). *Turbulence*. Cambridge University Press, UK.
Rashidi, M. and Banerjee, S. (1988). Turbulence structures in free surface channel flows. *Physics of Fluids*, 31 (9), 2491–2501.
Roy, A.G., Buffin-Bélanger, T., Lamarre, H. and Kirkbride, A.D. (2004). Size, shape and dynamics of large-scale turbulent flow structures in a gravel-bed river. *Journal of Fluid Mechanics*, 500, 1–27.
Torrence, C. and Compo, P.G. (1998). A practical guide to wavelet analysis. *Bulletin of the American Meteorological Society*, 79 (1), 61–78.

Yalin, M.S. (1977). *Mechanics of sediment transport*. 2nd ed. Pergamon Press, Oxford, England.

Yalin, M.S. (1992). *River mechanics*. Pergamon Press, Oxford, England.

Yalin, M.S. (2006). Large-scale turbulence and river morphology. In: *Proceedings of River Flow2006, 3th International Conference on Fluvial Hydraulics, Lisbon, Portugal, Sept. 6–8, 2006*. Ferreira, Alves, Leal & Cardoso (eds.). Taylor & Francis Group, London, 1243–1249.

Yalin, M.S. and da Silva, A.M.F. (2001). *Fluvial processes*. IAHR Monograph. IAHR, Delft, The Netherlands.

Yokosi, S. (1967a). The structure of river turbulence. *Bulletin of Disaster Prevention Research Institute,* Kyoto University, 17 (121), 1–29.

Yokosi, S. (1967b). Large scale turbulence in a river. *Bulletin of Disaster Prevention Research Institute,* Kyoto University, 10 (B), 1–8.

Chapter 6

Waves and currents: Hawking radiation in the hydraulics laboratory?

G.A. Lawrence[1], S. Weinfurtner[2], E.W. Tedford[1], M.C.J. Penrice[2] and W.G. Unruh[2]

[1] *Department of Civil Engineering, University of British Columbia, Vancouver, British Columbia, Canada*

[2] *Department of Physics and Astronomy, University of British Columbia, Applied Science Lane, Vancouver, Vancouver British Columbia, Canada*

ABSTRACT: This paper discusses experiments performed to test an analogy between Hawking radiation (the process by which black holes radiate energy) and the propagation of water waves against an adverse current. A streamlined obstacle was placed in a flume to create a region of high velocity. Long waves generated downstream of the obstacle were blocked by this region and converted to a pair of short waves. The group velocities of both the converted waves were downstream, but one of the converted waves retained an upstream phase velocity, whereas the other had a downstream phase velocity. These waves are shown to be analogous to Hawking radiation.

1 INTRODUCTION

When approached to participate in a memorial colloquium for my[1] long time friend and mentor, Gerhard Jirka, my first instinct was to present some unpublished research on selective withdrawal, the topic that brought Gerhard and I together in the 1980's. However, on second thought I realized that Gerhard would probably have been more interested in the results my recent collaboration with physicists, investigating the analogue between the interaction of surface water waves and currents, and the behavior of waves in the vicinity of black holes. I will begin with an outline of how this collaboration has evolved, followed by an overview of preliminary experiments we performed in the hydraulics laboratory to investigate the interaction of waves and currents. An attempt will then by made to briefly summarize the use of general relativity to obtain a mathematical analogy between the physics of black holes and hydraulics. The results of further laboratory experiments will then be analyzed in the context of black-hole physics.

2 BACKGROUND

We have all been exposed to the notion that nothing, not even light, can escape from a black-hole. Well, maybe not! Hawking (1974) proposed that, due to some sort of

[1] This paper is written in the voice of the first author.

quantum instability near their horizons, black holes emit a form of radiation, now known as Hawking radiation. Even in theory, the exact origin of this radiation is uncertain, and since it is too weak to be detected directly, physicists have sought other means of studying it. Unruh (1981) demonstrated a direct mathematical analogy between the physics of black holes and that of sound waves. Subsequently, Schutzhold and Unruh (2002) demonstrated that the analogy applied to water waves propagating against a current that could be varying in strength. They interested me in pursuing this analogy, and for several years Unruh and I could occasionally be found performing experiments in a flume in the hydraulics laboratory in the Department of Civil Engineering at the University of British Columbia. The experiments became more sophisticated, and intense, with the addition of three other researchers: a postdoctoral fellow in Physics, Dr. Silke Weinfurtner; a recent Civil Engineering PhD, Dr. Edmund Tedford; and a Physics undergraduate, Matt Pennrice. The results of these experiments appear in Weinfurtner *et al.* (2011); however, this paper was subject to a four-page limit and written with specialists in black hole physicists in mind. Here the goal is to present a more detailed overview in a manner that Gerhard Jirka would enjoy.

3 LABORATORY SETUP

We conducted a series of experiments in an open channel flume originally built to teach undergraduate students the principles of open channel hydraulics (Fig. 1). Streamlined obstacles were placed in the 6.22 m long, 0.154 m wide and 0.475 m deep flume. Surface waves were generated 2 m downstream of the obstacle by a vertically oscillating mesh that partially blocked the flow as it moved in and out of the water. Water height variations were measured and analysed using the same techniques as Tedford *et al.* (2009). The water surface was illuminated using laser induced fluorescence, and photographed with a high-resolution monochrome camera. Two sets of experiments were

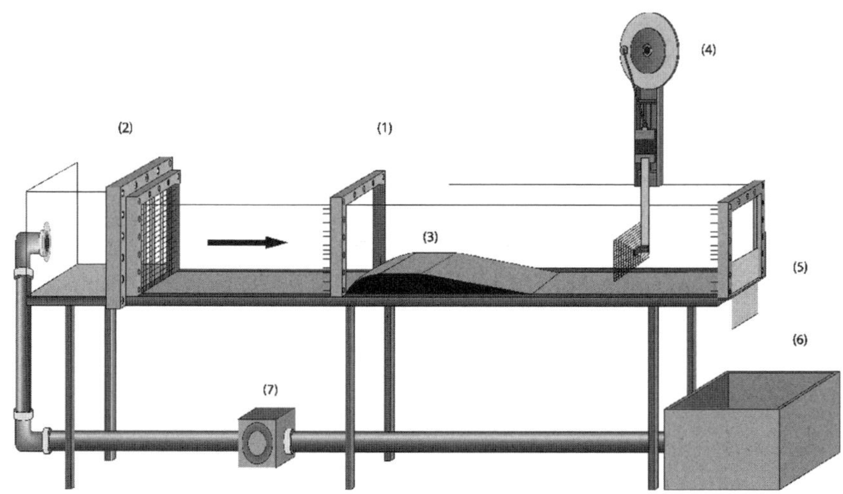

Figure 1 Experimental apparatus: (1) flume, (2) intake reservoir, (3) obstacle, (4) wave generator, (5) adjustable weir, (6) holding reservoir, and (7) pump and pump valve.

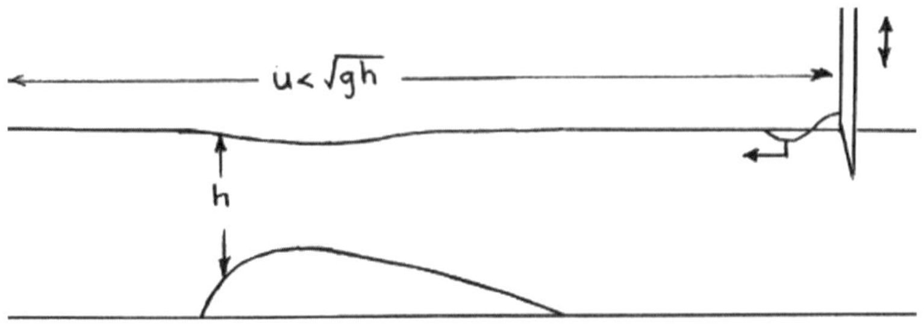

Figure 2 Schematic of subcritical flow over an obstacle. The Froude number $(\mathrm{Fr} \equiv u/\sqrt{gh})$ is less than 1 everywhere. What will happen to waves propagating upstream?

performed: the first set with a 1.0 m long and 0.05 m high obstacle, a two-dimensional flow rate $q \approx 0.02$ m²/s, and a downstream flow depth $h_o = 0.114$ m, subjected to a packet of waves of period $\tau \approx 1.5$ s; the second set was performed with a 1.55 m long and 0.106 m high obstacle, $q = 0.045$ m²/s, $h_o = 0.194$ m, subjected to trains of waves with periods ranging from 1.5 s–20 s, with corresponding still water wavelengths between 2.1 and 69 m. Before discussing the experiments in detail we will revisit some principles of open channel flow.

In the hydraulics of two-dimensional flow over an obstacle the most important parameter is the Froude number:

$$\mathrm{Fr} \equiv \frac{u}{\sqrt{gh}}, \tag{1}$$

which is the ratio of the cross-sectionally averaged flow velocity, u, and the phase speed of long surface waves, $\sqrt{gh}$, where g is the gravitational acceleration and h is the depth of flow. The flow depicted in Fig. 2 is said to be subcritical because $Fr < 1$ everywhere. Consider the behavior of waves generated downstream in such a flow. When asked what would the waves do, my first response (and that of everyone else that I have asked) was that the waves would propagate upstream past the obstacle. They would slow down as they passed over the obstacle, but not be arrested.

4 PRELIMINARY EXPERIMENTS ON THE INTERACTION OF WAVES AND CURRENTS

We conducted a preliminary experiment to examine the behavior of waves propagating upstream in a subcritical flow as depicted in Fig. 2. In this experiment a wave packet was generated downstream of the obstacle. The surprising result was that very little of the wave energy propagated upstream of the obstacle (see Fig. 3a) even though the maximum Froude number in the flow, $Fr_{max} \approx 0.44$. After a little head scratching we realized that this behavior is explained by noting that the Froude number is based on the long wave phase speed $\sqrt{gh}$, which is only the case if the wavelength is much greater than the depth of flow.

Relaxing the long wave assumption gives the phase speed:

$$c_p = \sqrt{\frac{g}{k}\tanh(kh)}, \tag{2}$$

where the wave number, $k = 2\pi/\lambda$. This result assumes that surface tension is not important (i.e., $k \ll \sqrt{pg/\sigma} = 370$, for water at 20 C). Taking (2) into consideration we can define a "finite wavelength" Froude number:

$$\mathbb{F} \equiv \frac{u}{c_\text{p}} = \frac{u}{\sqrt{\frac{g}{k}\tanh(kh)}}, \tag{3}$$

Note that:

$$Fr = \mathbb{F}\sqrt{\frac{\tanh(kh)}{kh}} < \mathbb{F}, \tag{4}$$

So for the flow depicted in Fig. 3, waves can be arrested above the lee side of the obstacle even when the flow is uncontrolled by the conventional definition ($Fr < 1$). In fact the situation is further complicated by the fact that for finite wavelengths the group velocity is less than the phase velocity. Also, the point at which the waves are arrested is a function of the wave number. Nevertheless, for a range incoming wavelengths there is a region of the flow that they cannot enter, analogous to what cosmologists call a white hole[2]. The location where the waves are arrested is called a white hole horizon, which is analogous to a hydraulic control. The main difference being that the position of the white hole horizon is a function of wavelength.

The dispersion relation for waves in the presence of a current is given by:

$$(\omega + uk)^2 = gk\tanh(kh), \tag{5}$$

where $\omega = 2\pi/\tau$, and τ is the wave period. The solution for positive ω and positive K is plotted in Fig. 3, and the detailed behavior of the waves in our experiment is illustrated by comparing this figure with the characteristics (space-time) diagram for our experiment.

Our wave packet was not uniform; the leading waves were longer with higher phase speed than the trailing waves. The leading waves managed to propagate over the obstacle; whereas, the remaining ones did not (Fig. 3a). The properties of these waves as they approached the obstacle are represented by "A" on Fig. 3; after the propagating halfway up the lee face of the obstacle their wavelength, phase speed and group velocity decreased (B). This process continued until they reached "C" where the group velocity vanished. Individual waves then continued to propagate upstream, but

2 From a theoretical perspective time reversal invariance leads to the equivalence of white and black hole horizons.

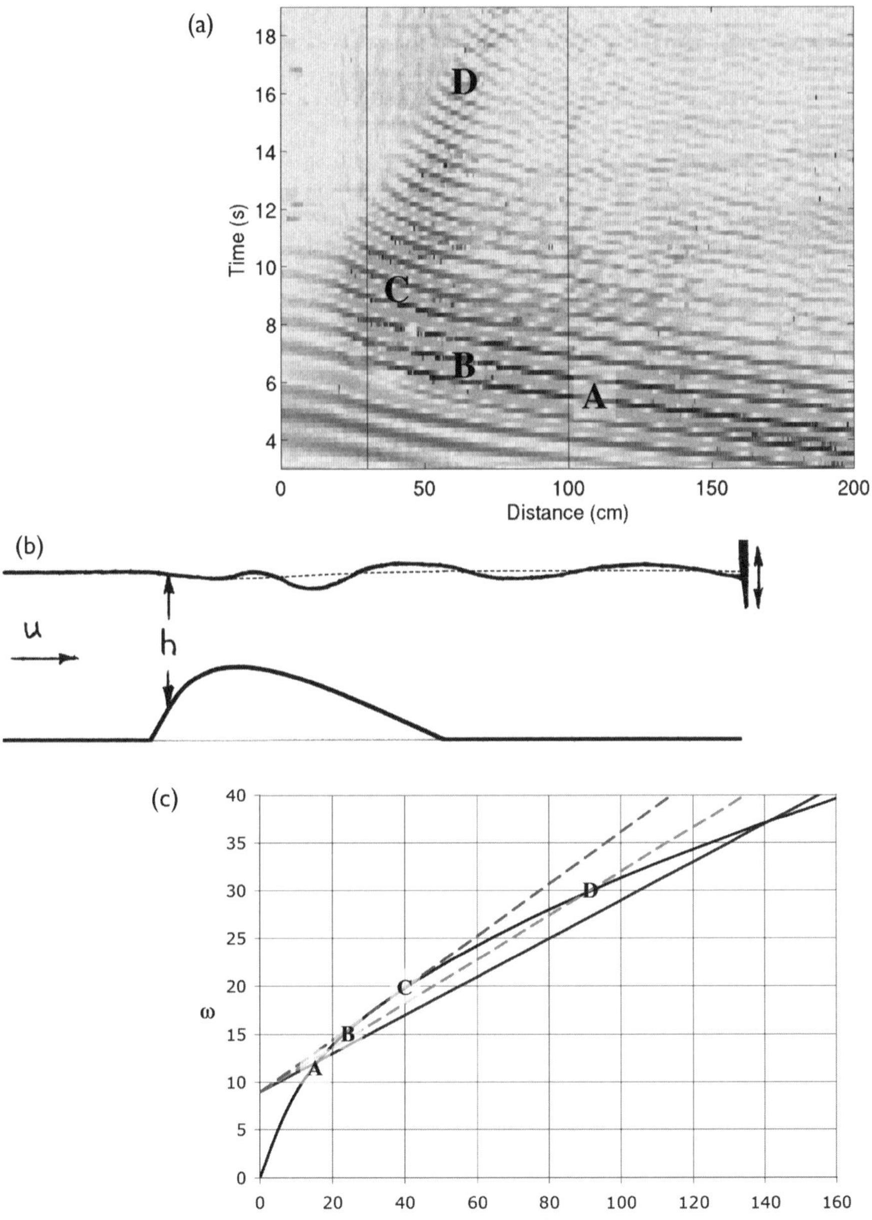

Figure 3 (a) Wave amplitude plotted as a function of position and time for a packet of waves propagating against subcritical flow over an obstacle. A white hole horizon arrests the waves just downstream of the crest of the obstacle. (b) Schematic of the flow. (c) Dispersion diagram. The curved line is $\sqrt{gk\ \tanh(kh)}$; the straight lines are $\omega + uk$ where u is the flow velocity away from the obstacle (bottom line), half wave up the obstacle (middle line) and near the crest (top line).

decayed rapidly, while the wave packet travelled downstream gradually losing energy to viscous dissipation. At "D" the waves were quite weak, but still detectable. In the process of propagating from "A" to "D" the waves were converted from relatively long waves with positive phase and group velocities, to short waves with positive phase velocity, but negative group velocity. Experiments such as these are not new; others, most notably Badulin *et al.* (1983), Suastika (2004) and Rousseaux *et al.* (2008) have performed similar, and more extensive, experiments. Badulin *et al.* (1983) even investigated further conversions involving capillary waves. Nevertheless, we present the above experiment as it provides context for our subsequent experiments investigating the analogy with Hawking radiation.

5 HAWKING RADIATION

Theure nature of Hawking radiation is still the subject of debate. The popular science description (Hawking, 1977) goes something like this. Particle-antiparticle pairs are continually generated throughout the universe, including near black hole horizons. Normally, these pairs annihilate each other soon after their formation. However, if a pair forms at a black hole horizon, one may fall into the hole while the other escapes as "Hawking radiation". The rate at which a black hole emits energy is normally too weak to be detected directly. However, Hawking radiation is inversely proportional a black hole's mass; consequently, there is the prospect that, in the absence of additional energy entering a black hole, its mass will radiate away at a rate that will increase as it shrinks. The black hole will ultimately disappear in a massive release of energy – hence the title of Hawking's 1974 paper "Black hole explosions".

It is probably more accurate to describe the particle-antiparticle pairs as field excitations (waves) with positive and negative norms (phase speeds), whose amplitudes α_f, β_f (Bogoliubov coefficients) are related by:

$$\frac{|\alpha_f|^2}{|\beta_f|^2} = \exp\left\{\frac{-2\pi\omega}{g_H}\right\} \tag{6a}$$

where ω is the frequency of the excitations, and $g_H = 10^{35}/M\, s^{-1}$ is the surface gravity of a black hole of mass M. Positive norm modes are emitted by the black hole, while negative ones are absorbed, effectively reducing its mass. Comparison of (6a) with the Boltzman-distribution yields a black hole "temperature".

$$T(K) = 6 \times 10^{-8} \frac{M_\odot}{M} \tag{6b}$$

where $M_\odot$ is a solar mass (mass of the sun). Equation (6b) leads to the notion that black hole evaporation (Hawking radiation) intensifies as a black hole shrinks; it also indicates that black hole evaporation would be difficult to observe directly (Hawking, 1975).

6 THEORETICAL ANALOGY BETWEEN WATER WAVES AND GENERAL RELATIVITY

In this section I will attempt to present Schutzhold and Unruh's (2002) analogy between water waves and general relativity in the context of the present study, and with readers more familiar with fluid mechanics than general relativity in mind.

Consider flow over an obstacle perturbed by gravity waves as depicted in Fig. 4.

Assuming the flow is incompressible and two-dimensional the continuity equation becomes:

$$\frac{\partial u}{\partial x} + \frac{\partial w}{\partial z} = 0, \tag{7}$$

If $h_B << 1$, where h_B is the depth of the background flow and λ is the wave-length of the gravity waves, we can make the shallow water assumption that $\partial u/\partial x$ is independent of z, where u and w are the horizontal (x) and vertical (z) components of velocity. Integrating (7) from $z = -h_B$ to $z = \eta$ yields:

$$h\frac{\partial u}{\partial w} + w(\eta) - w(-h_B) = 0, \tag{8}$$

where the depth of flow, $h = h_B + \eta$. Noting that:

$$w(\eta) = \frac{D\eta}{Dt} = \frac{\partial \eta}{\partial t} + u\frac{\partial \eta}{\partial x}, \text{and } w(-h_B) = -u\frac{\partial h_B}{\partial x},$$

(8) becomes:

$$h\frac{\partial u}{\partial x} + \frac{\partial \eta}{\partial t} + u\frac{\partial \eta}{\partial x} + u\frac{\partial h_B}{\partial x} = 0, \tag{9}$$

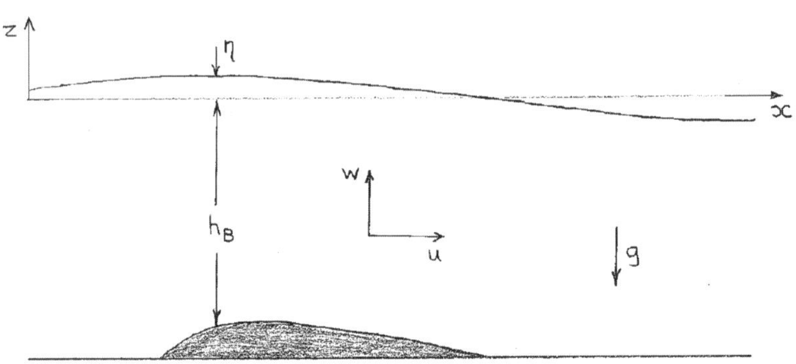

Figure 4 Definition sketch for analysis of water waves.

giving the Saint-Venant equation:

$$\frac{\partial h}{\partial t}+\frac{\partial (uh)}{\partial x}=0, \tag{10}$$

where the horizontal component of velocity $u = u_B + u'$, where u_B is the background velocity, and η and u' are the perturbations caused by the waves. Assuming $\eta << h_B$ we can rewrite the continuity equation as:

$$\frac{\partial \eta}{\partial t}+\frac{\partial (h_B u' + u_B \eta)}{\partial x}=0, \tag{11}$$

and Euler's equation at the free surface as:

$$-\frac{\partial \phi}{\partial t}+g\eta+\frac{1}{2}u^2=0, \tag{12}$$

assuming the flow is irrotational with velocity potential $\phi = \phi_B + \phi', u = -\partial\phi/\partial x$, and g is gravitational acceleration. Separating out the perturbation components yields:

$$\eta = \frac{1}{g}\left(\frac{\partial \phi'}{\partial t}+u_B\frac{\partial \phi'}{\partial x}\right). \tag{13}$$

Substituting (13) into (11) yields:

$$\frac{\partial}{\partial t}\left\{\frac{1}{g}\left(\frac{\partial \phi'}{\partial t}+u_B\frac{\partial \phi'}{\partial x}\right)\right\}+\frac{\partial}{\partial x}\left[h_B\left\{-\frac{\partial \phi'}{\partial x}\right\}+u_B\left\{\frac{1}{g}\left(\frac{\partial \phi'}{\partial t}+u_B\frac{\partial \phi'}{\partial x}\right)\right\}\right]=0, \tag{14}$$

or

$$\frac{\partial^2 \phi'}{\partial t^2}+\frac{\partial}{\partial t}\left(u_B\frac{\partial \phi'}{\partial x}\right)+\frac{\partial}{\partial x}\left(\{u_B^2-c^2\}\frac{\partial \phi'}{\partial x}\right)+\frac{\partial}{\partial x}\left(u_B\frac{\partial \phi'}{\partial t}\right)=0, \tag{15}$$

where $c=\sqrt{gh_B}$ is the wave speed. Equation (15) can be rewritten as:

$$[\partial_t, \partial_x]\begin{bmatrix} 1 & u_B \\ u_B & u_B^2-c^2 \end{bmatrix}\begin{bmatrix} \partial_t & \phi' \\ \partial_x & \phi' \end{bmatrix}=0 \tag{16}$$

It makes things simpler if we introduce a third dimension, with coordinate y on which nothing will depend; (16) then becomes:

$$\left[\partial_t\ \partial_x\ \partial_y\right]\begin{bmatrix} 1 & u_B & 0 \\ u_B & u_B{}^2-c^2 & 0 \\ 0 & 0 & c^2 \end{bmatrix}\begin{bmatrix} \partial_t & \phi' \\ \partial_x & \phi' \\ \partial_y & \phi' \end{bmatrix}=0 \tag{17}$$

We are now at the point where we can draw an analogy with general relativity. In general relativity the wave equation on a general curved space-time geometry is given by:

$$\frac{1}{\sqrt{|g|}}\partial_{\mu}\left(\sqrt{|g|}g^{\mu\upsilon}\partial_{\upsilon}\phi'\right)=0 \tag{18}$$

where the Greek indices run from 0–2, with 0 representing time, and 1–2 the two spatial dimensions (x and y), respectively, a repeated index is summed over, and $g^{\mu\nu}$ is the inverse of the "metric tensor" $g_{\mu\nu}$, and $\mathbf{g} \equiv \det(g_{\mu\nu})$. Equation (18), and variants of it, is central to the study of general relativity. With

$$g^{\mu\upsilon} = \frac{1}{c^4}\begin{bmatrix} 1 & u_B & 0 \\ u_B & u_B^{\,2} - c^2 & 0 \\ 0 & 0 & -c^2 \end{bmatrix}, \tag{19}$$

and

$$g_{\mu\upsilon} = c^2\begin{bmatrix} c^2 - u_B^2 & u_B & 0 \\ u_B & -1 & 0 \\ 0 & 0 & -1 \end{bmatrix}, \tag{20}$$

and $\mathbf{g} = c^8$, (17) and (18) are equivalent, and there is a direct analogy between fluid mechanics and general relativity.

Setting the upper right hand element of $g_{00} = 0$, yields:

$$\mathrm{Fr}^2 \equiv \left(\frac{u_B}{c}\right)^2 = 1 \tag{21}$$

where Fr is the Froude number, and (21) represents the condition for a hydraulic control in fluid mechanics, or for an event horizon in general relativity.

The above analogy relies on the shallow water assumption, which, as we have seen from the preliminary experiments, will be violated. Nevertheless, we felt that the analogy was sufficiently strong to warrant further experimentation with the goal of observing an analogue to Hawking radiation.

7 EXPERIMENTS IN SEARCH OF AN ANALOG TO HAWKING RADIATION

To investigate the possibility of simulating. Hawking radiation in a laboratory flume, we plot the dispersion relation (5) in the laboratory frame of reference (Fig. 5) for a flow typical of our experiments. We also consider the possibility of waves of negative k (waves with phase velocities in the opposite direction to the incoming

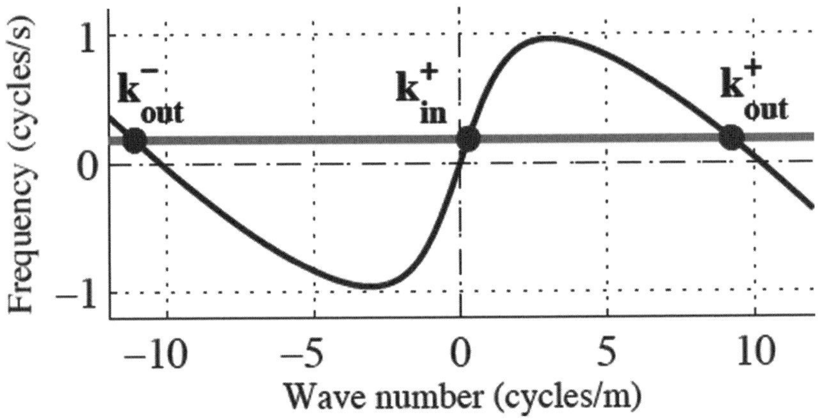

Figure 5 Dispersion relation for waves propagating against a flow typical of our experiments.A relatively long wave, k^+_{in}, propagating upstream, is blocked by the flow and converted to a pair of deep water waves (k^+_{out} and k^-_{out}) that are swept downstream. All waves have f = 0.185 HZ. (From Weinfurner *et al.*, 2011).

waves). Here, to be consistent with Weinfurtner *et al.*, 2011, we define frequency, $f \equiv 1/\tau$, and wave number, $K \equiv 1/\tau$.

For low frequencies, there are three possible waves, which we denote according to wave number. The first wave, k^+_{in}, is a shallow water wave with both positive phase and group velocities, and corresponds to the wave that we generate in our experiments. The second, k^+_{out}, has positive phase velocity, but negative group velocity. Both waves, k^+_{in} and k^+_{out}, are on the positive norm branch of the dispersion relation. The third, k^-_{out}, has both negative phase and group velocities, and it lies on the negative norm branch. It is this third wave that is the analogue of Hawking radiation. The first goal of our experiments was to detect the presence of this third wave. In the preliminary experiment depicted in figure 3, this wave was probably not present because the incoming wave frequency was too high, and even if it were present our preliminary detection methods would not have been sensitive enough to detect it.

Results of an experiment with incoming wave frequency f = 0.185 HZ are presented in figure 6. In this case we analysed images from exactly 18 cycles, measuring the free surface along approximately 2 m of the flow including the obstacle. To facilitate analysis of the surface wave data we redefined the spatial co-ordinate using, $\xi = \int dx/u_B(x)$, where x is the distance downstream of the crest of the obstacle. The ξ coordinate has dimensions of time, and its associated wave number κ has units of Hz (see We infurtner *et al.*, 2011). We then calculated the two-dimensional Fourier transformation as displayed in Fig. 6. Note that the amplitudes of the Fourier transform at frequencies above and below 0.185 Hz are very small, indicating that the noise level is small.

As anticipated, there are three peaks, one corresponding to the ingoing shallow water wavelength around κ = 0, and the other two corresponding to converted deep water waves peaked near $k^+_{out} = 9.7$ Hz and $k^-_{out} = -10.5$ Hz. The former is a positive norm and the latter a negative norm outgoing wave, an analogue of Hawking radiation.

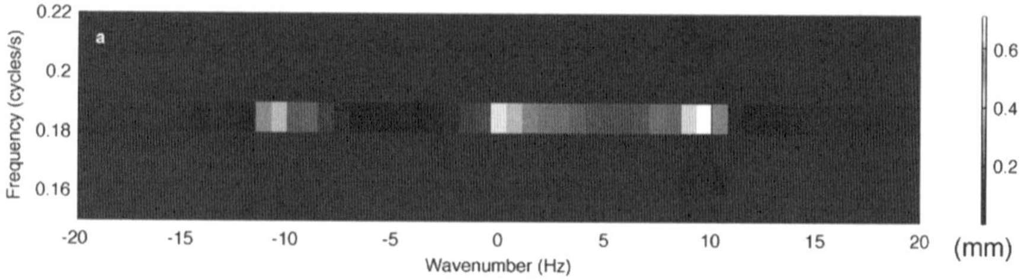

Figure 6 Demonstration of pair-wave conversion of an incoming frequency of 0.185 Hz.

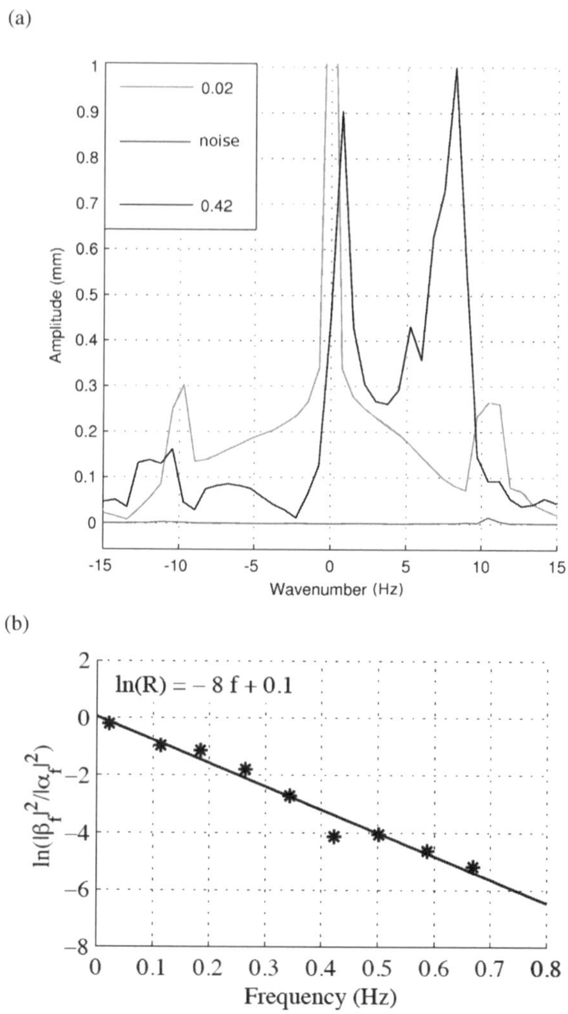

Figure 7 Amplitudes and thermal spectrum. (a) Absolute value of two ingoing frequency bands, and typical noise level (nearly horizontal line). (b) Natural log ratio of negative and positive norm components (stars). (From Weinfurner *et al.*, 2011).

Having detected wave-pair creation our focus shifted to determining whether or not the amplitudes of the waves obey (6). Our key results are presented in Fig. 7. Figure 7a shows the amplitude of the spatial Fourier transform at three selected ingoing frequencies. What is important to note, is that as the frequency increases, the ratio of the negative norm peak to positive norm peak decreases. Furthermore, the location of the positive norm peak moves slightly toward zero as the frequency increases, while the negative norm peak moves away from zero. This is to be expected from the dispersion plot, see Fig. 5. The almost horizontal line in Fig. 7a is the graph of the Fourier transform in the adjacent temporal frequency bands for the sample case of 0.185 Hz. This is a representation of the noise, and is a factor of at least 10 lower than the signals. Most importantly Figure 7b shows that the amplitudes of the waves do indeed satisfy (6) lending additional credence to the hypothesis that we have indeed observed an analogue to Hawking radiation.

8 SUMMARY

We have conducted a series of experiments to verify the stimulated Hawking process at a white hole horizon in a fluid analogue gravity system. These experiments demonstrate that the pair-wave creation is described by a Boltzmann-distribution, indicating that the thermal emission process is a generic phenomenon. It survives fluid-dynamical properties, such as turbulence and viscosity that, while present in our system, are not included when deriving the analogy. It is also robust against the vast alteration of the dispersion relationship in our system. The ratio is thermal despite the different dispersion relation in this model from that in the black hole derivation, increasing our trust in the ultraviolet-independence of the effect and that the thermal effect is a feature of the low frequency, long wavelength aspects of the physics. When this thermal emission was originally postulated by Hawking, it was believed to be a feature peculiar to black holes. Our experiments, and prior numerical work (Unruh, 1995; Corley & Jacobsen, 1999), demonstrate that this phenomenon seems to be ubiquitous, and not something that relies on quantum gravity or Planck-scale physics.

While our experiments measure only the stimulated emission from this white hole analogue, it has been known since Einstein (1917) that there is a very close relation between spontaneous and stimulated emission from a quantum system. Furthermore the time reversal invariance of the theory leads to the equivalence of black and white hole horizons. We believe our observations to be the most convincing demonstration of the Hawking process to date.

It would be exciting to measure the spontaneous emission from a black hole. While finding a small black hole to test the prediction directly is beyond experimental reach, such measurements might be achievable in other analogue models, like Bose Einstein condensates, or optical fibre systems (Dimopoulos, 2001; Jain *et al.*, 2007; Belgiorno *et al.*, 2010).

ACKNOWLEDGEMENTS

We thank Mauricio Richartz for his help during the initial stages of this project. WGU and GAL thank the Natural Sciences and Engineering research Council of Canada.

WGU thanks the Canadian Institute for Advanced research, and GAL thanks the Canada Research Chairs program. SW was supported by a Madam Curie Fellowship EMERGENT-2007-SW. We thank the Department of Civil Engineering for the use of the flume and experimental space. Their willingness to make do with other equipment and space made our experiments possible.

REFERENCES

Badulin, S.I., Pokazeyev, K.V. and Rozenberg, A.D. (1983). A Laboratory Study of the of Regular Gravity-Capillary Waves on Inhomogeneous Flows. Izvestiya, Atmospheric and Oceanic Physics 19, 10.

Belgiorno, F., Cacciatori, S.L., Ortenzi, G., Sala, V.G. and Faccio, D. (2010). Quantum radiation from superluminal refractive index perturbations. Phys. Rev. Lett. 104 1404003.

Corley, S. and Jacobson, T. (1999). Black hole lasers. Phys. Rev. D 59, 124011.

Dimopoulos, S. and Landsberg, G. (2001). Black Holes at the LHC. Phys. Rev. Lett. 87 161602.

Einurestein, A. Zur Quantentheorie der Strahlung. Phyis. Zeitschr. XVIII (1917).

Hawking, S.W. (1974). Black Hole Explosions. Nature 248, 30–31.

Hawking, S.W. (1975). Particle Creation by Black Holes. Commun. Math. Phys. 43, 199.

Hawking, S.W. (1977). The Quantum Mechanics of Black Holes. Scientific American 236, 34–40.

Jacobson, Theodore A. and Parentani, R. (2007). An echo of black holes. Scientific American 17, 12–19.

Jain, P., Bradley, A.S. and Gardiner, C.W. (2007). The Quantum de Laval Nozzle: stability and quantum dynamics of sonic horizons in a toroidally trapped Bose gas containing a superflow. Phys. Rev. A 76 023617.

Rouuresseaux, G., Mathis, C. Maissa, P., Philbin, T.G. and Leonhardt, U. (2008). Observation of negative-frequency waves in a water tank: a classical analogue to the Hawking effect? New Journal of Physics 10 053015.

Schützhold, R. and Unruh, W.G. (2002). Gravity wave analogs of black holes. Phys. Rev. D 66, 044019.

Suaurestika, I.K. (2004). Wave blocking PhD Thesis Technische Universiteit Delft, The Netherlands Online at http://repository.tudelft.nl/file/275166/201607.

Tedford, E.W., Pieters, R. and Lawrence, G.A. (2009). Symmetric Holmboe instabilities in a laboratory exchange flow. J. Fluid Mech. 636, 137–153.

Unruh, W.G. (1981). Experimental black hole evaporation. Phys. Rev. Lett. 46, 1351–1353.

Unruh, W.G. (1995). Sonic analogue of black holes and the effects of high frequencies on black hole evaporation. Phys. Rev. D 51, 2827–2838.

Weinfurtner, S., Tedford, E.W., Penrice, M.C.J., Unruh, W.G. and Lawrence, G.A. (2011). Measurement of stimulated Hawking emission in an analogue system. Physical Review Letters, 106, 021302 (see also arXiv:1008.1911v2 [gr-qc]).

Part 2

Fluvial hydraulics

Chapter 7

Numerical simulation of turbulent flow and sediment transport processes in arbitrarily complex waterways

Seokkoo Kang, Ali Khosronejad and Fotis Sotiropoulos
St. Anthony Falls Laboratory, Department of Civil Engineering, University of Minnesota, Minneapolis, Minnesota, USA

ABSTRACT: We review recent progress toward the development of a computational framework for simulating turbulence and sediment transport processes in real-life waterways with embedded hydraulic structures. The approach is based on the Curvilinear Immersed Boundary (CURVIB) method, which is extended to carry out LES and URANS and simulate free-surface effects and stream-bed morphodynamics in complex open channels. The predictive capabilities of the method are demonstrated by presenting results from the simulation of turbulence in a natural meander bend and in a straight flume with embedded rock structures.

1 INTRODUCTION

Natural streams and rivers are characterized by arbitrary geometrical complexity that spans a wide range of scales: from the scale of a stream or river meander, to the scale of transverse bathymetric variabilities, down to the scale of boulders, small rocks, and sand grains typically found in river beds. The presence of natural and/or man-made structures, such as riffles and pools, tree trunks, root wads, bridge foundations and stream restoration structures, and dynamically evolving boundaries (water surface and erodible bed) further add to the difficulty of the problem. These difficulties are daunting and render attempts to simulate turbulence in natural aquatic environments a rather challenging undertaking for even the most advanced numerical methods available today. Due to the enormous geometrical complexity of natural waterways and the wide variability of spatial and temporal scales of the various coherent structures, accurate numerical simulations require fine computational meshes, small time steps and numerical algorithms that are versatile and efficient enough to handle all underlying complexities. In this chapter we review recent progress toward the development of a numerical method that can tackle several of these challenges. We begin by presenting in the remainder of this section a literature review of all pertinent issues for simulating turbulent flow and transport processes in open-channel flows.

The first requirement for simulating complex flow physics occurring in the real-life streams is to be able to handle computational domains with arbitrary geometric complexity. In the last decade, sharp-interface immersed boundary (SIB) methods have become very popular for simulating flows in arbitrarily complex, multi-connected domains with moving immersed boundaries (see for example: Iaccarino & Verzicco, 2003; Gilmanov & Sotiropoulos, 2005; Ge & Sotiropoulos, 2007). In SIB methods a complex boundary is typically treated as a sharp interface immersed in a non-boundary-conforming background mesh. The immersed body is discretized with a separate surface mesh, which is used to identify the relative location or even track the motion, in the case of moving boundaries, of the immersed body within the fixed Cartesian background mesh. The presence of the solid boundary on the ambient flow is accounted for via suitable boundary conditions reconstruction techniques based on interpolation either along grid lines (Fadlun *et al.*, 2000; Iaccarino & Verzicco, 2003) or the local normal to the immersed boundary (Gilmanov and Sotiropoulos, 2005). For a recent review of SIB methods the reader is referred to Mittal & Iaccarino (2005). An important drawback of SIB techniques using a Cartesian background grid is encountered in internal problems involving flows in curved conduits or manifolds where a large number of grid points can be located outside of the region of interest and unnecessarily burden the computation. To circumvent this problem, Ge & Sotiropoulos (2007) introduced the curvilinear immersed boundary (CURVIB) method, which integrates SIB approaches with generalized curvilinear grids. Unlike standard SIB methods, which use Cartesian background grids, the CURVIB method employs a curvilinear mesh that can either exactly or approximately conform to the immersed boundary shape, thus, greatly minimizing wasted grid nodes. This method has been recently enhanced and extended to natural waterways (Kang *et al.*, 2011; Kang & Sotiropoulos, 2011a, b) and forms the basis for the numerical approach we describe in this book chapter.

Another major issue in the numerical simulation of turbulence in real-life waterways is that of turbulence modeling. Steady RANS models have been widely employed to simulate flow in a natural river reach with relatively coarse grids (Sinha *et al.*, 1998; Wilson *et al.*, 2003; Rodriguez *et al.*, 2004; Lai *et al.*, 2003; etc.). These models are efficient engineering simulation tools and have been shown to capture with reasonable accuracy the mean flow characteristics in natural rivers. However, they are inherently incapable of resolving the dynamics of large-scale coherent structures and their impact on turbulence production and scalar and particulate transport. Unsteady RANS (URANS) models, which solve the RANS equations in a time accurate manner, can in principle capture large-scale, organized vortex shedding and such models have also been applied to simulate turbulence in natural waterways. Ge & Sotiropoulos (2005), for instance, simulated the flow through a reach of the Chatachochee River near Cornelia, Georgia, with an embedded bridge foundation using an overset grid approach with the standard k–ε model in URANS mode. The model could capture large-scale vortex shedding from the bridge piers and the simulated mean velocity field was in good agreement with laboratory scale measurements. In spite of encouraging results, however, the model yielded an essentially steady horseshoe vortex system in the vicinity of the bridge foundation, which is in contrast with experimental observations of Devenport & Simpson (1990).

Large-eddy simulation (LES) models are computationally more demanding than URANS models but they can resolve the energetic coherent vortical structures that dominate flows in natural waterways. Hodges & Street (1999) carried out LES for turbulent open channel flow with presence of free surface, and Zedler & Street (2001) carried out LES for flow over periodic ripples to study flow dynamics of the sediment transport. McCoy *et al.* (2008) carried out LES for flow around multiple groynes in a straight, rectangular open channel. They obtained good agreement between computed and measured mean velocities and velocity variances. Recently Stoesser *et al.* (2010) carried out both steady RANS simulation and LES to simulate turbulent flow in a meandering open channel consisting of two 180° bends and compared the computed mean velocity profile with laboratory measurements. The computed results correctly reproduced the presence of the secondary cells in the bend. Kang *et al.* (2011) and Kang & Sotiropoulos (2011a) extended the CURVIB method to carry out LES of turbulent flow in a natural-like meandering streams. They employed computational grids sufficiently fine to directly resolve bed roughness elements of 10 cm in size and obtained good agreement between measurements and computations.

The presence of the free surface presents yet another major challenge in the numerical simulation of open channel flows because the air/water interface is dynamic and its motion is coupled in a non-linear manner with the instantaneous flow fields in both the air and water phases. To handle free surface effects a broad range of modeling approaches of varying degree of sophistication have been proposed in the literature. One- (cross-sectional averaged) and two-dimensional (depth-averaged) models (e.g., Molls & Chaudhry, 1995; Yoon & Kang, 2004) have been widely used for modeling free surface flows in open channels because of their computational simplicity and expedience and the fact that they greatly simplify the treatment of the air/water interface. Depth-averaged models, however, are based on overly simplifying assumptions that inherently limit their applicability to relatively simple flows and preclude their application to tackle turbulent free surface flows with complex hydraulic structures. A review of previous literature indicates that only a handful of studies have attempted to simulate 3D turbulent open channel flows (Hodges & Street, 1999; Yue *et al.*, 2003, 2005; Ramamurthy *et al.*, 2007). A free surface model capable of simulating 3D turbulent free surface flows in arbitrarily complex open channels with embedded hydraulic structures was developed recently by Kang & Sotiropoulos (2011b). This method is based on integrating the level set method of Osher & Sethian (1988) with the CURVIB computational framework of Kang *et al.* (2011) and will be discussed in detail in this chapter.

Finally, the simulation of sediment transport and scour of the stream bed presents yet another major challenge for numerical methods. The difficulties arise from the need to develop efficient algorithms that can capture the coupled, dynamic interaction of the turbulent flow with the streambed especially in the presence of embedded, geometrically complex, hydraulic structures, such as bridge foundations or stream restoration structures. A common feature among most of the previous numerical approaches for sediment transport modeling in open channels (Demuren & Rodi,1986; Rüther & Olsen, 2005; Khosronejad *et al.*, 2007, 2008) is that they employ structured, boundary-conforming, curvilinear meshes that are dynamically deformed to always remain fitted to the changing bed elevation. Consequently such methods require re-meshing, i.e., the generation of a new boundary-fitted curvilinear mesh, every time the bed elevation is modified as a result of local erosion and/or deposition processes. One difficulty with such an approach arises when the scour depth becomes locally too deep. Since the number of grid nodes in the vertical direction typically remains fixed, re-meshing will naturally result in coarser vertical grid spacing leading to reduction of numerical accuracy and deterioration of the bed change computation in subsequent times. Another potential difficulty with boundary fitted meshes could arise in regions where sediment deposition may become sufficiently large to raise the local bed elevation above the local water surface elevation causing an island to appear in the flow domain. Finally, another major challenge when dealing with boundary-conforming methods arises when attempting to generate good quality boundary-fitted meshes in arbitrarily complex, multi-connected domains, such as those encountered in natural waterways with embedded hydraulic structures. As an example, one can contemplate the difficulties that would be involved if the goal is to carry out coupled hydro-morpho dynamics simulations in a natural meandering stream with multiple in-stream structures, such as those commonly used in stream restoration projects (Nagata *et al.*, 2005). Khosronejad *et al.* (2011a) recently extended the CURVIB framework of Kang *et al.* (2011) to develop a new method for simulating open channel flows with mobile bed that has the potential to circumvent many of the aforementioned challenges. Because the requirement to conform the mesh to an arbitrarily complex boundary is eliminated in the CURVIB method, the method of Khosronejad *et al.* (2011a) does not, at least in principle, suffer from the previously discussed shortcomings of boundary-conforming methods and are thus well suited for simulating flow and transport processes in natural waterways with complex in-stream structures.

In this chapter we present the elements of a powerful computational framework for simulating turbulent flow and transport processes in arbitrarily complex open channels. The framework is based on the CURVIB method and is capable of carrying out LES or URANS calculation, simulate turbulent free-surface effects, and model coupled hydrodynamics and morphodynamics in natural waterways with arbitrarily complex embedded hydraulic structures. We also review recent results from the applications of the CURVIB method to a range of open channel flow problems that underscore the predictive capabilities and future potential of the method.

2 GOVERNING EQUATIONS

In this section, we describe the governing equations for flow, turbulence, free surface motion and sediment transport.

2.1 Flow equations

The equations governing the instantaneous, resolved flow field for three-dimensional, incompressible, turbulent flow are the time (Reynolds) or spatially averaged, for URANS and LES, respectively, continuity and Navier-Stokes equations. The governing equations are first formulated in Cartesian coordinates x_i (where $i = 1, 2, 3$) and then transformed fully (both the velocity vector and spatial coordinates are expressed in curvilinear coordinates) in generalized, curvilinear coordinates ξ^i as required by the CURVIB method (Ge & Sotiropoulos, 2007; Kang *et al.*, 2011). For the sake of generality, and to facilitate the subsequent discussion of the level-set approach we employ to develop the free-surface algorithm, we present the equations in the two-fluid formulation that is appropriate for the level-set formulations.

Let ϕ denote the time (Reynolds) or spatially averaged level set function, which measures the distance to the closest fluid interface and is positive in the water region, negative in the air region and zero at the interface (Osher & Sethian, 1988). Assuming for the time being that this function is known (see Section 2.3 for the equations governing the evolution of this quantity), the two-fluid, level set form of the incompressible, time-averaged or spatially-filtered Navier-Stokes equations read in compact tensor notation (where repeated indices imply summation) as follows ($i, j = 1, 2, 3$):

$$J\frac{\partial U^j}{\partial \xi^j} = 0, \tag{1}$$

$$\begin{aligned}\frac{1}{J}\frac{\partial U^i}{\partial t} = \frac{\xi_l^i}{J}\Bigg\{&-\frac{\partial}{\partial \xi^j}(U^j u_l) + \frac{1}{\rho(\phi)}\frac{\partial}{\partial \xi^j}\left(\mu(\phi)\frac{g^{jk}}{J}\frac{\partial u_l}{\partial \xi^k}\right)\\ &-\frac{1}{\rho(\phi)}\frac{\partial}{\partial \xi^j}\left(\frac{\xi_l^j p}{J}\right) - \frac{1}{\rho(\phi)}\frac{\partial \tau_{lj}}{\partial \xi^j} + G_i + \frac{1}{\rho(\phi)}H_i\Bigg\},\end{aligned} \tag{2}$$

where $\xi_l^i = \partial\xi^i/\partial x_l$ are the transformation metrics; J is the jacobian of the geometric transformation; u_i is the ith component of the velocity vector in Cartesian coordinates; $U^i = (\xi_m^i / J)\, u_m$ is the contravariant volume flux; $g^{jk} = \xi_l^j \xi_l^k$ are the components of the contravariant metric tensor; ρ is the density; μ is the dynamic viscosity; p is the pressure; τ_{ij} is the sub-grid stress tensor for LES models or the Reynolds stress tensor for RANS models; G_i is the acceleration due to gravity; and H_i is the surface tension force.

The equations governing single phase simulations (water phase only with rigid-lid free-surface) are recovered from Eq. (2) by setting G and H equal to zero and setting the density and viscosity as constants.

2.2 Turbulence closure equations and wall modeling

In this section we present the governing equations for the LES and URANS turbulence models and also discuss the approach we adopt to model the near-wall flow.

2.2.1 *Turbulence model*

The filtered (or Reynolds averaged) Navier-Stokes equations in the LES (or URANS) models are obtained by decomposing the velocity into resolved (or time-averaged) and unresolved (temporally fluctuating) components and integrating the Navier-Stokes equations over the spatial (or temporal) filter (Sagaut, 1988; Pope, 2000). As a result, sub-grid stress (or Reynolds stress) terms appear in the momentum equations (Eq. (2)), which are modeled using the eddy viscosity model:

$$\tau_{ij} - \frac{1}{3}\tau_{kk}\delta_{ij} = -2\mu_t \overline{S}_{ij}, \tag{3}$$

where the over bar denotes the grid filtering (or Reynolds averaging) operation, and $\overline{S}_{ij}$ is the filtered (or time-averaged) strain-rate tensor.

For the LES model, we employ the Smagorinsky sub-grid scale (SGS) model (Smagorinsky, 1963) to compute the eddy viscosity:

$$\mu_t = \rho C_s \Delta^2 \left|\overline{S}\right|, \tag{4}$$

where C_s is the Smagorinsky constant, Δ is the filter size, and $|\bar{S}| = \sqrt{2\bar{S}_{ij}\bar{S}_{ij}}$. The box filter (Sagaut, 1988) is employed in the present model. We employ the dynamic Smagorinsky model (Germano *et al.*, 1991) as a sub-grid model in which the model constant C_s evolves in spaces and time as function of the flow. The optimal value of C_s is selected to minimize the mean square error between the resolved stress at the grid filter and the test filter (Germano *et al.*, 1991).

To close the URANS equations and compute the eddy viscosity, we employ the k–ω model (Wilcox, 1988) and the shear stress transport (SST) model (Menter *et al.*, 2003).

For more details of the turbulence modeling the reader is referred to Kang *et al.* (2011).

2.2.2 *Wall modeling*

In high Reynolds number flow simulations, applying the no-slip boundary condition at the wall is often impractical as it results in excessively fine meshes and long computational times. To address this difficulty in this work we employ the wall model proposed in Cabot & Moin (2000) and Wang & Moin (2002). The model solves the boundary layer equation in the following form:

$$\frac{1}{\rho}\frac{\partial}{\partial l}\left((\mu + \mu_t)\frac{\partial u_s}{\partial l}\right) = \frac{1}{\rho}\frac{\partial p}{\partial s} + \frac{\partial u_s}{\partial t} + \frac{\partial (u_l u_s)}{\partial s}, \tag{5}$$

where l and s indicate the directions normal and tangential to the wall, respectively. By neglecting the right hand side of the Eq. (5), one obtains the equilibrium stress balance model (Wang & Moin, 2002):

$$\frac{1}{\rho}\frac{\partial}{\partial l}\left((\mu + \mu_t)\frac{\partial u_s}{\partial l}\right) = 0. \tag{6}$$

The eddy viscosity is given by the mixing length model with the near-wall damping as

$$\mu_t = \mu\kappa l^+ (1 - e^{-l^+/19})^2, \tag{7}$$

where κ is the von Karman constant (0.4), $l^+ = \rho u_\tau l/\mu$ and u_τ is the wall shear velocity. Eq. (6) is integrated from the wall to the second off-wall node to obtain the tangential velocity at the first off-wall node.

For rough wall modeling, we employ the following wall function (Khosronejad *et al.*, 2007):

$$\frac{u_s}{u_\tau} = \begin{cases} l^+ & l^+ \le l_0^+ \\ (\ln l^+)/\kappa + (B - \Delta B) & l^+ > l_0^+ \end{cases}, \tag{8}$$

where $l_0^+ = 11.53$, $B = 5.2$, and ΔB is a function of bed roughness function which is formulated in terms of the roughness Reynolds number as follows:

$$\Delta B = \begin{cases} 0 & k_s^+ < 2.25 \\ [B - 8.5 + \ln(k_s^+)/\kappa]\sin[0.4258(\ln(k_s^+) - 0.811)] & 2.25 \le k_s^+ < 90, \\ B - 8.5 + \ln(k_s^+)/\kappa & k_s^+ \ge 90 \end{cases} \tag{9}$$

where k_s is the effective roughness height of wall boundary and $k_s^+ = \rho u_\tau k_s/\mu$.

2.3 Free surface model

For a two-fluid formulation, the density and viscosity of the fluid in the Navier-Stokes equations (Eq. (2)) are not constant. Rather, the fluid properties vary as function of ϕ, transitioning smoothly across the interface from the respective values in the water phase to those in the air phase, as follows:

$$\rho(\phi) = \rho_{air} + (\rho_{water} - \rho_{air})h(\phi), \tag{10}$$

$$\mu(\phi) = \mu_{air} + (\mu_{water} - \mu_{air})h(\phi), \tag{11}$$

where $h(\phi)$ is the smoothed Heaviside function (Osher & Fedkiw, 2002) given as follows:

$$h(\phi) = \begin{cases} 1 & \phi < -\varepsilon, \\ 1/2 + \phi/2\varepsilon + \sin(\pi\phi/\varepsilon)/2\pi & -\varepsilon \le \phi \le \varepsilon, \\ 0 & \varepsilon < \phi, \end{cases} \tag{12}$$

where ε is a tunable parameter that determines the thickness of the numerical smearing of the interface (Osher & Fedkiw, 2002). The value of ε is usually set to be equal to the length of one to two grid spacings. The smoothing of the interface achieved by introducing the Heaviside function renders the density, velocity and pressure fields continuous across the interface and prevents potential numerical instabilities.

The surface-tension force H_i in Eq. (2) is calculated by the following equation (Sussman *et al.*, 1994):

$$H_i = -\sigma\Gamma(\phi)\tilde{\delta}(\phi)\frac{\partial\phi}{\partial\xi^j}\frac{\partial\xi^j}{\partial x_i}, \tag{13}$$

where $\tilde{\delta}(\phi) = \partial h/\partial\phi$ is the smoothed delta function, which is the derivative of the smoothed Heaviside function (Eq. (12)), σ is the surface tension, and Γ is the local curvature of the interface defined as follows:

$$\Gamma = \nabla \cdot \left(\frac{\nabla\phi}{|\nabla\phi|}\right). \tag{14}$$

Although the surface tension term is incorporated in the governing equations, it is neglected in all free surface simulations presented in this work because the interface curvature in open channel flows is negligible.

In order to model the motion of the free surface interface, the level set function ϕ is computed by the level set method proposed by Osher & Sethian (1988). The governing equation for the time-averaged motion of the interface of two immiscible fluids (e.g., air and water) curvilinear grids is written as

$$\frac{1}{J}\frac{\partial \phi}{\partial t} + U^j \frac{\partial \phi}{\partial \xi^j} = 0. \tag{15}$$

Due to the spatial filtering or Reynolds averaging of the advection equation for ϕ, second order correlation term involving fluctuations of ϕ and u_i appears in the right hand side of the equation. This term has been neglected from the above equation assuming that the fluctuations of the level set function are small and their overall contribution to the mean flow is negligible.

While Eq. (15) accurately advects the zero level set according to the given velocity field (Osher & Sethian, 1988), the level set function away from the gas/liquid interface is not necessarily a distance function, which should always satisfy the relation of $|\nabla\phi| = 1$. Keeping the level set function as a distance function, at least within the several grid cells near the interface, is very important in the level set method. This is because the gradient of ϕ can become large if the distance function is not preserved, which leads to a loss of accuracy. In order to preserve the distance function, Sussman & Fatemi (1999) suggested solving the following reinitialization equation:

$$\frac{\partial \phi}{\partial \tau} + S(\phi_0)\left(|\nabla\phi| - 1\right) = \lambda\tilde{\delta}(\phi)|\nabla\phi|, \tag{16}$$

with

$$\lambda = -\frac{\int_\Omega \tilde{\delta}(\phi)S(\phi_0)(1 - |\nabla\phi|)d\Omega}{\int_\Omega \tilde{\delta}^2(\phi)|\nabla\phi|\,d\Omega}, \tag{17}$$

where Ω is the volume of the individual grid cell, $\partial(\ )/\partial\tau$ is a pseudo-time derivative, which is driven to zero during every physical time-step through iterations, ϕ_0 is the level set function at the beginning of the pseudo-time iteration, and S is the smoothed sign function given by:

$$S(\phi_0) = \begin{cases} 1 & \phi_0 \geq \varepsilon, \\ -1 & \phi_0 \leq -\varepsilon, \\ \phi_0/\varepsilon - \sin(\pi\phi_0/\varepsilon)/\pi & \text{otherwise.} \end{cases} \tag{18}$$

2.4 Sediment transport model

In this section we outline the basic framework of the morphodynamic model developed by Khosronejad *et al.* (2011a) and present the mathematical equation that governs the temporal variation of the sediment/water interface.

We consider the case for which all sediment transport occurs in bed-load mode within a live layer of thickness δ_{BL} (set equal to $2d_{50}$) just above the bed, which is located at elevation z_b above a datum level. The key idea of the method is to adopt the CURVIB framework (see below) and discretize the sediment/water interface with an unstructured triangular mesh, which is embedded within a background mesh used to discretize the channel. The dynamic deformation of the sediment/water interface is governed by the sediment continuity equation, the so-called Exner-Polya equation (Paola & Voller, 2005), which reads as follows:

$$(1-\gamma)\frac{\partial z_b}{\partial t} = -\nabla \cdot \mathbf{q}_{BL} \tag{19}$$

where γ is sediment material porosity, z_b is the bed elevation, and $\mathbf{q}_{BL}$ is the bed-load flux vector. Eqn. (28) is solved in a fully-coupled manner with the flow equations using a fluid-structure interaction approach as described below.

We discretize the Exner-Polya equation on the unstructured, triangular grid used to discretized the sediment/water interface using a finite volume method. Using the divergence theorem, Eq. (19) can be discretized at a given triangular element as follows:

$$(1-\gamma)\frac{\partial z_b}{\partial t} = -\frac{1}{A_b}\sum_{i_e=1,2,3}[\mathbf{q}_{BL}.\mathbf{n}]_{i_e}, \tag{20}$$

where A_b is horizontal projected area of a triangular cell, the summation is carried out over the three edges of the triangular element ($i_e = 1, 2, 3$), ds is the vector along a cell edge with length equal to the edge length, $\mathbf{n}(= \mathrm{ds} \times \delta_{BL})$ is the unit vector normal to the cell face (see Figure 1 for a schematic description). The bed-load flux vector $\mathbf{q}_{BL}$ is calculated as follows:

$$\mathbf{q}_{BL} = \Psi \| ds \| \delta_{BL}\mathbf{u}_{BL}, \tag{21}$$

where ψ is the local sediment concentration on the bed, $\| ds \|$ is the length of edge i_e, and $\mathbf{u}_{\mathrm{BL}}$ is the flow velocity vector parallel to the bed surface at the edge of the bed-load layer. The calculated velocity vector and bed shear stress (and correspondingly the sediment concentration) are projected onto the cell centroid at the edge of the bed-load layer by applying the law of the wall. The bed load flux at the cell face is then calculated using the GAMMA scheme (for more details see Khosronejad *et al.* (2011a)).

The bed concentration at each cell center on the bed is computed as follows:

$$\Psi = 0.015\frac{D_{50}}{\delta_{BL}}\frac{T^{1.5}}{D_*^{0.3}} + \Psi^*, \tag{22}$$

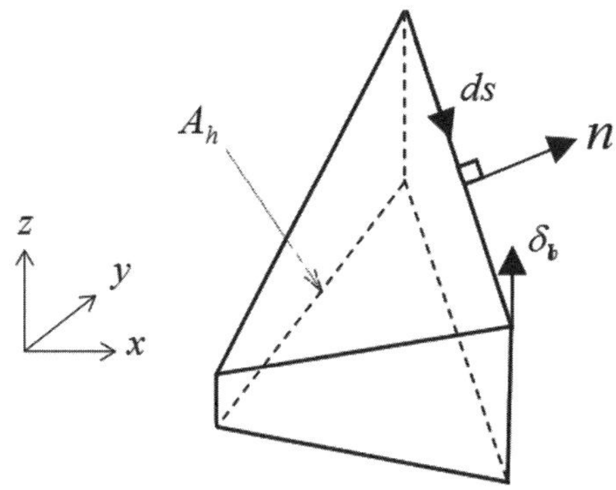

Figure 1 Schematic of a triangular element in the bed-load layer and definition of the cell area and unit normal vector.

where

$$D^* = D_{50}\left[\frac{(\rho_s - \rho)g}{\nu^2}\right]^{1/3}, \tag{23}$$

$$T = \frac{\tau_* - \tau_{*cr}}{\tau_{*cr}}, \tag{24}$$

$$\Psi^* = \frac{\Delta t}{A_h \delta_{BL}}\left[q_b . n\right]_{i_e}^* . \tag{25}$$

In the above equations ρ and ρ_s are the fluid and sediment densities, g is the gravitational acceleration, T is the non-dimensional excess bed shear stress (Van Rijn, 1993), τ_{*cr} is the critical bed shear stress calculated by taking into account both the longitudinal and lateral bed slopes, and τ_* is the bed shear stress. The first and second terms on right hand side of Eq. (22) represent the equilibrium sediment concentration from the deterministic equation of Van Rijn (Van Rijn, 1993) and non-equilibrium bed sediment concentration, respectively. The non-equilibrium part of bed sediment concentration is calculated based on the net sediment-flux into the bed cell at the previous time step. In other words, the actual bed sediment concentration is computed as a combination of an equilibrium concentration and the net sediment flux into the bed cell of volume of $A_h\delta_{BL}$. The characteristic length for sediment to adjust from non-equilibrium to equilibrium transport is assumed to be equal to the characteristic length of flow field. The second term of right hand side of Eq. (25) represents the divergence of the bed-load transport at the previous time level.

In order to prevent the bed slope from exceeding the sediment material angle of repose, a mass-conservative sand-slide algorithm has been. After solving the Exner-Polya equation and computing the new bed elevation, the entire bed is swept (for all the bed cells) to identify bed cells at which the maximum bed slope is larger than the angle of repose. The aforementioned local slope correction procedure is then applied only to the so identified bed cells. Because a single application of this algorithm may not be sufficient to identify and correct all cells with excess slope, the algorithm is applied iteratively until all slopes are smaller than the material angle of repose. The bed sweeping sequence during each interaction of this algorithm is alternated between the upstream-to-downstream and downstream-to-upstream directions. For more details the reader is referred to Khosronejad *et al.* (2011a).

3 THE CURVIB FRAMEWORK FOR OPEN CHANNEL FLOWS

The curvilinear immersed boundary (CURVIB) method (Ge & Sotiropoulos, 2007) was originally proposed for flow applications in which a moving immersed boundary is embedded in a background domain that can be efficiently discretized with a boundary-fitted curvilinear mesh. The CURVIB framework, however, is also ideally suited for simulating flows in natural meandering streams with arbitrarily complex bed topography and in-stream structures.

A schematic illustrating the application of the CURVIB method for a meandering stream is shown in Figure 2. The immersed boundaries (bathymetry and in-stream structures) are discretized with an unstructured triangular surface mesh and embedded in a curvilinear domain that outlines but does not exactly coincide with the meandering outline of the stream but has a regular (prismatic) cross-section that contains fully the actual stream everywhere. For fixed-bed simulations the stream-bed bathymetry mesh is held fixed throughout the simulation but for coupled morphodynamic simulations the mesh, which now discretizes the dynamic sediment/water interface, evolves in time by solving the Exner-Polya Equation (29) as described above. The details of the algorithm for handling boundary conditions at the sediment/water interface for this case are discussed in Section 4.4 below.

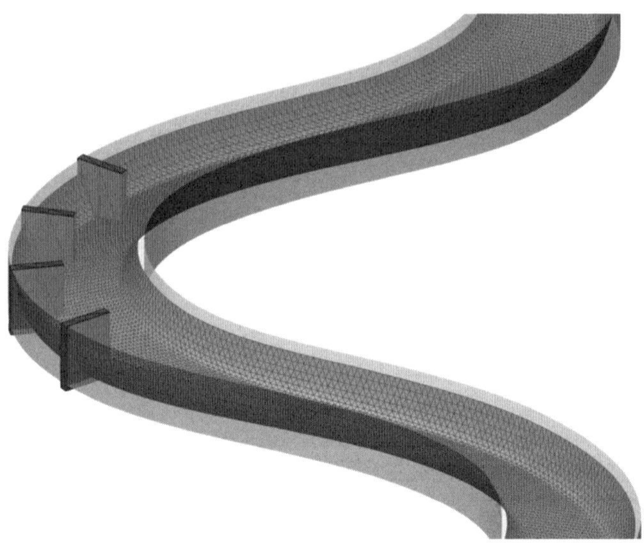

Figure 2 Schematic description of applying the CURVIB method for the natural stream.

The CURVIB method employs wall normal interpolation for calculating the velocity components at near-wall grid nodes - referred to as immersed boundary (IB) grid nodes - under the hybrid staggered/non-staggered grid layout proposed by Gilmanov & Sotiropoulos (2005). Figure 3 illustrates the wall normal interpolation for a velocity component or a turbulence quantity. At a given node (*B* in Figure 3) immediately adjacent to the immersed boundary, a straight line is drawn normal to the nearest wall (*A* in Figure 3) until it intersects with the grid line (or plane in 3D) which connects two (three in 3D) neighboring internal nodes. The point *C* in Figure 3 is called the interception point. The values of flow variables at the point *C* are obtained by linear interpolation from computed values at internal grid nodes α and β. Gilmanov & Sotiropoulos (2005) used the above approach to compute the velocity at the immersed boundary (IB) nodes in a manner such that the no-slip boundary condition is exactly satisfied at the wall. They used linear and quadratic interpolation which for turbulent flows would work well only if the grid spacing in the vicinity of the immersed boundary is sufficiently fine to resolve the laminar sublayer region and directly impose the no-slip condition on the solid immersed boundary. When the mesh is not sufficiently fine (as is often the case for high Reynolds number turbulent flows), however, the wall model to be described in the next section is used to reconstruct the boundary conditions.

3.1 Velocity boundary conditions for the Navier-Stokes equations

As discussed above, the no-slip boundary condition is employed when the near wall grid spacing is fine enough to resolve the viscous sublayer. Assuming the velocity component at the solid wall (point *A*) is u_i^A, which is zero when the wall is not moving, and its distribution along the points *A*, *B* and *C* is linear, the Cartesian velocity components at the IB node is given as follows:

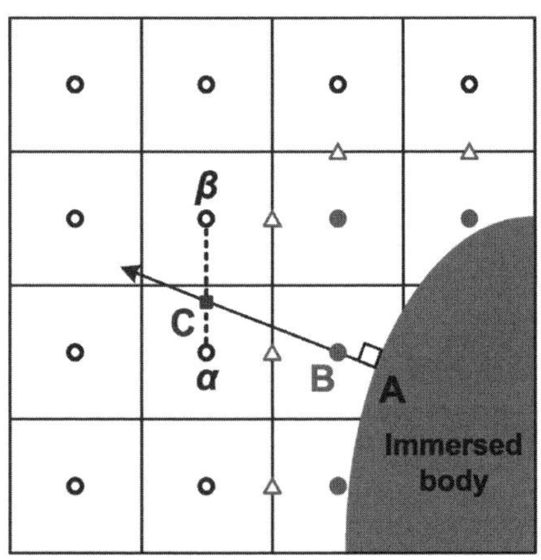

Figure 3 Schematic description of the wall normal interpolation at the IB node (filled circles: IB nodes, open circles: internal nodes, filled square: interception node, triangles: cell faces where the volume flux is stored).

$$u_i^B = \frac{\overline{AB}}{\overline{AC}}(u_i^C - u_i^A) + u_i^A, \tag{26}$$

where i = 1, 2, and 3, and $\overline{AB}$ and $\overline{AC}$ denote the distance between the points A and B and the points A and C, respectively.

When the near wall grid spacing is not fine enough to resolve the viscous sublayer, the velocity boundary condition is obtained either by the wall modeling approach described in Section 2.2.2 above or using standard wall functions. To implement the wall model in the context of the CURVIB method, we integrate the boundary layer equations (see Eq. (6)) from the surface of the wall to the second off-wall node (C in Figure 3) to obtain the tangential velocity component at the first off wall node (B in in Figure 3):

$$u_i^B = \frac{\int_0^{\delta_b} (\mu + \mu_t(l))^{-1} dl}{\int_0^{\delta_c} (\mu + \mu_t(l))^{-1} dl}(u_i^C - u_i^A) + u_i^A, \tag{27}$$

where δ_b and δ_c denote the distance from the wall to the point B and C, respectively. The normal velocity component at the IB nodes is obtained by the wall normal linear interpolation method. The above procedure is employed both for RANS and LES models.

3.2 Boundary conditions for the k and ω equations

The low Reynolds number wall boundary conditions for k and ω are as follows:

$$k = 0, \tag{28}$$

$$\omega = \frac{6\mu}{\rho\beta d^2}, \tag{29}$$

where d is the distance from the nearest wall at the first off-wall grid which is equal to $\overline{AB}$ in Figure 3. The above boundary conditions are adopted when the first off wall grid point is located inside the viscous sublayer. To impose the boundary condition given by Eq. (28) at the wall, the linear interpolation along the line normal to the wall is employed. Assuming $k = 0$ at the node A one obtains the value of k at the IB node (B in Figure 3):

$$k^B = \frac{\overline{AB}}{\overline{AC}} k^C. \tag{30}$$

The high Reynolds number wall boundary conditions for k and ω are (Wilcox, 1993):

$$k = \frac{u_\tau^2}{\sqrt{\beta^*}}, \tag{31}$$

$$\omega = \frac{u_\tau}{\sqrt{\beta^* \kappa d}}, \tag{32}$$

where $\kappa = 0.4$ is the von Kármán constant, and u_τ is the wall shear velocity. The above boundary conditions are adopted when the first off wall grid point is located inside the logarithmic layer. The boundary conditions given by Eqs. (29), (31) and (32) are directly imposed at the IB node.

3.3 Boundary conditions for the level set function

In order to implement the level set method in the context of the CURVIB method, we need to reconstruct boundary conditions for the level set function (ϕ) at the IB nodes. This is accomplished by setting the gradient of the level set function in the direction along an appropriate curvilinear grid line to be zero. For instance, when the grid nodes (i, j, k) and $(i + 1, j, k)$ are the fluid and IB nodes, respectively, the following relation is assumed:

$$\phi_{i+1,j,k} = \phi_{i,j,k}, \tag{33}$$

which implies zero Neumann boundary condition along the corresponding grid line. The above equation at the IB node is used as the boundary condition for adjacent fluid node for solving the level set and reinitialization equations.

4 NUMERICAL METHODS

The governing equations presented in Section 2 are discretized with second-order accurate, three-point central finite differences (unless otherwise indicated) on the hydrid staggered/non-staggered grid proposed by Gilmanov and Sotiropoulos (2005). In this section, we present the numerical procedures we adopt for solving these equations.

4.1 Solution of Navier-Stokes equations

To solve the URANS or LES version of the flow equations (Eqs. (1) and (2)), we employ the fractional step method proposed by Ge & Sotiropoulos (2007). During the first step, the momentum equations (Eq. (2)) are discretized in a fully implicit manner using second-order backward differencing in time:

$$\frac{1}{J}\frac{3\mathbf{U}^* - 4\mathbf{U}^n + \mathbf{U}^{n-1}}{2\Delta t} = RHS(\mathbf{U}^*, \mathbf{u}^*) \tag{34}$$

where n denotes the time level and RHS is the right hand side of Eq. (2). Both advection and diffusion terms in the right hand side of Eq. (2) are discretized using, three-point central, second-order accurate finite-differencing. A Jacobian-free Newton's method is employed to solve Eq. (34) (see Kang *et al.*, 2011 for details).

The intermediate velocity field $\mathbf{U}^*$ obtained by solving Eq. (34) is not divergence-free and needs to be corrected to satisfy the continuity equation. This is accomplished by formulating and solving the following Poisson equation for the pressure increment (or pressure correction) $\Pi = p^{n+1} - p^n$:

$$-J\frac{\partial}{\partial \xi^i}\left\{\frac{1}{\rho(\phi)}\frac{\xi_l^i}{J}\frac{\partial}{\partial \xi^j}\left(\frac{\xi_l^j \Pi}{J}\right)\right\} = \frac{3}{2\Delta t}J\frac{\partial U^{i,*}}{\partial \xi^i} \tag{35}$$

A major challenge for carrying out reach-scale, high-resolution simulations of turbulence in natural waterways stems from the large aspect-ratio of the resulting computational domains due to the large disparity in flow depth the stream wise and lateral dimensions of the reach. Large domain and grid cell aspect-ratios induce numerical stiffness and could dramatically deteriorate the convergence rate of powerful iterative algorithms. This is an especially critical issue for the solution of the Poisson equation that needs to be satisfied to machine zero during every time step in order to satisfy the incompressibility condition for the velocity field. To improve the convergence of the Poisson equation in large aspect-ratio grids and domains, we employ the algebraic multi grid (AMG) as the pre conditioner for the GMRES method (Saad & Schultz, 1986). This approach was proven to improve the convergence of the Poisson equation on highly stretched and distorted grids by several orders of magnitude (Kang *et al.*, 2011).

Following the solution of Eq. (35), the pressure and contravariant volume fluxes are updated as follows:

$$p^{n+1} = p^n + \Pi \tag{36}$$

$$U^{i,n+1} = U^{i,*} - J\frac{2\Delta t}{3}\frac{1}{\rho(\phi)}\frac{\xi_l^i}{J}\frac{\partial}{\partial \xi^j}\left(\frac{\xi_l^j \Pi}{J}\right) \tag{37}$$

The computer code is parallelized using the message passing interface (MPI) to take full advantage of massively parallel computational platforms.

4.2 Solution of URANS turbulence closure equations

Third-order WENO scheme (Jiang & Shu, 1996) and second-order central differencing scheme are used for the spatial discretization of advection and diffusion terms in k and ω equations, respectively, and the second-order backward differencing scheme is used for the time integration. The equations of k–ω and SST models are solved at every time step after the velocity fields are obtained by the previously described fractional step method. A fully implicit Jacobian-free Newton's method is employed for solving these equations in order to enhance numerical stability. This implicit method alleviates the numerical instability caused by the presence of stiff source terms in the k and ω equations. After computing the values of k and ω at all nodes, the values of eddy viscosity are calculated at the center of the cell and are subsequently interpolated to the cell face.

4.3 Solution of level set equations

The equation for the interface motion (Eq. (15)) is discretized in space by the third-order WENO scheme (Jiang & Shu, 1996), and the discrete equation is advanced in time using the second-order Runge-Kutta method. The spatial derivatives in mass conserving reinitialization equation (Eq. (16)) are discretized using the second-order ENO scheme proposed by Sussman *et al.* (1998).

4.4 Solution of Exner-Polya equation

To simulate coupled hydro-morphodynamic interactions we employ a partitioned fluid-structure interaction (FSI) approach (Borazjani *et al.*, 2008). That is, we partition the problem into the fluid and sediment domains and solve the governing equations for the flow and the bed morphodynamic equation (Eq. (19)) separately in each domain accounting for the interaction of the two domains by applying boundary conditions at the sediment/water interface.

More specifically, to solve the equations governing the flow boundary conditions on the bed, i.e., bed location and velocity, are specified while to solve the bed change equation (Eq. (19)) the velocity components and bed shear stress from the flow domain are required to calculate the sediment fluxes on the right hand side (see Khosronejad *et al.*, (2011a) for details).

Two types of FSI coupling (loose and strong coupling) are possible depending on whether the boundary conditions for each domain are obtained from the previous or current time level (Borazjani *et al.*, 2008). The domains are loosely coupled if the boundary conditions at the interface are obtained from the domain solutions from the previous time level (explicit in time) while they are strongly coupled if the interfacial boundary conditions are obtained iteratively from the domain solutions at the current time level (implicit in time). The implicit integration in the strong-coupling approach is accomplished by carrying out a number of sub-iteration steps until the coupled fluid/structure solution converges to a stable solution at every physical time step. The loose- and strong-coupling methods could differ significantly in their respective convergence rates and computational costs. Strong-coupling has better stability properties due to the implicit treatment of boundary conditions. Loose coupling, on the other hand, is computationally desirable due to the fact that no sub-iterations are required at each time level to converge the solution (Borazjani *et al.*, 2008).

The loose-coupling flow-bed interaction method we employ is summarized as follows. Assuming the position and velocity of the bed and the flow field to be known at time level n:

1. Obtain the flow field at time level $n + 1$ by solving the flow equations with the known position and velocity of the bed from time level n.
2. Calculate the boundary conditions, i.e., bed load flux vector, for the Exner-Polya equation from the solution of the flow field at time level $n + 1$ as follows:
   a. Find the bed shear stress and flow velocity components at the top of bed load layer (at a distance of δ_{BL} above bed);
   b. For the given sediment material and bed shape, compute the critical bed shear stress;

c Compute the sediment concentration at the center of each triangular cell on the bed using Eq. (22);
d Calculate the normal component of the bed load flux at each cell face edge using a conservative formulation;

3 Find the new bed elevation at each cell center at time level $n + 1$ by advancing Eq. (20) in time using the following first order, backward-difference scheme. Once the bed level at the cell centers has been updated, the related bed change for the vertices of a cell is computed by weighted interpolation among neighboring cells.
4 Apply the previously described sand-slide model to ensure that the local bed slopes nowhere exceed the material angle of repose.
5 Modify the shape of the sediment/water interface and calculate the bed velocity at the vertices of the cells to provide boundary conditions for the flow field computation at next time step.

The above algorithm is iterated until the steady state solution for bed surface is reached. The steady state condition is established when the infinity norm of the relative change of bed elevation between two successive time steps is less than 1 percent over one flow-through period. For more details, the reader is referred to Khosronejad *et al.* (2011a).

5 APPLICATIONS

In order to demonstrate the predictive capabilities and overall potential of the computational framework, we apply it herein to carry out: 1) LES of flow in a fixed-bed field scale meandering stream; 2) RANS calculation of free surface flow over a cross vane structure; and 3) RANS calculation of scour around a rock vane structure mounted on a mobile-bed open channel.

5.1 Turbulent flow in a fixed-bed field-scale meandering stream

To demonstrate the ability of the computational model to carry out high resolution simulations of flow in a natural stream with arbitrarily complex bathymetry, we apply it to carry out LES of turbulent flow in the meandering stream currently installed in the St. Anthony Falls Laboratory Outdoor StreamLab (OSL), University of Minnesota, Minneapolis, MN, USA (see Figure 4). For this case, the stream-bed is held fixed and the free-surface is treated as a slopping rigid lid (see below for details).

We consider the bankfull flow of the OSL; the Reynolds number (Re), Froude number (Fr), and flow rate (Q) of the bankfull flow condition were $Re = 10^5$, $Fr = 0.4$, and $Q = 2.85 \times 10^{-1}$ m^3/s, respectively, where Re and Fr are based on the flow depth and bulk velocity at the inlet. The mean sediment feed rate at the inlet was approximately 4 kg/min, and the median grain size of the sediment fed into the channel was 0.7 mm. The bed materials in the riffles consist of gravel ranging from 10 cm to 15 cm in length, which is large enough to withstand the maximum bed shear stress and as a result the riffle topography was mostly fixed. The bed in the pool was initially constructed with a flat sand bed and was allowed to evolve naturally toward

Figure 4 St. Anthony Falls Laboratory Outdoor StreamLab at University of Minnesota, Minneapolis, MN, USA.

a quasi-equilibrium state, with point bars and deep pool regions along the inner and outer banks, respectively.

Three-dimensional mean velocity and turbulence statistics measurements in the OSL were obtained using acoustic Doppler Velocimetry. Bed and water surface topography in the OSL were collected on a 1 cm horizontal grid at sub millimeter vertical accuracy using instruments mounted to a separate channel-spanning portable carriage, the position of which was registered using the total station. A laser distance sensor was used for collecting subaerial bank topography, while a pulser and ultrasonic submersible transducer system documented subaqueous topography. Mean water surface elevation measurements were sampled over a centimeter-scale spaced grid using an ultrasonic distance sensor.

The OSL was first run with the given discharge for several days until the bed in the pool region does not change significantly over time. The measured bed topography is shown in Figure 5. During the time of the measurements a long-term morphological change was observed in some regions of the pool but this was not accounted in the time-averaged bed topography. The bed geometry is discretized with 236,370 triangular elements as required by the CURVIB method (see Kang *et al.*, (2011) for details). The size of the bathymetry mesh is small enough to resolve individual bed roughness in the riffle zones. The body-fitted curvilinear grid that discretizes the rectangular meandering channel (see Kang *et al.*, (2011) for details) that contains entirely the natural OSL bathymetry consists of 2001 × 241 × 101 grid points in stream wise, transverse, vertical directions, respectively, and which approximately follows the meander of the channel. The total number of the grid points is about 49 million. The grid spacing in the streamwise, transverse and vertical directions is about 2.0 cm, 1.5 cm and 6 mm, respectively. The horizontal and vertical grid spacings are fine enough to directly resolve large roughness elements in the riffle region with size in the range of 10 cm to 15 cm. The grid nodes are uniformly distributed in the vertical direction and

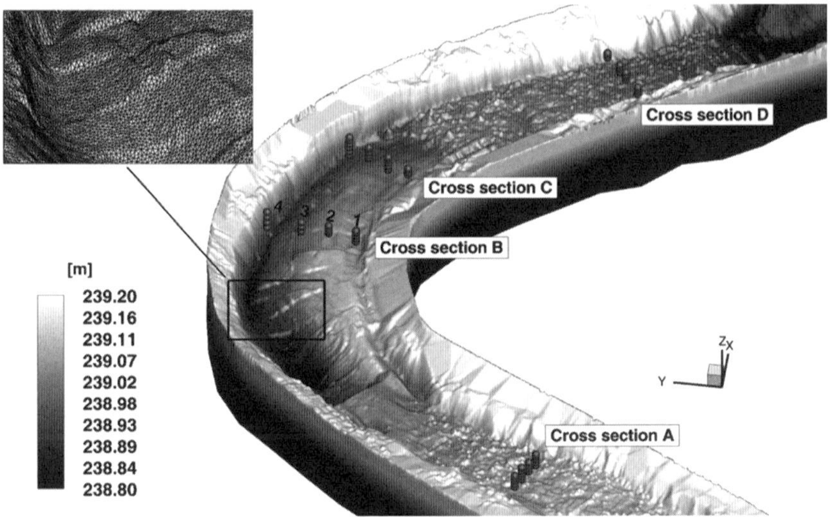

Figure 5 The SAFL Outdoor StreamLab bathymetry obtained from high-resolution measurements. The streambed is discretized using 236,370 triangular elements and treated as an immersed boundary. The contour levels denote bed elevations. The symbols and numbers mark velocity measurement locations. The flow direction is from bottom to top.

an orthogonal curved grid is used in the horizontal plane. In summary, the 30 m long, 3.5 m wide and 0.6 m high background computational grid, which surrounds the channel, is filled with 49 million cells with the size of 2.0 cm × 1.5 cm × 6 mm. The stream wise, transverse and vertical grid spacing correspond to approximately 400, 300 and 140 wall units, respectively, based on the measured mean bed shear stress. The measured time-averaged free surface elevation is used to prescribe the water surface, and no-flux and free-slip boundary conditions are employed for the instantaneous velocity field. The unsteady turbulent inflow condition, which was extracted from the separate LES solving the fully developed open channel flow with stream wise periodicity, is specified at the inlet. The separate LES for the generation of the inlet velocity boundary condition was carried out before running the main simulation.

The computational time step was chosen as 0.002 s, which corresponds to CFL (Courant-Friedrichs-Lewy) number of 0.6, and the computation was carried out by using 120 CPUs on a supercomputer. The simulation was first run until the total kinetic energy in the computational domain reached steady state. The simulation was continued for 105 time steps, which corresponds to about 6.5 flow through times for a 30 m long computational domain, and the data collected during this simulation interval were averaged to calculate the mean flow and turbulence statistics.

In Figure 6 we plot contours of the mean stream wise velocity component at the water surface. Figure 6 show that as the high-momentum flow from the first riffle enters the pool it narrows forming a strong jet-like flow structure at the water surface. This jet is seen to approach the outer bank of the bend, reaching the apex where the thin high-velocity core breaks down and starts diffusing laterally to cover a larger area of the outer bank before it encounters the second riffle. The high-velocity jet

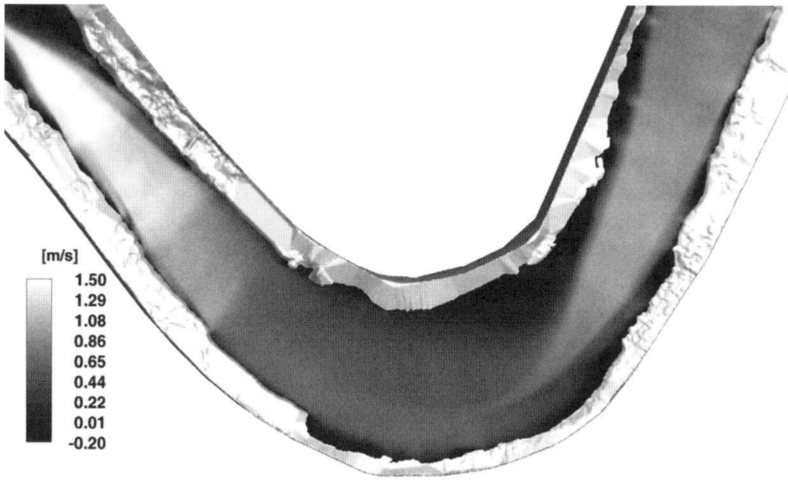

Figure 6 Mean streamwise velocity contours at the water surface (flow is from right to left).

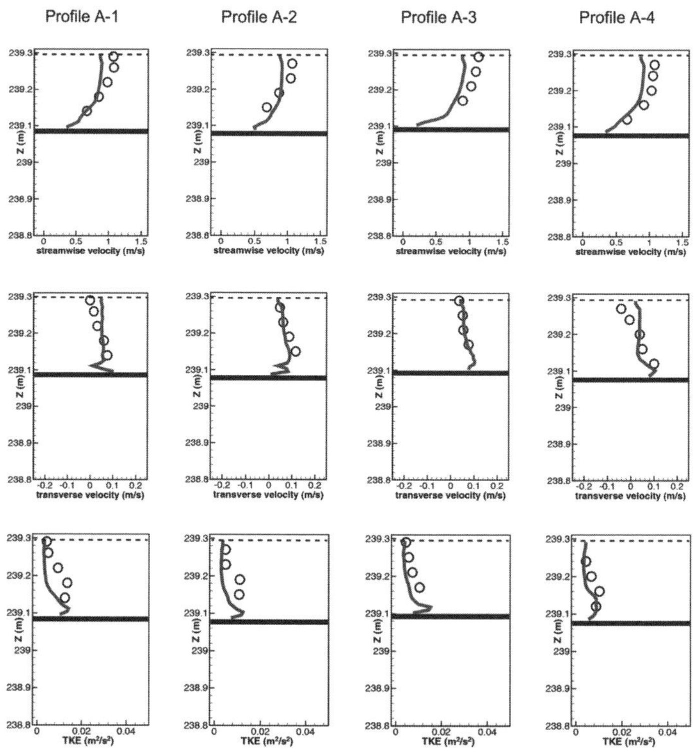

Figure 7 Comparisons of mean streamwise velocity, mean transverse velocity, and TKE with the measurements at the cross section A (solid line, LES; symbols, measurements). The locations of the cross section and each profile within a cross section are defined in Figure 5. The dashed line and thick solid line denote the locations of free surface and bed, respectively.

appears to become wider in the near-bed plane and is positioned above the thalweg of the bend. The emergence of such high-velocity core near the bed along the outer bank and the thalweg is expected to accelerate stream bank and streambed erosion as is known to occur in meander bend flows. An extensive discussion of the rich flow phenomena and mechanisms uncovered by these simulations can be found in Kang and Sotiropoulos (2011).

Figure 7 compares the computed mean velocity and TKE profiles with the measurements at the cross section A (see Figure 5). Total number of measurement points at the four cross sections is 116, but only the comparisons of the four profiles are shown. The first and second letters of the captions in Figure 7 denote locations of the cross section and the profile, respectively. It is seen that the simulations are in good overall agreement with the measurements. The overall level of agreement between experiments and simulations is particularly encouraging when one considers the uncertainties inherent in our numerical model and the fact that no attempt was made to calibrate the model in any way.

5.2 Free surface flow in an open channel with complex in-stream structure

We consider turbulent flow in a straight open channel with an embedded rock structure, namely a cross vane typically used in stream restoration projects for preventing stream bank erosion and increasing flow diversity (Rosgen, 2001). Experiments for this case were carried out in a St. Anthony Falls Laboratory (SAFL) flume. The flume is 0.9 m wide and 12 m long, and the tailgate weir is adjusted to control the downstream water elevation. The mean water surface profile at several locations was measured using Massa ultrasonic distance sensor. While the sidewalls are smooth, the channel bed is rough with the equivalent sand grain roughness of 7×10^{-3} m. The cross vane is installed at 6 m downstream from the inlet of the flume. The flow discharge is 3.81×10^{-2} m^3/s, which corresponds to a Reynolds number of 4.08×10^4 and a Froude number of 0.19 based on the mean flow depth (0.17 m) and velocity (0.24 m/s). The 3D geometry of the cross vane in the flume was scanned using the high-resolution laser topography scanner at sub-millimeter resolution. The measured topography was used to reconstruct the IB mesh for the numerical simulation.

The computational domain is 6 m long, 0.9 m wide and 0.32 m high, and it is discretized with of $191 \times 151 \times 121$ nodes in x, y and z directions, respectively. The unstructured IB surface mesh used to discretize the surface of the cross vane consists of 21,970 triangular elements. The background grids are stretched along all three spatial directions. The size of the grid spacing near the side walls ($y = 0$, $y = 0.9$ m) and the channel bed ($z = 0$) are 6.8×10^{-4} m and 1.5×10^{-3} m, respectively. The size of the vertical grid spacing near the free surface is 10^{-3} m. The maximum grid spacing in the transverse, vertical and streamwise directions is 1.35×10^{-2} m, 1.8×10^{-2} m and 1.2×10^{-1} m, respectively. The computational grid near the cross vane is shown in Figure 8. While the discharge of water at the inlet ($x = 3$ m) is fixed in time, the free surface at the inlet is allowed to move in the vertical direction. To preserve a constant water discharge for temporally varying free surface elevation, at each time step the velocity at the inlet is adjusted so that the volume flux of the water matches the measured discharge (3.81×10^{-2} m^3/s). At the outlet boundary ($x = 9$ m), the level

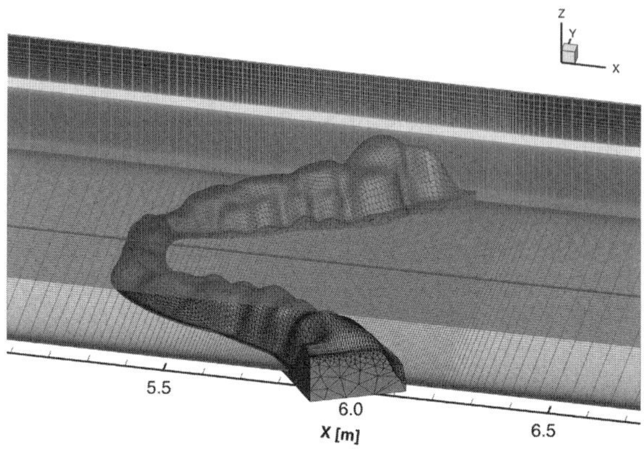

Figure 8 Background computational grids near the cross vane and an immersed boundary mesh for the cross vane discretized with triangular elements. Flow is from left to right.

set function corresponding to the measured free surface elevation (z = 0.172 m) is prescribed. The reference elevation is located at the bed of the outlet ($z = 0$). The noslip velocity boundary condition is applied at the sidewalls, and the following rough wall model is used at the channel bed. The density and dynamic viscosity of the water are set to 1,000 kg/m^3 and 10^{-3} kg/(m·s), and those of the air are set to 1.2 kg/m^3 and 1.8×10^{-5} kg/(m·s), respectively. The acceleration due to gravity is set to $g = 9.8$ m/s^2. The time step used for the computation is $\Delta t = 5 \times 10^{-4}$ s, and the value of ε (see Eq. (12)) is set to 1.5×10^{-3} m. The computation was run for 100,000 time steps.

Figure 9 shows the contour plots of the computed streamwise velocity and the free surface elevation at the water surface. We can observe that the mean streamwise velocity in the center region of the wake of the cross vane is significantly increased due to the effect of the rock structures. We also see the increased free surface elevation in the upstream of the structure and the regions of the low free surface elevation directly above the rocks near the sidewalls. These low free surface elevation demonstrate the ability of the method to resolve complex local effects as the flow transitions from subcritical (Fr < 1) to supercritical (Fr > 1) as a result of the very shallow flow depth near the sidewalls of the channel to which the rocks are attached. In Figure 10 and Figure 11, we compare the computed water surface profile with the one measured along transverse and streamwise directions, respectively. Both the computed transverse and streamwise water surface profiles show very good agreement with the measured profiles, thus clearly establishing the ability of the model to correctly predict the 3D free surface deformation due to the presence of an arbitrarily complex hydraulic structure.

5.3 Local scour around hydraulic structures

In this section, we demonstrate the predictive capabilities of the sediment transport model by simulating local scour around a rock vane structure embedded in a straight open channel flume.

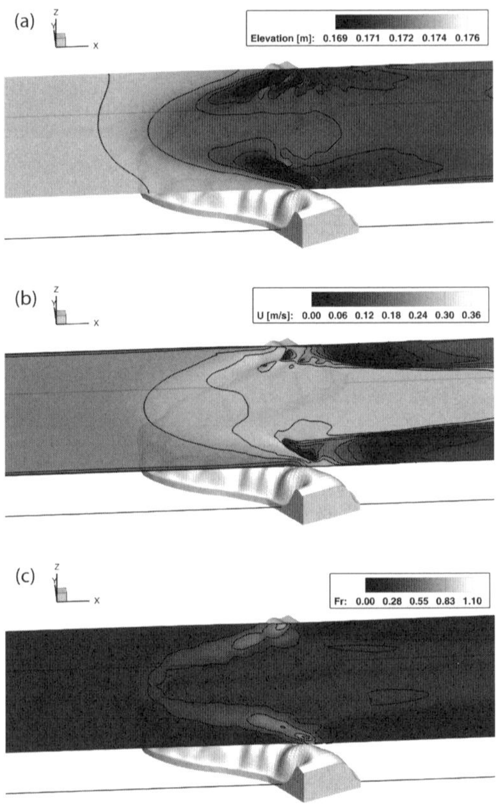

Figure 9 Contour plots of the computed (a) streamwise velocity, (b) free surface elevation, and (c) local Froude number at the water surface. Flow is from left to right.

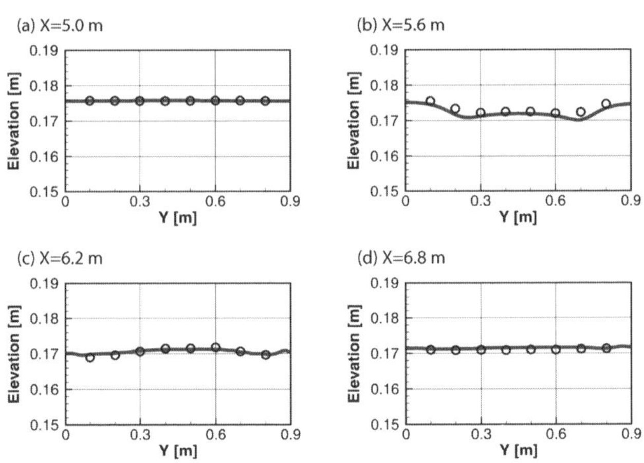

Figure 10 Comparison of the computed and measured transverse water surface profile (symbol: measurements, solid line: computation).

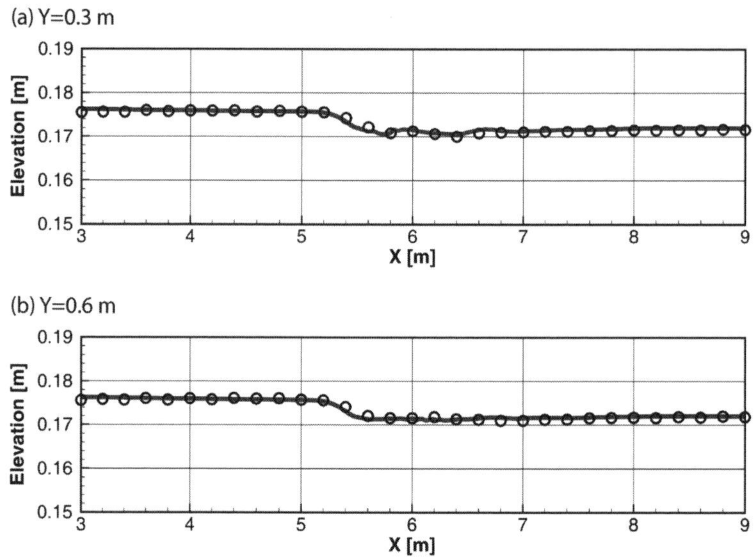

Figure 11 Comparison of the computed and measured transverse water surface profile (symbol: measurements, solid line: computation).

A flume experiment was carried out in a 15 m long, 0.9 m wide, and 0.6 m deep laboratory flume. A single-armed rock vane structure was installed on an 18 cm thick sand bed with d_{50} = 1.8 mm. There was no sediment feeding from the inlet to ensure a clear water scour condition. The rock-vane was installed at an angle of 20 degrees with respect to the oncoming flow and the tip of the structure was located at approximately one-third the channel width, 0.3 m, and the structure length was approximately 0.7 m. The rock vane structure was designed so that the top elevation of the bank attached rock was equal to that of the free surface elevation. The mean inflow velocity and flow depth are 0.36 m/s and 0.17 m, corresponding to a Reynolds number of approximately 60,000.

This calculation was carried out using the k–ω version of the URANS model (Kang *et al.*, 2011) with a Cartesian background mesh which consists of 401 × 121 × 61 grid nodes in stream wise, span wise and vertical directions, respectively. The computational grid is stretched in longitudinal and transverse directions so that grid nodes are clustered near the rock-vane structure. The mobile bed is discretized with unstructured grids consisting of 5,541 triangular elements and the rock-vane structure is covered by 188 triangle elements. Figure 12 shows a perspective view of the background grid for flow domain, the bed and the rock vane discretized with triangular grids. The time step, non-dimensionalized by the flow depth and bulk velocity, is set to 0.012, which corresponds to a maximum CFL number equal to 1.

Figure 13 compares the computed and measured bed topography 59 minutes after the flume experiment began. The contour levels denote the bed elevation around the rock-vane structure. At this time, both the experiment and computation did not reach the equilibrium condition yet, which implies that the bar in the downstream of

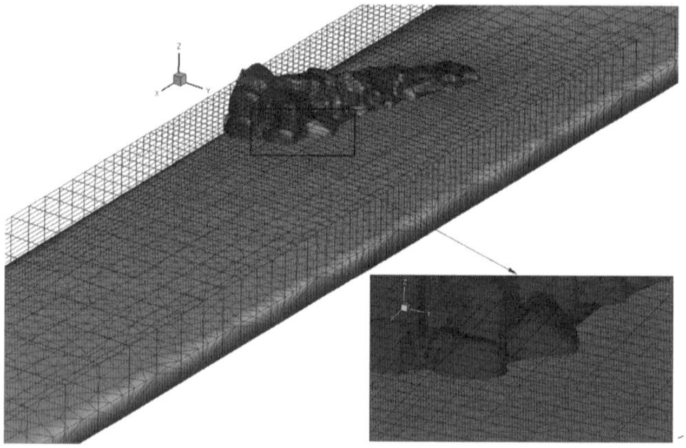

Figure 12 Computational domain for sediment transport modeling for the rock vane structure. The mobile bed and the rock vane structure are discretized with unstructured triangular grids and treated as a sharp-interface immersed body embedded in the background flow domain discretized with structured grids.

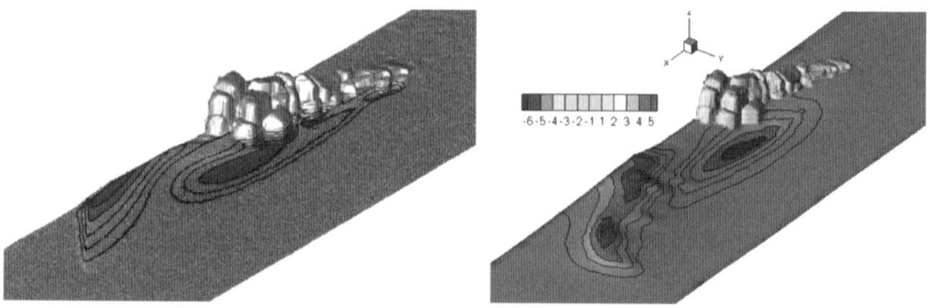

Figure 13 Measured (left) and computed (right) bed topography around rock-vane structure after 59 minutes. Flow is from top right to bottom left. Bed topography is not at equilibrium condition.

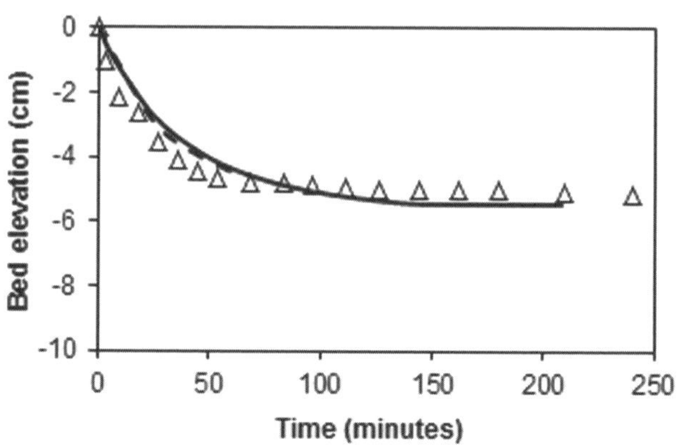

Figure 14 Computed (solid line) and measured (triangles) time evolution of the maximum scour depth in the channel.

the vane is still moving. Overall patterns of scour and deposition in the downstream of the structure predicted by the computational model are in good agreement with the measurements. The location of the maximum scour hole is also predicted with reasonable accuracy.

Figure 14 compares the measured and computed time evolution the maximum scour depth. The maximum equilibrium scour depth predicted by the simulation is approximately 5.7 cm, which is within 7% of the measured value (5.3 cm). Overall, the computed trend is in good agreement with the measurements.

6 SUMMARY

We presented our recent progress toward developing a versatile computational framework capable of simulating turbulent flow and sediment transport processes in real-life rivers and streams with embedded arbitrarily complex hydraulic structures. This framework is based on the CURVIB method (Ge & Sotiropoulos, 2007), which has recently been extended to carry out LES and URANS in waterways with arbitrarily complex bathymetry including free-surface effects and stream-bed erosion and scour (Kang *et al.*, 2011; Kang & Sotiropoulos 2011a, b Khsoronejad *et al.*, 2011a, b). We demonstrated the potential and predictive capabilities of the various elements of the proposed framework by applying it to carry out: 1) LES of flow in a fixed-bed field scale meandering stream; 2) RANS calculation of free surface flow over a cross vane structure; and 3) RANS calculation of scour around a rock vane structure. All of these simulations showed good agreement with measurements and demonstrated the ability of the method to handle arbitrarily complex nature and man-made geometries.

Whereas our simulations to date have yet to test and demonstrate the capabilities of the integrated computational framework (coupled turbulence-resolving, free-surface and morphodynamic simulations), the various components of the method have been developed into a unifying computational framework with this goal in mind. We are currently in the process of demonstrating the capabilities of our method in this regard by carrying out coupled LES with dynamic free-surface deformation and stream-bed erosion. The model is also being extended to incorporate suspended sediment transport. We will be reporting these results in a series of future papers in which we will hope to demonstrate the predictive capabilities of the proposed modeling framework as a powerful tool for carrying-out simulation-based research in real-life environmental hydraulics applications.

ACKNOWLEDGEMENTS

This work was supported by NSF grants EAR-0120914 (as part of the National Center for Earth-Surface Dynamics) and EAR-0738726; a grant from Yonsei University, South Korea; and a grant from National Cooperative Highway Research Program. Computational resources were provided by the Minnesota Supercomputing Institute.

REFERENCES

Armenio, V. and Piomelli, U. (2000). A Lagrangian mixed subgrid-scale model in generalized coordinates. *Flow, Turbulence and Combustion*. 65 (1), 51–81.

Borazjani, I., Ge, L. and Sotiropoulos, F. (2008). Curvilinear immersed boundary method for simulating fluid structure interaction with complex 3D rigid bodies. *Journal of Computational Physics*. 227, 7587–7620.

Cabot, W. and Moin, P. (2000). Approximate wall boundary conditions in the large-eddy simulation of high Reynolds number flow. *Flow, Turbulence and Combustion*. 63, 269–291.

Demuren, A.O. and Rodi, W. (1986). Calculation of flow and pollutant dispersion in meandering channels. *Journal of Fluid Mechanics*. 172 (1), 63–92.

Devenport, W.J. and Simpson, R.L. (1990). Time-dependent and time-averaged turbulence structure near the nose of a wing-body junction. *Journal of Fluid Mechanics*. 210, 23–55.

Fadlun, E.A., Verzicco, R., Orlandi, P. and Mohd-Yusof, J. (2000). Combined immersed-boundary finite-difference methods for three-dimensional complex flow simulations. *Journal of Computational Physics*. 161 (1), 35–60.

Ge, L. and Sotiropoulos, F. (2005). 3D unsteady Rans modeling of complex hydraulic engineering flows. I: Numerical model. *Journal of Hydraulic Engineering*. 131(9), 800–808.

Ge, L. and Sotiropoulos, F. (2007). A numerical method for solving the 3D unsteady incompressible Navier-Stokes equations in curvilinear domains with complex immersed boundaries. *Journal of Computational Physics*. 225 (2), 1782–1809.

Germano, M., Piomelli, U., Moin, P. and Cabot, W.H. (1991). A dynamic subgrid-scale eddy viscosity model. *Physics of Fluids A*. 3 (7), 1760–1765.

Gilmanov, A. and Sotiropoulos, F. (2005). A hybrid Cartesian/immersed boundary method for simulating flows with 3D, geometrically complex, moving bodies. *Journal of Computational Physics*. 207 (2), 457–492.

Hirt, C.W. and Nichols, B.D. (1981). Volume of fluid (VOF) method for the dynamics of free boundaries. *Journal of Computational Physics*. 39 (1), 201–225.

Hodges, B.R. and Street, R.L. (1999). On simulation of turbulent nonlinear free-surface flows. *Journal of Computational Physics*. 151 (2), 425–457.

Iaccarino, G. and Verzicco, R. (2003). Immersed boundary technique for turbulent flow simulations. *Applied Mechanics Reviews*. 56 (3), 331–347.

Jiang, G.S. and Shu, C.W. (1996). Efficient implementation of weighted ENO schemes. *Journal of Computational Physics*. 126 (1), 202–228.

Kang, S., Lightbody, A., Hill, C. and Sotiropoulos, F. (2011). High-resolution numerical simulation of turbulence in natural waterways. *Advances in Water Resources*. 34 (1), 98–113.

Kang, S. and Sotiropoulos, F. (2011a). Flow phenomena and mechanisms in a field-scale experimental meandering channel with a pool-riffle sequence: Insights gained via numerical simulation. *Journal of Geophysical Research*. 116, F03011.

Kang, S. and Sotiropoulos, F. (2011b). Numerical modeling of 3D turbulent free surface flow in complex open channels. *Advances in Water Resources*. under review.

Khosronejad, A., Kang, S., Borazjani, I. and Sotiropoulos, F. (2011a). Curvilinear immersed boundary method for simulating coupled flow and bed morphodynamic interactions due to sediment transport phenomena. *Advances in Water Resources*. 34 (7), 829–843.

Khosronejad, A., Kang, S. and Sotiropoulos, F. (2011b). Experimental and computational investigation of local scour around bridge piers. *Advances in Water Resources*. In press.

Khosronejad, A., Rennie, C., Salehi, A. and Townsend, R. (2007). 3D numerical modeling of flow and sediment transport in laboratory channel bends. *Journal of Hydraulic Engineering*. 133 (10), 1123–1134.

Khosronejad, A., Salehi, A., Rennie, C. and Gholami, I. (2008). Three dimensional numerical modeling of sediment release in a water reservoir. *Journal of Hydraulic Research*. 46 (2), 209–223.

Lai, Y.G., Weber, L.J. and Patel, V.C. (2003). Nonhydrostatic three-dimensional model for hydraulic flow simulation. I: Formulation and verification. *Journal of Hydraulic Engineering*. 129 (3), 196–205.

McCoy, A., Constantinescu, G. and Weber, L.J. (2008). Numerical investigation of flow hydrodynamics in a channel with a series of groynes. *Journal of Hydraulic Engineering*. 134 (2), 157–172.

Menter, F.R., Kuntz, M. and Langtry, R. (2003). Ten years of industrial experience with the SST turbulence model. Turbulence, *Heat and Mass Transfer* 4, 625–632.

Mittal, R. and Iaccarino, G. (2005). Immersed boundary methods. *Annual Review of Fluid Mechanics*. 37 (1), 239–261.

Molls, T. and Chaudhry, M.H. (1995). Depth-averaged open channel flow model. *Journal of Hydraulic Engineering*. 121 (6), 453–465.

Nagata, N., Hosoda, T., Nakato, T. and Muramoto, Y. (2005). Three-dimensional numerical model for flow and bed deformation around river hydraulic structures. *Journal of Hydraulic Engineering*. 131 (12), 1074–1087.

Osher, S. and Fedkiw, R. (2002). *The level set method and dynamic implicit surfaces*. Springer-Verlag, New York.

Osher, S. and Sethian, J.A. (1988). Fronts propagating with curvature-dependent speed: Algorithms based on Hamilton-Jacobi formulations. *Journal of Computational Physics*. 79 (1), 12–49.

Paola, C. and Voller, V.R. (2005). A generalized Exner equation for sediment mass balance. *Journal of Geophysical Research*. 110, F04014.

Pope, S.B. (2000). Turbulent Flows. Cambridge Univ. Press. Cambridge, UK.

Rosgen, D.L. (2001) The cross vane, W-weir and J-hook structures: Their description, design and application for stream stabilization and river restoration. *Proceedings, Wetland Engineering and River Restoration Conference CD-ROM*, ASCE, Reston, VA.

Ramamurthy, A.S., Qu, J. and Vo, D. (2007). Numerical and experimental study of dividing open-channel flows. *Journal of Hydraulic Engineering*. 133 (10), 1135–1144.

Rodriguez, J.F., Bombardelli, F.A., Garcia, M.H., Frothingham, K.M., Rhoads, B.L. and Abad, J.D. (2004). High-resolution numerical simulation of flow through a highly sinuous river reach. *Water Resources Management*. 18 (3), 1573–1650.

Rüther, N. and Olsen, N.R.B. (2005). Three-dimensional modeling of sediment transport in a narrow 90° channel bend. *Journal of Hydraulic Engineering*. 131 (10), 917–920.

Saad, Y. and Schultz, M.H. (1986). GMRES: a generalized minimal residual algorithm for solving nonsymmetric linear systems. *SIAM Journal on Scientific Computing*. 7, 856–869.

Sagaut, P. (1988). Large eddy simulation for incompressible flows. Springer, Berlin.

Sinha, S.K., Sotiropoulos, F. and Odgaard, A.J. (1998). Three-dimensional numerical model for flow through natural rivers. *Journal of Hydraulic Engineering*. 124 (1), 13–24.

Smagorinsky, J.S. (1963). General circulation experiments with the primitive equations. *Monthly Weather Review*. 91, 99–164.

Stoesser, T., Rüther, N. and Olsen, N.R.B. (2010). Calculation of primary and secondary flow and boundary shear stresses in a meandering channel. *Advances in Water Resources*. 33 (2), 158–170.

Sussman, M. and Fatemi, E. (1999). An efficient, interface-preserving level set redistancing algorithm and its application to interfacial incompressible fluid flow. *SIAM Journal on Scientific Computing*. 20, 1165–1191.

Sussman, M., Fatemi, E., Smereka, P. and Osher, S. (1998). An improved level set method for incompressible two-phase flows. *Computers & Fluids*. 27, 663–680.

Sussman, M., Smereka, P. and Osher, S. (1994). A level set approach for computing solutions to incompressible two-phase flow. *Journal of Computational Physics*. 114, 146–159.
Van Rijn, L.C. (1993). Principles of sediment transport in rivers, estuaries, and coastal seas. Aqua Publications.
Wang, M. and Moin, P. (2002). Dynamic wall modeling for large-eddy simulation of complex turbulent flows. *Physics of Fluids*. 14 (7), 2043–2051.
Wilcox, D.C. (1988). Reassessment of the scale-determining equation for advanced turbulence models. AIAA Journal. 26, 1299–1310.
Wilson, C.A.M.E., Boxall, J.B., Guymer, I. and Olsen, N.R.B. (2003). Validation of a three-dimensional numerical code in the simulation of pseudo-natural meandering flows. *Journal of Hydraulic Engineering*. 129 (10), 758–768.
Yoon, T.H. and Kang, S-.K. (2004). Finite volume model for two-dimensional shallow water flows on unstructured grids. *Journal of Hydraulic Engineering*. 130 (7), 678–688.
Yue, W., Lin, C.-L. and Patel, V.C. (2003). Numerical simulation of unsteady multidimensional free surface motions by level set method. *International Journal for Numerical Methods in Fluids*. 42 (8), 853–884.
Yue, W., Lin, C.-L. and Patel, V.C. (2005). Large eddy simulation of turbulent open-channel flow with free surface simulated by level set method. *Physics of Fluids*. 17 (2), 025108.
Zedler, E.A. and Street, R.L. (2001). Large-eddy simulation of sediment transport: Currents over ripples. *Journal of Hydraulic Engineering*. 127 (6), 444–452.

Chapter 8

Morphodynamic equilibrium of tidal channels

G. Seminara, M. Bolla Pittaluga and N. Tambroni
Department of Civil, Environmental and Architectural Engineering, University of Genova, Genova, Genoa, Italy

ABSTRACT: We explore the problem of morphodynamic equilibrium of tidal channels, bounded seaward by a tidal sea and shoaling landward as observed in coastal lagoons and estuaries. These channels are typically landward converging, meandering and bounded by tidal flats periodically flooded by the tidal wave. We then attempt to provide an answer to the following questions. How is equilibrium defined for tidal channels? Do tidal channels have an equilibrium length with a predictable bed profile? Why are channels typically converging? What fundamental differences exist between lagoon channels and estuaries? We identify three distinct cases. The first (*coastal*) case concerns the 'short' tidal channels observed in coastal wetlands and lagoons: their distinct feature is the absence of a fluvial supply of fresh water and sediments. This case has been fully explored. In particular, Seminara *et al.* (2010) showed that rigorous conditions of static equilibrium exist and require that the sediment flux must vanish at each instant throughout the tidal cycle. The equilibrium length is proportional to the inlet depth and decreases as convergence, roughness or tidal amplitude increase. These channels satisfy the so called O'Brien law. Results have been substantiated by detailed laboratory measurements of Tambroni *et al.* (2005*a*). The second (*fluvial*) case concerns the transition of a river into a tidal channel characterized by fairly 'small' tidal oscillations. We derive a perturbation solution for flow and bed topography showing that equilibrium arises from a balance between the aggrading effect of channel divergence and the opposite effect of the residual sediment flux driven by tide propagation. The third (*estuarine*) case concerns the transition of a river into a tidal channel characterized by fairly 'large' tidal oscillations. We derive a numerical solution for flow and bed topography able to describe conditions intermediate between the two limit cases discussed previously. Results show that the model is able to describe a wide class of settings: ranging from fluvial dominated estuaries to tidally dominated estuaries, where the equilibrium profile tends to the 'coastal' profile, with some correction needed in order for the hydrodynamics to accommodate the fluvial transport.

1 INTRODUCTION

We are concerned in this paper with the morphodynamics of erodible channels. This is a subject which has received considerable attention in the field of fluvial hydraulics, such that, to some extent, many fundamental issues may be considered as fairly settled (see, e.g., Seminara, 2010). The morphodynamics of tidal channels is not equally understood: the reader interested in a recent assessment of the subject is referred to Swart & Zimmermann (2009). Below, we wish to smake some further progress, focusing on the equilibrium of channels bounded seaward by a tidal sea and shoaling landward as observed in coastal lagoons and estuaries. These channels have various distinct features: they are typically landward converging (Friedrichs & Aubrey, 1996), meandering and bounded by tidal flats periodically flooded by the tidal wave (Figure 1).

A number of questions arise. The first set of questions concerns the paradigm of morphodynamic equilibrium, starting from the most fundamental one: how is equilibrium defined and what mechanisms control the establishment of an equilibrium state? Do tidal channels have an equilibrium length and a predictable bed profile? Why are channels typically converging? What fundamental differences exist between lagoon channels and estuaries? Some of these issues have been tackled by recent research efforts and are essentially settled. This is the case of the 'short' tidal channels typically observed in coastal wetlands and lagoons (Figure 1), their distinct feature being the absence of a fluvial supply of fresh water and sediments.

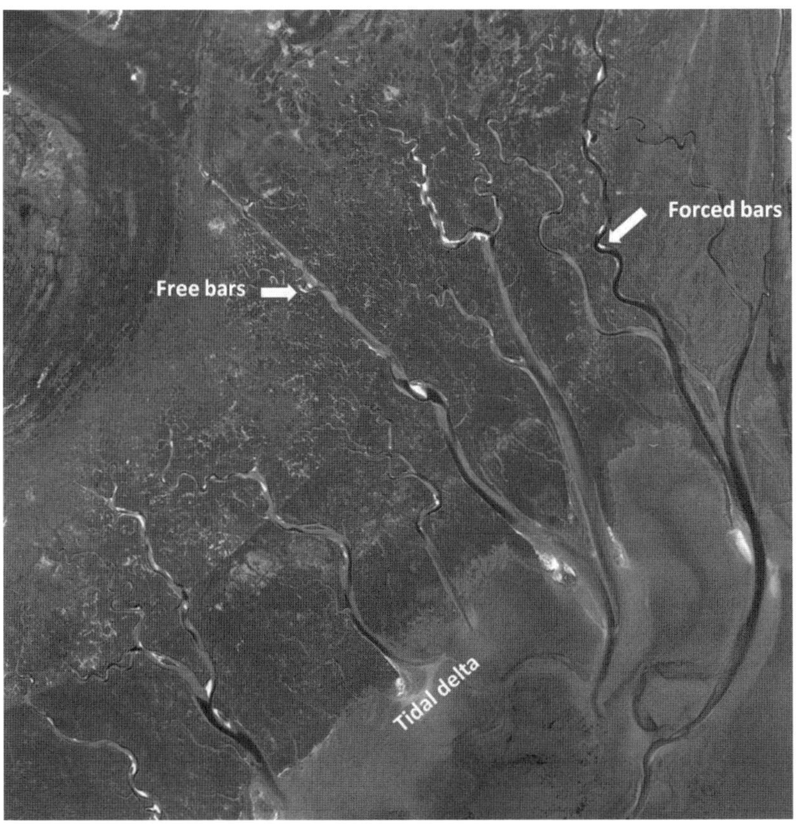

Figure 1 An aerial image of tidal channels bounded by tidal flats (Skallingen, Denmark). Courtesy of Aart Kroon.

Some of the available knowledge on this problem is reported in Section 3. Next, we investigate in Section 4 the other extreme case, that we call 'the fluvial case', where a river transitions into a tidal channel characterized by fairly 'small' tidal oscillations: hence, the fluvial discharge of water and sediments dominates over the tidal transport. Finally, in Section 5 we consider the intermediate case, that we call 'the estuarine case', where the tidal channel is characterized by fairly 'large' tidal oscillations.

An extension of the above problems concerns the hydrodynamic-morphodynamic interaction between channels and flats and between flats and salt marshes. This is a complex issue, which involves a number of further mechanisms, most notably wind driven sediment resuspension in tidal flats and the role of marshes as sinks of minerogenic sediments and sources of organic sediments. Understanding and modeling this phenomenon is crucial in order to predict the impact of sea level rise on the future of coastal wetlands and lagoons. Its importance has attracted in the last few years some intense research activity, falling in the realm of bio-morphodynamics. Space does not allow to review this subject here.

A third set of questions which would deserve attention concerns the paradigm of morphodynamic instability. We would preliminarily like to know whether the equilibrium profiles discussed in this paper are stable to 1-D perturbations. Next, it would be of interest to ascertain why, when and how bedforms (of both large and small scale) develop in tidal channels: while field observations of such features have been known from the geomorphological literature for a long time, attempts to provide theoretical explanations of these observations (Seminara & Tubino, 2001; Blondeaux & Vittori, 2011), reproduce bar formation in the laboratory (Tambroni *et al.*, 2005*a*) or through numerical simulations (Hibma *et al.*, 2004) have been proposed only recently and leave various fundamental issues as yet unsettled. A related set of questions concerns the paradigm of plan form instability. Why and when do tidal channels meander? Do tidal meanders exhibit features similar to their fluvial counterparts? In particular, do they exhibit a preferential (seaward or landward) skewness? Do they migrate? Do they amplify to the extent that neck or chute cutoff may occur? Some of these questions have been addressed (Marani *et al.*, 2002; Solari *et al.*, 2001) but a number of as yet unsolved problems still await to be properly tackled. Space will not allow to treat these subjects.

Below we then focus on the first issue, namely the one concerning the morphodynamic equilibrium of tidal channels. In the next Section we formulate the mathematical problem governing equilibrium. We then solve this problem analytically for the coastal (Sect. 3) and fluvial (Sect. 4) cases. The estuarine case is treated numerically in Sect. 5. The paper is concluded with some discussion of the future developments needed in order to substantiate our theoretical findings and relax some of the assumptions employed in the present work.

2 MORPHODYNAMIC EQUILIBRIUM OF A TIDAL CHANNEL: GENERAL FORMULATION

Let us consider a tidal channel and refer it to a longitudinal coordinate x^* with origin at the channel inlet and pointing landward as depicted in Figure 2. The channel width B^* is assumed to vary exponentially in the landward direction according to the following relationship:

$$B^* = B_0^*[b_u + (1-b_u)\exp(-x^*/l_b^*)] \tag{1}$$

where l_b^* is the channel convergence length, B_0^* is the width at the inlet and b_u is a 'dimensionless river width', scaled by the inlet width, asymptotically reached upstream. Note that the coastal case differs from the fluvial and estuarine cases because no river discharge is supplied from upstream, hence the river width b_u vanishes. We will examine this particular case in Section 3. At present, let us assume that some constant 'formative' river discharge per unit width q_u^* is supplied from upstream and the flow carries a constant sediment flux per unit width q_{su}^* in equilibrium with the fluid flux. Note that the notion of formative discharge of a river has been the subject of longstanding speculations in the geomorphological literature (Wolman & Miller, 1960; Carling, 1988; Emmett & Wolman, 2001) and can now be set on a firmer rational basis (Luchi *et al.*, 2011). Let us then denote by S, U_u^* and D_u^* the river slope, the modulus of the uniform cross sectionally averaged velocity of the fluvial stream and its associated uniform flow depth. One then readily finds that:

$$F_u^2 = \frac{S}{C_{fu}}; \; q_{su}^* = \frac{0.05C_{fu}^3[U_u^*]^5}{(s-1)^2 g^2 d^*} \tag{2}$$

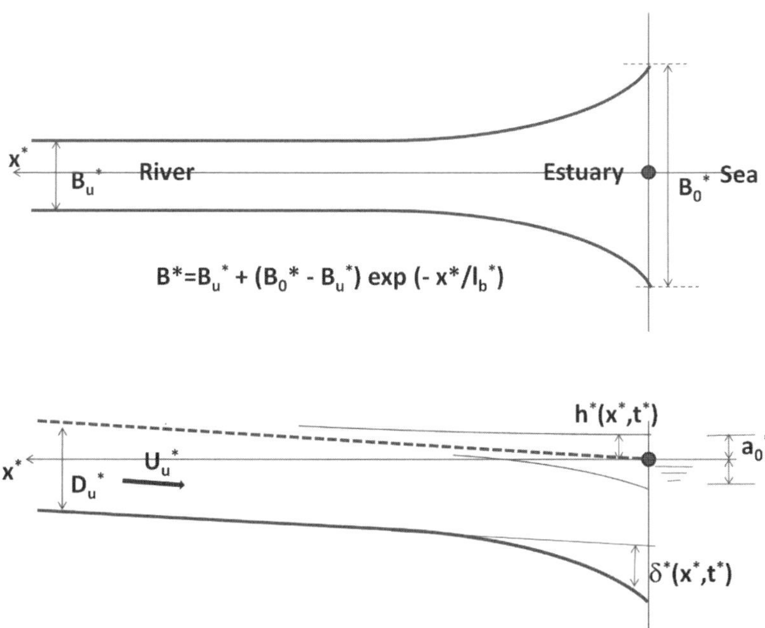

Figure 2 Sketch and notations for an exponentially converging tidal channel merging upstream into a fluvial channel.

Note, that we have assumed the bed to be cohesionless and have employed Engelund and Hansen's predictor (Engelund & Hansen, 1967) for the total sediment discharge per unit width, denoting by F_u, C_{fu}, d^* and s Froude number, friction coefficient, grain size and relative sediment density. In order to identify the appropriate spatial scale for the general case where both fluvial supply and tidal forcing play a role, let us assume that dissipation must balance gravity in the momentum equation: it turns out that the gradient of free surface elevation must be of order $\mathcal{O}(F_u^2 C_{fu}) = \mathcal{O}(S)$. Hence, the following scaling naturally arises from the above assumption:

$$x^* = \frac{D_u^*}{S} x; \quad t^* = \omega^{*-1} t; \\ h^* = D_u^* h; \quad D^* = D_u^* D; \quad U^* = U_u^* U, \tag{3}$$

with ω^* angular frequency of the tidal wave. The conservation equations for the fluid phase in their 1-D dimensionless forms then read:

$$\lambda \frac{\partial D}{\partial t} + \frac{\partial q}{\partial x} + \beta(x) q = 0, \tag{4}$$

$$\lambda F_u^2 \frac{\partial}{\partial t}\left(\frac{q}{D}\right) + \frac{F_u^2}{2} \frac{\partial}{\partial x}\left(\frac{q^2}{D^2}\right) + \frac{\partial h}{\partial x} + \frac{q|q|}{D^{10/3}} = 0. \tag{5}$$

where $q \equiv UD$ and the classical Strickler's relationship has been employed for the friction coefficient in the momentum equation. Note that a number of parameters emerge: $\beta(x)$ and λ measure the relative rate of width reduction in the upstream direction and the effect of local inertia relative to convective transport. They read:

$$\beta(x) = -\lambda_b \frac{\exp(-\lambda_b x)}{a + \exp(-\lambda_b x)}; \qquad \lambda = \frac{\omega^* D_u^*}{S U_u^*} \tag{6}$$

where:

$$a = \frac{b_u}{1 - b_u}; \qquad \lambda_b = \frac{D_u^*}{l_b^* S} \tag{7}$$

Note, that the above formulation is not appropriate for the coastal case, when the parameter a vanishes, the scaling quantity U_u^* becomes meaningless and variations of free surface elevation scale with the amplitude of the forcing tide. The appropriate scaling for the coastal case is discussed in Section 3.

The Equations (4) and (5) must be supplemented by appropriate boundary conditions. At the channel mouth the forcing oscillation of the free surface must be imposed:

$$h|_{x=0} = \epsilon f(t). \tag{8}$$

where $f(t)$ describes the temporal dependence of the forcing tide and the parameter ϵ has the form:

$$\epsilon = \frac{a_0^*}{D_u^*} \tag{9}$$

The landward boundary condition imposes that the stream must asymptotically merge into the uniform fluvial stream, hence:

$$U\big|_{x\to\infty} \to -\updownarrow \tag{10}$$

In order to complete the formulation we must allow for the channel bottom to evolve due to the effects of erosion and deposition. We then need to introduce the governing equations for the solid phase. Denoting by η^*, $\bar{c}$, p and q_s^* the instantaneous, laterally averaged, bed elevation at a given cross section, the volumetric concentration of the solid phase averaged over the cross section, the sediment porosity and the total flux of sediments per unit width respectively, the 1-D dimensionless form of Exner equation (Exner, 1925), which is a statement of mass conservation of the solid phase, reads:

$$\lambda\tau\left[(1-p)\frac{\partial\eta}{\partial t} + D\frac{\partial\bar{c}}{\partial t}\right] + \frac{\partial q_s}{\partial x} + \beta(x)q_s = 0 \tag{11}$$

where p is sediment porosity and the following dimensionless quantities have been defined:

$$\begin{aligned} \eta^* &= \eta D_u^*; & \tau &= \frac{U_u^* D_u^*}{\sqrt{(s-1)gd^{*3}}}; \\ B^* &= B_0^* b; & q_s^* &= q_s\sqrt{(s-1)gd^{*3}}. \end{aligned} \tag{12}$$

Note that τ is the ratio between the water supply and the scale for sediment supply, a large number which determines the slow character of morphodynamic evolution.

Using the Exner equation, we can now clarify the notion of morphodynamic equilibrium: a tidal channel is in *instantaneous morphodynamic equilibrium* provided η as well as $\bar{c}$ do not vary in time at any cross section. Hence, strictly, this condition would require that the instantaneous total sediment flux in the cross section should keep constant everywhere throughout the channel. In the coastal case, no sediment flux is supplied upstream, hence equilibrium requires that the instantaneous sediment flux must vanish everywhere: this is a condition of 'static' equilibrium and, in particular, no exchange of sediments is allowed with the sea. In the fluvial and estuarine cases, the constant sediment flux discharged from the river at formativeconditions must keep constant throughout the estuary.

A weaker form of equilibrium is achieved when the absolute bed elevation as well as the average concentration do not experience any net variation in a tidal cycle (*tidally averaged morphodynamic equilibrium*), hence:

$$\eta_{eq}(x)\,|_t^{t+2\pi} = \bar{c}(x)\,|_t^{t+2\pi} = 0 \tag{13}$$

Recalling the Equation (11), we must again distinguish the coastal case, where the condition (13) implies that the net sediment flux in a tidal cycle must vanish everywhere, from the fluvial and estuarine cases, where equilibrium is achieved provided the net sediment flux in a tidal cycle is constant throughout the tidal channel and equal to the constant sediment flux discharged from the river. In fact, integrating the Equation (11) both in time and in space and using the constraints (13), with some reduction achieved using the Engelund and Hansen predictor, one finds:

$$\exp\left[-\int_x^{\infty} \beta(x)dx\right] - 1 = < D^{-1/2}U^5 > \tag{14}$$

the symbol <> denoting the operation of averaging over a tidal cycle. Though the above conditions will never be exactly met as the external forcing to the system (namely tide oscillations and river discharge) undergo fluctuations, however the search for equilibrium under steady forcing conditions allows one to define a reference asymptotic state of conceptual relevance.

Finally, *long term equilibrium* is also affected by sea level rise and soil subsidence: equilibrium then requires that the bed elevation *relative to the rising sea level* should not experience a net variation in a tidal cycle. This is a condition of *dynamic equilibrium* where a small net sediment flux in a tidal cycle must be allowed (or added to the fluvial supply) in order for the channel to accrete and counteract sea level rise and subsidence (Nichols & Boon, 1994).

3 THE COASTAL CASE: NO SUPPLY OF FRESH WATER

Let us first note that no assumption has been made so far on the size of the various parameters introduced in the last section. We now examine the coastal case, which meets the following limit conditions:

$$\epsilon \to \infty; \quad F_u^2 \to 0. \tag{15}$$

The existence of a state of equilibrium of tidal channels in the coastal case has been the subject of a number of recent and less recent investigations (de Swart & Zimmermann, 2009), starting from the works of Friedrichs & Aubrey (1996) and Schuttelaars & de Swart (1996). It has been established that equilibrium is associated with a shoaling bed profile with bed elevation increasing from the inlet to a shoreline which develops landward. Let us summarize the scientific basis of this result.

3.1 Formulation

As previously pointed out, the appropriate scaling of the governing equations differs from the fluvial case. The natural scaling length for the longitudinal coordinate l^* is now the finite, as yet unknown, length of the channel; the scaling depth is chosen to coincide with the inlet depth D_i^*; the natural scale for variations of free surface elevation is the amplitude of tidal oscillations a_0^* and the flow speed is scaled by the flow speed $\left(\epsilon\sqrt{gD_i^*}\right)$ characteristic of small amplitude inviscid tidal waves. Hence, we write:

$$x^* = l^* x; \quad h^* = a_0^* h; \quad D^* = D_i^* D; \quad U^* = \epsilon\sqrt{gD_i^*}\,U; \tag{16}$$

The governing equations then read:

$$\lambda\frac{\partial D}{\partial t} + \epsilon\frac{\partial (UD)}{\partial x} - \epsilon\lambda_b (UD) = 0, \tag{17}$$

$$\lambda\frac{\partial U}{\partial t} + \epsilon U\frac{\partial U}{\partial x} + \frac{\partial h}{\partial x} + R\frac{U\,|U|}{D^{4/3}} = 0. \tag{18}$$

Here, λ is the ratio between the channel length and the inviscid tidal wavelength, R is a dimensionless parameter measuring the importance of friction relative to gravity and λ_b weighs as usual the relative effect of channel convergence:

$$\lambda = \frac{l^*\omega^*}{\sqrt{gD_i^*}}; \quad R = \frac{l^*\varepsilon C_{fi}}{D_i^*}; \quad \lambda_b = \frac{l^*}{l_b^*}. \tag{19}$$

The appropriate boundary condition at the channel mouth is again the forcing oscillation of the free surface:

$$D\big|_{x=0} = 1 + \epsilon h(x,t)\big|_{x=0} = 1 + \epsilon f(t). \tag{20}$$

On the contrary, the landward boundary condition requires more care: in fact, the shoreline, located at $x_{sh}^*(t^*)$, is a moving boundary through which no relative flux may occur, hence:

$$U\big|_{x=x_{sh}} = \frac{\lambda}{\epsilon}\dot{x}_{sh}; \quad D\big|_{x=x_{sh}} = 0. \tag{21}$$

In the early work of Schuttelaars & de Swart (1996) the correct landward boundary condition was simply replaced by a condition of vanishing fluid flux set at $x = 1$. This approach ignored the inner flow region where the bed emerges during part of the tidal cycle: this region, where the flow depth scales with the tidal amplitude a_0^*, behaves as a boundary layer where rescaling is necessary. The mathematical formulation is completed by the 1-D form of the evolution equation of the bed interface in the form of Exner's equation which takes a dimensionless form similar to that presented in Section 2.

3.2 The asymptotic equilibrium solution of Seminara *et al.* (2010)

The above problem is amenable to analytical treatment for 'short' tidal channels (channel length much smaller than the tidal wavelength) subject to low amplitude tides, conditions typically met in Venice lagoon. Hence, Seminara *et al.* (2010) set:

$$\epsilon << 1; \quad \lambda = \Lambda\epsilon; \quad R = r\epsilon; \quad \lambda_b \sim \mathcal{O}(1); \tag{22}$$

with Λ and r $\mathcal{O}(1)$ quantities. The reader is referred to Seminara *et al.* (2010) for details of the analysis. Here, it suffices to recall that, in the latter work, Seminara *et al.* (2010) used the method of matched asymptotic expansions (Nayfeh, 1973), whereby the problem was solved distinctly in an outer region consisting of the bulk of the channel and in the inner region adjacent to the moving shoreline. Outer and inner solutions were then appropriately matched. Let us summarize the main results of that analysis.

1 The first result of the theory is the proof that *rigorous conditions of static equilibrium require that the sediment flux must vanish at each instant throughout the tidal cycle*, i.e., the stress at the bottom must not exceed the threshold conditions for sediment motion. In fact, any sediment transport law establishes a non linear relationship between the sediment flux and the flow speed: for an arbitrary temporal dependence of the forcing tide, yet periodic with zero mean, a residual sediment flux will generally arise from non linearity of the transport law, unless transport vanishes identically throughout the tidal cycle. In the real world exact static equilibrium will hardly be reached for several reasons: the external (astronomical, atmospheric and climatic) forcing is random; equilibrium is approached asymptotically at a rate decreasing as the residual sediment flux decreases (see Section 3.3); channels exchange sediments with adjacent flats. This is why a slightly different criterion for equilibrium has been proposed in the literature: the condition that, during a tidal cycle, the velocity attained throughout the channel must be spatially constant and of the same order of magnitude as the critical velocity for sediment motion (Friedrichs & Aubrey, 1996; Pritchard & Hogg, 2003 and Waeles *et al.*, 2004). It is worth mentioning at this stage, that, in a recent paper (Toffolon & Lanzoni, 2010), a slightly different approach was followed to investigate the equilibrium of some lagoon channels which end close to the divide between adjacent sub-basins: these channels are also shoaling but, apparently, they keep permanently submerged at the inner end. The above Authors treat this end cross section as an impermeable boundary. They also treat the sediments as cohesive assuming equal values for the threshold stresses for sediment erosion and deposition. Under these conditions, the Authors find an equilibrium solution satisfying the condition of vanishing residual sediment exchange with the bed (i.e., of constant U_{rms}) which the Authors suggest to be fairly similar to the solution obtained imposing a constant value of the peak velocity throughout the channel. The latter analysis leads to results which may appear to leave the picture of the process substantially unchanged. On the contrary, it raises a conceptual problem

of some interest, namely: can equilibrium be rigorously reached (in a mathematical sense) with non vanishing sediment transport? We must honestly state that we feel that the latter analysis cannot be taken to be a proof that such an equilibrium may exist. For various reasons. Firstly, the above Authors use a somewhat artificial boundary condition. In particular, they force the channel to be permanently submerged and, as a result, the peak velocity has a discontinuity tending to infinity at incipient emersion. We doubt the bed would actually keep permanently submerged if the system were allowed to evolve freely. This never occurred in the numerical simulations performed by Lanzoni & Seminara (2002). Secondly, the observation that in some channels of Venice Lagoon, the landward end keeps permanently submerged is not a proof either. In fact, there is no observational evidence that these channels are indeed in equilibrium and, more notably, the end cross sections are essentially parts of the tidal flats, which are known to be subject to the strong resuspensive effect of the winds frequently blowing on Venice Lagoon: this effect might be the cause of the permanent submersion of those channels. Finally, a possible objection which might be leveled to the static equilibrium criterion is that it requires the existence of a threshold condition for sediment motion. However, any transport law which does not contain such a threshold (e.g., Engelund-Hansen formula) predicts extremely low values of sediment flux as the Shields stress tends to vanish. As a result, though an exact equilibrium can never be reached, as the Shields stress decreases to very low values, departure from equilibrium is so weak that the time scale of the transient process becomes extremely large and equilibrium may be taken to be effectively reached.

2 Imposing the static equilibrium constraint, Seminara *et al.* (2010) were able to predict the *equilibrium profile and the equilibrium length of tidal channels* in closed form. At the leading order of approximation, they read:

$$D_0(x)^{7/6} = \frac{l_b^*}{l_\infty^*}\left\{1 - \exp\left[\frac{l_{eq0}^*}{l_b^*}(x-1)\right]\right\}, \tag{23}$$

where l_{eq0}^* is the equilibrium length of the actual channel and l_∞^* is the equilibrium length of a non converging channel. These quantities read:

$$l_{eq0}^* = -l_b^* \log\left(1 - \frac{l_\infty^*}{l_b^*}\right); \quad l_\infty^* = \frac{U_{cr}^*}{\epsilon\omega^* \mid \dot{f}_{max} \mid}. \tag{24}$$

Concavity of the bed profile increases as l_b^* decreases, but a weak concavity persists also in the limit $l_b^* \to \infty$ $(\lambda_b = 0)$. Note that the equilibrium channel shortens as the convergence length, the channelroughness or the tidal amplitude increase, all the other parameters being fixed. Also note that the equilibrium length is proportional to the inlet depth: this finding confirms a famous Venetian say, 'Gran Laguna fa gran porto', meaning that a great lagoon needs a deep inlet.

3 A 'short' microtidal channel at equilibrium satisfies the so called O'Brien law: this is a well known power law relationship with exponent 0.85 between the so called tidal prism P^* and the mean cross sectional area Ω_0^*, empirically established by O'Brien

(1969) for tidal inlets. Recently, it has been claimed, on the basis of both numerical simulations (Lanzoni & Seminara, 2002) and field observations (D'Alpaos *et al.*, 2009), that an identical relationship would hold for tidal channels. The theory of Seminara *et al.* (2010) predicts the validity of a power law relationship with exponent $6/7 = 0.857$, thus providing a theoretical substantiation of those findings.

4 The effect of tidal flats on channel equilibrium has only been superficially investigated so far. In fact, in the works of Seminara *et al.* (2010) and Toffolon & Lanzoni (2010) only the influence of flats on the channel hydrodynamics was accounted for and was shown to lead to comparatively shorter channels at equilibrium. However, shallow flats are obviously able to exchange with the channels sediments resuspended by sufficiently strong winds. This mechanism, fairly common in Venice lagoon, tends to erode the flats unless flood currents carry enough sediments to balance the loss experienced by the flats during the ebb phase. Whether or not the system may reach equilibrium under these strongly asymmetric conditions is an as yet unsolved problem. Hence, the findings of both Seminara *et al.* (2010) and Toffolon & Lanzoni (2010) will require some deeper understanding.

The results described above were also satisfactorily compared with laboratory observations performed by Tambroni *et al.* (2005*a*), as shown in Figure 3.

3.3 Transient evolution towards equilibrium

The transient process whereby, starting from an initially horizontal bed profile, equilibrium is reached has been the subject of numerical simulations (Lanzoni &

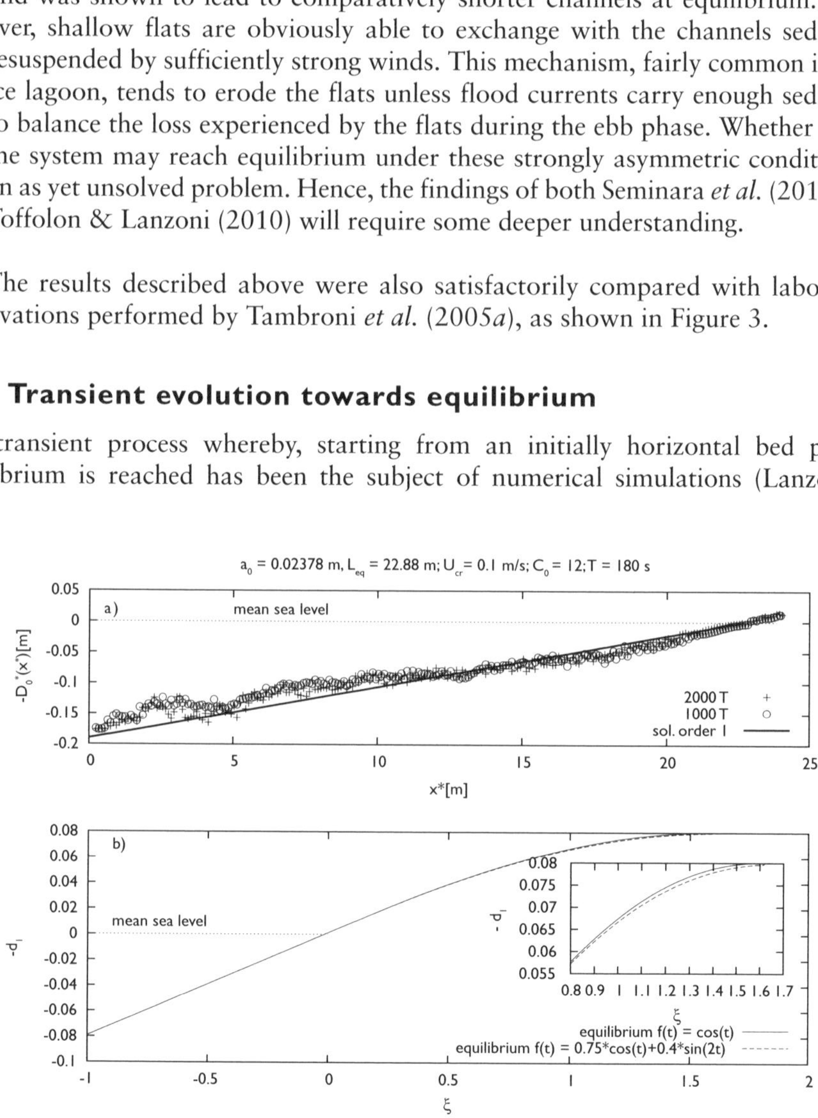

Figure 3 a) Comparison between the bed profile predicted by Seminara *et al.* (2010) and the laboratory observations of Tambroni *et al.* (2005*a*); b) Bed profile in the inner region predicted by Seminara *et al.* (2010).

Seminara, 2002; Todeschini *et al.*, 2008) and laboratory observations (Tambroni *et al.*, 2005*a*). The picture arising from these investigations is as follows: an aggradational wave originating from the inlet migrates landward leading eventually to bed emersion and the formation of a shore, i.e., tidal channels choose their own length adjusting to the asymptotic equilibrium condition of vanishing sediment flux at any time throughout the tidal cycle in any cross section of the channel. The net tidally averaged volume of sediments exchanged between basin and channel is found to be positive and grows in time (the channel imports sediments) in the very initial stage of the process, but it soon reaches a maximum and then decreases monotonically (the channel exports sediments) tending asymptotically to some constant negative value corresponding to the final equilibrium state (no further import or export of sediments). An interesting feature, which will deserve further analysis, is the observation that shore formation is associated with a process of reflection that the bottom wave undergoes when it reaches the landward boundary. Finally, Lanzoni & Seminara (2002) also showed that the validity of O'Brien relationship is asymptotically approached as the channel tends to equilibrium. The work of Todeschini *et al.* (2008) has clarified the role of the length of the initial channel chosen in the numerical simulations, showing that an initial length smaller than some threshold value determines the final equilibrium state: this is the case of 'short' channels typically observed in coastal lagoons and wetlands. Essentially, in the light of the work of Seminara *et al.* (2010), this result implies that the size of a coastal lagoon is determined by the inlet depth, which is in turn determined by the size of the inlet. In other words, it is the inlet depth which ultimately controls the length of channels issuing from the inlet. On the contrary, if the initial length exceeds the above threshold, the morphological evolution of the channel leads to sediment deposition, the bed emerges and a barrier for tidal propagation forms setting the landward end of the channel (Figure 4): in this case, the length of the channel is intrinsically tied to its plan form shape and to tidal forcing.

3.4 Multiple equilibria?

An interesting issue has been brought to the attention of the scientific community by the work of Schuttelaars & de Swart (2000). These Authors investigated the role of externally imposed overtides on the occurrence of channel equilibrium and found that multiple equilibria would be possible in the presence of a sufficiently

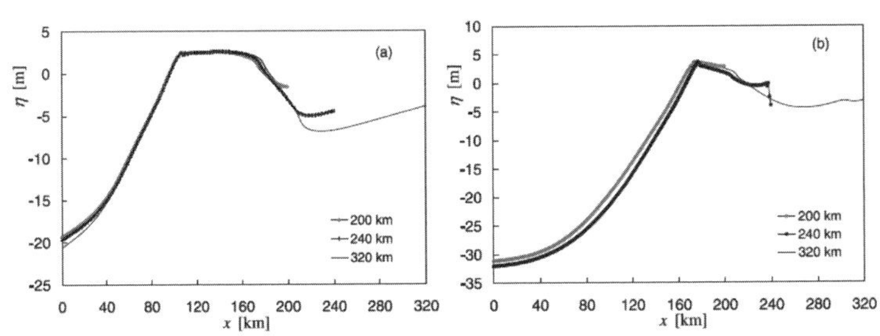

Figure 4 Equilibrium bottom profiles obtained by Todeschini *et al.* (2008) for different values of the length of the simulation domain: convergence length is 40 km (a) or 80 km (b) tidal amplitude is 2 m and initial depth is 10 m. Note that the equilibrium profile is practically independent of the length of the simulation domain.

strong M_4 component of the tidal forcing. Though the possibility of multiple states of morphodynamic equilibrium cannot be ruled out a priori, however we feel that the findings of Schuttelaars & de Swart (2000) may have arisen from the non-obvious boundary conditions imposed by these Authors. In fact, they impose the temporal dependence of the free surface elevation as well as the bed elevation at the inlet; moreover, at the landward end, they impose a vanishing bed elevation (relative to the MSL) but allow for a non vanishing velocity. The rationale behind this approach is not clear to us. According with the theory of Seminara *et al.* (2010) the length of the channel at equilibrium is tied with the flow depth at the inlet: if the length is given, then the depth at the inlet must be allowed to vary in order to ensure equilibrium; alternatively, if the the depth at the inlet is assigned then the channel length cannot be also fixed.

4 THE FLUVIAL CASE: DOMINANT SUPPLY OF FRESH WATER

We next examine the fluvial case, which is characterized by the limit conditions:

$$\epsilon \to 0; \quad F_u^2 \sim F\epsilon \tag{25}$$

with F an $\mathcal{O}$ (1) quantity. These assumptions make the resulting problem amenable to analytical treatment. The details of the analysis cannot be reported here and will be the subject of a forthcoming paper (Seminara *et al.*, 2011). Here, we introduce the reader to the theoretical formulation of the problem and present some preliminary results.

4.1 Perturbation expansion

We take advantage from the conditions (25) and expand the solution in powers of ϵ in the form:

$$(q, D, h, U) = (q_0, D_0, h_0, U_0) + \epsilon(q_1, D_1, h_1, U_1) + \epsilon^2(q_2, D_2, h_2, U_2) + \mathcal{O}(\epsilon^3) \tag{26}$$

Substituting from (26) into the governing Equations (4), (5), (11) and equating likewise powers of ϵ one finds a sequence of differential problems.

$\mathcal{O}(\epsilon^0)$

The differential problem obtained at $\mathcal{O}(\epsilon^0)$ describes the steady effect of downstream widening on the fluvial stream. This problem is readily solved in closed form to find:

$$q_0 = -\exp\left[-\int_\infty^x \beta(x)dx\right] = -\frac{a}{a + \exp(-\lambda_b x)}; \tag{27}$$

$$D_0 = q_0^{8/11}; \; h_0 = \int_0^x q_0^{-14/33} dx$$

$\mathcal{O}(\epsilon)$

The differential problem obtained at $\mathcal{O}(\epsilon)$ describes the propagation of a small amplitude tidal wave on the slowly varying fluvial stream. Its solution can be expressed in the form:

$$[q_1, h_1] = [0, h_{10}(x)] + [q_{11}(x), h_{11}(x)][\exp(it) + c.c.] \tag{28}$$

Moreover, denoting by $\delta_1(x)$ the contribution to the flow depth perturbation D_1 associated with the perturbation of bed elevation, the equilibrium constraint determines immediately the function $\delta_1(x)$ in the form:

$$\delta_1(x) = -h_{10}(x) \tag{29}$$

With some reduction, the function $h_{10}(x)$, which represents the $\mathcal{O}(\epsilon)$ correction of the effect of downstream widening on the fluvial stream, is found to satisfy a first order differential equation which is immediately solved with the boundary condition $h_{10}|_{x=0} = 0$ to give:

$$h_{10} = \frac{F}{2}\left[\left(\frac{a}{a+1}\right)^{6/11} - \left(\frac{a}{a+\exp(-\lambda_b x)}\right)^{6/11}\right] \tag{30}$$

Note that this solution implies that the uniform stream is displaced vertically downward, as:

$$h_{10}|_{x\to\infty} = \frac{F}{2}\left[\left(\frac{a}{a+1}\right)^{6/11} - 1\right] \tag{31}$$

$\mathcal{O}(\epsilon^2)$

The differential problem obtained at $\mathcal{O}(\epsilon^2)$ can be solved in the form:

$$\begin{aligned}[q_2, h_2] = {} & [q_{20}(x), h_{20}(x)] + [q_{21}(x), h_{21}(x)][\exp(it) + c.c.] \\ & + [q_{22}(x), h_{22}(x)][\exp(2it) + c.c.]\end{aligned} \tag{32}$$

We are only interested in the steady component of the solution, which affects the equilibrium topography of the tidal channel. Noting that the tidally averaged water flux per unit width q cannot differ from its leading order value q_0, it follows that q_{20} must vanish identically. Moreover, the equilibrium constraint at $\mathcal{O}(\epsilon^2)$ is readily found to impose the following condition:

$$\frac{D_2}{D_0} = \frac{20}{11}\frac{< q_1^2 >}{q_0^2} + \frac{143}{44}\frac{< D_1^2 >}{D_0^2} - 5\frac{< q_1 D_1 >}{q_0 D_0} \tag{33}$$

Performing a tidal average of the momentum equation at $\mathcal{O}(\epsilon^2)$ and using (33), we are left with a first order ordinary differential equation for $h_{20}(x)$ which can be solved to give:

$$h_{20} = \int_0^x \left[-\frac{167}{33} q_0^{-80/33} < q_1^2 > -\frac{65}{18} q_0^{-62/33} < D_1^2 > +10 q_0^{-71/33} < q_1 D_1 > \right] dx \tag{34}$$

Once the solution for h_{20} is known, then the perturbation of bottom elevation is readily derived from (33). It may be worth noting that, as $x \to \infty$, $D_2 \to 0$ hence $h_{20}\,|_{x\to\infty}$ tends to $(-\delta_2)\,|_{x\to\infty}$. These quantities do not vanish and turn out to be negative: they essentially set the position of the fluvial stream relative to the mean sea level.

4.2 The particular case of nonconvergent channels

It is of some interest to consider the particular case when channel convergence vanishes, a condition reinforced by setting $\lambda_b = 0$. Recalling the solution obtained in Section 4.1 one then readily finds:

$$q_0 = -1; \;\; D_0 = 1; \;\; h_0 = x \tag{35}$$

$$(q_1, h_1) = \left(\frac{i\lambda}{2\mu}, \frac{1}{2} \right) [\exp(-\mu x + it) + c.\,c.] \tag{36}$$

where

$$\mu = \frac{5}{3}\left[1 + \sqrt{1 + \frac{18}{25} i\lambda} \right] \tag{37}$$

As expected, the solution at $\mathcal{O}(\epsilon)$ vanishes in the limit as $x \to \infty$ in the absence of channel convergence. In fact, the effect of residual terms deriving from tidal oscillations appears firstly in the solution at second order, which is found to read:

$$h_{20} = \frac{\left[\frac{167}{66} \frac{\lambda^2}{|\mu|^2} + \frac{65}{36} + 5 \frac{\lambda}{|\mu|^2} Im(\mu) \right]}{\mu + \bar{\mu}} [\exp - (\mu + \bar{\mu})x - 1] \tag{38}$$

The reader will immediately note that, in the limit as $x \to \infty$, h_{20} does not vanish and turns out to be negative. As already pointed out, this quantity affects the position of the free surface of the fluvial stream relative to the meansea level. Finally, from (33) and (38) one finds:

$$\delta_2 = -\Delta_2 \exp - (\mu + \bar{\mu})x - h_{20}\,|_{x\to\infty} \tag{39}$$

where

$$\Delta_2 = \frac{\lambda^2}{|\mu|^2} \left[\frac{167}{66(\mu + \bar{\mu})} - \frac{10}{11} \right] + \left[\frac{65}{36(\mu + \bar{\mu})} - \frac{143}{88} \right] + \frac{\lambda Im(\mu)}{|\mu|^2} \left[\frac{5}{\mu + \bar{\mu}} - \frac{5}{2} \right] \tag{40}$$

As expected, in the limit as $x \to \infty$, δ_2 tends to $-h_{20}|_{x\to\infty}$: hence the depth of the fluvial stream in its uniform state is obviously unaffected by tidal oscillations.

4.3 Results for the fluvial case

Results for non convergent channels are reported in the Figure 5 for a given value of λ and various increasing values of the small parameter ϵ. It is apparent that the effect of tidal oscillations is to let the slope of the free surface decrease and the bottom slope

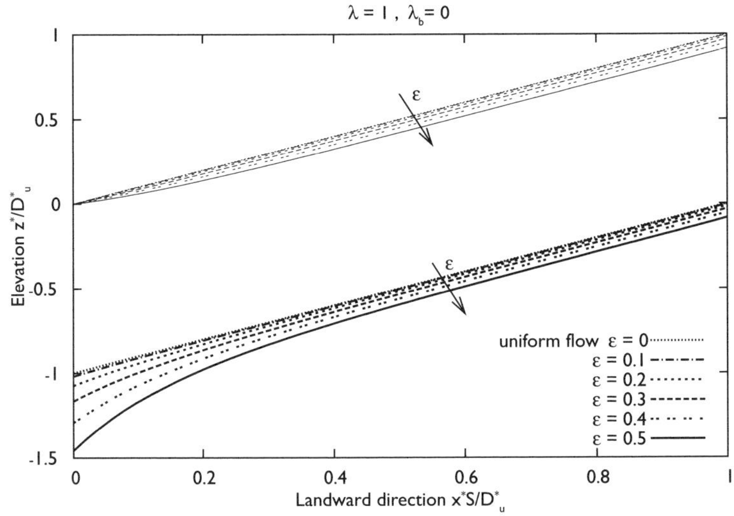

Figure 5 Bed and free surface profiles at equilibrium according to the present numerical solution for a nonconvergent channel. $\lambda = 1$.

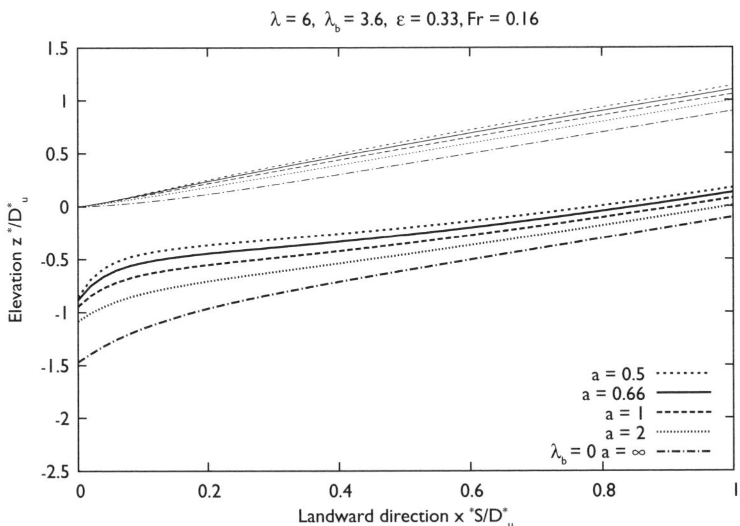

Figure 6. Bed and free surface profiles at equilibrium according to the present numerical solution for a convergent channel. $\lambda = 6$.

increase as the sea is approached. As a result, the flow depth increases, leading to a decreasing effect of dissipation. As expected, these effects are enhanced as ϵ increases. Results for convergent channels are reported in the Figure 6 for given values of λ and ϵ. They show that the effect of an increasing channel convergence (through an increasing value of either λ_b or b_u) is to induce increasing bed aggradation close to the inlet.

5 THE ESTUARINE CASE

We finally examine the estuarine case, which is typically characterized by the condition:

$$\epsilon \geq \mathcal{O}(1) \tag{41}$$

5.1 Numerical solution

The dimensionless governing differential equations presented in Section 2 have been solved numerically. We have employed a classical scheme (MacCormack, 1969) that is second order accurate both in time and space. In order to eliminate spurious oscillations that are generated close to discontinuities of the solution when linear schemes higher than first order are used (Godunov, 1959), we have employed a Total Variation Diminishing (TVD) algorithm that reformulates the MacCormack scheme in a nonlinear fashion (Garcia–Navarro & Alcrudo, 1992). In order to ensure the stability of the numerical scheme, the time step was computed such to meet the Courant-Friedrichs-Lewy constraint. The choice of the boundary conditions also deserve some comments. In fact, the MacCormack scheme was employed to compute the solution marching in time at every computational node except for the first and last nodes, where boundary conditions were prescribed. However, as the number of boundary conditions is not sufficient to define the solution completely, we have employed the method of characteristics to close the problem. Analysis of the governing equations shows that, irrespective of the Froude number, the celerity of two characteristics is invariably negative (i.e., perturbations propagate downstream) and one is positive. When the flow velocity was negative (dominated by the river), two boundary conditions, namely those involving the flow discharge and the sediment flux, were imposed at the landward boundary and the third condition was determined through the compatibility equation associated with the upstream propagating eigenvalue. Conversely, when the flow velocity was positive (dominated by the tide during the flood phase), only one boundary condition, on flow discharge, was imposed at the landward boundary whereas the other variables were determined from the compatibility equations associated with the two eigenvalues propagating downstream. Similar considerations are valid for the seaward node where the free surface elevation was prescribed. During the ebb phase the flow is negative (as in the fluvial case) and the two unknown variables are determined from the two compatibility equations associated with the eigenvalues propagating downstream. During the flood phase, sediment flux was also prescribed assuming that it was in equilibrium with local hydrodynamics conditions. The remaining variable was evaluated through the compatibility equation associated to the upstream propagating eigenvalue.

5.2 Results for the estuarine case

Results for non convergent channels are reported in the Figure 7 for different values of the small parameter ϵ. The plots report also the analytical solution valid for small values of ϵ. Note, that the analytical solution deviates increasingly from the numerical solution as ϵ increases. Again, the effect of tidal oscillations is to reduce the free surface slope and enhance the bottom slope close to the inlet. When ϵ reaches $\mathcal{O}(1)$ values, the free surface in the 'estuarine' region is practically horizontal, as found in the 'coastal' case.

Results for convergent channels are reported in the Figure 8 for a given value of the small parameter ϵ. It turns out that the effect of channel convergence at equilibrium is to induce bed aggradation close to the inlet, an effect which is enhanced as λ_b increases. In order to appreciate the time scale of the transient process leading to morphodynamic equilibrium, we plot in Figure 9a the bed elevation throughout the estuary for a non-convergent channel at different times and, in Figure 9b, the corresponding values of the spatial distribution of the net sediment. It is apparent that the temporal scale of the process is of the order of decades and that the final equilibrium is characterized by a constant value of the net sediment flux equal to the sediment flux supplied upstream.

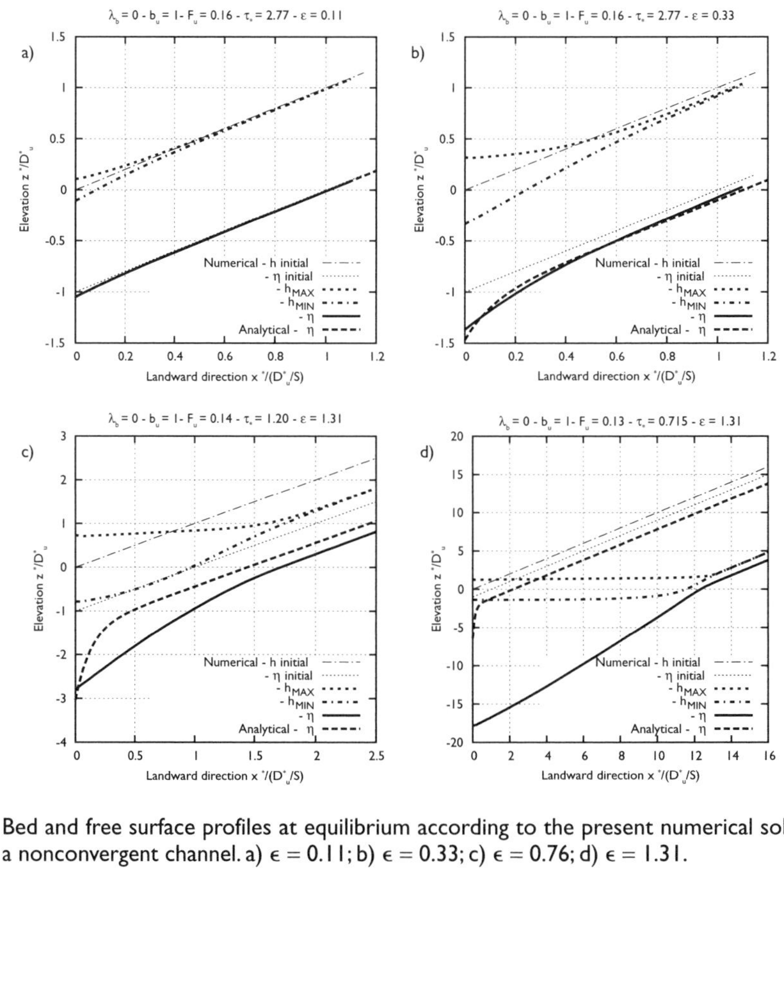

Figure 7 Bed and free surface profiles at equilibrium according to the present numerical solution for a nonconvergent channel. a) $\epsilon = 0.11$; b) $\epsilon = 0.33$; c) $\epsilon = 0.76$; d) $\epsilon = 1.31$.

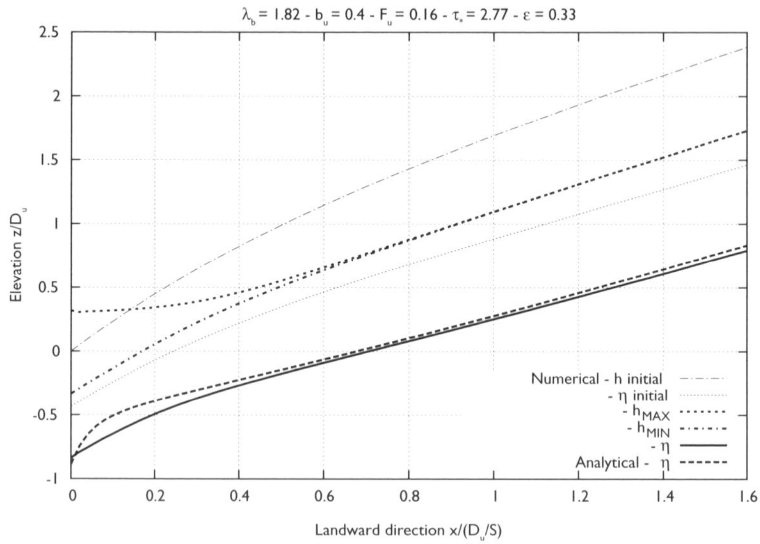

Figure 8 Bed and free surface profiles at equilibrium according to the present numerical solution for a convergent channel. $\lambda_b = 1.82$, $b_u = 0.4$, $\epsilon = 0.33$.

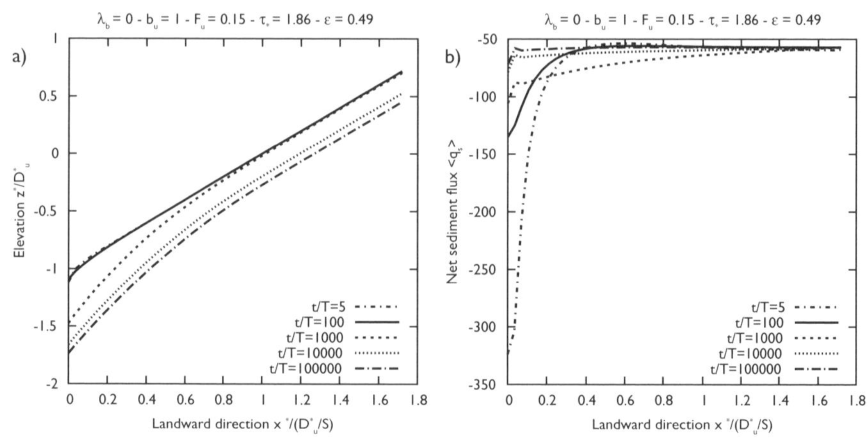

Figure 9 a) Temporal evolution of the bed throughout the estuary for a nonconvergent channel and b) spatial distribution of the net sediment flux. $\epsilon = 0.49$.

6 CONCLUDING REMARKS

The general theoretical framework presented in this paper appears to be able to describe consistently a continuum of environmental settings ranging from fluvial dominated estuaries to tidally dominated estuaries. The effect of tidal oscillations is to let the slope of the free surface decrease and the bottom slope increase approaching the sea, these effects being enhanced as ϵ increases. Eventually the equilibrium profile tends to the 'coastal' profile, with some increasingly minor correction needed in order for the hydrodynamics to accommodate the fluvial transport. Results need be substantiated on the basis of field observations as well as, possibly, laboratory experiments: an exercise that has successfully been completed so far only for the coastal case. We are presently trying to extend such a comparison to the fluvial and estuarine cases. Results will be reported in Seminara *et al.* (2011).

Here, we wish to point out some of the limitations of the present work. Firstly, we have assumed the sediment to be cohesionless, a condition that is not invariably satisfied in tidal settings, where the presence of mud and possibly clay is not uncommon. The role of cohesion may be readily incorporated using the classical Partheniades-Krone formulation. While this may alter results quantitatively, it is unlikely that qualitatively unexpected features may arise.

A second major assumption concerns stratification: the estuary has been assumed to be well mixed. This is typically true for coastal wetlands (the coastal case) and microtidal estuaries (the fluvial case): it does not necessarily fit the case of macrotidal estuaries. The interaction between stratification and sedimentation adds a number of further ingredients to the problem, which may indeed result in qualitatively different results. Further research will be needed to clarify this issue.

A third major problem which will require clarification concerns the very notion of formative discharge for a tidal channel. While this issue does not arise in the coastal case, it requires to be settled in the fluvial and estuarine cases. We may expect that the formative conditions in the fluvial case are associated with low return period floods (say 1 year floods) as typical of rivers (Luchi *et al.*, 2011). Whether or not the same criterion applies to estuarine conditions is a matter that will need to be substantiated on the basis of field observations interpreted in the light of appropriate unsteady morphodynamic simulations of estuaries subject to sequences of floods of variable intensities.

Among the many further issues that are left open, of special interest is the question of why tidal channels are typically converging. The answer to this question is qualitatively straightforward (see also Seminara *et al.* (2001)). In fact, let us assume for the sake of simplicity that the equilibrium of the channel cross section is established through a mechanism of bank stability similar to the fluvial mechanism. We know since the early work of Schumm (1960) that the equilibrium of alluvial rivers is such that the aspect ratio of the channel cross section depends on the amount of clay contained in the banks: more cohesive banks leading to narrower channels. Hence, if the banks of the tidal channel have a fairly uniform composition, the channel width at equilibrium must decrease in the landward direction in order to maintain the same aspect ratio with the decreasing value of the channel mean depth associated with the equilibrium bed profile imposed by the tidal flow. This picture needs to be quantitatively substantiated and the actual exponential character of channel convergence will have to be ascertained and clarified.

ACKNOWLEDGMENTS

The present work has been funded by MIUR (Prin 2008 Project - Eco-morphodynamics of tidal environments and climate change). N. Tambroni holds a 'Ricercatore' permanent position at the University of Genova, temporally funded by Thetys S.p.A.

REFERENCES

Blondeaux, P. and Vittori, G. (2011). Dunes and alternate bars in tidal channels. *Journal Fluid Mechanics*, 670, 558–580.

Bolla Pittaluga, M. (2003). Long term morphodynamic equilibrium of tidal channels, PhD Thesis, University of Padua (Italy).

Carling, P. (1988). The concept of dominant discharge applied to two gravel–bed streams in relation to channel stability thresholds. *Earth Surface Processes and Landforms*, 13(4), 355–367.

Dalrymple, R.W. and Rhodes, R.M. (1995). Estuarine dunes and bars. G.M.E. *Perillo, Geomorphology Sedimentology of Estuaries, Developments in Sedimentology*, 53, 359–422.

D'Alpaos, A., Lanzoni, S., Marani, M. and Rinaldo, A. (2009). On the tidal prism–channel area relations. *Journal of Geophysical Research*, 115, F01003, doi:10.1029/2008 JF001243.

Defina, A. (2000). Two–dimensional shallow flow equations for partially dry areas, *Water Resources Research*, 36(11), 3251–3264.

de Swart, H.E. and Zimmermann, J.T.F. (2009). Morphodynamics of tidal inlet systems. *Annual Review of Fluid Mechanics*, 41, 203–229.

Emmett, W.W. and Wolman M. G. (2001). Effective discharge and gravel–bed rivers. *Earth Surface Processes and Landforms*, 26, 1369–1380.

Engelund, F. and Hansen, E. (1967). A Monograph on Sediment Transport in Alluvial Streams. Copenhagen: Danish Technical Press.

Exner, F.M. (1925). Uber die wechselwirkung zwischen wasser und geschiebe in flussen. *Acad. Wissenscaften, Wien*, 134, 165–180.

Friedrichs, C.T. and Aubrey, D.G. (1994). Tidal propagation in strongly convergent channels. *Journal of Geophysical Research*, 99, 3321–3336.

Friedrichs, C.T. and Aubrey, D.G. (1996). Uniform bottom shear stress and equilibrium hypsometry of intertidal flats, in *Mixing in Estuaries and Coastal Seas, Coastal and Estuarine Studies*, 50, 405–429.

Garcia–Navarro, P. and Alcrudo, F. (1992). 1–D Open Channel Flow Simulations using TVD MacCormack scheme. *Journal of Hydraulic Engineering, ASCE*, 118(3), 1359–1372.

Godunov, S.K. (1959). A Finite Difference Method for the Computation of Discontinuous Solutions of the Equations of Fluid Dynamics. *Matematicheskii Sbornik*, 47(89), 271–306.

Hibma, A., Schuttelaars, H.M. and de Vriend, H.J. (2004). Initial formation and long–term evolution of channel–shoal patterns. *Continental Shelf Research*, 24, 1637–1650.

Lanzoni, S. and Seminara, G. (2002). Long–term evolution and morphodynamic equilibrium of tidal channels. *Journal of Geophysical Research*, 107 (C1), 3001.

Lanzoni, S., Siviglia, A., Frascati, A. and Seminara, G. (2006). Long waves in erodible channels and morphodynamic influence. *Water Resources Research*, 42, W06D17, doi:10.1029/2006 WR004916.

Luchi, R., Bolla Pittaluga, M. and Seminara, G. (2011). Modeling the morphodynamic response to width variations with an application to the Magra River (Italy). *Proc. River, Coastal and Estuarine Morphodynamics: RCEM2011*, Bejing, 4–8 September 2011.

Mac Cormack, R.W. (1969). The Effect of Viscosity in Hypervelocity Impact Cratering, *AIAA Paper*, 69–354, Cincinnati, Ohio, 1959.

Marani, M., Lanzoni, S., Zandolin, D., Seminara, G. and Rinaldo, A. (2002). Tidal meanders. *Water Resources Research*, 38(11), 1225, doi:10.1029/2001 WR000404.

Meyer–Peter, E. and Müller, R. (1948). Formulas for bedload transport. *Conf. of Internat. Ass. of Hydraul. Res., Stockholm, Sweden.*

Nayfeh, A. (1973). In Perturbation Methods (ed. H. Riffert and H. Herold). John Wiley and Sons.

Nichols, M.M. and Boon, J.D. (1994). Sediment transport processes in coastal lagoons. In *Coastal Lagoon Processes (ed. B. Kjerfve)*, vol. 60, 157–209. Elsevier Science.

O'Brien, M.P. (1969). Equilibrium flow areas of inlets on sandy coasts. *Journal Waterways Harbors Division ASCE*, 95, 43–52.

Pritchard, D. and Hogg, A.J. (2003). Cross–shore sediment transport and the equilibrium morphology of mudflats under tidal currents, *Journal of Geophysical Research*, 108(C10), 3313, doi:10.1029/2002 JC001570.

Schumm, S.A. (1960). The shape of alluvial channels in relation to sediment type, *U.S. Geol. Surv. Prof. Pap.*, 352B.

Schuttelaars, H.M. and de Swart, H.E. (1996). An idealized long term morphodynamic model of a tidal embayment. *European Journal of Mechanics B/Fluids*, 15, 55–80.

Schuttelaars, H.M. and de Swart, H.E. (2000). Multiple morphodynamic equilibria in tidal embayments. *Journal of Geophysical Research*, 105, 105–118.

Seminara, G. (2010). Fluvial Sedimentary Patterns, *Annual Review of Fluid Mechanics*, 42, 43–66.

Seminara, G. and Tubino, M. (2001). Sand bars in tidal channels. Part 1. Free bars. *Journal Fluid Mechanics*, 440, 49–74.

Seminara, G., Lanzoni, S., Bolla Pittaluga, M. and Solari, L. (2001). Estuarine patterns: an introduction to their morphology and mechanics. In *Lecture Notes on Geomorphological Fluid Mechanics*, Springer Verlag, Berlin, 455–499.

Seminara, G., Lanzoni, S., Tambroni, N. and Toffolon, M. (2010). How long are tidal channels?. *Journal Fluid Mechanics*, 643, 479–494.

Seminara, G., Bolla Pittaluga, M., de Capitani di Vimercate, C. and Tambroni, N. (2011). *In preparation.*

Solari, L., Seminara, G., Lanzoni, S., Marani, M. and Rinaldo, A. (2001). Sand bars in tidal channels. Part 2. Tidal meanders. *Journal Fluid Mechanics*, 451, 203–238.

Tambroni, N., Bolla Pittaluga, M. and Seminara, G. (2005*a*). Laboratory observations of the morphodynamic evolution of tidal channels and tidal inlets, *Journal of Geophysical Research*, 110, F04009, doi:10.1029/2004 JF000243.

Tambroni, N., Stansby, P.K. and Seminara, G. (2005*b*). Modeling the morphodyanmics of tidal inlets, *Flooding and Environmental Challenges for Venice and its Lagoon: State of Knowledge*, Cambridge, CUP, *Ed. C.A. Fletcher and T. Spencer*, 379–389.

Todeschini, I., Toffolon, M. and Tubino, M. (2008). Long–term morphological evolution of funnelshape tide–dominated estuaries. *Journal of Geophysical Research*, 113, C05005, doi:10.1029/2007 JC004094.

Toffolon, M. and Lanzoni, S. (2010). Morphological equilibrium of short channels dissecting the tidal flats of coastal lagoons. *Journal of Geophysical Research*, 115, F04036, doi:10.1029/2010 JF001673.

Waeles, B., Le Hir, P. and Jacinto, R.S. (2004). Modelisation morphodynamique cross–shore d'un estran vaseux. *Comptes Rendus Geoscience*, 336, 1025–1033.

Wolman, M.G. and Miller, J.P. (1960). Magnitude and frequency of forces in geomorphic processes. *Journal of Geology* 68, 54–74.

Chapter 9

Flow structure and sustainability of pools in gravel-bed rivers

S.E. Parkinson[1], *P. Goodwin*[2] *and D. Caamaño*[3]
[1] *US Bureau of Reclamation, Pacific Northwest Region, Boise, Idaho USA*
[2] *Center for Ecohydraulics Research, University of Idaho, Boise, Idaho, USA*
[3] *Department of Civil Engineering, Universidad Católica de la Santísima Concepción, Concepción, Chile*

ABSTRACT: Pool-riffle channel morphology in gravel-bed streams creates a range of micro-habitats that are important for maintaining ecological diversity. This study is motivated by the need to restore sustainable spawning and over-wintering habitat for salmon as a means of mitigating for dam construction. The velocity reversal hypothesis (Gilbert, 1914, Keller 1971, 1972) has been used as a potential mechanism for the sustainability of pools. Velocity reversal implies that the velocities are smaller through the pool than across the riffle at low and intermediate flows. At high discharges this condition becomes reversed and velocities through the pool may exceed the velocities across the riffle. Recent criteria have been developed that utilize the bathymetric characteristics of the channel to ascertain whether or not velocity reversal will occur. However, it is unclear whether the velocity reversal criterion predicts pool sustainability under all flow conditions and if limitations to this approach exist. The role of a high velocity core in controlling sediment deposition and scour processes was investigated under different flow discharges in addition to conditions when velocity reversal may not recreate the original pool riffle morphology.

The flume results of this study demonstrate that although velocity reversal provides a useful indicator of the persistence of pool features, bed shear stress and sediment flux reversal do not occur at the same discharge or location as the velocity reversal. The study also demonstrated the differential transport capacity through the pool and riffle as well as conditions when these features are lost beyond self-recovery.

1 INTRODUCTION

Gravel-bed rivers are an important habitat for salmonids, which are at risk throughout the Northwest region of North America and are the focus of extensive environmental legislation and litigation. The pool-riffle morphology in gravel-bed rivers is particularly important as the diversity in physical habitat in these reaches is critical for spawning and other life stages (Stanford *et al.*, 2005, Tilman, 1998). The river cross sectional shapes, at any location along the watercourse, are a function of the flow, the quantity and character of the sediment in movement through each section, and the character or composition of the materials making up the bed and banks of the channel (Leopold *et al.*, 1964; Knighton, 1998; Federal Interagency Stream

Corridor Restoration Working Group, 1998). Pool-riffle sequences are one of the possible river reach morphologies created by these interactions. Many restoration projects attempt to restore pool morphology in a sustainable and minimum maintenance manner.

A number of hypotheses have been presented to explain the formation and self maintenance of pool riffle sequences. The reversal hypothesis has been evaluated with field measurements, laboratory investigations and numerical simulations. The reversal hypothesis defines the condition where a high discharge event creates a higher velocity condition within the pool feature than across the riffle. Reversals in near-bed velocity and shear stress have been more commonly reported. Average cross sectional velocity reversals have been demonstrated to occur under limited conditions (Booker 2001, Thompson 2011). These reversals require that a more rapid increase in cross sectional area with increasing discharge of the riffle relative to the pool has to occur. However, this one dimensional simplification does not fully describe the three dimensional physical processes occurring between the pool and riffle. Recent findings have shown that constriction of flow through the pool, either caused by point bar geometry or effective area reduction due to a recirculating eddy, creates a narrow core of higher velocity water.

This flow convergence was suggested by MacWilliams *et al.* (2006) as the pool formation mechanism in both free-formed and forced pools. Caamaño *et al.* (2010) further defined this jet structure as comprising velocities greater than 90% of the maximum depth-averaged velocity observed within a cross section. Numerical simulations of this flow structure also illustrated a change in orientation as well as a change in intensity and location of vertical and horizontal eddies. These non-uniform flow effects created by this high velocity core structure have the capacity to influence the local transport and sediment routing dynamics through the pool.

Self maintenance of pool-riffle sequences are dependent on the differential routing and scouring of sediment. In order to assess the response of a pool to sediment pulses during high flows or artificially high sediment loads created by landuse practices or wildfire, it is necessary to measure the differential sediment transport between the pool and riffle. However, this is very difficult and hazardous to do in the field at the high flow conditions when the geomorphically significant sediment transport is occurring. Therefore, to investigate the sediment transport characteristics of a pool-riffle sequence, a physical model was constructed.

This model facilitated the visualization of physical processes and detailed velocity structure observed in field data and in numerical simulations. This model was built to represent a pool-riffle sequence located on the Red River within the Red River Wildlife Management Area. The purpose of utilizing this specific reach was to validate the results observed in the three-dimensional numerical modeling as well as qualifying the conceptual model processes proposed in the self-maintenance of pool-riffle sequences (Caamaño *et al.*, 2010).

2 STUDY SITE

The headwaters of the Red River originate within the Clearwater Mountain Range in north-central Idaho (lat. 458 45', long. 1158 24'), USA. The Red River joins the

American River to become the South Fork Clearwater River. This particular study reach is a meandering gravel-bed channel that flows through an unconfined 4.5 km long meadow situated at an elevation of 1,280 m and bounded by forested mountains (Klein *et al.,* 2007) within the Red River Wildlife Management Area (RRWMA) (Figure 1).

The study site illustrated in Figure 2 is the Lower Red River Meadow where restoration activities were conducted in several reaches starting in 1993. The pool-riffle sequence selected for this study was free formed (not forced by external flow obstructions such as logs or boulders; Montgomery *et al.*, 1995) during the first high flow event immediately after the final restoration phase of the RRWMA in 2000. For approximately 10 years, the geometry of these features has remained relatively consistent since project completion. This specific pool-riffle sequence is situated downstream of a straight, 50 m long run and curves into a double pool-riffle sequence with bankfull discharge at 16.62 m^3/s and an average bank top width of 11.25 m, before opening to a wider stream section. The upstream run has a symmetrical, almost rectangular cross section that allows a uniform flow distribution in the stream before entering the first pool. The first pool-riffle sequence has a deep residual pool depth followed by a coarse riffle, whereas the second pool is shallower, with a finer-grained riffle (Figure 2 and Table 1). Figure 3 illustrates the first riffle feature at both low flow and high flow conditions.

Figure 1 Location of Red River Wildlife Management Area, north central Idaho.

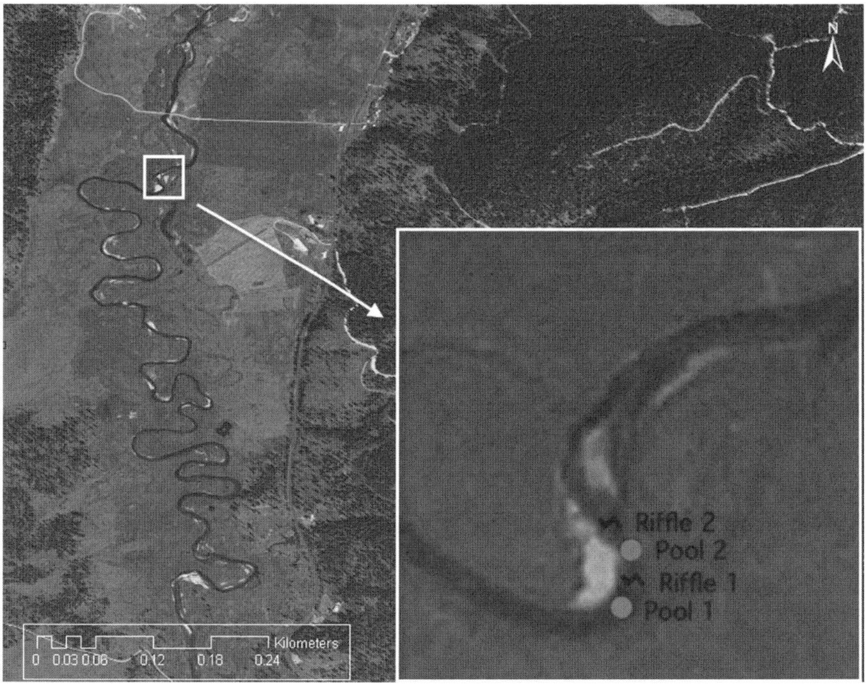

Figure 2 Location of the study reach (inset) relative to the RRWMA restoration efforts.

Table 1 Grain size distribution, modified from Caamaño *et al.* (2010).

Bedform	*Pool 1*	*Riffle 1*	*Pool 2*	*Riffle 2*
d_{16}	12 mm	38 mm	19 mm	26 mm
d_{50}	42 mm	65 mm	46 mm	47 mm
d_{84}	74 mm	94 mm	82 mm	80 mm
d_{90}	82 mm	105 mm	90 mm	85 mm

Figure 3 Riffle 1 at low flow and high flow conditions, pool 2 and riffle 2 immediately downstream, adapted from Caamaño *et al.* (2009).

This specific site was selected for the detailed field data and numerical model output available for this reach. The site provides an opportunity to compare the numerically modeled flow structure (Caamano, 2008) with the laboratory observations, in addition to measuring the sediment transport behavior in the laboratory. The numerical model was calibrated against extensive field data, although there are no field observations for discharges above 80% of the bankfull discharge. Since most of the morphological changes occur at discharges at or greater than bankfull, the numerical simulations were used to scale the physical model.

3 METHODOLOGY

3.1 Model scale computations

A Froude scaled physical model of the pool riffle sequence was constructed. A surface was created in ArcMap of the reach survey points. This surface was queried for bathymetric elevations every two meters for physical model construction. Figure 4 illustrates a 30 m by 60 m boundary drawn around the first pool-riffle sequence. The geometric length ratio (L_R) between the model and prototype is:

$$L_R = \frac{L_P}{L_M} = \frac{30}{2} = 15 \qquad (1)$$

L_P = prototype or field length dimension (m)
L_M = model length dimension (m)
L_R = geometric length ratio.

The pool and riffle cross sections are shown in Figure 5. The locations of these cross sections capture both the deepest point within the pool and across the riffle.

In order for a physical model to be representative of the processes being evaluated, similarity between the significant model and prototype processes (or field conditions) must be retained, although this is problematic for the sediment transport processes. The computations and assumptions that defined the representation of the modeled pool riffle sequence in the flume are summarized in Table 2. Turbulent flows were maintained in the model at all simulated discharges. The initiation of motion and sediment transport were analyzed using several approaches but the Shields criteria and the Meyer-Peter Mueller (MPM) sediment transport equations, shown in Table 2, were used to scale the particle sizes for the bathymetry used in this experiment. A lightweight artificial material was used in the experiments to represent bedload transport. Scaling the natural bedload sediment transport with lightweight artificial material resulted in larger particles that are not subject to cohesion and can be tracked visually. Nylon material was used, with a specific gravity of 1.15. Transport and initiation of motion were scaled on the basis of the average shear stress for the pool cross section which was estimated from the numerical model. Once the particle size was estimated, based on Shield's parameter for initiation of motion in the pool, two additional particles sizes were used to represent a larger and smaller diameter particle.

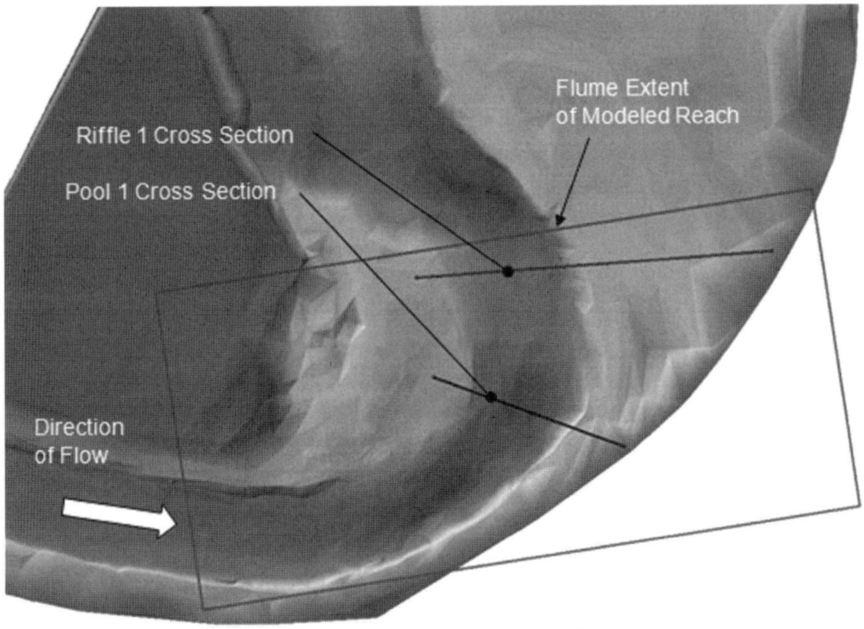

Figure 4 Generated surface using survey data with pool-riffle features identified within the modeled reach extent.

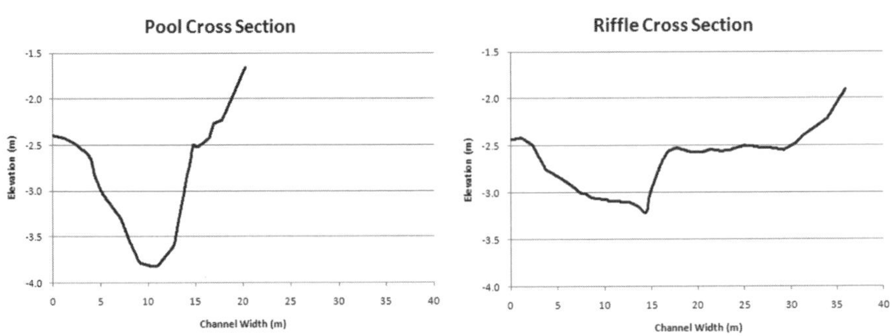

Figure 5 Plotted pool and riffle cross sections.

Table 3 presents the physical dimensions of the prototype pool riffle sequence and the modeled dimensions in the flume based on the geometric scaling ratio. The bank-full dimensions were based on numerical model results.

3.2 Model construction

Model construction of the pool riffle bathymetry utilized natural sediments ranging in size from 3 mm to 8 mm and represented the scaled d_{50} to d_{90} material sampled in the pool. Observations during the past decade have shown that the larger sediment on the armored bed and bars move infrequently. Field measurements of bedload material

Table 2 Scaling parameters and ratios.

Criterion	*Equation*		*Scaled ratio*
Froude number	$Fr = \frac{u}{\sqrt{gh}}$		$Fr_R = Fr_p/Fr_m = 1.0$
Reynolds number	$Re = \frac{uh}{v}$	(733,000 prototype) (15,500 model)	$Re_R = Re_p/Re_m = 47.3$
Shear stress	$\tau_b = \rho ghS$		$\tau_{bR} = \tau_{bp}/\tau_{bm} = 15.0$
Initiation of motion (Shields parameter)	$\theta_{cr} = 0.055$ (prototype material) $\theta_{cr} = 0.013D_*^{0.29}$ (model material)		$\Theta_{crR} = \Theta_{crp}/\Theta_{crm} = 1.0$
Dimensionless particle parameter	$D_* = \left[\frac{(s-1)g}{v^2}\right]^{1/3} d_m$		$D_{*R} = D_{*p}/D_{*m} = 2.6$
Dimensionless bed-load transport rate (MPM)	$\Phi_b = 8(\mu\theta - 0.047)^{1.5}$		$\Phi_{bR} = \Phi_{bp}/\Phi_{bm} = 1.0$
Volumetric bed-load transport rate (MPM)	$q_b = \left(\sqrt{(s-1)gd_m^{1.5}}\right)\Phi_b$		$q_{bR} = q_{bp}/q_{bm} = 5.1$
Specific gravity	$S_s = \frac{\rho_s}{\rho_w}$		$s_{s,R} = s_{s,p}/s_{s,m} = 2.3$

Table 3 Prototype and model dimensions at bankfull discharge.

	Calculated physical parameters	
	Prototype (Field)	Model (Flume)
Geometric scaling	30 m width	2 m width
pool		
Maximum depth	1.64 m	0.11 m
Average cross sectional velocity	1.35 m/s	0.35 m/s
riffle		
Maximum depth	0.81 m	0.05 m
Average cross sectional velocity	1.29 m/s	0.33 m/s
sediment		
Pool $d_{transported}$ (computed size for movement)	17 mm 13 mm 8 mm	13 mm 10 mm 6 mm

transport taken with a Helley-Smith sampler indicate that the bed load transport material is considerably finer than the material found on the bed.

Due to the expected sensitivity of the results to bathymetry, particular care was taken in the accurate construction of the model. A grid frame was constructed within the flume representing 2 m spacing in the field and each intersection representing an elevation data point extracted from the ArcMap generated surface (Figure 6).

Figure 6 Flume construction of pool-riffle sequence and final configuration and location of cross sections.

Metal stakes were scored at 1 cm intervals to measure gravel depth and placed in a single row at each grid intersection. Once shaping and contouring of the bathymetry to the correct depths was completed, the grid points were painted and the stakes moved to the next row. Gravel was smoothed between each grid point and row to maintain a smooth transition between elevations. Once the entire grid was shaped, the elevations were verified prior to the model runs using an acoustic depth profiler and adjusted as necessary. The approximate bankfull location of the water surface is painted on the model channel for documentation and reference purposes.

3.3 Model scenarios

3.3.1 Objectives

Numerical modeling of the pool-riffle sequence demonstrated a velocity reversal at bankfull conditions. At this discharge rate, the cross sectional velocity across the pool was greater than the riffle. The objectives of the physical modeling are to understand the spatial distribution of sediment flux reversal and the patterns of sediment deposition and erosion at a range of river discharges. These observations will then be used to assess the sustainability conceptual model postulated in Caamaño *et al.*, 2010.

The scenarios in the flume were developed with the following objectives:

1 Characterize the high velocity core formation at bankfull flows and greater, and validate the numerical modeling observations
2 Quantify when shear stress reversal occurs relative to velocity reversal

3 Quantify the point when sediment flux reversal occurs relative to velocity reversal
4 Characterize the formation of the high velocity core flow structure within the pool and how this changes when the pool begins to fill with sediment
5 Observe self-sustaining mechanisms within the pool-riffle sequence.

Field observations and the predictions of numerical models indicate that significant bedload transport does not occur until the bankfull discharge is exceeded. The discharges used in the model are:

1 110-percent bankfull discharge ($1.1Q_{BKF}$)
2 120-percent bankfull discharge ($1.2Q_{BKF}$)
3 135-percent bankfull discharge ($1.35Q_{BKF}$)
4 145-percent bankfull discharge ($1.45Q_{BKF}$).

3.3.2 High velocity core visualization

A dye tracer was employed to record the flow characteristics within the pool. Fluorescein dye was used with black light to enhance the visual contrast of the dye with the background. The grid points and bankfull line were visible under the fluorescent lights and provided dye release reference points for comparison between discharge scenarios. The dye injections were recorded to assess the presence and location of the flow features near and within the pool. The flow structure obtained from these images is compared qualitatively to the numerical model. The dye releases were made from an assembly consisting of a metal rod, syringe, and surgical tubing. It was configured such that the dye was released horizontally, into the flow stream, without interfering with the flow structure. Dye releases were made by hand to mimic local velocity conditions.

3.3.3 Artificial sediment releases

The artificial sediment was sized to be mobile at bankfull conditions and be representative of bedload sediment transport observed at the site. These sediments are either placed in the run upstream of the pool to observe the transport of bedload sediments into and through the pool, or the sediments were placed in the pool to observe the recovery or non-recovery of the pool morphology. Sediment movement was recorded as the flume was brought up to steady state discharge conditions and maintained for a defined period of time.

4 RESULTS

4.1 Velocity characteristics

Dye tracer evaluation of the four discharge scenarios was performed. For presentation purposes, conditions representative of $1.1Q_{BKF}$ and $1.45Q_{BKF}$ are shown to demonstrate the extremes observed. Conditions observed at $1.2Q_{BKF}$ and $1.35Q_{BKF}$ followed the same trend between the discharges and, therefore, draw similar conclusions.

Figure 7 illustrates dye released at the pool's deepest location for two discharges. The left photograph depicts $1.1Q_{BKF}$ while the photo on the right depicts $1.45Q_{BKF}$ conditions. The near bed flow direction is directly across the point bar at the $1.1Q_{BKF}$ discharge. Even though dye is released in the pool thalweg, it is diverted over the point bar to the left. A change in direction of the near-bed high velocity core flow was observed as the modeled flows were progressively increased. At the highest modeled discharge, $1.45Q_{BKF}$, the flow direction is maintained along the line of the pool thalweg and continues over the pool tailout toward the riffle.

The visualizations demonstrate consistent flow patterns and behaviors observed in numerical model output or field measurements (Booker *et al.*, 2001, Caamaño *et al.*, 2010). In these discussions, the term "jet" is used loosely to define the high velocity core structure generated by the physical attributes of the pool. Caamaño *et al.* (2010) defined it as velocities greater than 90% of the maximum depth-averaged velocity observed within a cross section. Caamaño *et al.* (2010) described the Coanda effect, the lateral pressure gradient across the pool, forcing the jet across the point bar, inside the bend (left photo, Figure 7) which was observed in the numerical simulations and at the field site. At higher flows, the high velocity core or jet migrates to align through the pool thalweg and toward the outer bank (right image).

Dye releases were made to identify and assess the eddy on the outside of the bend and high velocity core at different discharge conditions, features that had been observed from field measurements and numerical simulations (Figure 8). Thompson *et al.* (1996) stated that these recirculating eddies, or wake zones, reduce the effective cross-sectional area of the pool, thus contributing to the area of higher velocity in the pool.

Figure 9 illustrates the influence of the high velocity core on the eddy feature in these tracer studies. The images on the left are at $1.1Q_{BKF}$ and the images on the right are $1.45Q_{BKF}$. The deepest part of the pool is marked with an "X" for reference. In the top images, dye is released upstream of the pool, near the bottom of the channel. The direction of the jet, as seen in Figure 9, migrates toward the outer bank as discharges increase. Superimposed is an arrow indicating the location of the eddy for comparison with the lower photographs. The lower photographs show the dye release in the eddy with the high velocity core or 'jet' direction superimposed. As discharges increased,

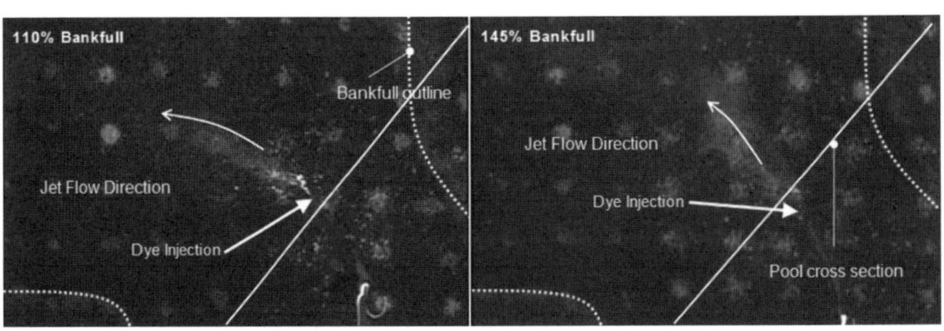

Figure 7 Dye tracer visualization of flow direction at 1.1 times and 1.45 times bankfull conditions, respectively.

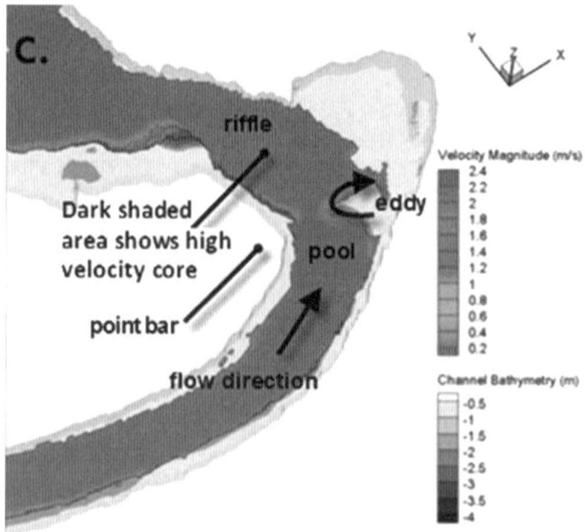

Figure 8 Numerical simulation of the high velocity core flow location of eddy feature (adapted from Caamaño, 2008).

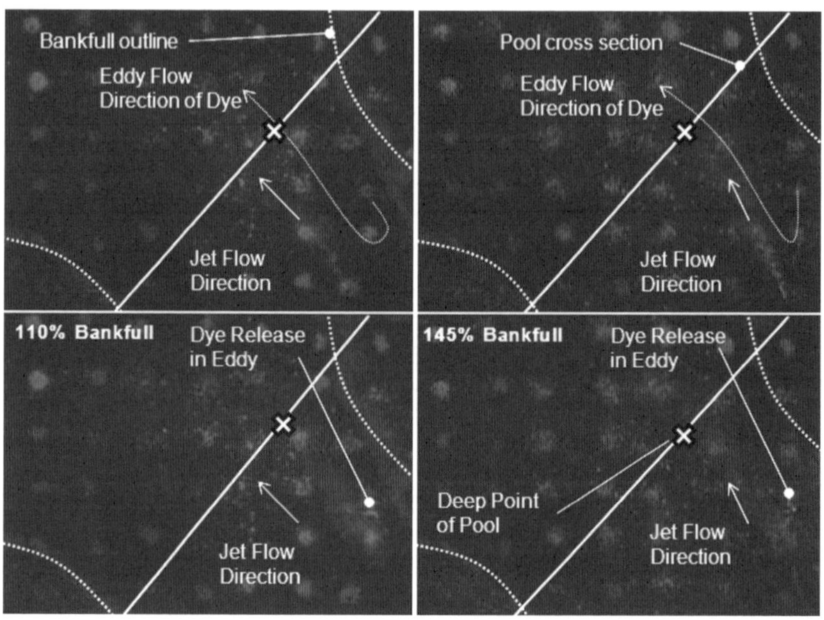

Figure 9 Stream bed dye releases upstream of the pool and within the eddy feature at $1.1Q_{BKF}$ and $1.45Q_{BKF}$.

the eddy feature remained present but became narrower in width and moved farther upstream. The persistence of the eddy feature at these high flows prevents the jet from impinging directly against the outer bank, thus reducing the opportunity for bank erosion.

The results of the tracer studies provide a qualitative validation of the field observations and numerical simulations for this pool-riffle sequence.

4.2 Shear stress reversal

In these experiments, the shear stress was evaluated through the initiation of motion of artificial sediment particles. The particle motion was observed for each series of constant discharges over 30 minutes. This period was selected based on scaling the duration of typical peak flood flows at the RRWMA. The $1.2Q_{BKF}$ discharge was insufficient to scour the sediments in the deepest part of the pool, although a few particles did move onto the riffle. However at $1.45Q_{BKF}$, most of the deposited sediment was scoured from the bottom of the pool and was either deposited on the riffle or was transported out of the study reach. This implies that for this pool-riffle sequence, the shear stress reversal occurs between 1.2 and $1.45Q_{BKF}$. The dye studies indicate that this was also the flow range where the high velocity core realigned from across the point bar to follow the thalweg. The scour of sediments from the deepest part of the pool and subsequent deposition onto the downstream riffle demonstrate the process that sustains the pool and builds the riffle crest (Figure 10). The black dotted line shows the deepest part of the pool and the initial location of the sediments and the solid black line highlights the area of particle deposition.

Table 4 summarizes the fate of the marked particles in the deepest part of the pool for each discharge.

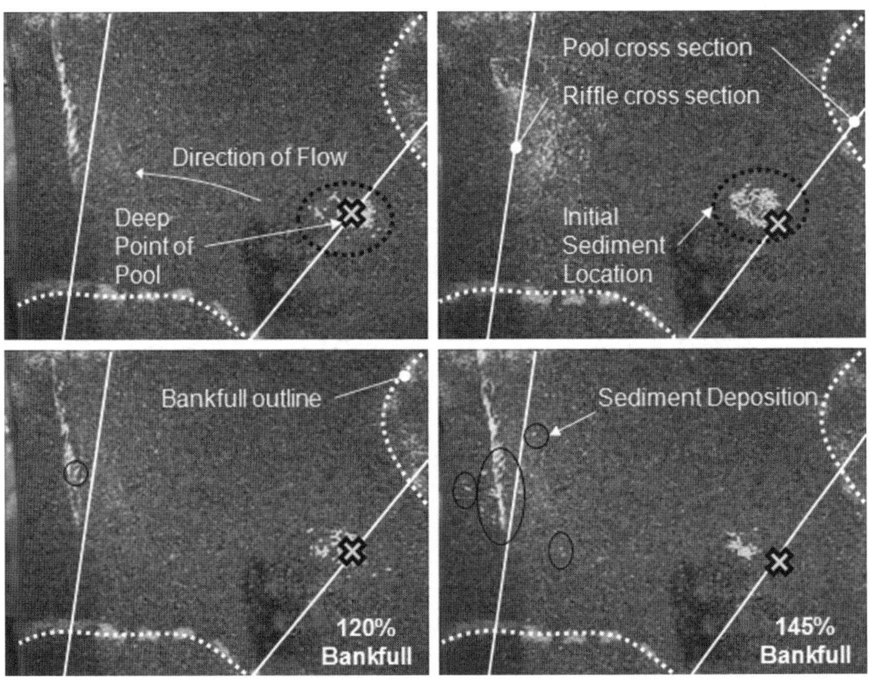

Figure 10 Sediment scour from pool at $1.2Q_{BKF}$ and $1.45Q_{BKF}$.

Table 4 Scour and deposition of particles from the pool thalweg.

	Shear stress results		
	Deposition location		
Discharge rate	Pool or pool tail out	Riffle	Particles moving out of reach
$1.1Q_{BKF}$	0%	0%	0%
$1.2Q_{BKF}$	99%	1%	0%
$1.35Q_{BKF}$	58%	33%	9%
$1.45Q_{BKF}$	40%	26%	34%

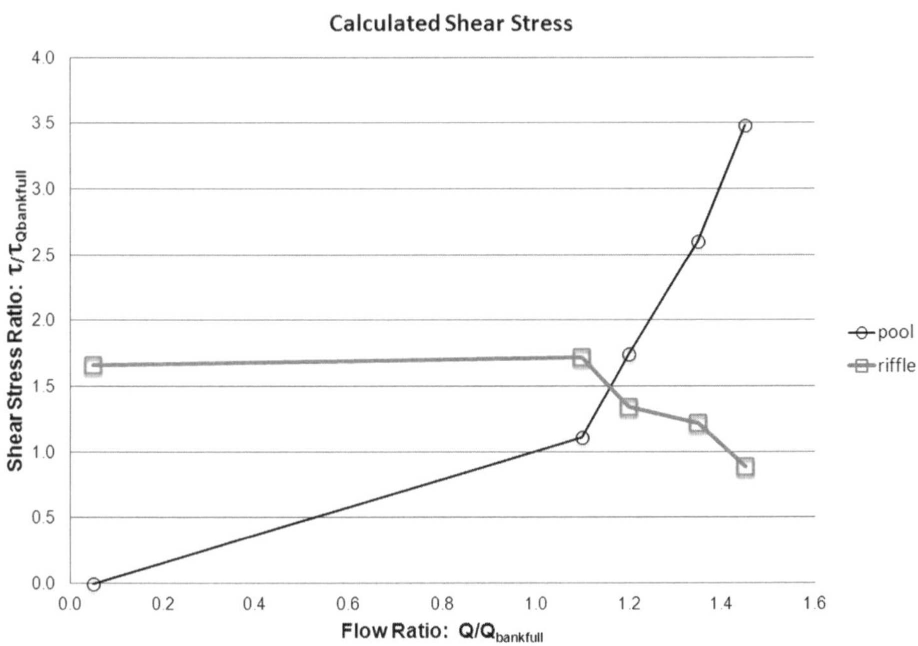

Figure 11 Estimate of normalized average shear stress reversal from the physical model.

The average shear stress through the pool and riffle were estimated from the water surface elevation slope observed in the physical model and the results are normalized with the bed shear stress calculated at $Q_{1.1BKF}$ (Figure 11). The small scale of the model resulted in some inaccuracy in the estimation of water surface slope, but the trends and relative differences between discharges are consistent. The estimates of low flow shear stress were made using field measurements.

The trend toward shear stress reversal was also evident in the numerical simulations (Caamaño, 2008). The shear stress was computed utilizing both the quadratic stress law and the logarithmic velocity relationship. Both methodologies indicated a reduction of the riffle shear stress between the medium flow and bankfull flow modeled. That trend also indicated that reversal should occur at discharges greater than bankfull. The decrease in riffle shear stress as the discharge increases could be attributable to downstream control affecting this upstream pool-riffle sequence (De Almeida & Rodriguez, 2011, Caamaño, 2008). This indicates that the occurrence of shear stress reversal could be influenced by both downstream control and the local bathymetry.

4.3 Sediment flux reversal

These scenarios assess sediment flux reversal by quantifying the relative difference between the pool and riffle bed load transport. The artificial sediment particles were aligned across the run, just upstream of the pool and simulate the fate of bed-load moving through the reach at high discharges (Figure 12). Flow was gradually increased until the target discharge was achieved and then held constant for 30 minutes. For all of the modeled discharges, none of the mobilized particles deposited within the pool. The sediments were transported through the pool and either deposited on the riffle, point bar, or continued out of the study reach (Table 5). Another observation was that as the modeled discharges increased, deposition on the point bar occurred higher up the bank. In addition, more sediment was deposited on the

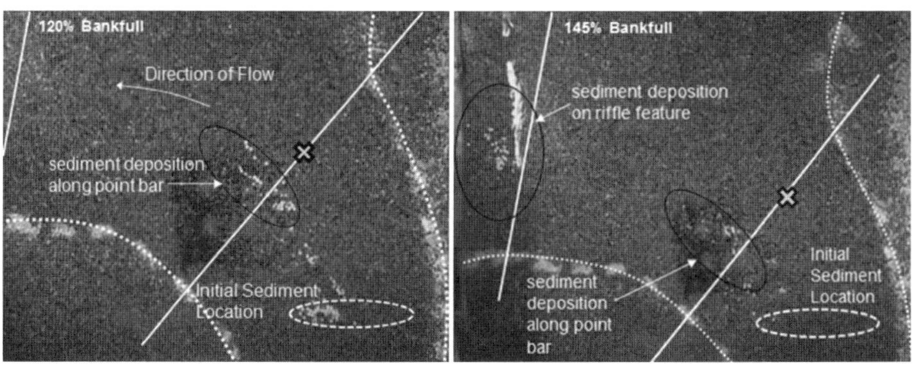

Figure 12 Transport of artificial sediment at $1.2Q_{BKF}$ and $1.45Q_{BKF}$ discharges.

Table 5 The deposition of incoming particles in the study reach at high flows.

	Sediment flux results		
	Deposition location		
Discharge rate	*Point bar*	*Riffle*	*Out of system*
$1.1Q_{BKF}$	0%	0%	0%
$1.2Q_{BKF}$	94%	4%	2%
$1.35Q_{BKF}$	85%	11%	4%
$1.45Q_{BKF}$	55%	29%	17%

riffle than carried downstream. These results demonstrate that at discharges higher than bankfull, sediment flux, or bedload transport, through the pool is greater than across the riffle. Mobilization of the bedload sediments did not occur until $1.2Q_{BKF}$, therefore, sediment flux reversal may occur near bank full for smaller sediments than those represented in these experiments.

These sets of experiments provided validation of the sediment flux reversal occurring for this pool-riffle sequence. The numerical model results of this sequence also indicated that near bankfull, reversal would occur although there is considerable uncertainty in the sediment transport formula used for simulating localized differential transport (Caamano, 2008). These results also show the incoming sediment particles are building the point bar and riffle, not filling the pool.

4.4 Pool infill

The preceding scenarios described the scour and deposition trends of sediments for the current channel morphology. Although this morphology has been stable for the past decade, it is possible that under a different sequence of annual hydrographs that sediment could have accumulated within the pool through either the lack of large flows capable of scouring the pool or an overload of sediment to the system due to changing landuse practices or the consequences of wildfire in the basin. These scenarios evaluate the ability of the pool to recover following the accumulation of significant sediment in the pool. The sedimentation in the pool assumed that the sediment delivery to the study reach had exceeded the bedload transport capacity of the pool for some period of time. The visualizations of the flow characteristics were made to assess the potential recovery mechanisms. The pool feature was in-filled to approximately 90-percent of the residual pool depth to compare the numerical simulations of the same condition (Caamaño *et al.*, 2010).

4.4.1 *Flow characteristics*

The dye tracer technique was used to compare pool features under in-filled and original conditions. Dye was released across the surface of the pool feature from river left to right at a flow condition of $1.1Q_{BKF}$. For existing conditions, the eddy on the outer bank is very pronounced (Figure 13, left image). When the pool is filled with sediments, the eddy is no longer present. Without the eddy feature, the dye directly impacts the outer bank. This validates the numerical model output observations under the similar discharge conditions.

When dye was released upstream of the pool, the distinct high velocity core is more diffuse in the infilled pool (Figure 14, right image). Even with remnants of the point bar in place, there is not enough constriction of the flow to create the high velocity core or any large eddies.

4.4.2 *Pool recovery*

Artificial sediment particles were placed on top of the pool infill area at the former deepest part of the pool to assess their mobility and potential for pool recovery by scour. Very little movement of the particles occurred and the few particles that did

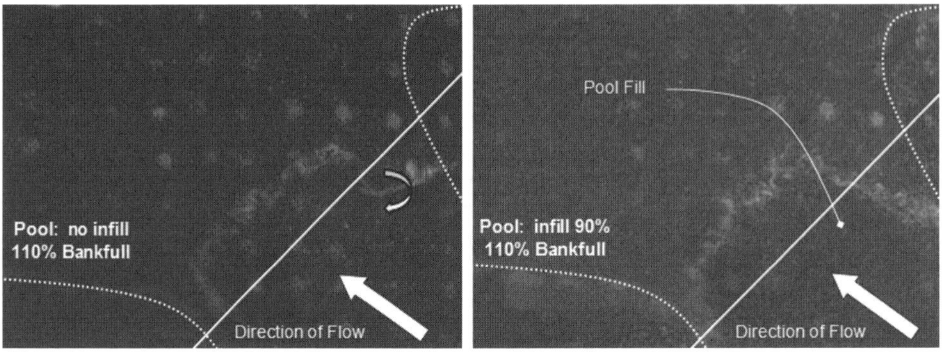

Figure 13 Dye tracer comparison between original pool structure and infilled condition.

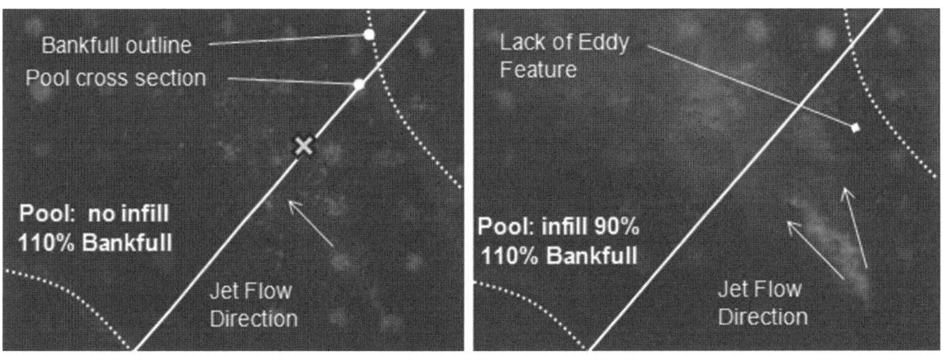

Figure 14 Dye tracer comparison of jet formation between original pool structure and infilled condition.

move were transported downstream and were not deposited preferentially on the riffle or point bar. Figure 15 illustrates the location of the sediments after 15 minutes at $1.45Q_{BKF}$. Similar to the numerical model results, at bankfull conditions, shear stress reversal was not observed and this condition potentially indicates a situation where pool recovery may not be possible.

However, if bedload transport occurs, there may be an alternative mechanism for recovery. Sediments were placed across the run, just upstream of the in-filled pool section. The discharge of $1.45Q_{BKF}$ was held constant for 30 minutes. The sediments were deposited toward the upstream end of the point bar and began to elongate the bar (Figure 16). Under these circumstances, if the point bar develops at a faster rate than erosion of the outer bank, the semblance of a jet feature may gradually redevelop. Eventually the point bar will constrain the flows and possibly scour the sediments that had filled the pool or will create a constriction capable of forming a new pool. The key factor would be whether the outer bank would erode faster or keep pace with the build out of the point bar. If the bank erodes such that the reduction in pool width compared to the riffle cannot be generated, then the pool will be unable to recover.

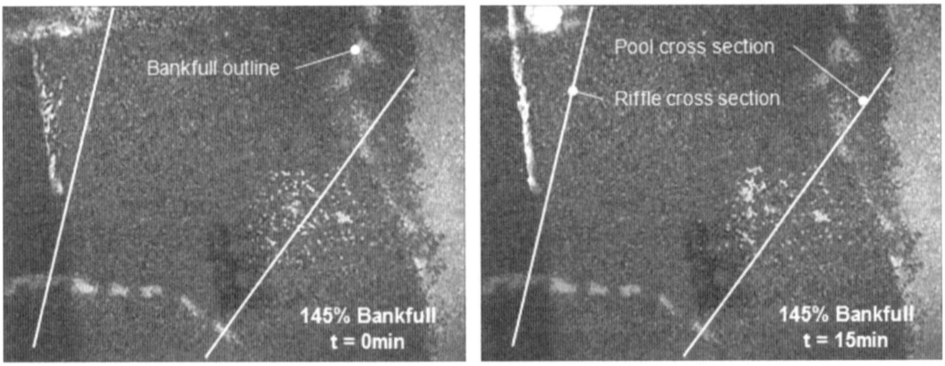

Figure 15 Artificial sediment movement on top of pool in-fill section at $1.45Q_{BKF}$.

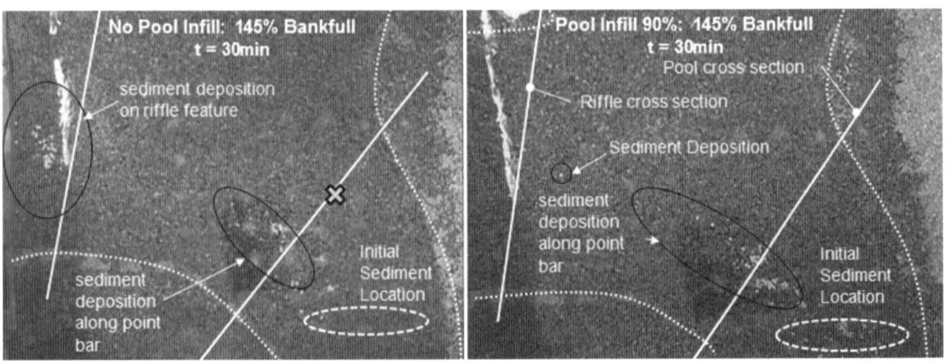

Figure 16 Comparison of sediment flux between the original pool configuration and the infilled pool.

5 CONCLUSION

5.1 Flume model

Dye tracer analysis qualitatively validated the general behavior of the high velocity core feature predicted by numerical studies. As discharges increased above bankfull, the jet direction migrated from the inner bank, across the point bar towards the outer bank. The existence of an eddy feature on the outer bank was present at flows greater than bankfull. The eddy prevented direct impact of the high velocity core against the outer bank. As flows increased above bankfull, the jet orientation begins to align through the pool thalweg and the sediments begin to scour out of the pool. These sediments deposit in the pool tail out, the riffle, or carried further downstream. This phenomenon was also witnessed in a forced pool configuration (MacVicar & Roy, 2010) where flow acceleration due to the constriction mobilized the sediments in the center of the pool and over the exit slope.

The artificial sediments representing the coarse bedload placed upstream of the pool were not mobile until roughly $1.2Q_{BKF}$. When mobilized, none of the sediments deposited within the pool, most likely due to entrainment in the jet turbulence (MacVicar & Roy, 2007) and carried around the deep part of the pool. Some of the sediments were carried out of the study reach while some of the entrained sediment deposited directly on the downstream riffle. Deposition also occurred along the point bar as sediment was routed over the side bar near the pool entrance. MacVicar & Roy (2010) also documented this sediment deposition characteristic at their forced pool-riffle study site. As the flows increased, deposition of the sediments occurred higher up on the bar. It is also interesting to note that deposition percentage on the riffle was greater than the percentage of sediments carried out of the study reach. Based on these observations, it can be surmised that sediment flux reversal occurs at flows greater than bankfull conditions and the transport characteristics observed are the primary processes for sustaining the pool morphology.

The pool feature was filled with pea-gravel bed material to approximate the numerically modeled 90-percent infill condition (Caamaño *et al.*, 2010). The dye tracer analysis validated the numerical model output and demonstrated that the eddy feature was no longer present to protect the outer bank. In addition, the lack of a defined high velocity core or jet was also apparent in the visualizations. Artificial sediment deposited across the in-filled pool did not mobilize at any of the modeled discharges. This indicates that the high velocity core structure plays a role in pool maintenance. The scenario simulating the fate of bedload material being transported into the study reach when the pool is filled indicated the possibility of elongating the point bar. If the building of the point bar occurs at a faster rate than outer bank erosion, it is possible that the constriction necessary to create the pool riffle morphology could develop over time, but this process would be slower than scouring the pool fill material.

5.2 Conceptual model

A conceptual model for self-sustaining pool-riffle features was developed in Caamaño *et al.* (2010) utilizing field data and other observations (Dietrich & Smith, 1984, MacWilliams *et al.*, 2006, Pasternack *et al.*, 2008). The preliminary results from this physical modeling exercise further clarify the various principles postulated in the conceptual model.

1 The physical model demonstrated that near bankfull conditions, a sediment flux reversal occurred, preserving the pool feature and depositing material along the point bar or riffle. This assumption is predicated on the fact that the sediment flux transport of the pool exceeds that of the sediment delivery to the study reach.
2 If the pool was not aggraded to a point where a jet structure was no longer present, local transport capacity was exceeded and the pool could scour. Higher than bankfull flows were necessary to redirect the high velocity core from the point bar to the pool thalweg. At all the modeled discharges, the eddy feature was present and provided a level of protection to the outer bank.
3 A scenario modeled in the flume considered the altered flow structure and scouring ability if sediment accumulated in the pool and reduced the residual pool depth to

10% of the existing bathymetry. In this condition the well-defined high velocity core was no longer present. The eddy feature was also gone and the high velocities directly impacted the outer bank.

4 Artificial sediments placed in the in-filled pool were not mobilized under any of the modeled discharges. It was also shown that sediment transported into the study reach is routed across the side bar upstream of the pool. While the sediments were not entrained and carried to the riffle and deposited, they did create an elongated point bar. Therefore, if a flow constriction could begin to form near the top of the pool, and if the point bar formed faster than the rate of outer bank erosion, then a jet could gradually develop. If this occurred, it is possible that the pool feature could re-form.

ACKNOWLEDGMENTS

The original concept of the study was developed by Dr. Gerhard Jirka and the authors with the intent of supporting collaborative research between the University of Idaho and Karlsruhe Institute of Technology through exchange visits. These preliminary results of the physical model study are presented as a tribute to Dr. Jirka and his commitment to exploring innovative applications of fluid mechanics to environmental issues, for supporting young professionals and fostering international collaboration. The authors would like to acknowledge the support of Reclamation's Pacific Northwest Region through Reclamation's Science and Technology Program (Grant ID 4362) to investigate pool-riffle sustainability mechanisms. Reclamation's Columbia Snake Salmon Recovery Office (CSRO) receives federal monies to improve and restore salmonid habitat impacted by the construction of federal dams. These mitigation efforts include the design of specific habitat features for several lifestages of salmonids in various streams or rivers identified throughout the Northwest. This publication was made possible by the NSF Idaho EPSCoR Program and by the National Science Foundation under award numbers EPS-0814387 and BES-9874754) as well as the Chilean Fondecyt Project N° 11100399.

REFERENCES

Booker, D.J., Sear, D.A. and Payne, A.J. (2001). Modeling three-dimensional flow structures and patterns of boundary shear stress in a natural pool-riffle sequence. *Earth Surface Processes and Landforms*, 26, 553–576.

Caamaño, D., Goodwin, P. and Buffington, J.M. (2010). Flow structure through pool-riffle sequences and a conceptual model for their sustainability in gravel-bed rivers. *River Res. Applic.* [Online] Available from doi:10.1003/rra. 1463 [Accessed May 2011].

Caamaño, D., Goodwin, P., Buffington, J.M., Liou, J.C. and Daley-Laursen, S. (2009). A unifying criterion for velocity reversal hypothesis in gravel-bed rivers. *Journal of Hydraulic Engineering*. Asce. 135 (1), 66–70.

Caamaño, D. (2008). The velocity reversal hypothesis and implications to the sustainability of pool-riffle bed morphology (Doctoral dissertation). Boise, Idaho, University of Idaho.

De Almeida, G.A.M. and Rodriguez, J.F. (2011). Understanding pool-riffle dynamics through continuous morphological simulations, *Water Resour. Res.*, 47. W01502.

Dietrich, W.E. and Smith, J.D. (1984). Bed Load Transport in a River Meander, *Water Resour. Res.*, 20 (10), 1355–1380.

Federal Interagency Stream Corridor Restoration Working Group, (1998). Stream Corridor Restoration: Principles, Processes, and Practices. By the Federal Interagency Stream Restoration Working Group (FISRWG)(15 Federal agencies of the US government). GPO Item No. 0120-A; SuDocs No. A 57.6/2:EN 3/PT.653. ISBN-0-934213-59-3. http://www.nrcs.usda.gov/technical/stream_restoration/

Garcia, M.H. (2008). "Sediment Transport and Morphodynamics". *Sedimentation Engineering: Processes, Measurements, Modeling, and Practice.* No. 110. ASCE. New York.

Gilbert, G.K. (1914). "The transportation of debris by running water." *US Geological Survey Professional Paper* 86, Washington, DC.

Keller, E.A. and Florsheim, J.L. (1993). "Velocity-reversal hypothesis: A model approach." *Earth Surface Processes and Landforms*, 18, 733–740.

Klein, L.R., Clayton, S.R., Alldredge, J.R. and Goodwin, P. (2007). Long-term monitoring and evaluation of the Lower Red River Meadow Restoration Project, Idaho, U.S.A. *Restoration Ecology*, 15, 2, 223–229.

Knighton, D. (1998). Fluvial Forms & Processes: A new perspective, Arnold, London.

Leopold, L.B., Wolman, M.G. and Miller, J.P. (1964). *Fluvial Processes in Geomorphology*, Freeman, San Francisco.

Lisle, T.E. (1982). "Effects of aggradation and degradation on riffle-pool morphology in natural gravel channels, Northwestern California." *Water Resources Research*, 18 (6), 1643–1651.

MacVicar, B.J. and Roy, A.G. (2007). Hydrodynamics of a forced riffle pool in a gravel bed river: 1. Mean velocity and turbulence intensity. *Water Resour. Res., 43.* W12401.

MacVicar, B.J. and Roy, A.G. (2007). Hydrodynamics of a forced riffle pool in a gravel bed river: 2. Scale and structure of coherent turbulent events. *Water Resour. Res., 43.* W12402.

MacWilliams, M.L., Jr., Wheaton, J.M., Pasternack, G.B., Street, R.L. and Kitanidis, P.K. (2006). Flow convergence routing hypothesis for pool-riffle maintenance in alluvial rivers, *Water Resour. Res.*, 42. W10427

Montgomery, D.R., Buffington, J.M., Smith, R. Schmidt, K. and Pess, G. (1995). Pool spacing in forest channels. *Water Resour. Res.*, 31:1097–1105.

Pasternack, G.B., Bounrisavong, M.K. and Parikh, K.K. (2008). Backwater control on riffle-pool hydraulics, fish habitat quality, and sediment transport regime in gravel-bed rivers. *Journal of Hydrology* 357: 1–2: 125–139.

Pugh, C.A. (2008). "Sediment Transport Scaling for Physical Models." *Sedimentation Engineering: Processes, Measurements, Modeling, and Practice.* No. 110. ASCE. New York, N.Y.

Richards, K.S. (1976). The morphology of riffle-pool sequences. *Earth Surface Processes and Landforms*, 1, 71–88.

Stanford, J.A., Lorang M.S. and Hauer, F.R. (2005). The Shifting Habitat Mosaic of River Ecosystems. *Verh. Internat. Verein. Limnol.*, 29. 123–136.

Thompson, D.M. (2011). The velocity-reversal hypothesis revisited, *Progress in Physical Geography,* 35: 123.

Tilman, D. (1999). The ecological consequences of changes in biodiversity: a search for general principles. *Ecology*, 80, 1455–1474.

van Rijn, L.C. (1993). *Principles of Sediment Transport in Rivers, Estuaries and Coastal Seas.* Amsterdam, Aqua Publications.

Wilkinson, S.N., Keller, R.J. and Rutherfurd, I.D. (2004). "Phase-shifts in shear stress as an explanation for the maintenance of pool-rifle sequences." *Earth Surface Processes and Landforms*, 29, 737–753.

Wohl, E.E. (2000). *Mountain Rivers*, American Geophysical Union, Water Resources Monograph, Washington, DC.

Chapter 10

Drag forces and flow resistance of flexible riparian vegetation

A. Dittrich[1], J. Aberle[1] and T. Schoneboom[1,2]
[1] Leichtweiß-Institute for Hydraulic Engineering and Water Resources, Technische Universität Braunschweig, Braunschweig, Germany
[2] INROS LACKNER AG, Bremen, Germany

ABSTRACT: Hydraulic resistance of riparian vegetation depends on the drag exerted on the plants and hence on plant specific parameters and foliage. However, until today, most studies related to the estimation of flow resistance of riparian emergent vegetation have been carried out under idealised conditions using cylindrical roughness elements, regular element spacing, etc. Although these studies contributed significantly to the current understanding of the relevant processes, there are still many open questions with regard to the influence of flexibility and foliage on drag forces, flow resistance and the flow field. This paper discusses various issues related to these questions using data from experiments carried out with flexible artificial and natural small scale vegetation elements in a laboratory flume and with natural trees in a towing tank facility. The comparison of drag forces on artificial and natural plants is used to introduce a concept for the comparison of the resistance behaviour of flexible artificial plants with their natural counterparts. The data are also used to discuss the spatial variability of the drag forces within a vegetation array and to highlight the effect of foliage on drag forces. It is shown and that foliage contributes significantly to total drag predominantly at low velocities while the wooden parts of the trees contribute more to total drag at larger velocities. The differences between the drag force - velocity relationships for flexible plants and rigid bluff bodies are evaluated in terms of the recently defined Vogel exponent. It is shown that this exponent is implicitly incorporated in an existing approach for the determination of flow resistance of emergent riparian vegetation, which is validated using the acquired data.

1 INTRODUCTION

Recent interest in river restoration, environmental flood management, and the application of bioengineering techniques has stimulated interdisciplinary research in environmental hydraulics. An important aspect in this research area is the development of sustainable river management strategies which are in accordance with both ecology and flood management. The key for the development of such strategies is the better understanding of the complex interaction between flow and vegetation as vegetation exerts a major control on the conveyance capacity of rivers but is, at the same time, also an integral part of riverine ecosystems.

Despite significant research work in recent years (see overviews in DVWK, 1991, Finnigan, 2000, Schnauder & Moggridge, 2009, Folkard, 2011) there are still many open questions related to the complex interaction between flow and vegetation from both ecological and engineering perspectives (e.g., Järvelä *et al.*, 2006, Nikora, 2010). For example, vegetation has often been modeled in a simplified way by rigid cylinders without considering peculiarities associated with natural plants such as plant structure, flexibility, and foliage. Today it is known that such a simplification is inappropriate for modeling hydraulic resistance of flexible natural vegetation (e.g., Järvelä, 2004, Folkard, 2011). Moreover, it has been shown that the hydraulic resistance of natural plants depends on the resistance associated with different scales (e.g., stem scale, leaf scale, canopy scale, depth scale, reach scale; Nikora 2010), the plants ability to streamline with the flow (i.e., biomechanical properties, Fathi-Maghadam & Kouwen, 1997), and foliage (e.g., Vogel, 1994, James *et al.*, 2008, Wilson *et al.*, 2008).

The complexity of the problem requires detailed experimental studies in which various hydraulic and plant parameters are determined. However, experiments with living vegetation in a laboratory environment (e.g., Freeman *et al.*, 2000, Stephan & Gutknecht, 2002, Järvelä, 2004, and 2006) or in the field (e.g., Sukhodolov & Sukhodolova, 2010) are sophisticated. Field studies have the advantage that plants grow in their natural environment but are characterised by heterogenous boundary conditions that are difficult to control (e.g., discharge, water level). Laboratory studies have the advantage that they can be carried out with controlled boundary conditions (e.g., discharge, water depth, uniform flow) and that sophisticated and sensitive measurement techniques can be deployed. However, studying the flow resistance of natural trees in the laboratory depends on the available flume size and corresponding experiments are therefore often carried out with parts of trees or small trees (e.g., Järvelä, 2004, Wilson *et al.*, 2008, Västilä *et al.*, 2011). Moreover, the artificial environment in hydraulic laboratories affects the living conditions of the plants and hence (potentially) their biomechanical properties. As a consequence various studies have been carried out with artificial plants to overcome difficulties with plant nursing and to eliminate the variability of individual plants within patches or arrays (e.g., Ikeda & Kanazawa, 1996, Ciraolo *et al.*, 2006, Okamoto & Nezu, 2009). The majority of these studies focused on aquatic vegetation while investigations with flexible artificial riparian vegetation are rare (e.g., Schoneboom, 2011). However, until today there is no guidance available how artificial vegetation elements can be chosen and unambiguously compared to the natural target species. Choosing artificial vegetation only on the basis of geometrical similarity with the prototype vegetation does not guarantee adequate reflection of biomechanical properties and of the resistance behaviour of the natural plants (Järvelä, 2006).

The objective of the present paper is to highlight current knowledge with regard to drag forces on flexible leafy riparian vegetation and associated hydraulic resistance. Using data from experiments carried out in the hydraulic laboratory of the Leichtweiß-Institute for Hydraulic Engineering and Water Resources at the Technische Universität Braunschweig and in the framework of a Hydralab III project, various issues will be emphasised ranging from the presentation of a concept for the comparison of artificial leafy plants with their natural counterparts, the effect of leaves on the flow resistance, and the hydraulic resistance of arrays of flexible vegetation arrays in emergent conditions.

2 BACKGROUND

2.1 Single rigid and flexible plants

The hydrodynamic drag on a plant of arbitrary shape consists of viscous and pressure drag whose proportions depend on the shape of the plant (e.g., Hoerner, 1965) as well as biomechanical properties. Based on conceptual considerations, Nikora (2010) hypothesized that submersed highly flexible aquatic vegetation elements mainly experience viscous drag while the major component of drag on a riparian vegetation element such as a tree is the pressure or form drag (Finnigan, 2000). Theoretically, the total drag F_D can be computed by integrating both shear stress and pressure distribution over the whole plant surface (e.g., Jirka, 2007). However, such a computational approach is sophisticated for complex shaped bodies and therefore there is no other way than to rely on measurements (Nakayama, 1999) when investigating the drag forces on plants.

In general, drag is expressed according to

$$F_D = \frac{1}{2}\rho C_D A u_R^2 \tag{1}$$

where ρ = fluid density, C_D = drag coefficient, A = reference area, and u_R = reference velocity. In investigations with single elements it is often assumed that u_R corresponds to cross-sectionally averaged flow velocity u_m and that the reference area A corresponds to the frontal projected area A_p of the body. The dimensionless C_D-value depends on body shape, Reynolds number, and on the surface roughness of the body. C_D-values for simply shaped rigid bodies such as disks, plates, or cylinders can be found in hydraulic text-books. Equation (1) further shows that a quadratic relationship exists between F_D and u_m for situations where A and C_D are constant (e.g., for bluff bodies and a constant water depth).

In many studies related to the hydraulic resistance of vegetation, plants have been conceptualised as rigid cylinders (e.g., Li & Shen, 1973, Nepf, 1999) for which the projected area A_p corresponds to the product of the cylinder diameter d and the cylinder height l (submerged conditions) or to the product of d and the water depth h (emergent conditions). For a rigid cylinder in emergent conditions $1.0 \leq C_D \leq 1.2$ for $800 < Re_d < 2 \cdot 10^5$, where Re_d denotes the cylinder Reynolds-number $Re_d = u_m d/\nu$, with ν = kinematic fluid viscosity (Schlichting & Gersten, 2006). As a consequence, a constant value of $C_D = 1.0$ or 1.2 is often used as a global drag coefficient in practical applications.

Flexible plants show, in contrast to rigid elements, a significantly different resistance behaviour due to their deformation under flow action. The streamlining of a flexible plant, shown exemplarily in Figure 1 for an artificial plant which is described in more detail in Section 3.1, is accompanied by a height and breadth-contraction of the plant and hence with a reduction of both A_p and C_D with increasing flow velocity. The streamlining reflects the efforts of the plant to reduce flow separation and hence pressure drag (Nikora, 2010).

Estimating C_D for flexible plants from experimental data using Equation (1), the result depends on the definition of both the characteristic area A (e.g., frontal projected

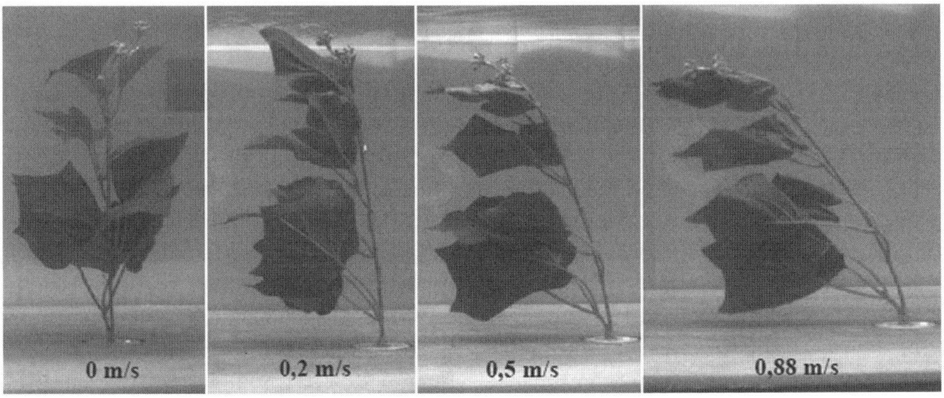

Figure 1 Longitudinal deformation of a flexible leafy vegetation element in an open channel flow at different flow velocities.

area A_p, product of lateral projection of the plant and plant characteristic diameter, or total wetted surface area of the plant) and the reference velocity u_R (undisturbed velocity integrated over the plant surface area and u_m; see Statzner *et al.*, 2006). The dependence of C_D on the chosen definition of A and u_R aggravates the comparability of experimental results from different studies. In fact, these parameters have often been defined differently highlighting problems with experimental design, data analysis and interpretation in studies of drag acting on plants (Statzner *et al.*, 2006).

The difference in the resistance behaviour of flexible and rigid plants has been expressed in the literature in a more general way in terms of the Vogel exponent b describing the deviation of the F_D-u_m relationship from the quadratic law (de Langre, 2008):

$$F_D \sim u_m^{2+b} \tag{2}$$

A value of $b = -1$ was suggested for flexible leafy trees in various wind-tunnel investigations (e.g., Cullen, 2005, de Langre, 2008) indicating a linear relationship between F_D and u_m. Similar results were reported in studies carried out in water flows (e.g., Fathi-Maghadam & Kouwen, 1997, Armanini *et al.*, 2005, Wilson *et al.*, 2010). Taking into account Equation (1), a value of $b = -1$ suggests $A_pC_D \sim u_m^{-1}$. However, the results of Oplatka (1998) indicate that b depends on the flexural rigidity of the plants. For a fully stiff tree, Oplatka (1998) obtained $b = 0$, for a partially stiff tree $b = -0.36$ and for a fully flexible tree $b = -1$. Thus, although $b = -1$ has been concluded in various studies, the linear model may not be a priori applicable, indicating the need for further research. This issue will be discussed in more detail in Section 4.

2.2 Plant communities

Although single trees are not uncommon on floodplains, they do not contribute significantly to the total hydraulic resistance at larger scales which depends predominantly on the resistance caused by plant communities and hence, besides on plant characteristics, on planting pattern (e.g., Hirschowitz & James, 2009, Dittrich & Aberle, 2010) and wake flow characteristics affecting the turbulent flow structure (e.g., Nepf, 1999).

The total hydraulic resistance at the reach scale can be obtained by applying the superposition principle (e.g., Petryk & Bosmajian, 1975, Yen, 2002):

$$f = f' + f'' \tag{3}$$

where f' = surface friction factor and f'' = friction factor due to form resistance. For a cylinder array and emergent conditions, the latter corresponds to:

$$f'' = 4mdh\langle C_D \rangle \tag{4}$$

where $m = (a_x a_y)^{-1}$ denotes the number of cylinders per unit area, a_x and a_y denote the longitudinal and transverse spacing of the elements, respectively, and $<C_D>$ defines the drag coefficient of the array.

The flow in a multicylinder array is influenced by the decay and spread characteristics of wakes forming at upstream cylinders. Various studies showed that, dependent on the cylinder arrangement and spacing, $<C_D>$ may differ substantially from the drag coefficient of a single isolated cylinder C_D (e.g., Li & Shen, 1973, Lindner, 1982, Nepf, 1999, Schoneboom *et al.*, 2011). Therefore, the assumption of a constant $<C_D>$-value for different cylinder densities and arrangements of e.g., $<C_D> = 1.5$ (DVWK, 1991), $<C_D> = 1.05$ (Stone & Shen, 2002), or the estimation of $<C_D>$ in analogy to an isolated cylinder by assuming $u_R = u_m$ is a simplification which is, strictly speaking, not correct.

The wake flow characteristics and hence the associated differences between the approach velocity u_R of an individual cylinder within the array and the cross-sectionally averaged flow velocity u_m, can be taken into account by a computational approach developed by Lindner (1982). This approach, representing an enhancement of the approach developed by Li and Shen (1973), allows the determination of $<C_D>$ and considers the effects of cylinder arrangement and density. It was recently validated by Schoneboom *et al.* (2011) on the basis of direct drag force measurements within a multicylinder array.

One of the first approaches for the calculation of the flow resistance of flexible riparian vegetation was developed by 'Fathi-Maghadam and Kouwen (1997). According to this approach, the friction factor f'' can be determined from

$$f'' = 4C_D\left(\frac{A_M}{A_b}\right) \tag{5}$$

where C_D = drag coefficient of a plant, A_b = ground area occupied by a plant and A_M = momentum absorption area of a plant which is closely related to the one-sided area of leaves and stems. Based on theoretical and dimensional considerations 'Fathi-Maghadam and Kouwen (1997) linked the right hand side of Equation (5) with the flexural rigidity of the plants, which is defined as the product of Young's elasticity modulus and the cross-sectional moment of inertia of the trees. Kouwen and Fathi-Moghadam (2000) extended this approach later by defining a species-related vegetation index. However, as plant biomechanical properties and the required empirical coefficients are difficult to determine, Järvelä (2004) developed a modified version of Equation (5) for emergent riparian vegetation, i.e., for $H \geq h$ (where H = plant height). This approach is based on the leaf-area index LAI and species-specific coefficients which can be obtained from hydraulic experiments:

$$f'' = 4C_{D\chi} LAI \left(\frac{u_m}{u_\chi} \right)^{\chi} \frac{h}{H} \tag{6}$$

where LAI = leaf area index = one-sided (upper) leaf area of the plants per unit area; $C_{D\chi}$ = species-specific drag coefficient, χ = vegetation parameter which accounts for the effects of plant deformation in a flow and is unique for a particular species, and u_χ = lowest velocity used in the determination of χ which is used for the normalization of the relationship.

The parameter h/H represents a scaling parameter for LAI. It is based on the assumption that the leaves are uniformly distributed over the height of the vegetation and enables the application of Equation (6) for different partial submergences. Järvelä (2004) determined the empirical coefficients $C_{D\chi}$ and χ for different species using own experiments and data of Fathi-Maghadam and Kouwen (1997). It is worth mentioning that Equation (6) complies with Equation (2), i.e., rearranging Equation (6) it can be shown that $F_D \sim u_m^{2+\chi}$. Thus, the χ-coefficients are identical to the Vogel-exponent b.

3 DATA

The following sections mainly focus on drag forces and hydraulic resistance of flexible leafy elements. Various aspects introduced in Section 2 will be highlighted and discussed using experimental data recently obtained from experiments carried out in the hydraulic laboratory of the Leichtweiß-Institut für Wasserbau (LWI), Technische Universität Braunschweig and in the framework of the Transnational Access Activities EU Hydralab III scheme. The corresponding data sets are described below.

3.1 LWI-Experiments

The LWI-experiments were carried out in a 32 m long, 0.6 m wide and 0.4 m deep tilting flume. The discharge Q during the experiments was controlled by a valve and measured by an inductive flow meter. Water depth was adjusted by a tailgate located in a distance of 25 m to the flume inlet and measured by 10 piezometers installed along the flume length. The bed roughness consisted of a rubber mat with 3 mm high pyramidal shaped roughness elements. More details on the experimental setup can be found in Schoneboom (2011).

Drag forces on vegetation elements were measured using an innovative drag force measurement system (DFS) consisting of multiple sensors which represent an enhancement of the drag force sensor used by Wilson *et al.* (2008). The design of the individual sensors (Figure 2a), which is described in detail in Schoneboom *et al.* (2008) and Schoneboom (2011), is based on the employment of a series of strain gauges configured as two Wheatstone full bridges to a cantilever in bending. The compact design of the individual sensors allowed the installation of up to 10 sensors below the flume bed in the 1.5 m long test section which was located in a distance of 15.2 m to the flume inlet (Figure 2b). This setup ensured that the DFS did not disturb the flow and that the individual sensors could be easily rearranged to match the investigated plant patterns. Hence it was possible to simultaneously measure drag forces on 10 plants with high accuracy (±0.02 N) and temporal resolution (up to 1613 Hz). In the experiments, the sampling rate was restricted to 200 Hz and drag forces were measured for a sampling duration of 60 s. The drag force measurements were repeated three times to ensure repeatability. The frontal projected areas A_p of the vegetation elements attached to the sensors were estimated from the analysis of digital photographs taken with a submersible mini-camera (Figure 2c).

In this paper we use a subset of the data set described in Schoneboom (2011). These data were collected in experiments carried out with natural and artificial flexible plants, different densities (isolated plants and arrays with spacing 15 × 15, 20 × 20, and 30 × 30 cm^2 corresponding to plant densities of 11.1, 25, and 44.4 plants/m^2), and in-line (L) and staggered (S) arrangements (Figure 3).

Most of the experiments were carried out with artificial flexible elements which were used to guarantee the plants' persistence with time in order to ensure comparability of the results. The 23 cm high artificial plants (see Figures 1 and 2), commercially available off the shelf and described in detail in Schoneboom and Aberle (2009), are composed of a 3 mm thick coated wire stem, a blossom, and four branches with three leaves each. The leaves are made of fully flexible dyed textile and the one-sided leaf area varies between 14.32 to 57.6 cm^2 with a total cumulative leaf area of 373.57 cm^2.

The array experiments were carried out with 18.5 m long canopies to ensure fully developed flow conditions in the test section which was located 9.25 m downstream

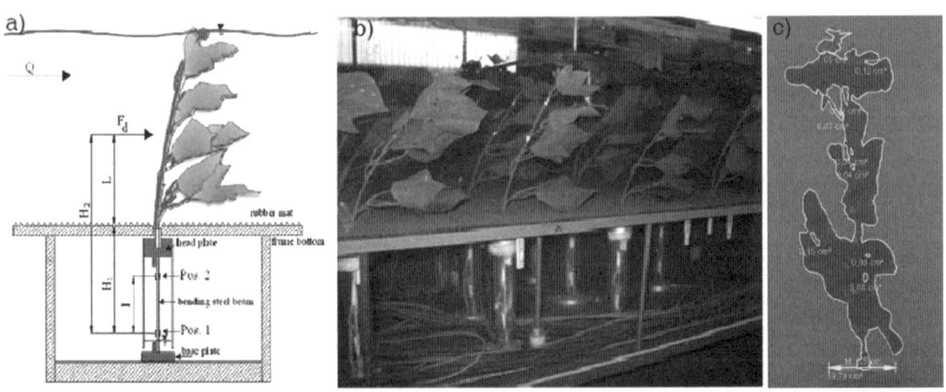

Figure 2 a) Design-sketch of the DFS, b) view of the test section with DFS mounted below the flume bed, and c) digital photograph of a streamlined vegetation element used for the evaluation of the projected area.

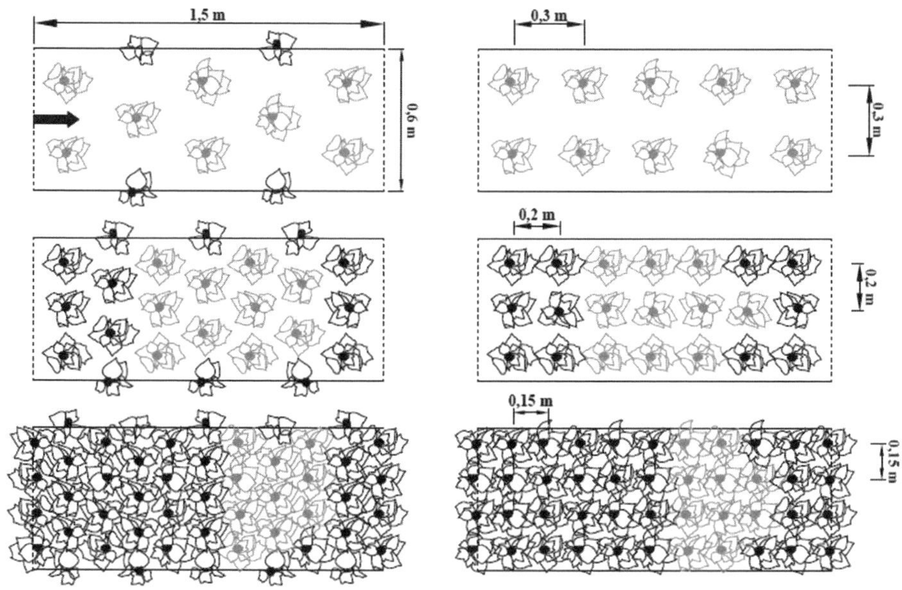

Figure 3 Investigated plant arrangements in the test section. The left column shows the staggered (S) and the right column the in-line (L) arrangements for densities of 11.1, 25, and 44.4 plants/m^2 (from top to bottom). The highlighted plants were attached to the DFS-sensors.

from the canopy leading edge. In case of the densest vegetation pattern, the canopy length was reduced to a total length of 10 m starting at a distance of 10 m from the flume inlet due to the limited number of available vegetation elements. For the staggered pattern, some plants had to be placed close to the flume wall (Figure 3). Thus, to ensure a constant plant density and a homogeneous distribution of the leaves, six of the twelve leaves were removed from these plants. In addition, the plant blossom was removed from every second 'half-plant'. The additional stem in every other row affects the overall flow resistance only marginally due to the low stem diameter of 3 mm (Schoneboom, 2011).

The flow and drag force measurements were carried out with steady uniform flow conditions and, in most cases, with just submerged vegetation elements, i.e., the water level was adjusted to the height of the deflected plants.

3.2 Hydralab-Experiments

The LWI-data are complemented by data from experiments carried out at the 320 m long, 12.5 m wide and 6.5 m deep CEHIPAR ship canal facility in El Pardo, Madrid in the framework of an Hydralab III project (see Wilson *et al.*, 2010, Xavier *et al.*, 2010 for more details on the experimental setup). The experiments were carried out with natural trees sampled in the surrounding of Madrid in March and April 2008. For the towing experiments, the sampled trees were attached upside-down to a dynanometer suspended beneath a carriage located above the canal. The carriage moved along canalside railings with the velocity controlled to an accuracy of 1 mm/s. The forces on the dynamometer were monitored in real time to an accuracy of 0.0098 N. In order to discuss the influence of foliage on the drag forces exerted by individual trees, we will present data from tests carried out with natural alders in foliated and defoliated conditions. This data complements the LWI-experiments which were carried with artificial small scale plants.

4 RESULTS AND DISCUSSION

4.1 Comparison of hydraulic resistance of artificial and natural plants

The geometrical similarity of artificial plants with their natural counterparts (e.g., leaf and plant shape, plant structure) does, as outlined in the introduction, not guarantee that the resistance behaviour of the natural plants is adequately reflected. Acknowledging this fact, Schoneboom (2011) compared additionally F_D-u_m relationships as well as the lever arm L of various plants of similar height. The lever arm L corresponds to the distance of the resulting drag force application point to the flume bed and can be obtained from the drag force sensor data based on geometrical considerations (Figure 2a). For flexible plants, L is expected to decrease with streamlining and hence with the reduction of the deflected plant height. As a consequence, L contains implicit information on flexibility and vertical biomass distribution as these two parameters define the momentum absorption area. Smaller differences in foliation (i.e., the momentum absorption area) between plants can be taken into account by normalizing F_D with the one-sided leaf area A_Σ, of the respective plants, i.e., by comparing F_D/A_Σ-u_m relationships.

Figure 4 shows results from experiments carried out with isolated plants in the LWI-flume: an artificial element (see Figure 1), a young leafy twig of a natural poplar with relatively large flexibility, and an artificial and natural willow twig, respectively. Figure 4a reveals a good agreement of both the F_D/A_Σ-u_m and L-u_m relationship for the

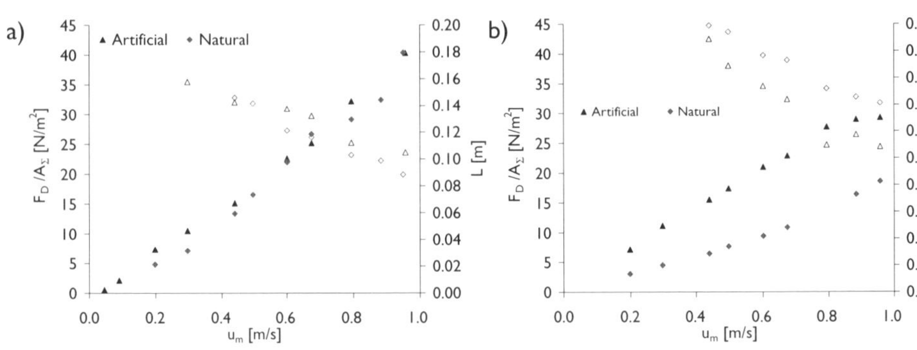

Figure 4 Comparison of the resistance behaviour of a) artificial and natural poplar and b) artificial and natural willow (full symbols denote drag forces and open symbols denote lever arm). The lever-arm values for low velocities are not shown due to the limited accuracy of L for low loads (see Schoneboom, 2011 for details).

artificial element with the respective curves of the natural poplar. On the other hand, the willow curves show larger deviations (Figure 4b) indicating that the artificial willow does not adequately reflect the biomechanical properties of its natural counterpart. The good agreement of the artificial element (referred to as artificial poplar in the following) with the natural poplar was a strong indicator for Schoneboom (2011) that the resistance behaviour of these plants is similar. This assumption was justified by further tests carried out with natural poplars in which different stages of foliation were simulated by successively removing leaves from the plants. Based on the well agreement of the corresponding F_D/A_Σ-u_m curves with the curves obtained in similar tests with the artificial poplar (see Schoneboom & Aberle, 2009), Schoneboom (2011) concluded that the behaviour of the artificial poplar resembles the behaviour of young natural poplars. The concept developed by Schoneboom (2011) can be seen as an innovative approach for the assessment of the resistance behaviour of an artificial flexible element in comparison with its natural counterpart.

The data shown in Figure 4 can also be used to discuss the F_D-u_m relationships. The visual inspection of Figure 4 suggests, at a first glance, a linear relationship and fitting a straight line yields linear squared regression coefficients $R^2 \geq 0.98$ for all four curves. On the other hand, fitting a power law according to Equation (2) $b = -0.66$ and -0.63 are obtained for the artificial and natural poplar, respectively, and $b = -1.09$ and $b = -0.85$ for the artificial and natural willow, respectively. For all power laws, R^2-values were in the same order of magnitude as for the linear fit, i.e., $R^2 \approx 0.98$. The b-values for the poplars are almost identical (showing once more the good agreement between the natural and flexible poplar) and are significantly different from -1. This means that the assumption of a linear F_D-u_m relationship is not justified by the power-law analysis for the investigated poplars. On the other hand, the exponents for the willows are closer to -1 but show at the same time larger scatter confirming the identified differences in the resistance behaviour.

4.2 Contribution of foliage to the total drag of natural trees

The artificial poplar was characterised by a large leaf area compared to its stem area. Hence it was difficult to isolate the contribution of the drag on the stem to total drag using this plant. However, this issue can he highlighted using the Hydralab data set introduced in Section 3. Figure 5a presents the obtained F_D-u_m relationships for both the foliated and defoliated alder trees. The corresponding curves show a clear separation of the drag forces exerted on the foliated and defoliated specimen, respectively. We note that the drag force variability within the foliated and defoliated data set is related to differences in foliage and plant structure of the individual specimen.

In order to investigate the effect of foliage, its contribution to the total drag was estimated for identical towing velocities by subtracting the measured drag forces in the defoliated condition $F_{D,d}$, (i.e., the drag force component of the wooden parts of the plant) from the measured total drag of the foliated tree $F_{D,t}$. The resulting difference $F_{D,f} = F_{D,t} - F_{D,d}$ represents the contribution of the foliage to total drag and the ratio $F_{D,r} = F_{D,f}/F_{D,d}$ enables the investigation of the proportions of the drag due to foliage and the wooden parts (Schoneboom, 2011).

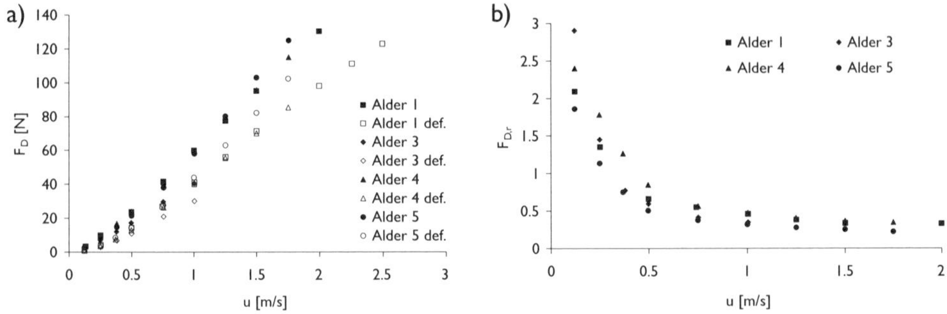

Figure 5 a) Drag force F_D as a function of the carriage velocity *u* for alders in foliated and defoliated conditions and b) Drag force ratio $F_{D,r}$ as a function of towing velocity.

Figure 5b shows $F_{D,r}$ as a function of the towing velocity and reveals that foliage contributes up to 75% to the total drag at lower towing velocities ($F_{D,r} > 1$ or $u \leq 0.25$ m/s). The influence of foliage decreases with increasing flow velocity due to streamlining of the leaves and $F_{D,r}$ approaches an approximately constant value of $F_{D,r} \approx 0.33$ for $u > 1$ m/s. Hence, at larger velocities the leaves of the investigated plants contribute approximately 25% to the total drag while the wooden parts contribute approximately 75%. It may be hypothesised that the deformation of the leaves has reached its maximum when this threshold ($F_{D,r} \approx 0.33$) is reached.

Wilson *et al.* (2008) found in an investigation with an ivy-twig that foliage contributed approx. 60% to total drag while Armanini *et al.* (2005) reported a contribution of 40% for a salix specimen. The present results fit well within this broader range. On the other hand, Västilä *et al.* (2011) found in an investigation with arrays of natural poplar twigs that foliage contributed up to 75% to the to total resistance, even at large velocities. The difference between the results of Västilä *et al.* (2011) and Figure 5 may be related to differences in flexural rigidity of the small-scale plants (twigs) and full-scale trees (which were still in the growing period), different flow characteristics (multi-plant-arrangement in turbulent flow and towing tests with single plants in still water), and different leaf properties (see Albayrak *et al.*, 2010 for this topic). This issue remains to be investigated in further studies.

The observed contribution of the wooden parts and foliage to total drag can also be discussed in the context of the F_D-u relationships shown in Figure 5a. A closer inspection of the individual relationships for $u \leq 0.5$ m/s indicates different gradients compared to larger velocity ranges. A similar behaviour was identified by Wilson *et al.* (2010) and Xavier *et al.* (2010) investigating the F_D-u relationships of willows within the same towing experiments. Acknowledging the different gradient in the low velocity range and assuming a linear F_D-u relationship, the corresponding velocity range was called trans-flexing zone in these studies. The existence of such a trans-flexing zone may also be inferred from Figure 4, as the corresponding F_D-u relationships would not pass through the origin if linearly extrapolated.

Furthermore, Figure 5b shows that, at low velocities, the contribution of the leaves to total drag is more significant than the contribution of the wooden parts.

Visual observations during the experiments indicated that tree deformation was not significant for $u \leq 0.5$ m/s, which is also in agreement with the analysis of the b-exponent for the defoliated trees in this range resulting in $b \approx 0$ for all four trees. This exponent indicates that the trees behaved as a rigid body. On the other hand, $b \approx -0.5$ was obtained for the same velocity range for the foliated trees suggesting that the difference in b is associated with the streamlining of the foliage. Although Equation (2) does, strictly speaking, not justify the definition of a trans-flexing zone as a power law passes through the origin, these results still indicate the existence of such a zone at low velocities. It seems that the resistance behaviour in the trans-flexing zone is dominated by the leaves and to a less extent by the wooden parts of the plant, which behave almost as a rigid body.

The wooden parts start streamlining at larger velocities, which becomes apparent from the F_D-u relationships of the defoliated trees for $u > 0.5$ m/s deviating from the quadratic relationship identified for the lower velocities ($b \approx -0.5$). The corresponding b-value for the foliated trees also decreases slightly in this velocity range ($b \approx -0.65$). It can therefore be concluded that the smaller exponent is associated with the streamlining of the wooden parts. Consequently, this velocity range may be defined, in accordance with Wilson *et al.* (2010) and Xavier *et al.* (2010), as flexing-zone. In this context it is worth mentioning that it would also be possible to fit a linear relationship with a high degree of accuracy to the data so that $b \approx -1.0$. However, as outlined in the previous section, the assumption of such a linear relationship should be substantiated by a theoretical background in future studies.

4.3 Hydraulic resistance of arrays with flexible leafy vegetation

The following section presents results from the experiments carried out with arrays of the artificial poplar and different boundary conditions. The corresponding data are used to highlight drag force variability in arrays, the impact of plant arrangement, and the investigation of hydraulic resistance.

4.3.1 Drag force variability

Figure 6 presents the time averaged drag forces F_D of the individual artificial poplars which were attached to the drag force sensors. These data were obtained in experiments carried out with the lowest array density (30 × 30 cm) and various flow velocities for the staggered (full symbols) and in-line (open symbols) arrangement. The figure, which is representative for all investigated array patterns, reveals a large variability of the measured drag forces for both arrangements, with the observed maximum variation of the drag force within the canopy of more than 50% (Schoneboom *et al.*, 2010). Visual observations during the experiments showed that the vegetation elements did reconfigure differently although the individual elements were identical and great care was given to plant arrangement during the experimental setup to ensure that all canopy elements were similarly aligned. The variability is also partly caused by the wake flow structure in the canopy which contributes to the differences between the staggered and in-line arrangement in Figure 6. Nonetheless, the large drag force variability for both arrangements resulted in the conclusion that the estimation of the

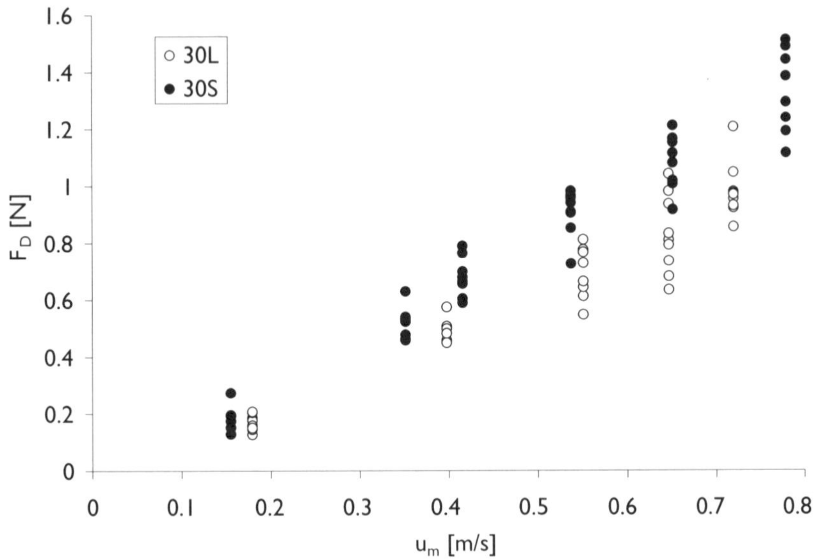

Figure 6 Drag forces F_D on individual plants for the 30 L arrangement (open circles) and 30 S (full circles) as function of u_m. Each circle indicates a DFS element.

spatially averaged drag forces from measurements with a single test plant is prone to errors (Schoneboom *et al.*, 2010).

4.3.2 *Plant arrangement*

The spatially averaged drag forces $\langle F_D \rangle$ are shown as a function of velocity in Figure 7 for all plant arrangements and densities together with the F_D-u_m curve of the artificial poplar. The separation of the data points shows that larger spatially averaged drag forces were exerted on the plants in staggered arrangement than on the plants arranged in-line. Furthermore, the figure indicates that plant density did not play a crucial role in the present experiments.

The influence of the plant arrangement is related to the wake flow characteristics indicating, as for multicylinder arrays, the importance of the flow structure within canopies. For the data in Figure 7, $\langle F_D \rangle$ is consistently 1.22 times larger for the staggered than for the in-line arrangement. A similar trend was obtained for $\langle F_D \rangle$-measurements in a multicylinder array and by applying the Lindner (1982) approach (Schoneboom *et al.*, 2011). However, the development of an adequate computational approach for flexible vegetation is difficult (if not impossible) as the wake flow structure depends on the properties of individual plants, plant reconfiguration, plant density and arrangement.

Figure 7 further shows that the F_D-u_m relationship of the individual artificial poplar almost coincides with the relationship of the in-line arrangement. This somewhat unexpected result may be explained by smaller differences in the experimental procedure. The tests with the individual plant were carried out with a constant water

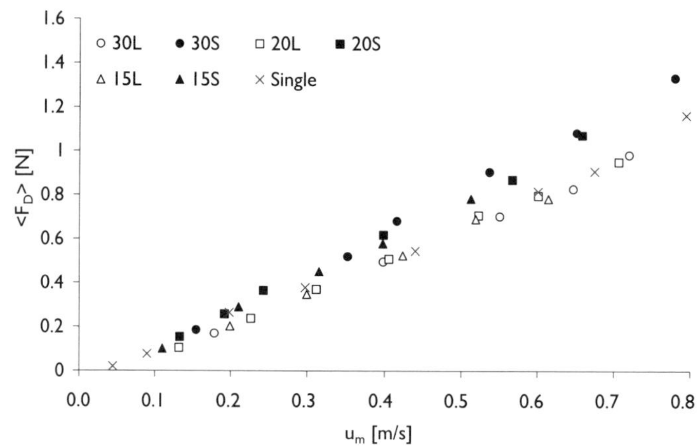

Figure 7 $<F_D>$-u_m relationship for the experimental LWI-vegetation array tests as well as for an isolated vegetation element as a function of cross-sectionally averaged velocity.

depth of h = 0.25 m while the array tests were carried out with just submerged flow conditions for which the water depth was, at large velocities, approximately h = 20 cm. Hence, taking the height reduction of the single plant into account, the 'real' approach velocity of the plant would be somewhat smaller than the cross-sectionally averaged velocity. Schoneboom (2011) therefore inferred that the F_D-u_m relationship for the isolated plant should lie between the curves for the in-line and staggered arrangements when these circumstances are considered.

The analysis of the F_D-u_m curves using Equation (2) resulted in b = –0.74 and –0.73 for the staggered and in-line arrangement, respectively. These values were somewhat lower than b = –0.65 for the isolated plant. The relatively small deviation of Δb = 0.08 may partly be explained by the different number of observations which can have an effect on the regression results. Figure 7 also suggests that the investigated plant densities may be interpreted as sparse. This conclusion arises from the fact that the plant density had no significant influence on the $<F_D>$-u_m relationships and is underpinned by visual observations during the experiments in which mutual 'physical' plant interference was not observed. Inasmuch mutual plant interference becomes significant and masks over the influence of the plant arrangement should be explored in further research (e.g., Aberle *et al.*, 2010). Based on the results in Figure 7, Schoneboom (2011) concluded that the spatially averaged drag force in sparse arrays may be estimated from experiments with an isolated plant given that the effect of the arrangement is considered.

4.3.3 *Frontal projected area and drag coefficient*

Figure 8 presents the spatially averaged plant projected area $<A_p>$ of the plants attached to the drag force sensors as well as the corresponding mean drag coefficients $<C_D>$ (obtained using Equation 1) as a function of the mean velocity. The reduction of $<A_p>$ with increasing flow velocity is more pronounced at low velocities which

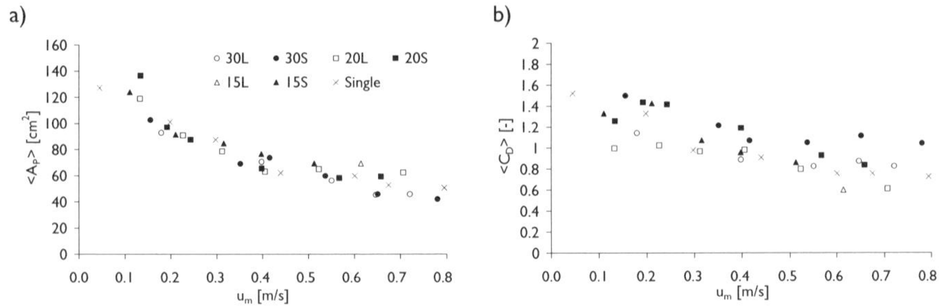

Figure 8 Spatially averaged values of a) $<A_p>$ and b) $<C_D>$ as a function u_m. For comparison, the data of the individual plant ate also displayed.

becomes apparent from the lager gradients d$<A_p>$/du_m (Figure 8a). Figure 8 therefore confirms the conclusion in Section 4.1.2 that leaf deformation is more pronounced at lower velocities. Moreover, it was found that, for larger velocities, the plant projected area may be as low as 15% of the one-sided leaf area (Aberle *et al.*, 2011).

Compared to the F_D-u_m relationships shown in Figure 7, the $<A_p>$-curves collapse on a single line and are therefore independent of plant arrangement. This result can partly be explained with the measurement accuracy as one would expect minor differences in $<A_p>$ due to the influence of the wake flow pattern (and hence differences in the approach velocity of individual plants). In the analysis, $<A_p>$ was determined from the analysis of A_p-snapshots of the plants attached to the DFS (Figure 2c; up to three snapshots per plant) and was found to provide a reasonable accurate estimate of time averaged A_p values because plant motion was not pronounced. The analysis procedure was identical for both the staggered and in-line arrangements so that the associated error resulting from the processing of the digital photographs (e.g., Sagnes, 2010) can be considered as constant. On the other hand, the differences in the approach velocity for both arrangements will be relatively small but can be expected to still result in small deviations in A_p. Such small differences are difficult to quantify with the applied procedure (see scatter in the data) due to the heterogeneity associated with the individual plants and to some extent due to the resolution of the digital photographs (see Figure 2c).

The functional relationships of the spatially averaged $<C_D>$ values with flow velocity are presented in Figure 8b showing clearly that plant deformation is accompanied by a reduction of the drag coefficient. However, in contrast to the $<A_p>$-u_m curve, the $<C_D>$-curves are separated with regard to plant arrangement. This separation follows directly from the results shown in Figure 7 and Figure 8a. Bearing in mind that $<A_p>$ showed no such separation with arrangement, the separation in Figure 8b may be amplified to some extent if a dependency of $<A_p>$, which could not be resolved from the measurements, exists.

4.3.4 *Hydraulic resistance of flexible plant arrays*

The aforementioned difficulties related to the estimation of $<A_p>$ and $<C_D>$, and as a consequence of the product $<A_p>$ and $<C_D>$, show the need for surrogate measures in order to unambiguously parameterise the drag force equation (Equation 1) and/or of the friction factor f''. As discussed in Section 2.2, the approach proposed by Järvelä (2004) is appealing for this purpose as it uses the LAI as surrogate measure for $m <A_p>$ combining the single sided leaf area and the plant density into a single parameter. Thus, using this approach it is not necessary to determine the frontal projected area for different flow velocities. Moreover, the uncertainties related to $<A_p>$ and $<C_D>$ are lumped into the species specific drag coefficient $C_{D\chi}$ and the vegetation parameter χ, which can be determined from hydraulic measurements.

Figure 9 shows the friction factor f'' normalized with LAI as a function of u/u_χ for the staggered ($u_\chi = 0.11$ m/s) and the in-line setup ($u_\chi = 0.13$ m/s; note that $h/H = 1$ for all data points). The shape of the curves is similar to the shape of the f''-u_m relationships shown in Järvelä (2006) and Västilä *et al.* (2011) for natural flexible vegetation. Figure 9 shows distinct differences between the in-line and staggered setups which are associated with the aforementioned different $<F_D>$-u_m curves (Figure 7). The calibration of Equation (6) resulted in $C_{D\chi} = 0.50$ and $\chi = -0.74$ for the staggered and $C_{D\chi} = 0.34$ and $\chi = -0.73$ for the in-line arrangement, respectively. Both relationships were fitted with a high degree of accuracy ($R^2 = 0.99$) and the reported χ-values are confirmed by the comparison with the b-exponents in Section 4.3.2. The obtained parameters are slightly different from the values reported in Schoneboom *et al.* (2010) due to a refinement of the analysis. Therefore, the values reported in Schoneboom *et al.* (2010) should be replaced for future references with the values provided above.

The species-specific values for the artificial poplar are also confirmed by the analysis of additional experiments (25 plants/m² in a staggered arrangement) in which the vertical distribution of the leaves (and hence LAI) was varied (Aberle *et al.*, 2011). The analysis of these data resulted in almost identical values $C_{D\chi} = 0.49$ and $\chi = -0.72$ ($u_x = 0.14$ m/s). It is worth mentioning that the $C_{D\chi}$ and χ -values for the individual data series reported in Aberle *et al.* (2011) showed some scatter which was related to

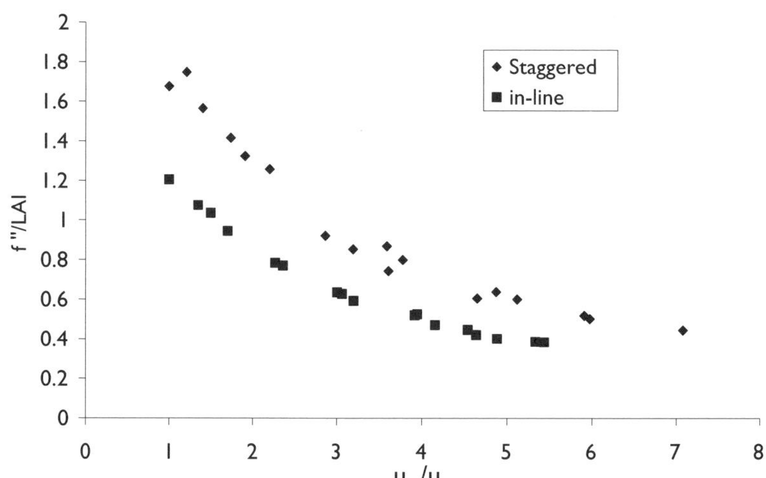

Figure 9 Friction factor f'' normalised with LAI plotted against the normalized flow velocity for the staggered and in-line setup.

the influence of the plant blossom and the limited number of available data points for the regression.

5 SUMMARY AND CONCLUSIONS

The present study highlighted various issued related to the drag forces on flexible riparian vegetation elements and the associated hydraulic resistance. These were illustrated using experimental data obtained from measurements with artificial small-scale as well as natural large scale plants.

Using data of artificial and natural plants of similar height, the concept developed by Schoneboom (2011) was introduced which is a useful tool to evaluate if the resistance behaviour of natural plants is adequately reflected by its artificial counterpart. The concept is based on a visual comparison of the plants and the additional analysis of F_D-u_m relationships and the lever arm L of the resulting drag force. The lever arm decreases with increasing plant deformation and inhibits hence implicit information on the momentum absorption area. This concept may be enhanced in future studies to develop scaling laws for vegetation, taking also into account allometric relationships.

The analysis of drag-force – velocity relationships of full-scale foliated and defoliated trees obtained in towing tank experiments was used to assess the contribution of foliage to total drag. It was shown that foliage contributes more significantly to total drag at lower velocities (up to 75%) and that, at larger velocities, the contribution of the wooden plant parts becomes more significant. Moreover, the analysis suggested that the contribution of the leaves to total drag reached a constant percentage at larger velocities in these tests. It was hypothesised that this fact can be explained with the maximum streamlining of the leaves.

The detailed analysis of the F_D-u curves from the towing tests further suggested that the wooden parts of trees may act as a rigid body below a certain threshold velocity. This observation substantiates the existence of the so called trans-flexing zone defined by Wilson *et al.* (2010), in which the total resistance is composed of the drag on the rigid tree parts and the streamlining leaves. Above this threshold, in the so called flexing zone, the wooden parts start to reconfigure significantly and the contribution of the leaves to total drag approaches the aforementioned constant percentage.

The analysis of drag force measurements on up to 10 artificial flexible vegetation elements in a vegetation array showed that the spatially heterogeneous flow field and hence the wake flow structure has a significant effect on the F_D-u_m relationship of the individual plants. The observed F_D-variability indicated that care should be taken to estimate the spatially averaged drag force from drag force measurements with a single plant. On the other hand, the analysis of the spatially averaged drag forces indicated that they may be, for sparse arrays, estimated from measurements with a single isolated plant. However, such estimates must be corrected for the influence of the array arrangement as drag forces exerted on the vegetation elements were found to be larger for staggered than for the in line-arrangements which was associated with wake flow characteristics.

The investigation of the obtained drag force – velocity relationships for both single elements and vegetation arrays indicated that the data could be described by both a power law and a linear relationship with similar accuracy. The deviation from the quadratic law expected for rigid bodies results from the reconfiguration of the plants which is associated with their efforts to reduce pressure drag. While the linear relationship has been suggested in various investigations, it still requires theoretical justification. On the other hand, the Vogel-exponents b resulting from the power law approach allow for a broader quantification of the flexibility. Moreover, b is equivalent to Järveläs' (2004) vegetation parameter χ which accounts for plant deformation in the corresponding approach. Therefore, the reported χ-values in the present study as well as in the literature, which are larger than –1, also contradict the existence of a universal linear relationship between the drag force and velocity for flexible vegetation elements.

The Järvelä (2004) approach was found to be practicable approach for the quantification of flow resistance of emergent riparian vegetation. Although it requires species-specific parameters, it eliminates to some extent uncertainties associated with the estimation of the plant projected area and shape-dependent C_D-values as the leaf area index is used as a surrogate measure for the frontal projected area. The presented analysis demonstrated that the required species-specific parameters can be obtained from hydraulic measurements and that similar values are obtained for the artificial poplars in experiments with different boundary conditions. Moreover, the recent study of Antonarakis *et al.* (2010) also showed the applicability of the Järvelä (2004) approach for field applications as novel measurement techniques such as Terrestrial Laser Scanning enable the determination of LAI in the field.

In the present paper we focused mainly on drag forces exerted on flexible vegetation elements and bulk considerations. These are only a few aspects in the challenging field of flow-vegetation interaction. Our analyses showed the need for future investigations of the spatially heterogeneous flow field as well as biomechanical properties of the plants. From a hydraulic point of view, the double-averaging methodology, which is based on the temporal and spatial averaging of the Navier-Stokes equations (e.g., Nikora *et al.*, 2007a and 2007b) provides a theoretical background for such investigations. Moreover, Nikora (2010) showed that this approach provides an appropriate framework for studying flow-biota coupling and integration (i.e., up-scaling) of physical interactions and mass-transfer processes.

ACKNOWLEDGEMENTS

This research was conducted under contract AB 137/3-1 from DFG (Deutsche Forschungsgemeinschaft). The authors acknowledge valuable discussions with Juha Järvelä and the fruitful collaboration with of C.A.M.E Wilson, P. Xavier, H.-P. Rauch, W. Lammeraner, and H. Thomas in the Hydralab III project.

REFERENCES

Aberle, J., Järvelä, J., Schoneboom, T. and Dittrich, A. (2010). Flow resistance of rigid and flexible emergent vegetation revisited. In: *Proc. 1st European IAHR Congress*, 4–6 May 2010, Edinburgh, UK. Papers on memory stick.

Aberle. J., Schoneboom, T. and Rhiner, B. (2011). Influence of leaf mass on drag forces in vegetated flows. In: *Proc. 34rd IAHR Congress*, 26 June – 1 July 2011, Brisbane, Australia. ISBN 978-0-85825-868-6, pp. 2744–2751.

Albayrak, I., Nikora, V., Miler, O. and O'Hare, M. (2010). Flow-plant interaction at a leaf scale: Effects of leaf shape and flexural rigidity. In: Dittrich, A., Koll, Ka., Aberle, J. and Geisenhainer, P. (eds.) *Proc. Int. Conf. on Fluvial Hydraulics River Flow 2010*, 8–10 September 2010, Braunschweig, Germany. Karlsruhe, Bundesanstalt für Wasserbau, pp. 253–260.

Antonarakis, A.S., Richards, K.S., Brasington, J. and Muller, E. (2010) Determining leaf area index and leafy tree roughness using terrestrial laser scanning. *Water Resour. Res.*, 46(W06510), doi:10.1029/2009WR008318.

Armanini, A., Righetti, M. and Grisenti, P. (2005). Direct measurement of vegetation resistance in prototype scale. *J. Hydraul. Res.*, 43 (5), 481–487.

Ciraolo, G., Ferreri, G.B. and La Loggia, G. (2006). Flow resistance of Posidonia oceanica in shallow water. *J. Hydraul. Res.*, 44 (2), 189–202.

Cullen, S. (2005). Trees and wind: a practical consideration of the drag equation velocity exponent for urban tree risk management. *J. Arboric.*, 31 (3), 101–113.

de Langre, E. (2008). Effects of wind on plants. *Annu. Rev. Fluid Mech.*, 40, 141–168.

Dittrich, A. and Aberle, J. (2010). Die Vegetation an Fliessgewässern aus dem Blickwinkel eines Hydraulikers. *Ingenieurbiologie/Genie Biologique*, 3 (2010), 37–42.

DVWK (1991). *Hydraulische Berechnung von Fließgewässern.* Merkblätter zur Wasserwirtschaft 220/1991. Hamburg/Berlin, Verlag Paul Parey.

Fathi-Maghadam, M. and Kouwen, N. (1997). Nonrigid, nonsubmerged, vegetative roughness on floodplains. *J. Hydraul. Eng.*, 123 (1), 51–57.

Finnigan, J.J. (2000). Turbulence in plant canopies. *Annu. Rev. Fluid Mech.*, 32, 519–571.

Folkard, A.M. (2011). Vegetated flows in their environmental context: a review. *Engineering and Computational Mechanics*, 164 (EM1), 3–24.

Freeman, G., Rahmeyer, W. and Copeland, R.R. (2000). *Determination of resistance due to shrubs and woody vegetation.* US Army Corps of Engineers Engineer Research and Development Centre, ERDC/ CHL TR- 00-25.

Hirschowitz, P.M. and James, C.S. (2009). Conveyance estimation in channels with emergent bank vegetation. *Water SA*, 35 (5), 607–614.

Hoerner, S. (1965) *Fluid-Dynamic Drag.* Brick Town, Hoerner.

Ikeda, S. and Kanazawa, M. (1996). Three-dimensional organized vortices above flexible water plants. *J. Hydraul. Eng.*, 122 (11), 634–640.

James, C.S., Goldbeck, U.K., Patini, A. and Jordanova, A.A. (2008). Influence of foliage on flow resistance of emergent vegetation. *J. Hydraul. Res.*, 46 (4), 536–542.

Järvelä, J. (2004) Determination of flow resistance caused by non-submerged woody vegetation. *Int. J. River Basin Manag.*, 2 (1), 61–70.

Järvelä, J. (2006). Vegetative flow resistance: characterization of woody plants for modeling applications. In: *Proc. World Environmental and Water Resources Congress*, 21–25 May 2006, Omaha, USA. American Society of Civil Engineers, Papers on CD-Rom.

Järvelä, J., Aberle, J., Dittrich, A. Schnauder, I. and Rauch, H.P. (2006). Flow-vegetation-sediment interaction: research challenges. In: Ferreira, R.M.L., Alves, E.C.T.L., Leal, J.G.A.B. and Cardoso, A.H. (eds.) *Proc. Int. Conf. on Fluvial Hydraulics River Flow* 2006, 6–8 September 2006, Lisbon, Portugal. Taylor & Francis, Vol. 2, pp. 2017–2026.

Jirka, G.H. (2007). *Einführung in die Hydromechanik.* Karlsruhe, Universitätsverlag Karlsruhe.

Kouwen, N. and Fathi-Moghadam, M. (2000). Friction factors for coniferous trees along rivers. *J. Hydraul. Eng.*, 126 (10), 732–740.

Li, R.M. and Shen, W. (1973). Effect of tall vegetations on flow and sediment. *J. Hydraul. Div.*, 99 (HY5), 793–814.

Lindner, K. (1982). *Der Strömungswiderstand von Pflanzenbeständen*. Mitt. Leichtweiß-Institut für Wasserbau No. 75, Braunschweig, Technische Universität Braunschweig.

Nakayama, A. (1999). *Introduction to Fluid Mechanics*. Oxford, Butterworth-Heinemann.

Nepf, H.M. (1999). Drag, turbulence, and diffusion in flow through emergent vegetation. *Water Resour. Res.*, 35 (2), 479–489.

Nikora, V. (2010). Hydrodamics of aquatic ecosystems: An interface between Ecology, Biomechanics and Environmental Fluid Mechanics. *River Res. Appl.*, 26, 367–384.

Nikora, V., McEwan, I., McLean, S., Coleman, S., Pokrajac, D. and Walters, R. (2007a). Double-Averaging concept for rough-bed open-channel and overland flows: Theoretical Background. *J. Hydraul. Eng.*, 133 (8), 873–883.

Nikora, V., McLean, S., Coleman, S., Pokrajac, D., McEwan, I., Campbell, L., Aberle, J., Clunie, D. and Koll, Ka. (2007b). Double-Averaging concept for rough-bed open-channel and overland flows: Applications. *J. Hydraul. Eng.*, 133 (8), 884–895.

Okamoto, T. and Nezu, I. (2009). Turbulence structure and "Monami" phenomena in flexible vegetated open-channel flows. *J. Hydraul. Res.*, 47 (6), 798–810.

Oplatka, M. (1998). *Stabilität von Weidenverbauungen an Flussufern*. Zürich, Diss. ETH Nr. 12575.

Petryk, S. and Bosmajian, G. (1975). Analysis of flow through vegetation. *J. Hydraul. Div.*, 101 (HY7), 871–884.

Sagnes, P. (2010). Using multiple scales to estimate the projected frontal surface area of complex three-dimensional shapes such as flexible freshwater macrophytes at different flow conditions. *Limnol. Oceanogr. Meth.*, 8, 474–483.

Schlichting, H. and Gersten, K. (2006). *Grenzschicht-Theorie*. 10. Auflage. Berlin, Springer.

Schnauder, I. and Moggridge, H.L. (2009). Vegetation and hydraulic-morphological interactions at the individual plant, patch and channel scale. *Aquat. Sci.*, 71, 318–330.

Schoneboom, T. (2011) *Sohlen- und Formwiderstand von durchströmter flexibler Vegetation*. Dissertation. Braunschweig, Technische Universität Braunschweig.

Schoneboom, T. and Aberle, J. (2009). Influence of foliage on drag force of flexible vegetation. In: *Proc. 33rd IAHR Congress*, 9–14 August 2009, Vancouver, Canada. Papers on CD-Rom.

Schoneboom, T., Aberle, J. and Dittrich, A. (2010). Hydraulic resistance of vegetated flows: Contribution of bed shear stress and vegetative drag to total hydraulic resistance. In: Dittrich, A., Koll, Ka., Aberle, J. and Geisenhainer, P. (eds.) *Proc. Int. Conf. on Fluvial Hydraulics River Flow 2010*, 8–10 September 2010, Braunschweig, Germany. Karlsruhe, Bundesanstalt für Wasserbau, pp. 269–276.

Schoneboom, T., Aberle, J. and Dittrich, A. (2011). Spatial variability, mean drag forces, and drag coefficients in an array of rigid cylinders. In: Rowinski, P. (ed.). *Experimental Methods in Hydraulic Research, Geoplanet: Earth and Planetary Sciences*. Vol. 1, pp. 255–265, doi: 10.1007/978-3-642-17475-9-18.

Schoneboom, T., Aberle, J., Wilson, C.A.M.E. and Dittrich, A. (2008). Drag force measurements of vegetation elements. In: *Proc. 8th International Conference on Hydro-Science and Engineering (ICHE) 2008*, Nagoya, Japan. Papers on CD-ROM.

Statzner, B., Lamoroux, N., Nikora, V. and Sagnes, P. (2006). The debate about drag and reconfiguration of freshwater macrophytes: comparing results obtained by three recently discussed approaches. *Freshwat. Biol.*, 51, 2173–2183.

Stephan, U. and Gutknecht, D. (2002). Hydraulic resistance of submerged flexible vegetation. *J. Hydrol.*, 269, 27–43.

Stone, B.M. and Shen, H.T. (2002). Hydraulic resistance of flow in channels with cylindrical roughness. *J. Hydraul. Eng.*, 128 (5), 500–506.

Sukhodolov, A., and Sukhodolova, T. (2010). Case Study: Effect of submerged aquatic plants on turbulence structure in a lowland river. *J. Hydraul. Eng.*, 136 (7), 434–446.

Västilä, K., Järvelä, J., Aberle, J. and Schoneboom, T. (2011). Vegetative drag in natural, foliated plant stands. In: *Proc. 34rd IAHR Congress*, 26 June – 1 July 2011, Brisbane, Australia. ISBN 978-0-85825-868-6, pp. 2978–2985.

Vogel, S. (1994) *Life in moving fluids: the physical biology of flow*. 2nd edition, Princeton, Princeton University Press.

Wilson, C.A.M.E., Hoyt, J. and Schnauder, I. (2008). Impact of foliage on the drag force of vegetation in aquatic flows. *J. Hydraul. Eng.* 134 (7), 885–891.

Wilson, C.A.M.E., Xavier, P., Schoneboom, T., Aberle, J., Rauch, H.-P., Lammeranner, W., Weissteiner, C. and Thomas, H. (2010). The Hydrodynamic Drag of Full Scale Trees. In: Dittrich, A., Koll, Ka., Aberle, J. and Geisenhainer, P. (eds.) *Proc. Int. Conf. on Fluvial Hydraulics River Flow 2010*, 8–10 September 2010, Braunschweig, Germany. Karlsruhe, Bundesanstalt für Wasserbau, pp. 453–459.

Xavier, P., Wilson, C.A.M.E., Aberle, J., Rauch, H.P., Schoneboom, T., Lammeranner, W. and Thomas, H. (2010). Drag force of flexible submerged trees. In: *Proc. Hydralab Closing Event*, 2–4 Februar 2010, Hannover, Germany.

Yen, B.C. (2002). Open Channel Flow Resistance. *J. Hydraul. Eng.*, 128 (1), 20–39.

Chapter 11

Flow-biota interactions in aquatic systems: Scales, mechanisms, and challenges

V. Nikora[1], S. Cameron[1], I. Albayrak[1], O. Miler[1], N. Nikora[1], F. Siniscalchi[1] M. Stewart[1] and M. O'Hare[2]
[1] *School of Engineering, University of Aberdeen, Aberdeen, UK*
[2] *Centre for Ecology and Hydrology, Penicuik, Midlothian, UK*

ABSTRACT: The paper outlines the current trends and challenges in studies of flow-biota interactions particularly focusing on freshwater aquatic systems. The multiple scales and a variety of mechanisms involving hydrodynamic, biomechanical and ecological processes differentiate these interactions from conventionally studied flow-structure interaction problems, and highlight the methodological and conceptual challenges to be resolved. The appearance of new promising measurement capabilities (e.g., use of high-resolution laboratory and field panoramic and stereoscopic PIV at a variety of scales), employment of advanced turbulence concepts in flow-biota considerations, accelerated advances in organism biomechanics, and developments in multi-scale descriptions of multi-component systems represent the current trends that are briefly discussed and illustrated with case studies. These trends reflect the emergence of a new inter-discipline subject area - *Hydrodynamics of Aquatic Ecosystems* - that can be defined as a study of flow-organism interactions in running waters with particular focus on relevant transport processes and mutual physical impacts occurring at multiple scales from the sub-organism scale to the organism patch mosaic scale.

1 INTRODUCTION

Although the importance of hydrodynamics for biophysical processes that determine biological communities in streams, estuaries, lakes and seas is widely recognised, the knowledge of hydrodynamic effects in flow-biota interactions remains very limited (e.g., Hart and Finelli, 1999; Biggs *et al.*, 2005; Statzner, 2008). A slow progress in the implementation of fluid mechanical concepts and approaches into ecological theories can be explained by a variety of reasons such as measurement difficulties at scales most relevant to organisms; poorly understood biomechanical properties of organisms that change significantly across species, scales, and environments; and the absence of a solid unifying interdisciplinary platform for integrating hydrodynamic, biomechanical and ecological considerations and their scaling up (or down) from the sub-organism scale to the patch mosaic scale. It should also be noted that the subject of flow-organism interactions lies at the borders between fluid mechanics, ecology, and biomechanics, i.e., at the discipline interfaces which are typically avoided by researchers. These negative factors, however, have been significantly weakened in recent years with the appearance of new promising measurement capabilities (e.g., use of high-resolution laboratory and field panoramic and stereoscopic PIV at a variety of scales), employment of advanced turbulence concepts in flow-biota considerations, accelerated advances in organism biomechanics, and developments in multi-scale descriptions of multi-component systems. This paper is an attempt to further enhance and promote mechanics of flow-biota interactions as an emerging research area at the interfaces between environmental fluid mechanics, aquatic ecology, and biomechanics. This new area, that may be called *Hydrodynamics of Aquatic Ecosystems*, bridges these disciplines together and can be defined as a study of flow-organism interactions in running waters with particular focus on relevant transport processes and mutual physical impacts (e.g., due to interplay between flow-induced forces and reaction forces generated by the organisms) occurring at multiple scales from the sub-organism scale to the organism patch mosaic scale comparable to the flow width (Nikora, 2010a, b). The focus of *Hydrodynamics of Aquatic Ecosystems* on the interfaces between fluid mechanics, ecology and biomechanics should help with elimination of existing knowledge gaps in the least understood facets of flow-biota interactions. The considerations in this paper mainly relate to hydrodynamic aspects of flow-biota interactions. Their association with biomechanical and ecological counterparts has been recently discussed in Nikora (2010a, b).

A variety of scale-dependent mechanisms involving hydrodynamic, biomechanical and ecological processes is a fundamental feature of flow-biota interactions and therefore in Section 2 we first draw attention to the scale issues that differentiate these interactions from conventionally studied flow-structure interaction problems. Then, in Section 3, we highlight current trends and challenges that researchers encounter in studying scale-dependent mechanisms of flow-biota interactions, followed in Section 4 by three recent case studies to illustrate these challenges. The centre of attention of this paper is on freshwater systems (i.e., streams and rivers) where, compared to marine systems, the knowledge of flow-biota interactions remains fairly modest. The material is unavoidably biased by the authors' studies for the last 10 years which were partly discussed in Nikora (2004, 2007, 2009, 2010a, b), Nikora and Nikora (2007, 2010), O'Hare (2007), Windsor *et al.* (2010a, b), Albayrak *et al.* (2011), Cameron (2011), Miler *et al.* (2011), and Siniscalchi *et al.* (2011), and other papers.

The *Hydrodynamics of Aquatic Ecosystems* has been considered by Professor Gerhard H. Jirka as an important emerging branch of Environmental Fluid Mechanics and the authors are deeply grateful for his encouragement, constant support, and inspiring discussions of this topic.

2 MULTI-SCALE MECHANICS OF FLOW-BIOTA INTERACTIONS

Physical interactions and ecologically relevant mass-transfer processes in aquatic ecosystems occur in a wide range of scales and depend on how characteristic physical scales match biological scales such as organism dimensions, patch/community dimensions, life cycles, and others. Furthermore, at least some of the physical and biological scales may not be independent and arise as a result of flow-biota interactions. These characteristic scales represent a variety of interaction-induced mechanisms, from the sub-organism scale to the system scale. For example, there is a growing body of evidence that the interplay between flow and biota is largely responsible for the aggregation of aquatic organisms into patches which, in turn, are often 'organised' into even larger aggregations called patch mosaics (e.g., Okubo and Levin, 2001). These multi-scale aggregations represent a form of organisms' 'self-organisation', in addition to the traditionally considered patchiness 'imposed' by heterogeneity of habitat properties and/or nutrient supply. The biological scales such as those of community patchiness are often reflected in the flow structure as physical scales of biota-induced hydrodynamic patterns (e.g., wakes behind individual organisms or their patches; see examples in the following sections).

The above conjecture suggests that physical scales in aquatic ecosystems represent two kinds of mechanisms, those induced by abiotic environment and therefore independent of ecosystem's biota, and those which are induced by flow-biota interactions. This conjecture can be illustrated using streams and rivers as an example. Flow variability in streams and rivers covers wide ranges of temporal and spatial scales, from milliseconds to many years and from sub-millimetres to tens of kilometers. This variability is largely driven by abiotic factors and can be conveniently summarised using velocity spectra showing how the energy of fluctuations is distributed across the scales (Figure 1, modified from Nikora, 2008). The low frequency (large periods) range in the frequency spectrum is formed by intra-annual and inter-annual hydrological variability while the high-frequency (small periods) range is formed by flow turbulence (Figure 1a). The low wave-number (large spatial scale) range in the wave-number spectrum is formed by morphological variability along the flow such as bars and/or meanders (Figure 1b). At smaller spatial scales (comparable to and less than the flow width) velocity fluctuations are due to turbulence. This 'turbulence' range of scales is most relevant to organisms as their own scales (up to the patchiness scale) typically fall within this range. Thus, the biota-induced mechanisms and associated scales are most likely to be observed at scales comparable to or smaller than the channel width. Although our focus is on the instream biota, it is worth noting that flow-biota interactions may also control larger or much larger scales such as those of channel meandering that may be significantly influenced by riparian and terrestrial vegetation (G. Parker, personal communication). The examples of the hypothesized biota-induced scales, for the case of aquatic vegetation, are shown in Figure 1 and are supported by laboratory and field experiments (e.g., Nepf, 1999; Naden *et al.*, 2006).

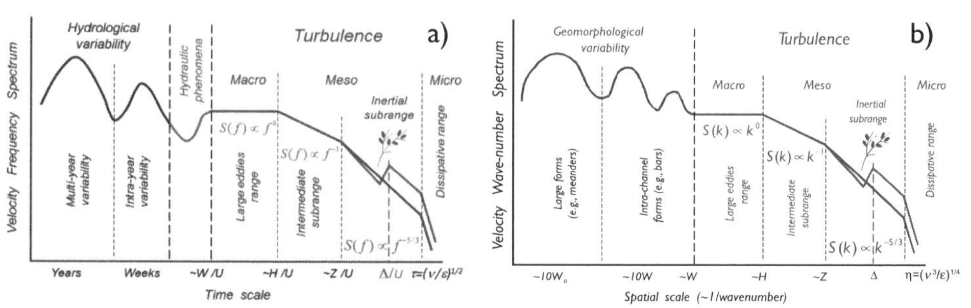

Figure 1 Schematised velocity spectra in rivers: (a) frequency spectrum; and (b) wave-number spectrum (W_0 and W = river valley and river channel widths, H = depth, Z = distance from the bed, Δ = plant scale, U = flow velocity, η = turbulence micro-scale).

The authors' experience suggests that within a wide range of scales (from the sub-organism scale to the patch mosaic scale comparable to the lateral flow size) there are distinct characteristic scales where flow-organism interactions and transport processes are scale-specific and interconnected. There are at least four such characteristic scales: sub-organism scale (or several sub-organism scales for an organism of complex morphology such as many aquatic plants), organism scale, patch scale, and patch mosaic scale (Nikora, 2010a). It is likely that the strength of interaction between the scales diminishes with increase in scale separation, although there may be certain long-range correlations between the scales.

In the following sections we will discuss research challenges in this subject area and then will illustrate them with examples of flow-vegetation interactions at three distinct scales: single leaf scale, single plant scale, and plant patch scale.

3 RESEARCH CHALLENGES

The conceptual suite of *Hydrodynamics of Aquatic Ecosystems* is likely to be always strongly linked to the original mother disciplines, i.e., fluid mechanics, biomechanics, and ecology. Fluid mechanics contributes to *Hydrodynamics of Aquatic Ecosystems* with concepts of boundary layers (BL), mixing layers (ML), wakes, and jets. Depending on the specific conditions these flow types may exhibit properties of two turbulence phenomena: coherent structures (CS) and/or eddy cascades (EC). In aquatic ecosystems, BL, ML, wakes, and jets are fundamental for characterisation of both (1) hydraulic habitats, as most aquatic communities live within BLs, MLs, etc; and (2) flow patterns around individual organisms that are often surrounded by BLs, MLs, or wakes generated at organism surfaces or within/around organism communities. The occurrence of these flow types in aquatic ecosystems, however, often deviates from their canonical forms, thus leading to *Challenge #1: What are the manifestations of the canonical flow types in aquatic ecosystems?* Figure 2 may illustrate this challenge for the case of aquatic plants which typically span a wide range of scales from a leaf scale to individual plant to the plant patch mosaic (i.e., an assemblage of plant patches of different shapes and sizes). Indeed, biological communities quite often are embedded in a superposition of interacting multi-scale BLs generated by a variety of boundaries including those introduced by the organisms themselves (e.g., flow-depth BL and leaf/stem BLs in Figure 2a).

Often, these BLs have limited thicknesses and small relative submergence of roughness elements, being often organised as a cascade of internal boundary layers (e.g., Nikora, 2010a). As a result, the conventional concepts and descriptions may not be always applicable and may require refinements (e.g., applicability of the conventional log velocity profile in low-submergence BLs is questionable and needs to be justified). The flow patterns, schematically summarised in Figure 2 for the case of aquatic plants, may include (Nikora, 2010b): (1) 'conventional' depth-scale shear-generated turbulence which may be significantly altered by the vegetation; (2) canopy-height-scale turbulence resulting from the Kelvin-Helmholtz instability (KHI) at the upper boundary of the vegetation canopy (known as the mixing-layer analogy (Raupach *et al.*, 1996); (3) generation of small-scale turbulence associated with flow separation from stems (i.e., von Karman vortices); (4) generation of

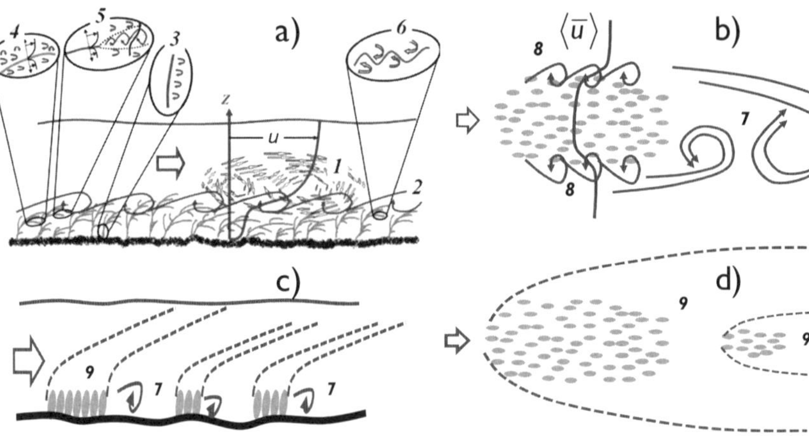

Figure 2 Hypothesised flow patterns in vegetated channels: side view at a patch scale (a); plan view at a patch scale (b); side view at a patch mosaic scale (c); and plan view at a patch mosaic scale (d).

small-scale turbulence within local boundary layers attached to leaf/stem surfaces; (5) generation of small-scale turbulence behind plant leaves serving as small 'splitter plates' that generate local leeward mixing layers, with subsequent turbulence production through the Kelvin-Helmholtz instability; (6) turbulence generation due to plant waviness at a range of scales (if biomechanical properties allow this); (7) generation of large-scale 3D and 2D turbulence associated with wakes and flow separation at a patch scale; (8) generation of 3D and 2D boundary layer and mixing layer turbulence at patch sides aligned with the flow; and (9) generation of interacting vertical and horizontal internal boundary layers at the patch mosaic scale. The patterns in Figure 2 are hypothesized based on conceptual consideration and some results from laboratory and field studies (e.g., Hondzo and Wang, 2002; Sand-Jensen, 2003; Poggi *et al.*, 2004; Ghisalberti and Nepf, 2006; Nikora and Nikora (2007, 2010); Nezu and Sanjou, 2008; Nikora, 2010a, b; Sukhodolov and Sukhodolova, 2010; Folkard, 2011; Rominger and Nepf, 2011; Nepf, 2012). The occurrence of these patterns and their exact manifestations (if any) are not yet clear. The identification and quantification of interrelationships between these patterns, as well as detection of their individual and combined roles in transport processes and drag generation for a range of biomechanical parameters represent *Challenge #2: What are the combined effects of canonical flow types in aquatic ecosystems?*

Figure 2 also highlights a possibility that biological communities may create unique flow types with specific properties making them distinctly different from the canonical flow types or their combinations, leading to *Challenge #3: Do flow-biota interactions create new unconventional flow types or patterns which are still waiting for identification?* To illustrate this point, one may look into a "mixing layer analogy" originally proposed for terrestrial vegetation canopies by Raupach *et al.* (1996) and further advanced in Finnigan (2000) and Finnigan *et al.* (2009). For the case of submerged aquatic vegetation, this analogy was first implemented by H. Nepf's Group (e.g., Nepf, 2012) and then used in a number of follow-up studies of flow-vegetation interactions (e.g., Poggi *et al.*, 2004; Nezu and Sanjou, 2008). These studies showed that large-scale ML eddies formed as a result of KHI at the canopy top may play a crucial role in mass and momentum exchange between canopy region and flow region above the canopy. Although the mixing layer analogy for aquatic vegetation has already been explored (see citations above), there are still a number of issues that require clarification. Some of them suggest that the mixing layer analogy may be a manifestation of a new flow type (i.e., not described previously) that exhibits unique properties absent in canonical flows. Examples include (1) the existence of a 'detached' logarithmic BL above a ML at the canopy top (i.e., ML may block access of BL eddies to the canopy layer thus 'detaching' BL eddies from the bed and destroying the conventional conditions for BL formation); (2) big difference between the convection velocity of large eddies at the canopy top and a local mean velocity, reported for both terrestrial and aquatic canopies (e.g., Finnigan, 2000), although for conventional mixing layers these two velocities should be equal or very close (Ho and Huerre, 1984); and (3) monami effect, i.e., wavy motions of a canopy top often observed in natural aquatic canopies, known as 'honami' for terrestrial canopies (Okubo and Livin, 2001). Using these features (especially the existence of the logarithmic layer above the mixing layer at the canopy top), we may further suggest that the hypothesized new flow type is a materialization of a generalized boundary layer concept where a solid boundary, required by a conventional concept, is replaced with a region of a spatially distributed momentum sink. In this conjecture, the momentum sink region plays the same physical role as the conventional boundary, i.e., it creates a downward vertical flux of momentum required for the appearance of the conventional logarithmic layer (e.g., Landau and Lifshitz, 2004). Within this generalized concept, the momentum sink mechanisms define the virtual position of the bed which is likely to correspond to the position of the resultant momentum sink, in analogy with the flat solid boundary. The momentum sink mechanisms may also influence the mechanisms of the momentum transport towards the sink region and, through this, the constants in the logarithmic formula. Applying this conjecture to the flow over a vegetation canopy (or a mussel bed) we may define the canopy as the momentum sink region while the mixing layer at the top of the canopy and wakes behind vegetation elements provide the mechanisms for the momentum sink. This generalized concept, where a solid boundary is replaced with a momentum sink surface or a spatial region, may also explain the existence of the logarithmic boundary layers reported for flows over beds of porous material and may be a useful tool in other research areas such as magnetohydrodynamics.

The multi-scale property together with the physical and biological complexity of boundary conditions in aquatic ecosystems highlight a need for an appropriate unifying framework that would serve for flow-biota coupling and integration (i.e., up-scaling) of physical interactions and mass-transfer processes. The integration of fluid mechanical, biomechanical, and ecological processes together and upscaling the effects of these processes from the sub-organism scale to the patch mosaic scale constitute another task awaiting to be completed. This task, *Challenge #4*, should lead to the development of the unifying framework expected to be (1) quantitative by nature; (2) capable of coupling fluid mechanical, ecological, and biomechanical processes in a reasonably rigorous way; (3) a convenient and rigorous tool for upscaling small-scale flow-organism interactions to a larger scale (e.g., from the organism scale to patch or patch mosaic scale); (4) suitable as a basis for mathematical modelling and computer simulations; and (5) appropriate for guiding field and laboratory studies and data interpretation and generalization (Nikora, 2010b). The spatially filtered (but instantaneous in time domain) hydrodynamic, transport, and biomechanical equations, which couple flow and organisms together through a rigorous spatial averaging operation (over local volume or area in the plane parallel to the mean bed surface) may serve as a potential candidate for such a framework. The coupled spatially filtered equations can be derived for both fluid (considering organisms as embedded media) and organisms (considering fluid as embedded media). The flow and 'organism' equations are linked by the interface terms describing physical interactions and/or exchange of substances (e.g., the same term describing transport of nutrients through organisms' surfaces will be included in both 'flow' and 'organism' equations, but with opposite signs). The 'instantaneous' equations can also be simultaneously spatially- and time-averaged to produce the double-averaged (in time and space) coupled equations for fluid and organisms. The double-averaged equations for the fluid phase have been originally proposed to describe flow properties within and above terrestrial canopies (e.g., Finnigan, 2000 and references therein; see also Nikora *et al.*, 2007 for derivation and detailed discussion). The instantaneous spatially averaged equations and double-averaged equations explicitly contain important (although still unconventional) terms such as form-induced stresses and fluxes, and for the flow region with embedded organisms, form and viscous drag terms, wake and waving production terms (e.g., energy production due to the wake effects behind mussels or due to mobile interfaces such as plants), and source/sink terms describing interface transport and heterogeneous reactions (e.g., sediment 'breathing' or nutrient uptake by aquatic organisms). The spatial averaging methodology is conceptually close to the Large-Eddy Simulation (LES) philosophy, which is currently actively used in turbulence research. In relation to organisms, the spatial averaging approach can be supplemented with the homogenisation techniques such as those developed in composite materials mechanics (e.g., Torquato, 2002).

4 CASE STUDIES OF FLOW-VEGETATION INTERACTIONS AT MULTIPLE SCALES

The challenges highlighted in the previous section can be illustrated with three examples which reflect the recent trends in this research area, particularly (1) multi-scale considerations, (2) the appearance of new measurement capabilities (e.g., use of high-resolution laboratory and field panoramic and stereoscopic PIV at a variety of scales), and (3) employment of advanced turbulence concepts in flow-biota considerations. The examples include unique experiments on flow-plants interactions at a patch scale (field PIV study, Section 4.1), individual plant scale (laboratory PIV study, Section 4.2), and leaf scale (laboratory PIV study, Section 4.3). Although the results presented below are still preliminary they highlight a number of hydrodynamic features unknown before, and define the focus for further steps.

4.1 Flow-vegetation interactions at a patch scale

This example represents a world-first application of stereoscopic PIV in a field experiment to study flow-vegetation interactions. At the core of the in-situ stereoscopic PIV system developed in our group is a rigid sub-assembly that holds a pair of cameras and laser sheet forming optics at a fixed position relative to a glass bottomed boat shaped structure (Figure 3).

The 'boat' incorporates water prisms (e.g., Prasad, 2000) to minimise refraction and internal reflections at the air-glass-water interface. The cameras (Dalsa 4M60, 2352 × 1728 pixels, 60 frames per second, Nikon 60 mm lens, f/5.6) are mounted at an angle of approximately 30 degrees from vertical and are adjusted according to the Scheimpflug principle to ensure that the vertical light sheet is entirely within the field of view of the cameras. The sub assembly is designed to be configured and calibrated in the laboratory, so that the only adjustment required in the field is to carefully position the structure so that the bottom of the 'boat' just skims the water surface, minimising any disturbance to the flow. The 'boat' assembly is supported by a carriage that traverses a 7.5 m long Aluminium extrusion truss bridge. The carriage also supports a computer system for continuous direct-to-disk recording of images from a pair of cameras, and the laser system (100 mJ per pulse Nd:YAG, Oxford Laser Neo 50/100 PIV) which connects to the laser sheet forming optics via an articulated arm. The truss bridge is mounted on abutment rails, and can be traversed over a 0.5 m range in the streamwise direction. The entire bridge, carriage, and 'boat' sub-assembly can be installed at a field site by a 6-person team in around 8 hours.

The in-situ stereoscopic PIV system was deployed on the Urie River, near the town of Inverurie and around 25 minutes drive from Aberdeen City. The selected site featured a rich variety of aquatic plants including *Myriophyllum, Potamogeton, Callitriche*, and *Ranunculus genera* along with various aquatic mosses. At the measurement location, the river was 12.9 m wide and 0.385 m deep, the flow rate was 2.7 m^3/s and the water surface slope was $1.5 \pm 0.4 \times 10^{-3}$. The Reynolds number based on flow depth and mean velocity was 152×10^3 and the Froude number was 0.28. The gravel bed of the river had an estimated D_{50} of 35 mm and featured occasional boulders and sandy patches interspersed with patches of aquatic plants. A *Ranunculus*

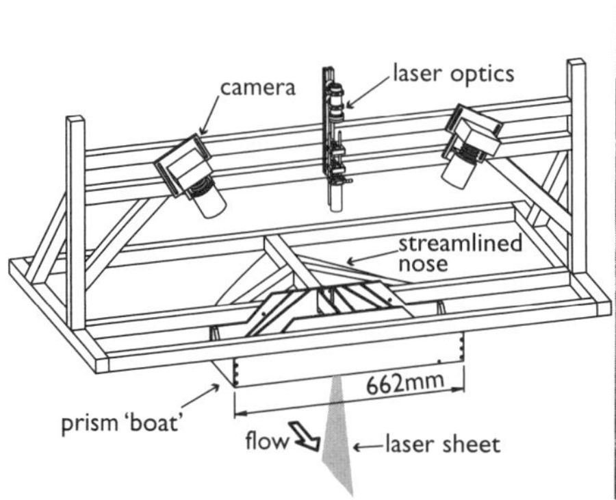

Figure 3 In situ stereoscopic PIV setup: prism 'boat', camera, and laser optics sub assembly (left), and deployed in the Urie River, Scotland (right).

penicilatus plant, 2.8 m from the river bank, and near the middle of a 50 m long approximately straight section of the river was selected for the study. A set of three PIV records were collected, each three minutes long and incremented in the streamwise direction by 130 mm resulting in a total measurement coverage of 400 mm in the streamwise direction and 320 mm in the vertical direction. The recording rate was 30 image pairs per second, but some randomly distributed image pairs were found not to be viable due to a technical issue, resulting in an average recording rate of 20 image pairs per second.

Mean flow velocity streamlines, time averaged streamwise velocity ($\overline{u}$), and turbulent kinetic energy ($0.5[\overline{u'u'} + \overline{v'v'} + \overline{w'w'}]$, where $\overline{u'u'}$, $\overline{v'v'}$ and $\overline{w'w'}$ are the variances of the streamwise, transverse, and vertical velocity components respectively) illustrate the main characteristics of flow field around the *Ranunculus* plant (Figures 4 and 5).

An inflection point in the mean velocity profile behind, and near the top level of the plant indicates a sink of momentum associated with pressure or viscous drag forces induced by the plant (Figure 5). Immediately behind the plant, the mean flow velocity is significantly reduced, but the flow has not separated, and no recirculation zone forms. In contrast, the small rock near x = 250 mm clearly exhibits flow separation (Figure 4) – a signature of pressure drag. The absence of flow separation from the plant may indicate a viscous drag mechanism as was suggested in Nikora

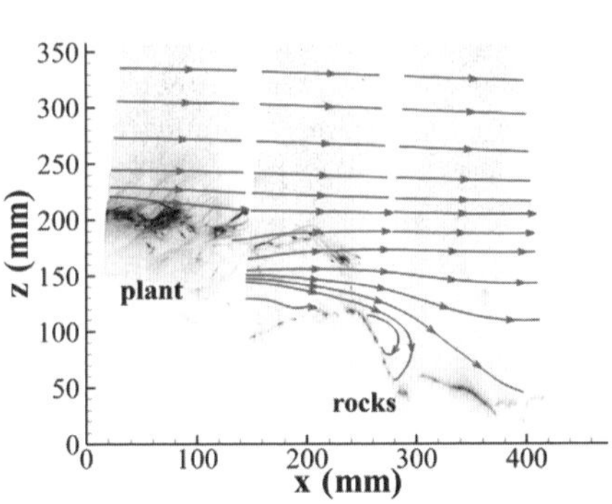

Figure 4 PIV images (inverted gray levels) from 3 measurement planes with mean flow streamlines.

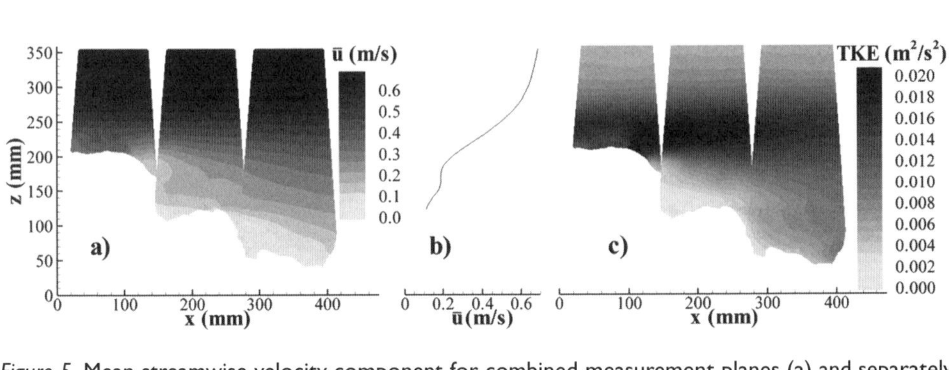

Figure 5 Mean streamwise velocity component for combined measurement planes (a) and separately for x = 210 mm (b); turbulent kinetic energy (c).

and Nikora (2007) and Nikora (2010a). However, separation and pressure drag may still occur at the scales of individual plant stems or leaves not visible in Figure 4.

The turbulent kinetic energy distribution has a maximum near the level of the top of the plant and decays rapidly both towards the river bed and the free surface. This localised region of high turbulent kinetic energy may be associated with flow instability in the wake of the plant such as the Kelvin Helmholtz instability (KHI) mechanism suggested for plant canopies by Raupach *et al.* (1996). The applicability of KHI in the environment of high background turbulence, and for a single plant is however still unclear and other mechanisms of energy production driven by the large mean velocity gradients in the plant wake may be possible. Fluctuations of the plant may also contribute significantly to the production of turbulent kinetic energy, and therefore it is worth examining potential interactions between plant and fluid velocity fluctuations.

Plant velocity fluctuations were extracted from PIV images using a cross correlation based method. Rectangular interrogation regions were used with one side extended to cover the entire plant cross section, but narrow in the streamwise direction so that plant fluctuations could be estimated as a function of position along the plant. By combining the velocities measured from both cameras, plant velocities in the transverse (v_p) and vertical (w_p) directions were separated. The power spectral density of plant velocity fluctuations (Figure 6a) reveals a characteristic frequency around 1 Hz. This frequency may reflect the passage of periodic flow structures such as those generated by KHI, or alternatively might be interpreted as a natural frequency of the plant. Aquatic plants such as *Ranunculus*, however, have very low flexural rigidity and it is unlikely that resonance phenomena such as that observed in terrestrial crops (e.g., de Langre, 2008), could have a significant influence on the plant velocity spectrum. It is interesting to note that fluid velocity fluctuations measured behind the plant (Figure 6b), where the turbulent kinetic energy distribution has a local maximum, also exhibit increased energy near 1 Hz. While this suggests a possible coupling between plant and fluid fluctuations, it is not immediately clear whether plant fluctuations are generating the fluid fluctuations or vice versa. Figures 6a, b also indicate that transverse and vertical velocity fluctuations are nearly isotropic, both for the plant, and for the fluid in the wake. This suggests that the turbulent energy production mechanisms in the shear layer created by the plant are substantially three-dimensional, which may relate to the multistage breakdown of vortices generated by KHI due to secondary instabilities (Finnigan, 2000; Finnigan *et al.*, 2009).

Normalised covariance estimates

$$R_{v_p,v} = \overline{v'_p\, v'} / \left(\overline{v'^2_p}\ \overline{v'^2}\right)^{0.5} \text{ and } R_{w_p,w} = \overline{w'_p w'} / \left(\overline{w'^2_p}\ \overline{w'^2}\right)^{0.:} \qquad (1)$$

between fluid and plant velocity fluctuations show that a region above the plant, where fluid and plant motion is strongly correlated, is surprisingly narrow (Figure 6c). The thinness of this region indicates the flow plant interaction is a local phenomenon and probably not associated with depth scale flow structures. Overall, the preliminary data analysis is consistent with periodic flow instability, generated in the high shear region near the top of the plant, with a characteristic frequency of 1 Hz which causes the plant to fluctuate at the same frequency. Further analysis of the data, which

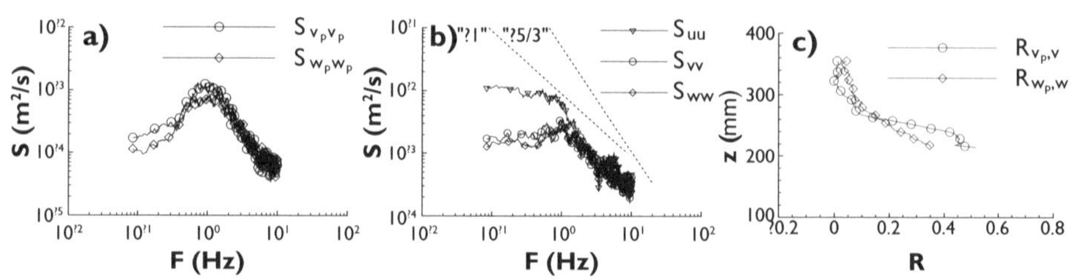

Figure 6 Power spectral density of plant velocity fluctuations (a) and fluid velocity fluctuations in the wake of the plant at x = 184 mm, z = 225 mm (b); and normalised covariance between fluid and plant velocity fluctuations (c).

is underway, may reveal other possible interaction mechanism between the plant and the flow.

4.2 Flow-vegetation interactions at a scale of individual plant

A set of laboratory experiments were conducted to further evaluate the potential mechanisms of flow plant interaction in a controlled environment at a smaller scale. The experiments were completed in the 1.18 m wide, 20 m long Aberdeen Open Channel Facility (AOCF). The entire bed of the channel was covered by a single layer of 16 mm diameter glass spheres in a hexagonal close packed arrangement. Two plant species were tested, *Ranunculus penicillatus* and *Myriophyllum alterniflorum*, with each plant consisting of six 300 mm long stems which were anchored to a drag measurement device using a small cable gland. Preliminary results are presented below for experiments conducted with a 120 mm flow depth, a flow rate of 0.081 m^3/s, a bed surface slope of 1.5×10^{-3}, a Reynolds number of 69×10^3 (based on flow depth and mean velocity), and a Froude number of 0.53. Detailed measurements of the flow field in a 3-dimensional region surrounding the plant were carried out using panoramic and stereoscopic particle image velocimetry. The panoramic PIV mode, employed in this set of experiments, uses four cameras in a side-by-side arrangement to measure two velocity components (u and w) over an 840 mm flow region in the streamwise direction and covering the entire 120 mm flow depth. The panoramic measurement plane is aligned with the centre of the plant and covers the 300 mm plant length and 540 mm of the plant wake. Stereoscopic PIV was used in a cross flow configuration to measure all three velocity components over a measurement area of 340 mm in the transverse direction and the entire 120 mm flow depth. The stereoscopic measurement plane was deployed in four different streamwise locations, one upstream and three downstream of the plant to capture both the approach flow and the plant wake.

Mean streamwise velocity combined for all measurement planes has similar characteristics for both the *Ranunculus* and *Myriophyllum* plants (Figure 7). Immediately behind the plant, a wake region forms which extends over an area approximately equal to the cross section of the plant. In the centre of the wake, at the first transverse measurement plane behind the plant (x = 361 mm), the velocity deficit $(\overline{u}_{us} - \overline{u}_{ds}) / \overline{u}_{us}$ is 0.28 and 0.37 for *Ranunculus* and *Myriophyllum* respectively ($\overline{u}_{us}$ and $\overline{u}_{ds}$ are the

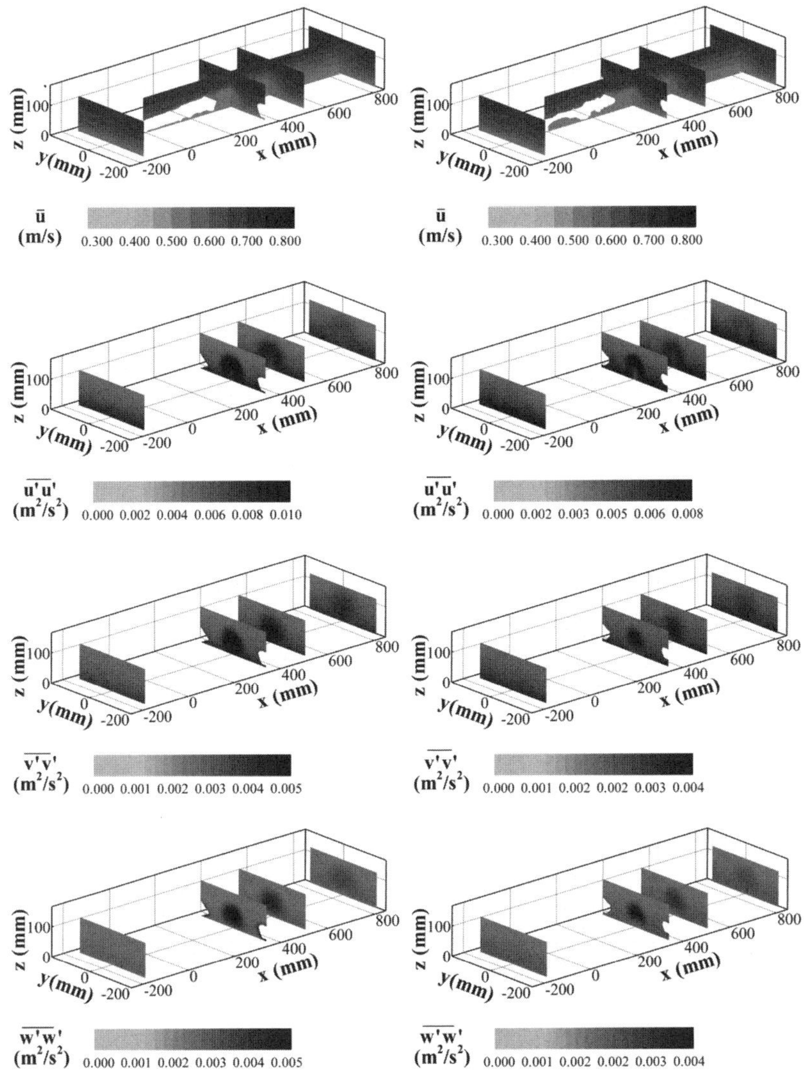

Figure 7 Mean streamwise velocity and velocity variances for *Myriophyllum* (left column) and *Ranunculus* (right column) from laboratory plant scale experiments.

upstream of the plant and downstream mean velocities measured at corresponding y and z coordinates). At the furthest downstream measurement plane (x = 832 mm) the velocity deficit is reduced to 0.09 for *Ranunculus* and 0.10 for *Myriophyllum*, but the wake has expanded to cover a much larger area (defined by the zero crossing of the velocity deficit), including the entire flow depth.

The velocity variances in the wakes of the plants have rather distinct characteristics for each component (Figure 7). The streamwise velocity variance forms a horseshoe-shaped pattern of comparatively large variance compared to the surrounding flow whereas the transverse and vertical variances form circular regions. The region of increased vertical velocity variance has a flat bottom, probably due to the dampening effect of the bed. The source of these unique patterns remains to be clarified, and a complete analysis of the per component turbulent kinetic energy budget would seem a useful starting point.

Frequency spectra calculated for velocity fluctuations in the wake of the *Ranunculus* and *Myriophyllum* plants indicate that turbulent energy is increased across a broad range of scales relative to the same flow without a plant present (Figure 8). The largest difference between plant and no-plant spectra occur for frequencies near 2 Hz which might, similar to the field experiments, be interpreted as a characteristic frequency of either the plant motion, or some periodic turbulence production mechanism. The power spectrum of plant drag force fluctuations also indicates potential periodicity at similar frequencies to those observed for the wake velocity (Figure 9). Their origin and significance to the total drag force fluctuation is not immediately apparent. Analysis of the phase and magnitude relationships based on the synchronous wake velocity and drag force fluctuation measurements may provide further insight into the nature of the periodicity. The low frequency part of the drag force spectrum in Figure 9 is most likely formed by large-scale upcoming turbulence.

4.3 Flow-vegetation interactions at a leaf scale

Further experiments were conducted to study potential mechanisms of flow-plant interactions at even smaller scale of a single plant leaf. Small plant leaves (less than 30 mm long) were towed through stationary water at speeds between 0.1 and 0.8 m/s

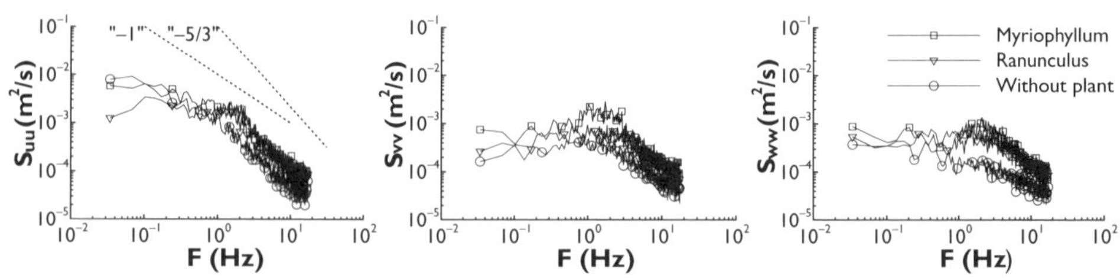

Figure 8 Power spectral density of fluid velocity fluctuations in the wake of the plant at x = 371 mm, y = 0 mm, z = 40 mm.

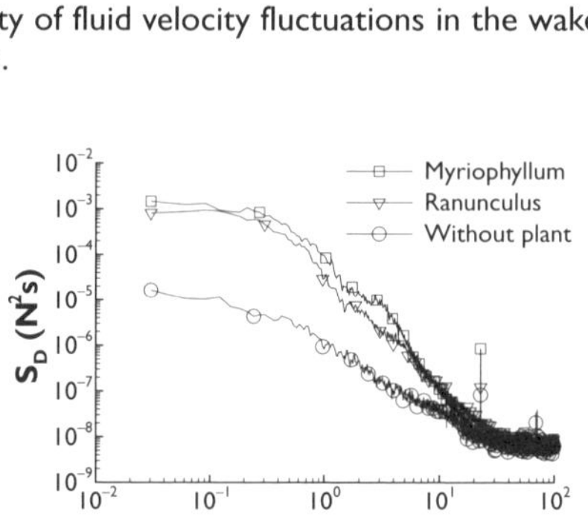

Figure 9 Power spectral density of drag force fluctuations.

using the AOCF precision instrumental carriage with synchronous high resolution PIV and drag force measurements. Leaf drag forces were measured using a MEMSCAP AE-801 sensor element which was mounted inside a 2.4 mm diameter glass tube to shield it from the flow as illustrated in Figures 10a, b. The glass tube was mounted inside a 6.0 mm diameter stainless steel tube with a tapered nozzle to streamline the diameter transition. The sensing element of the AE-801 was extended by a short length of 0.2 mm diameter tungsten wire which had sufficient rigidity to transmit the load between plant leaves and the sensor element. Plant leaves were attached to the end of the wire using a small amount of cyanoacrylate adhesive and were a sufficient distance from the glass tube to avoid any interaction with the tube wake. PIV measurements were made in a horizontal plane, level with the tip of the drag measurement device using two-component, two-camera panoramic PIV. A pair of rhomboid prisms were used to form a stable optical interface at the water surface and displace the optical axes of the cameras so that their respective fields of view (30 mm in the streamwise direction) overlapped by a small amount. This overlap could not be achieved without the prisms because the width of the camera body (90 mm) did not allow the cameras to be positioned close enough to each other. A stainless steel frame streamlined and supported the prisms and fixed the drag measurement device at the correct position 50 mm below the lower surface of the prisms. This distance was sufficiently large that the boundary layer created by the prism arrangement did not interfere with the leaf. Leaves from four plant species were tested, including *Myriophyllum alterniflorum*, *Ranunculus penicillatus*, *Elodea canadensis*, and *Cryptocoryne undulata* (Figure 10c). Each leaf was towed over a distance of 13.4 m, and several repeats were made to ensure at least 60 s of data was collected at each of the towing speeds.

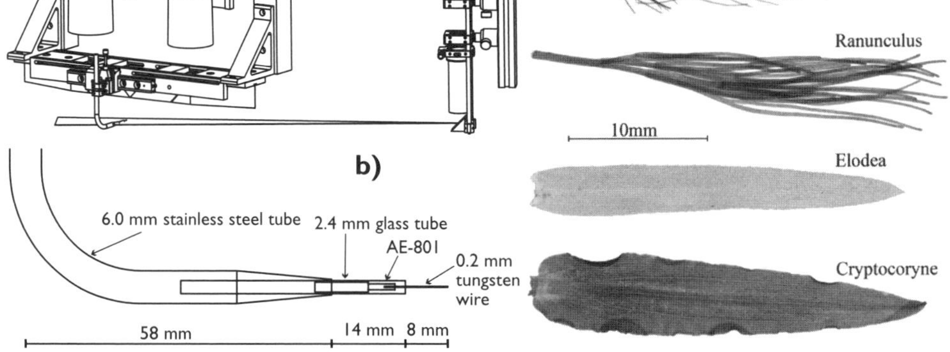

Figure 10 Particle image velocimetry (a) and drag force measurement setup (b) in laboratory leaf scale experiments; photographs of the leaves used for the experiments (c).

Measured mean leaf drag forces ($\overline{F_D}$) as function of the towing velocity (u_t) are illustrated in Figure 11 for the four tested plant species, and for the drag device without a leaf attached. For simply shaped bodies such as cylinders the scaling of the drag force with velocity provides an indication of the balance between viscous ($F_D \propto u_t$) and pressure drag ($F_D \propto u_t^2$). For a cylinder shaped object, such as the tungsten wire of the drag measurement device, the transition from viscous to pressure dominated drag is expected at a Reynolds number around 100 (0.5 m/s for the 0.2 mm diameter wire). For aquatic plant leaves, however, interpreting the balance between viscous and pressure drag is not as straightforward because there are several potentially competing mechanisms at play. Firstly, flexible leaves can reconfigure themselves depending on the flow velocity to become more streamlined. We observed this phenomenon for all of the tested leaves, including *Elodea* and *Cryptocoryne* which had an initially curved profile and tended to straighten with increasing velocity. Secondly, for the *Ranunculus* and *Myriophyllum* leaves, the flow is able to penetrate through the leaf structure more effectively at higher velocities due to the thinner boundary layers forming on the individual leaf elements. This is reflected by the reduced wake velocity deficit observed for larger tow velocities (Figure 12), and is likely to increase the drag force as more leaf elements are exposed to the flow.

Finally, all of the tested leaves were observed to oscillate slightly as they were towed through the water which may have an effect on the mean drag force. The origin of these vibrations is not yet clear, but they may be related to the periodic vortex production that was observed for the *Myriophyllum* and *Cryptocoryne* leaves (Figure 13). For the *Myriophyllum* leaf, the vortices suddenly appear some distance downstream of the leaf and are likely to be the result of the Kelvin Helmholtz instability of the inflectional velocity profile in the wake. It is interesting to note that the vortices only appear on one side of the leaf, probably due to the asymmetry of the leaf structure. For the *Cryptocoryne* leaf, the periodic vortices appear immediately at the leaf tip and are found in both clockwise and counter-clockwise rotating forms as indicated by the signed swirling strength $\omega\lambda_{ci} / |\omega|$ (Christensen and Wu, 2005), where $\omega = \partial v / \partial x - \partial u / \partial y$ is the vertical component of the vorticity vector, and λ_{ci} is the imaginary component of the complex conjugate eigenvalue of the two-dimensional

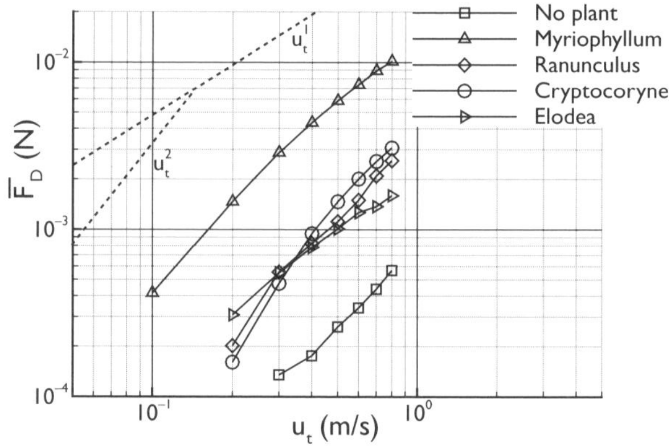

Figure 11 Mean drag force versus towing velocity from laboratory leaf scale experiments.

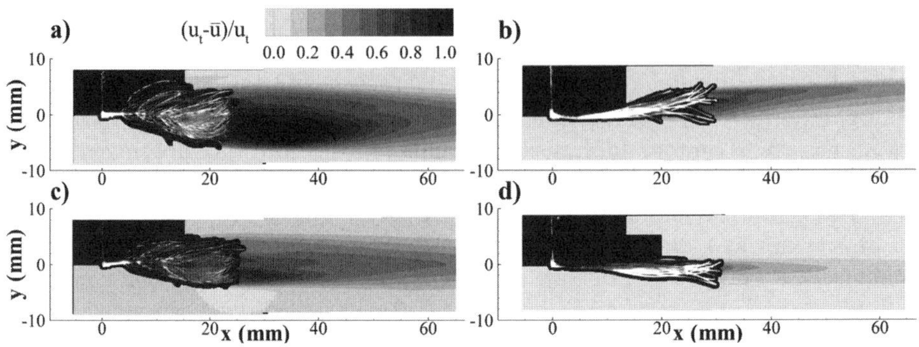

Figure 12 Wake velocity deficit for *Myriophyllum* (a, c) and *Ranunculus* (b, d). Wake velocity deficits for towing velocity of 0.2 m/s (a and b) and 0.8 m/s (c and d) are shown.

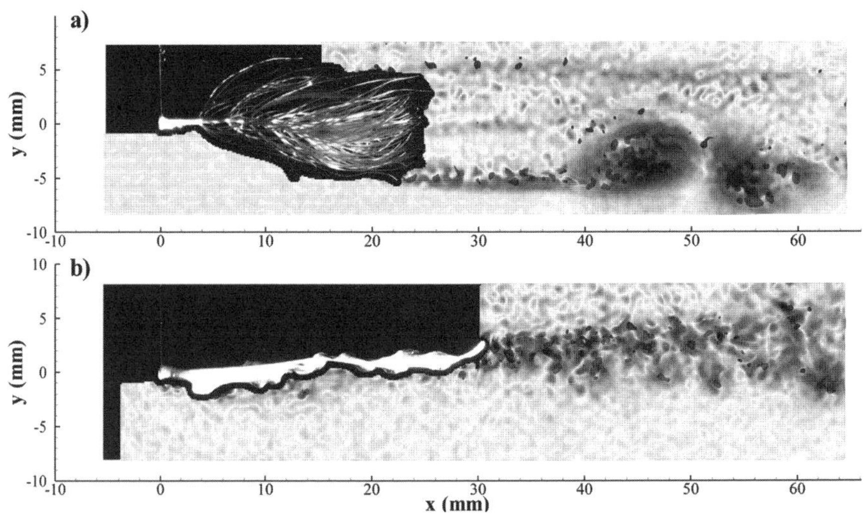

Figure 13 Snapshots of instantaneous velocity fluctuation vectors and contours of signed swirl strength for (a) *Myriophyllum* and (b) *Cryptocoryne*.

velocity gradient tensor. These vortices may also result from the KHI in the wake of the leaf, but alternatively, could be generated by unsteady flow separation from the corrugated leaf perimeter. Further analysis of the data may provide further insight into the nature of these periodicities and their potential interaction with the leaf motion.

5 CONCLUSIONS

Addressing the challenges described in the paper and further exploration of the effects highlighted in the case studies should help in eliminating multiple knowledge gaps at the borders between fluid mechanics, ecology and biomechanics, i.e., areas where the probability of new discoveries is highest. In this respect, *Hydrodynamics of Aquatic Ecosystems* will provide a missing platform for developing process-based models, to replace current approaches such as diffusion-type approximations, often operating with coefficients disconnected to the underlying processes and actual organisms (Okubo and Levin, 2001). The knowledge on specific mechanisms of flow-biota interactions will also enhance capabilities of large-scale models based on complex systems approaches which are currently under active development and may have significant practical relevance. *Hydrodynamics of Aquatic Ecosystems* promises not only step changes in the current understanding of our aquatic environments but also responds to the growing demands for the advanced knowledge in numerous applications, including civil and environmental engineering (e.g., stream restoration design), resource management (e.g., definition and determination of 'environmental flows' for regulated rivers), aquaculture (e.g., optimal design for aqua-farms), and bio-security (e.g., control of invasive species or transport of pathogens). It will also provide a solid biophysical basis for eco-hydraulics which has been formed as an applied research area based on largely empirical or semi-empirical approaches. Finally, the integration of methodologies of fluid mechanics, ecology and biomechanics and the focus on the interfaces between these disciplines creates the strong possibility of major breakthroughs not only in the understanding of aquatic ecosystems but also beyond it, with benefits for as diverse fields as design of bio-mimicking devices, fluid-structure interactions, and the adaptive evolution concept, among others.

ACKNOWLEDGMENTS

The authors are grateful to the Organizing Committee of the Gerhard H. Jirka Memorial Colloquium for their kind invitation to present this paper. Stimulating discussions of this topic with J. Aberle, B. Biggs, S. Coleman, P. Davies, J. Finnigan, A. Folkard, D. Goring, D. Hart, C. Howard-Williams, G.H. Jirka, I. Jowett, N. Lamouroux, S. Larned, S. McLean, G. Parker, S. Rice, T. Riis, M. Righetti, P. Sagnes, B. Statzner, A. Sukhodolov, T. Sukhodolova, A. Suren, and S. Thrush are greatly acknowledged. The work was partly supported by the Leverhulme Trust, Grant F/00 152/Z "Biophysics of flow-plants interactions in aquatic systems."

REFERENCES

Albayrak, I., Nikora, V., Miler, O. and O'Hare, M. (2011). Flow-plant interactions at a leaf scale: effects of leaf shape, serration, roughness and flexural rigidity. *Aquatic Sciences*, DOI 10.1007/s00027-011-0220-9.

Biggs, B.J.F., Nikora, V. and Snelder, T. (2005). Linking scales of flow variability to lotic ecosystem structure and function. *River Res. Appl.*, 21, 283–298.

Cameron, S.M. (2011). PIV algorithms for open-channel turbulence research: accuracy, resolution and limitations. *J. Hydro-Environment Res.* (in print).

Christensen, K.T. and Wu, Y. (2005). Visualization and characterization of small-scale spanwise vortices in turbulent channel flow. *Journal of Visualisation*, 8 (2), 177–185.

de Langre, E. (2008). Effects of wind on plants. *Annual Rev. Fluid. Mech.*, 40, 141–168.

Finnigan, J.J. (2000). Turbulence in plant canopies. *Annu. Rev. Fluid. Mech.*, 32, 519–571.

Finnigan, J., Shaw, R. and Patton, E. (2009). Turbulence structure above a vegetation canopy. *J. Fluid Mech.*, 637, 387–424.

Folkard, A. (2011). Vegetated flows in their environmental context: a review. *Proceedings of the Institution of Civil Engineers: Engineering and Computational Mechanics*, 164 (EM1). ISSN 1755-0777.

Ghisalberti, M. and Nepf, H. (2006). The structure of the shear layer in flows over rigid and flexible canopies. *Environmental Fluid Mechanics*, 6, 277–301.

Hart, D.D. and Finelli, C.M. (1999). Physical-biological coupling in streams: The Pervasive effects of flow on benthic organisms. *Annu. Rev. Ecology System.*, 30, 363–395.

Ho, C.-M. and Huerre, P. (1984). Perturbed free shear layers, *Ann. Rev. Fluid Mech.*, 16, 365–424.

Hondzo, M. and Wang, H. (2002). Effects of turbulence on growth and metabolism of periphyton in a laboratory flume, *Water Resour. Res.*, 38 (12), 1–9.

Landau L.D. and Lifshitz, E.M. (2004). *Fluid Mechanics*. Amsterdam, Elsevier.

Miler, O., Albayrak, I., Nikora, V. and O'Hare, M. (2011). Biomechanical properties of aquatic plants and their effects on plant–flow interactions in streams and rivers. *Aquatic Sciences*, DOI 10.1007/s00027-011-0188-5.

Naden, P.S., Rameshwaran, P., Mountford, O. and Robertson, C. (2006). The influence of macrophyte growth, typical of eutrophic conditions, on river flow velocities and turbulence production. *Hydrological Processes*, 20, 3915–3938.

Nezu, I. and Sanjou, M. (2008). Turbulence structure and coherent motion in vegetated canopy open-channel flows. *J. Hydro-Environment Res.*, 2, 62–90.

Nepf, H. (1999). Drag, turbulence and diffusivity in flow through emergent vegetation. *Water Resour. Res.*, 35 (2), 479–489.

Nepf, H. (2012). Flow and transport in regions with aquatic vegetation. *Annu. Rev. Fluid. Mech.*, 44, 123–142.

Nikora, V. (2004). Hydrodynamic effects in aquatic ecosystems. In: *Hydraulics in Water Engineering: Proc. of 8th Conf.* Brisbane, Australia, The Institution of Engineers (keynote address, full text in CD ROM Proceedings, pp. 1–12.

Nikora, V. (2007). Hydrodynamics of aquatic ecosystems: spatial-averaging perspective. *Acta Geophysica*, 55 (1), 3–10.

Nikora, V., McEwan, I., McLean, S., Coleman, S., Pokrajac, D. and Walters, R. (2007). Double averaging concept for rough-bed open-channel and overland flows: Theoretical background, *J. Hydraul. Eng.*, ASCE, 133 (8), 873–883.

Nikora, V. (2008). Hydrodynamics of gravel-bed rivers: scale issues. In: Habersack, H., Piegay, H. and Rinaldi, M (eds.) *Developments in Earth Surface Processes*, v. 11, Gravel-Bed Rivers VI. Berlin, Springer. pp. 61–81.

Nikora, V. (2009). Flow-organism interactions in running waters: interplay between flow mechanics, biomechanics, and ecology. In: Lopez-Jimenez, P.A., Fuertes-Miquel, V.S., Iglesias-Rey, P.L., Lopez-Partino, G., Martinez-Solano, F.J. & Palau-Salvador, G. (eds.) *Environmental Hydraulics. Theoretical, Experimental, and Computational Solutions: Proc. of 1st Intern. Workshop on Environmental Hydraulics.* Valencia, Spain, CRC Press, pp. 11–14.

Nikora, V. (2010a). Hydrodynamics of aquatic ecosystems: an interface between ecology, biomechanics and environmental fluid mechanics. *River Res. Appl.*, 26, 367–384, DOI: 10.1002/rra.1291.

Nikora, V. (2010b). Hydrodynamics of aquatic ecosystems: Current state, challenges and prospects. In: Mallinson G.D. & Cater J.E. (eds.) *Proc. of 17th Australsian Fluid Mechanics Conference*, Auckland, New Zealand, December (Flash Disk).

Nikora, N. and Nikora, V. (2007). A viscous drag concept for flow resistance in vegetated channels. In:Di Silvio, G. & Lanzoni, S. (ed.) *Harmonizing the demands of art and nature in hydraulics: Proc. of 32nd Congress of IAHR*, Venice, Italy, Coirdila (CD ROM).

Nikora, N. and Nikora, V. (2010). Flow penetration into the canopy of the submerged vegetation: definitions and quantitative estimates. In: Dittrich, A., Koll, Ka., Aberle, J. & Geisenhainer, P. (eds.) *River Flow 2010: Proc. of 5th International Conference on Fluvial Hydraulics River Flow 2010, Braunschweig, Germany*, v.1. Karlsruhe, Germany, BAW. pp. 437–444.

Okubo, A. and Levin, S.A. (2001). *Diffusion and Ecological Problems. Modern Perspectives*, Springer, New York.

Poggi, D., Porporato, A., Ridolfi, L., Alberston, J.D. and Katul, G.G. (2004). The effect of vegetation density on canopy sublayer turbulence, *Boundary-Layer Meteorol.*, 111, 565–587.

O'Hare, M.T., Hutchinson, K. and Clarke, R.T. (2007). The drag and reconfiguration experienced by five macrophytes from a lowland river. *Aquatic Botany* 86, 253–259.

Prasad, A.K. (2000). Stereoscopic Particle Image Velocimetry. Experiments in Fluids, 29 (2), 103–116.

Raupach, M.R., Finnigan, J.J. and Brunet, Y. (1996). Coherent Eddies and Turbulence in Vegetation Canopies: The Mixing Layer Analogy. *Boundary-Layer Meteorology* 78 (3–4), 351–382.

Rominger, J. and Nepf, H. (2011). Flow adjustment and interior flow associated with a rectangular porous obstruction. *J. Fluid. Mech.*, doi:10.1017/jfm.2011.199.

Sand-Jensen, K. (2003). Drag and reconfiguration of freshwater macrophytes. *Freshwat. Biol.*, 48, 271–283.

Siniscalchi, F., Nikora, V. and Aberle, J. (2011). Plant patch hydrodynamics in streams: mean flow, turbulence and drag forces. *Water Resour. Res.* (accepted)

Statzner, B. (2008). How views about flow adaptations of benthic stream invertebrates changed over the last century, *Intern. Rev. Hydrobiol.*, 93, 2008, 593–605.

Sukhodolov, A.N. and Sukhodolova, T.A. (2010). Case study: Effect of submerged aquatic plants on turbulence structure in a lowland river. *J. Hydraul. Eng.* 136 (7), 434–446.

Torquato, S. (2002). *Random Heterogeneous Materials*. Springer, New York.

Windsor, S.P., Norris, S., Cameron, S.M., Mallinson, G.D. and Montgomery, J.C. (2010). The flow fields involved in hydrodynamic imaging by blind Mexican cave fish (Astanax fasciatus). Part I: Open water and heading towards a wall. *Journal of Experimental Biology*, 213 (22), 3819–3831.

Windsor, S.P., Norris, S., Cameron, S.M., Mallinson, G.D. and Montgomery, J.C. (2010). The flow fields involved in hydrodynamic imaging by blind Mexican cave fish (Astanax fasciatus). Part II: Gliding parallel to a wall. *Journal of Experimental Biology*, 213 (22), 3832–3842.

Part 3

Small-scale phenomena

Chapter 12

Interaction of flows and particles at sub-micrometer scales

Daoyi Chen[1,2] *and Huan Huang*[2]
[1] *Division of Ocean Science and Technology Graduate School at Shenzhen, Tsinghua University Tsinghua Campus, The University Town, Shenzhen, P.R. China*
[2] *School of Engineering, University of Liverpool, Liverpool, UK*

ABSTRACT: The flow effects on Brownian motion of nanoparticle suspensions are measured by back scattering diffusing wave spectroscopy (CCD-DWS). At 1% weight concentration, a series of experimental studies were carried out for different flow velocities in two flow cells: 1 mm by 1 mm and 3 mm by 3 mm cross sections, respectively. First, in static conditions where only Brownian motions exist, the autocorrelation functions of natural particles were obtained. As flow velocity increased, it was found that the autocorrelation functions started to change slightly when the Peclet number ($Pe = U/k_0D$) reached about 29. Further flow velocity increase caused significant reduction in the slopes of the autocorrelation curves. The effects of Brownian motion persisted approximately up to Peclet number = 187. The above classifications clearly indicated the regions dominated by Brownian motion and flow motion respectively, as well as the transition.

1 INTRODUCTION

It is well known that Brownian motion was discovered by Robert Brown in 1827 when he made microscopic observations of the particles contained in the pollen of plants (Brown, 1827). This was long before the discovery of microorganisms like bacteria and their roles in water-borne diseases in the mid 1800s. Later, Einstein (1906) developed the theory of the Brownian motion. Bachelor (1976) investigated the effects of concentration on Brownian motion from the interactions of particles. The primary type of interaction is caused by hydrodynamic forces. When a particle moves in the medium, it will drag surrounding fluid and affect other particles nearby. Some theoretical formulas have been derived, by assuming hard spheres and considering the hydrodynamic interactions, in the form $D_{eff}/D_0 = 1 - 1.83\Phi$ where Φ is the volumetric concentration. It shows that the effective difusivity D_{eff} will be reduced due to hydrodynamic hindrance to the movement of particles.

The effects of flows on Brownian motion were studied after the development of nanoparticle sizing techniques based on laser applications such as dynamic light scattering (DLS) (see Wu *et al.*, 1990). The authors' purpose in revisiting this problem is due to practical needs in developing a probe for nanoparticle sizing for on-line applications where fluids are flowing through the probe. However, the results may stimulate further theoretical thinking beyond the practical applications.

Dynamic light scattering (DLS), also called quasi-elastic light scattering (QELS), has become an important tool for nanoparticle sizing in colloidal systems over the last 30 years (Berne & Pecora, 1976). However, the technique is usually used only in dilute systems due to the DLS theoretical models used to calculate the particle size; these DLS models are valid only in the single scattering regime. In the late 1980s, an extension of the principles of photon correlation spectroscopy to concentrated strongly-scattering media was introduced (Maret & Wolf, 1987; Pine *et al.*, 1988). This technique is known as diffusing-wave spectroscopy (DWS) (Weitz & Pine, 1993). The foundation of DWS has also been laid down in Chaikin *et al.* (1988). The propagation of light through a strongly scattering medium is treated as a diffusion process. Therefore, DWS is able to provide information about the local dynamics of particle dispersions at relatively higher concentrations. The concentration effects have been addressed further by introducing two-colour or 3-D optical arrangements to overcome the multi-scattering. Recently, we investigated the concentration effects together with the effects of laser power and a journal paper is in preparation (Huang *et al.*, 2011).

However, the DWS technique has been used mostly in the laboratory rather than as an on-line monitoring tool which requires working in moving fluids. Before a practically reliable monitoring tool can be developed, thorough understanding is needed of the flow effects on Brownian motions in relation to DWS measurement. There are some questions to be answered: when the flow starts to affect the Brownian motion; when the Brownian motion becomes negligible; and the correct formula to express the quantitative relationship between the measured autocorrelation functions and the parameters to be measured such as flow velocities and particle sizes.

So far, from the DWS measurement point of view, no literature has been found on the onset of the flow effects and the relative importance of Brownian motions in fluid flows. As to the flow effects on DWS, Wu *et al.* (1990) present a technique for measuring velocity gradients for laminar shear flow, using dynamic light scattering in the strongly multiple-scattering regime. Bicout and Maret (1994) used the same formula (2) as in Wu *et al.* (1990) for plane Couette flows. Wu *et al.* (1990)'s results have been regarded as classical, as quoted in Duncan and Kirkpatrick (2008).

In the last two decades, the DWS technique has been widely used in various applications and several examples are mentioned here. One potential application of DWS is for performing non-invasive measurements of the velocity of blood flow in blood vessels and detecting changes in blood volume in some biological tissues. Skipetrov and Meglinski (1998) used multiple scattering of laser radiation i in a laminar flow of scatterers in a cylindrical capillary embedded in the medium. G. Antar (1999) used laser scattering to measure small-scale turbulence. Li *et al.* (2005) conducted non-invasive detection of functional brain activity with near-infrared diffusing-wave spectroscopy. Ruis *et al.* (2008) used DWS to study the influence of shear on aggregation. Marze *et al.* (2008) studied aqueous foam slip and shear regimes determined by rheometry and multiple light scattering. Crassous *et al.* (2009) used DWS in a dilating scattering medium and found that an isotropic expansion of the material is equivalent to a contraction of the wavelength. In all studies mentioned above, flow effects are involved in various ways. However, the mechanisms behind the flow effects are yet to be understood.

The traditional PhotoMultiplier Tube PMT type of DWS setup is very time consuming because the data sampling time has to be long enough to reduce noise. Recently, CCD-DWS has become popular because hundreds of imaging sensor pixels can collect data simultaneously and only a relatively short time is required for each experiment (Harden & Viasnoff, 2001). The CCD-DWS could help the study of many more cases than previously.

In this paper, a series of experiments is carried out for different-sized nanoparticle suspensions at various flowrates. A CCD-DWS system has been setup and tested by the measurement of standard size polystyrene (PS) latex nanoparticles. We have chosen two channel cross sections (1 mm by 1 mm and 3 mm by 3 mm) for the aggregate formations. Autocorrelation function measurements of standard polystyrene latex particles at volume fraction = 0.9% were carried out with diameters 60, 100, 300, 500, 820, 1100 and 3200 nm. A new formula for the autocorrelation function is successfully produced to fit the experimental results. 4) The new formula has been tested with previous published datasets.

2 DWS METHOD AND EXPERIMENT SETUP

Under strong multiple-scattering conditions, photons execute random walks inside the turbid sample with the temporal autocorrelation function (ACF)

$$g_{1(r,t)} = \frac{\langle E(r,t)E^*(r,t)\rangle}{\langle |E(r,t)|^2\rangle} \tag{1}$$

of depolarized multiply-scattered radiation measured at point $\boldsymbol{r}$. The calculation of this time-averaged autocorrelation function involves the determination of the photon path length distribution function *P(s)* through the sample. In the continuum limit this is given by

$$G_1(t) = \int_0^\infty P(s)e^{-(2t/\tau_0)s/l^*}ds \tag{2}$$

where $\tau_0 = (k^2D_0)^{-1}$, $k = 2\pi n/\lambda$, D_0 is the diffusion coefficient of particles in the medium, n is the refractive index of the solvent, λ is the wavelength of the incident light, s is the length of photon trajectories and l^* is the transport photon mean-free-path. We measured the time-averaged intensity $\langle I\rangle$ and the normalized intensity autocorrelation function

$$G_2 \equiv \frac{\langle I(t_0)\, I(t_0+t)\rangle}{\langle I(t_0)\rangle \langle I(t_0+t)\rangle} - 1 \tag{3}$$

The intensity autocorrelation function can be approximated (Scheffold 2002)

$$G_2(t) - 1 = exp\left(-2\gamma\sqrt{6t/\tau_0}\right) \tag{4}$$

In this formula, γ is a constant depending on the boundary conditions in the photon diffusion approach. The typical γ value experimentally observed varies from 1.7 to 2.7 (Weitz & Pine, 1993; Scheffold 2002).

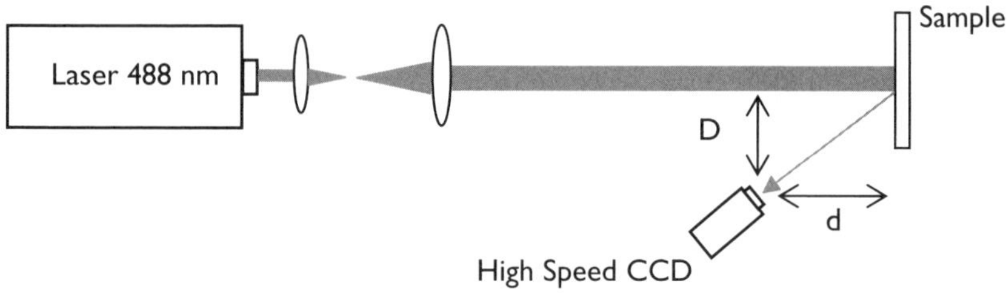

Figure 1 Schematic representation of diffusing wave spectroscopy setup.

A schematic diagram of the experimental apparatus is shown in Figure 1. Coherent laser radiation at a wavelength of $\lambda = 488$ nm, with power up to 2 W, was generated by an Argon Ion laser (Spectra Physics 2017). The laser beam was expanded to a spot with a diameter 1.5 mm approximately.

Two kinds of sample cell were applied in the experiments. The first group is made of microscopic glasses with a size of 1 * 1 * 50 mm. The observing windows of the cell are covered by 0.17 mm-thick slides. The other group is made from plastic tubes with a size of 3 * 3 * 180 mm. Slots are made on one of the walls for observation and also sealed by 0.17 mm-thick cover slides. Both have thickness L larger than 20 l^*. The optical window is located in the middle of the cells. This would give a length of 22.5 times channel width from the channel entry to the laser beam in the middle of the channel for boundary shear flows to develop in the 1 mm × 1 mm cell. Similarly, the 3 mm × 3 mm cell has a length of 31.7 times of the channel width for the boundary shear flows to develop. Syringe pumps were used to deliver the flows at precise rates.

The back-scattered light was detected directly by a high speed CCD camera (Photron Ultima APX system) that was placed approximately d = 14.5 cm from the sample cell and D = 9.5 cm from the laser beam. The resolution of the camera is 128 * 32 pixels with a frame rate of 100000 fps. This CCD was employed as a multi-speckle detector which enabled us to ensemble-average directly (Viasnoff *et al.*, 2002).

Polystyrene (PS) nanoparticle solutions used in experiment were ordered from Sigma-Aldrich with particle sizes ranging from 60 nm to 1100 nm and with 10% concentration by mass. For example, LB8 solution contains particles with average diameter (D) of 820 nm and has a size distribution of $\sigma/D = 1.7$ where σ denotes the standard deviation of the size distribution. The other solutions have a similar value of σ indicating a very narrow size distribution. In order to produce the required particle concentration, dilution with distilled water and stirring were carried out.

3 RESULTS AND DISCUSSION

3.1 Performance of the experimental system

To validate the experimental system and to study the effects of 300 nm and 500 nm particle solutions with different volume fractions, 0.9%, 1.8%, 3.6%, 5.4%, 7.2%

and 9% volume fractions were studied with laser power 40 mw, 80 mw, 160 mw, 240 mw, 320 mw, 640 mw, 800 mw, 960 mw, 1120 mw, 1200 mw and 1440 mw.

Based on our 132 experimental DWS runs, it was found that above about 640 mw, the laser power appeared to be adequate and the autocorrelation functions $g_2(t)$ display no difference as shown in Figure 2a.

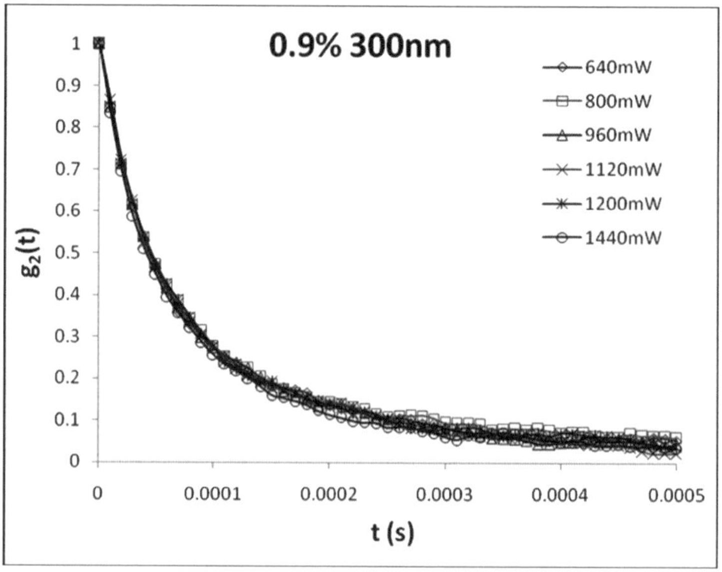

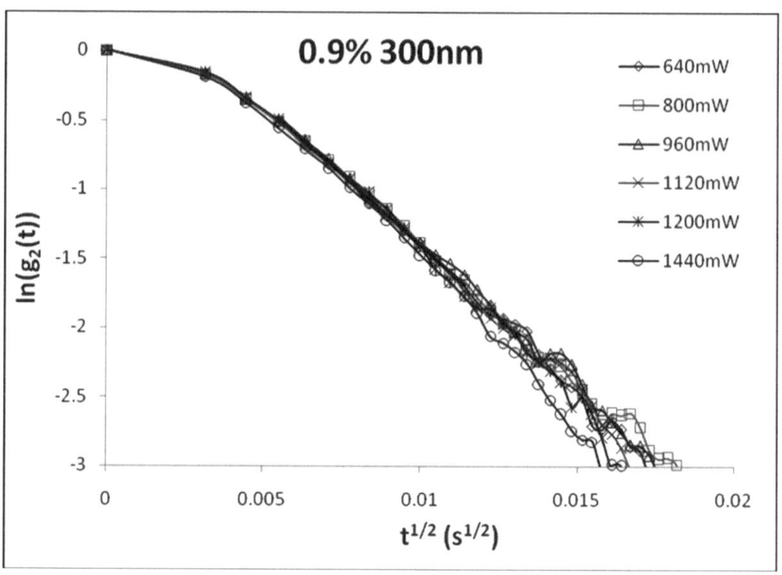

Figure 2 The normalised $g_2(t)$ for 0.9% volume fraction and 300 nm suspension, for high laser power (comparison for 640 mW to 1440 mW): a) plotting $g_2(t)$ against time t; b) $\ln(g_2(t))$ against $t^{1/2}$.

3.2 Particle sizing

As shown in Figure 3, the decay rate of the autocorrelation functions decreased from particle size 300 nm to 1100 nm. As we know, the Brownian motion contributes to the decay of the autocorrelation function. The strength of Brownian motion is weaker for larger particles, so the decay rate of $g_2(t)$ will tend to be smaller when particle size increased.

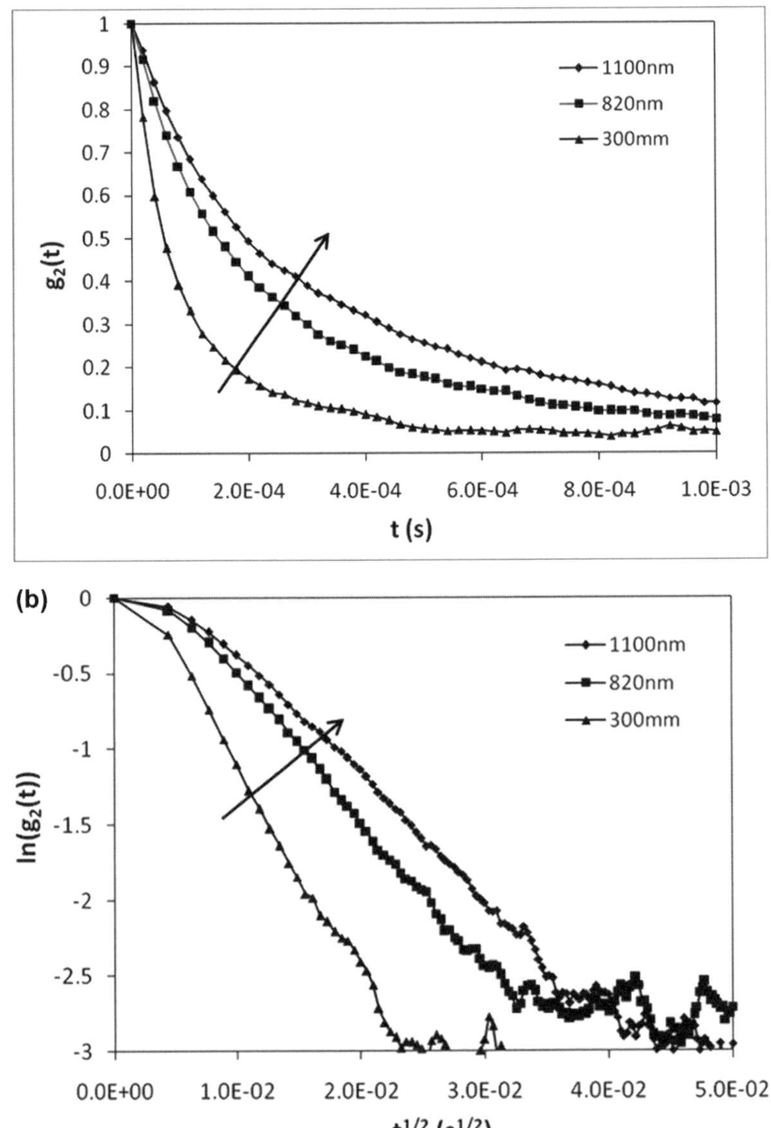

Figure 3 Intensity autocorrelation $g_2(t)$ from a typical DWS measurement in backscattering geometry for 0.3 μm, 0.82 μm and 1.1 μm PS particles respectively (volume fraction $\Phi = 1.8\%$): a) $g_2(t)$ vs. t; b) $\ln(g_2(t))$ vs. $(t)^{1/2}$, showing a linear relation over a broad range of decay times.

With the slopes obtained from the DWS experiments, the theory discussed earlier can be used to calculate the particle sizes. According to Equation (5), plotting $ln[g_2(t)]$ as a function of $t^{1/2}$ results in a straight line. The slope of the line is

$$S = -2\gamma\sqrt{6/\tau_0} \tag{5}$$

where $\tau_0 = 1/k^2D_0$, $k = 2\pi n/\lambda$ is the wave number in the scattering medium and D_0 is the diffusivity caused by Brownian motions. Therefore, the particle size can be calculated using the Stokes-Einstein relation (Einstein, 1906):

$$D_0 = \frac{K_B T}{6\pi\eta R} \tag{6}$$

where K_B is the Boltzmann constant, T is the temperature, η is the viscosity of the suspension and R is the radius of the particle. In our experiments, the temperature $T = 300$ K. Combining Equations (6) and (7), the particle size can be obtained:

$$R = \frac{4\gamma^2 k_B T k^2}{S^2 \pi\eta} \tag{7}$$

Here the suspension viscosity is equal to (Batchelor, 1976):

$$\eta_s = \eta_0(1 + 2.5\phi) \tag{8}$$

with η_0 equal to $1 * 10^{-3}$ Pa·s.

By using experimental data and Equation (8), the average value of $\gamma = 2.2$ was obtained with small variations for different particle size and these results agree well with those reported in previous experiments (Weitz & Pine, 1993; Scheffold, 2002). It proved that our back-scattering CCD-DWS system is reliable.

3.3 Onset of flow effects

An example is given for the autocorrelation curves of 1% 300 nm particles in a 3 * 3 mm flow cell. The static case serves as a reference to show the flow effects. The flowrate ml/m given in the figure is ml per minute.

As the flowrate increases, the slope of the autocorrelation function $\ln(g_2(t))$ increases. A 5% reduction in the slope is suggested as the dividing line for the onset of the flow effects. For this purpose, more tests were conducted in this flow rate range to give better estimation of the dividing line. To present the results, a new parameter is introduced here as

$$\mathrm{Pe} = \frac{\mathrm{U}}{\mathrm{kD_0}} \tag{9}$$

where U is the flow mean velocity, $k = 2\pi n/\lambda$ is the wave number in the scattering medium. and D_0 is the diffusivity caused by Brownian motions. This parameter is in the form of a Peclet number. It will be demonstrated later that Pe is directly proportional to a ratio τ_B/τ_s where τ_B is the time scale representing the Brownian motion and τ_s is the time scale representing the flow effects. It is interesting to note that Pe values for all cases are very close to a constant and the average is 29. As shown in Figure 5, the dividing line for the onset of the flow effects can be fitted by a constant Pe value of 29 (or $\tau_B/\tau_s = 1.79$).

3.4 Length scale Ut and Brownian effects at high flows

We would expect that when the flow rate increases to exceed a certain threshold value, the effect of Brownian motion may become relatively less important. A simple way to check the flow effects is to plot the $g_2(t)$ as a function of Ut which is a length scale. Similar to what has been shown in Fig. 4, at the high end of the flow rate cases, the curves tend to overlap each other, indicating that the flow effects become dominant and the Brownian effects gradually disappear. A fundamental difficulty in determining a dividing line for flow dominance is that the pure flow-effect limit is for the flow rate tending to infinity. From the practical application point of view, a dividing line is estimated to be Pe = 187 (or $\tau_B/\tau_s = 11.4$) for the 1 mm × 1 mm flow cell. The flow rates for the 3 mm × 3 mm flow cell were not sufficiently high to estimate

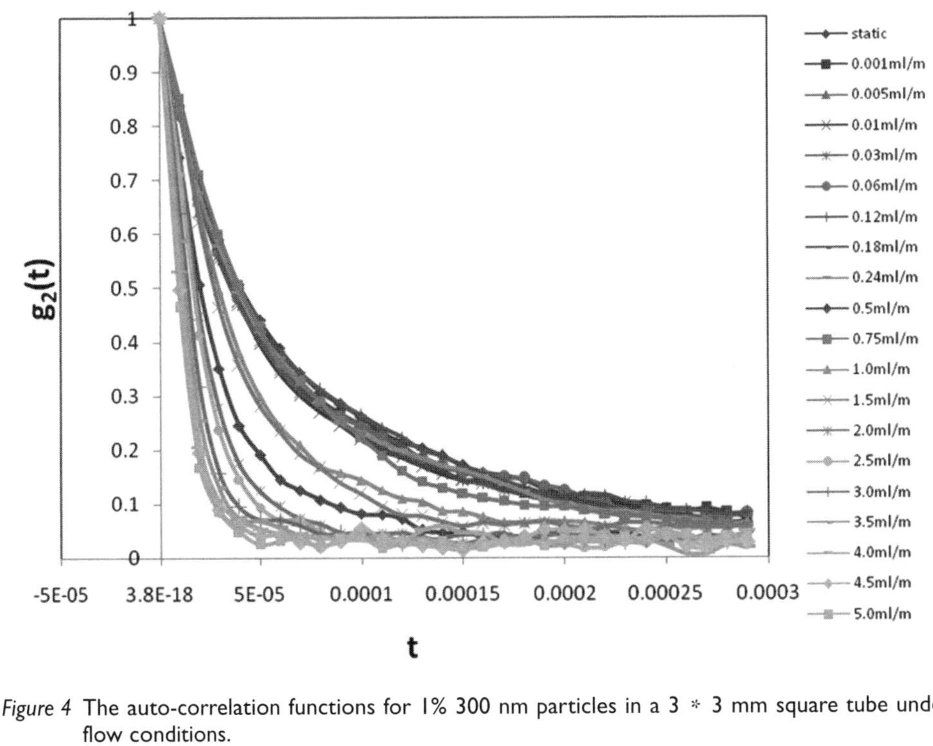

Figure 4 The auto-correlation functions for 1% 300 nm particles in a 3 * 3 mm square tube under flow conditions.

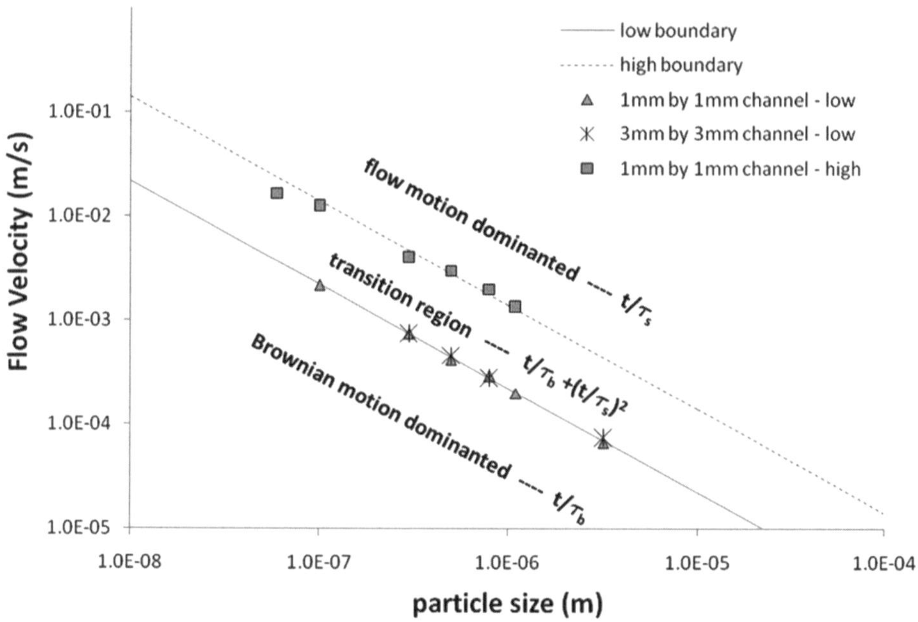

Figure 5 The diagram shows the dividing lines between Brownian and flow effects.

the dividing line. Compared with the lower dividing line (Pe = 29), which has been checked very carefully, it needs to be emphasized here that the upper dividing line (Pe = 187) is relatively rough, due to insufficient information, so it can only be used as a qualitative indication.

Figure 5 now provides a classification diagram for three different regions. When Pe < 29, the Brownian motion is dominant and flow has no or very little effect on the DWS autocorrelation functions. The autocorrelation can be expressed as a function of t/τ_B. When Pe > 187, the flow becomes dominant and the Brownian effect decreases to insignificance. The autocorrelation can be expressed as a function of t/τ_s. The region between these two extreme limits is the transitional region where both Brownian motion and the flow have important effects. Therefore, both τ_B and τ_s should appear in the formula for autocorrelation.

4 CONCLUSION

In conclusion, after carefully considering the effects of laser power, a CCD-DWS system was established successfully to conduct particle sizing at low concentrations. It can produce accurate results efficiently. It has been revealed that after a threshold value of Peclet number ($Pe = U/k_0D$) = 29, increasing flow rates lead to larger slopes of the autocorrelation curves, indicating the onset of the flow effects. The effects of Brownian motion persisted up to about Peclet number = 187. Beyond that value, it appeared that the flow effects dominate.

REFERENCES

Antar, G. (1999). Visible light scattering to measure small scale turbulence, *Review of Scientific Instruments,* 71 (1), 113–117.

Batchelor, G.K. (1976). Brownian diffusion of particles with hydrodynamic interaction, *J. Fluid Mechanics.* **74** (1), 1–29.

Berne, B.J. and Pecora, R. (1976). *Dynamic Light Scattering: With Applications to Chemistry*, New York, John Wiley.

Brown, R. (1828). A brief account of microscopical observations made in the months of June, July and August, 1827, on the particles contained in the pollen of plants; and on the general existence of active molecules in organic and inorganic bodies. *Philosophical Magazine 4,* 161–173.

Bicout, D. and Maret, G. (1994). Multiple light scattering in Taylor-Couette Flow, *Physica A*, 210, 87–112.

Chaikin, P.M., Pine, D.J., Weitz, D.A. and Herbolzheimer, E. (1988). *Physical Review Letters.* **60**, 1134–1137.

Crassous, J., Erpelding, M. and Amon, A. (2009). Diffusive Waves in a Dilating Scattering Medium, *Physical Review Letters,* 103 (1), 0139031–0139034.

Duncan, D.D. and Kirkpatrick, S.J. (2008). Can laser speckle flowmetry be made a quantitative tool?, *Journal of the Optical Society of America A*, 25 (8), 2088–2094.

Einstein, A. (1906) in: Fürth, R. and Cowper, Tr. A. D. (eds.). *Investigation on the theory of the Brownian movement*, New York, Dover.

Harden, J.L. and Viasnoff, V. (2001). Recent advances in DWS-based micro-rheology, *Current opinion in colloid & interface science,* 6 (5–6), 438–445.

Li, L., Dietsche, G., Iftime, D., Skipetrov, S.E., Maret, G., Elbert, T., Rockstroh, B. and Gisler, T. (2005). Noninvasive detection of functional brain activity with near-infrared diffusing-wave spectroscopy, *Journal of Biomedical Optics*, 10 (4), 044002.

Maret, G. and Wolf, P.E. (1987). Multiple light scattering from disordered media. The effect of Brownian motion of scatters, *Z. Phys. B*, **65**, 409–413.

Marze, S., Langevin, D. and Saint-Jalmes, A. (2008). Aqueous foam slip and shear regimes determined by rheometry and multiple light scattering, *Journal of rheology*, 52 (5), 1091–1111.

Pine, D.J., Weitz, D.A., Chaikin, P.M. and Herbolzheimer, E. (1988). Diffusion-wave spectroscopy, *Physical Review Letters*, **60** (12), 1134–1137.

Ruis, H.G., Venema, P. and Linden, E. (2008). Diffusing wave spectroscopy used to study the influence of shear on aggregation, *Langmuir*, 24 (14), 7117–7123.

Scheffold, F. (2002). Particle Sizing with Diffusing Wave Spectroscopy, *Journal of Dispersion Science and Technology*, **23** (5), 591–599.

Skipetrov, S.E. and Meglinski, I.V. (1998). Diffusing-wave spectroscopy in randomly inhomogeneous media with spatially localized scatterer flows, *Journal of Experimental and Theoretical Physics*, 86 (4), 661–665.

Viasnoff, V., Lequeux, F. and Pine, D.J. (2002). Multispeckle diffusing-wave spectroscopy: A tool to study slow relaxation and time-dependent dynamics, *Review of Scientific Instruments*, 73 (6), 2336–2344.

Weitz, D.A. and Pine, D.J. (1993). *Diffusing Wave Spectroscopy. In: Brown, W. (eds.). Dynamic Light Scattering: TheMethod and Some Applications*, Oxford, Oxford University Press.

Wu, X.-L., Pine, D.J., Chaikin, P.M., Huang, J.S. and Weitz, D.A. (1990). Diffusing-wave spectroscopy in a shear flow, *Journal of the Optical Society of America B,* 7 (1), 15–20.

Chapter 13

Environmental aspects of wastewater hydraulics

Willi H. Hager
Laboratory of Hydraulics, Hydrology and Glaciology (VAW), ETH Zurich, Zurich, Switzerland

ABSTRACT: Two basic sewer flow types are considered in this research, namely transitional and supercritical flows related to wastewater hydraulics. For transitional flows from the sub- to the supercritical regime as occur for instance on spillways, the free surface and velocity profiles are known to be continuous, whereas the bottom pressure profiles may become rapidly-varied, depending on the degree of curvature variation. A similar flow pattern is established at slope breaks from mild to steep slopes associated with sewer flows. The hydraulics of these are discussed both for the open rectangular as also the closed circular profiles. Because curvature effects are demonstrated to be small in terms of free surface effects, the hydraulic approach may be employed for the latter case, resulting in particular in an expression for the minimum tailwater sewer diameter to inhibit choking flow conditions.

The second problem presented relates to junction manholes under supercritical approach flow, for which again choking may become a high risk, associated with 'geysering' flow. These manholes include the through-flow and the bend manholes as particular cases, so that a general analysis is amenable, based on laboratory observations and a systematic data analysis. In contrast to standard knowledge, the maximum filling ratios of the approach flow and the discharge capacities of the three manhole configurations are detailed, along with a design basis that was successfully laboratory-tested. The results of both basic special manholes may thus be considered a significant advance in sewer hydraulics, by which undesirable sewage loss onto public space is prevented.

1 INTRODUCTION

Wastewater engineering currently counts among the most sponsored disciplines in civil engineering because of concerns in environmental pollution, water quality standards and city sanitation requirements. Given its multi-disciplinary background from civil, environmental and chemical engineering, among others, one might think that the current knowledge is so developed that few additional research is required. A look into the hydraulics of wastewater engineering reveals that only few standard books are available. Most of these were written decades ago, so that the recent book of this author may be counted among the few that are available. Wastewater hydraulics is essentially based on open channel hydraulics, with the particularity that flows are in closed conduits normally of circular cross-section, in which the interaction between air and water flows may become a concern, mainly under high-speed flow or if the pipe filling is close to full conduit flow. It is a basic requirement in wastewater hydraulics that flows are always under a free surface, to eliminate transitional effects from free surface to pressurized flow, except for the few sewer portions where the sewage is pumped. Usually, sewer flows are then assumed to have a maximum filling of some 85% of the pipe diameter, for which identical conditions as under full-filling occur. However, this simple requirement relates exclusively to uniform flow, but has to be redefined particularly for high-speed flows, because of additional complications.

It is well known that the transition from sub- to supercritical sewer flows occurs between a bottom slope of 0.2 and 0.3%, again under uniform flow conditions. Except for extremely mild-sloped reaches, sewer flow may therefore often be in the supercritical regime. Two particular aspects thus deserve attention: (1) Shockwaves as a response of any change of boundary conditions to the flow, including for instance bends, junctions, change of bottom slope, discharge addition or roughness variations, and (2) Hydraulic jumps whenever the supercritical flow regime cannot be maintained due to capacity limits, tailwater submergence or abrupt variations of either of the above mentioned parameters. Whereas shockwaves are no direct danger to a sewage system as long as the supercritical flow structures is maintained, hydraulic jumps may lead to so-called flow choking, namely the breakdown of the free surface flow due to a change to pressurized sewer flow. This scenario has been often observed in practice yet there are few steps that were initiated to counter such a breakdown of the system. Below, the junction manhole is considered in relation to choking, given its high relevance in applications, and recent design guidelines are presented by which a thorough supercritical flow in a sewer system may be guaranteed.

The first problem to be discussed relates to the slope change from mild to steep, a case that also caused numerous accidents in wastewater engineering. The basic mistake of the past was to assume that uniform flow is the only governing law to be considered. However, at these slope changes, the flow has to accelerate from a small to a higher velocity, so that choking again will be caused if there is not sufficient length for flow development. Based on an introduction in Chapter 2, the effects of streamline curvature and inclination in relation to these flows will be presented. Luckily, these effects may be demonstrated to be so small that the standard hydraulic approach may be used for these flows, thereby simplifying considerations to the theory based on hydrostatic pressure and uniform velocity distributions. In Chapter 3, the essence of the hydraulics of slope change manholes from mild to steep slopes will be elaborated, whereas chapter 4 deals with the junction manhole, including both bend and through-flow arrangements. The results will then be summarized in Chapter 5.

2 FLOWS FROM MILD TO STEEP SLOPES

2.1 Problem statement

A change of bottom slope from mild to steep involves particular flow characteristics in open channel flow. These flows are transitional from sub- to supercritical conditions; they are smooth and curvilinear-streamlined involving a continuous free surface profile, but a rapidly-varied bottom pressure profile (Fig. 1). The free surface profile $h = h(x)$ was analyzed by Massé (1938) using the singular point theory based on the gradually-varied open channel flow theory. Assuming hydrostatic pressure distribution this approach does not predict the two-dimensional (2D) flow features in a vertical plane associated with curvilinear-streamline flow. Although the singular point method approximates weakly-curved mild to steep slope transitions, no comparison with experimental data is available. At an abrupt slope break (Fig. 1a), the flow separates at the bottom kink (Rouse 1932, Weyermuller & Mostafa 1976), a feature beyond the present scope.

The inclusion of streamline curvature effects in the open channel flow equation of Boussinesq (1877) assumes a linear velocity distribution normal to the channel bottom resulting in a pseudo 2D approach. His mathematical development is commonly subjected to small streamline curvature (Hager & Hutter 1984). Interestingly, as discussed by Matthew (1995), its range of application may be higher than expected from the limited mathematical constraints. Other approximations for flow over curved channel bottoms using bottom-fitted coordinates include the perturbation approach of Dressler (1978) and the models by Berger and Carey (1998) or Dewals *et al.* (2006).

Herein, the transition from the horizontal to the steeply sloping rectangular channel reach is investigated to analyse the application range of the Boussinesq-type equation. Slope breaks with a rounded transition from the brink section to the tailwater channel slope are considered. Numerical results are compared with laboratory test data for transitions with a large downstream slope to investigate strong curvilinear gravity effects. Massé's (1938) analysis is also compared with the Boussinesq

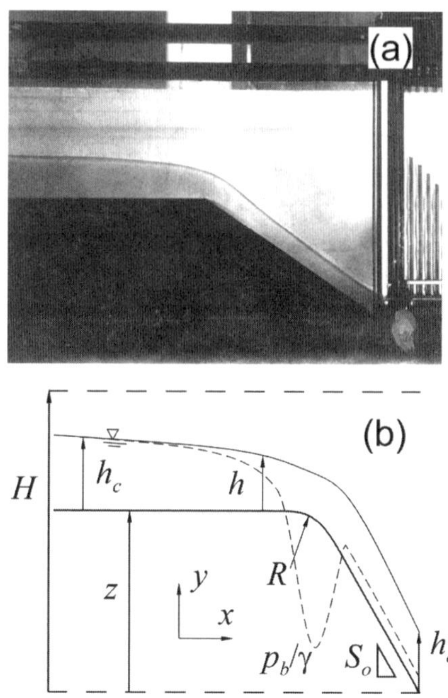

Figure 1 Transitional flow from mild to steep slopes (a) typical test arrangement (Rouse 1932) (b) definition sketch.

approach, as is a generalized momentum equation for curvilinear flows with results pertaining to the energy concept.

2.2 Governing equations

Matthew (1991) obtained for 2D, irrotational and incompressible free surface flow a second-order approximation for horizontal velocity u in the streamwise x-direction as

$$u = \frac{q}{h}\left[1 + \left(z'' - \frac{2h'z'}{h}\right)\left(\frac{2\eta - h}{2}\right) + \left(\frac{h''}{2h} - \frac{h'^2}{h^2}\right)\left(\frac{3\eta^2 - h^2}{3}\right)\right] \tag{1}$$

with q = discharge per unit width, h = flow depth measured vertically, $h' = dh/dx$, $h'' = d^2h/dx^2$, z = bottom elevation, $z' = dz/dx$, $z'' = d^2z/dx^2$, y = vertical coordinate and $\eta = y - z$. The vertical velocity profile $v(\eta)$ varies linearly from the channel bottom to the free surface as

$$v = \frac{q}{h}\left[z' + \frac{\eta}{h}h'\right] \tag{2}$$

Similar results to Equations (1) and (2) were obtained by Hager and Hutter (1984) and Hager (1985) yet by defining the flow depth as the vertical projection of an equipotential curve, or normal, rather than the vertical distance between the bottom and the free surface. At the latter ($\eta = h$) the pressure is atmospheric and the energy head H is given by

$$H = z + h + \frac{q^2}{2gh^2}\left(1 + \frac{2hh'' - h'^2}{3} + hz'' + z'^2\right) \tag{3}$$

similar to the equations of Fawer (1937), Matthew (1963), Hager and Hutter (1984) or Montes (1998). Equation (3) is a second-order differential equation for the free surface profile $h = h(x)$. For given H and prescribed boundary conditions, Equation (3) may be solved numerically. The velocity distributions $u(\eta)$ and $v(\eta)$ are then computed from Equations (1) and (2), and the pressure p distribution deduced from the Bernoulli equation as

$$\frac{p}{\gamma} = H - z - \eta - \frac{u^2 + v^2}{2g} \tag{4}$$

The bottom (subscript b) pressure profile $p_b = p(\eta = 0)$ is obtained from Equation (4), using Equations (1) and (2), as

$$\frac{p_b}{\gamma} = h + \frac{q^2}{2gh^2}\left(2hz'' + hh'' - h'^2 - 2z'h'\right) \tag{5}$$

Note that if $z'' = h'' = h' = 0$ the pressure distribution is hydrostatic and the pressure head is equal to the vertical flow depth h.

2.3 Boundary conditions

Test data of Hasumi (1931) and Westernacher (1965) indicate that the critical depth $h_c = (q^2/g)^{1/3}$ for parallel-streamlined flow establishes on the horizontal slope, at a distance of ~$3h_c$ upstream from the brink section, near the start of the circular arc transition, where hydrostatic pressure distribution prevails. Consider a Cartesian coordinate system (x, η) placed at the brink section, with a circular-shaped transition of radius R connecting the horizontal and the tailwater reaches (Fig. 1b). The upstream (subscript u) boundary condition $h_u = h(x_u = -3h_c) = h_c$ is used for computational purposes, thereby fixing the energy line on the horizontal bottom ($z = 0$) to $H = 3h_c/2$.

The downstream (subscript d) boundary condition is set where streamlines are nearly parallel to the channel bottom. From Equation (3), this condition reads (Hager 2010).

$$H = z + h + \frac{q^2}{2gh^2}\left(1 + z'^2\right) \tag{6}$$

The downstream flow depth h_d thus must satisfy Equation (6). Based on test data of Hasumi (1931) and Westernacher (1965) $x_d \cong +3h_c$.

2.4 Computational results

The computational results of Castro-Orgaz and Hager (2009) based on Equation (3) are favourably compared in Figure 2 with the test data of Hasumi (1931) for $S_o = 1$ and 1.732, and $R/h_c = 1.59$, 1 and 0.76 despite the highly-curvilinear flow. The model results also compare well with the measured free surface profiles of Westernacher (1965).

The brink depth $h_b = h(x = 0)$ (Fig. 3a) as a function of S_o for $R/h_c = 1$ using Equation (3) is compared in Figure 3b with the solution of the inverse form of the Laplace equation, according to which $h_b \approx 0.70$ for $S_o > 1$, whereas the extended Boussinesq equation yields $h_b \approx 0.68$. Test data of Mandrup Andersen (1975) and Weyermuller and Mostafa (1976) for small downstream slopes also corroborate these results.

The free surface brink slope h'_b is successfully compared with that of the Laplace equation in Figure 3c, with a limiting value of $h'_b = -0.27$ (15.1°), as compared with $h'_b = -0.31$ (17.22°) from the Boussinesq equation. The free surface slope at a free overfall (Matthew, 1995)

$$h'^2 = 3\left(1 - \frac{h_b}{h_c}\right)^3 \tag{7}$$

is also inserted, using the computed values for h_b in Equation (3). Accordingly, flows over a free overfall and the transition from mild to steep slopes behave hydraulically similar.

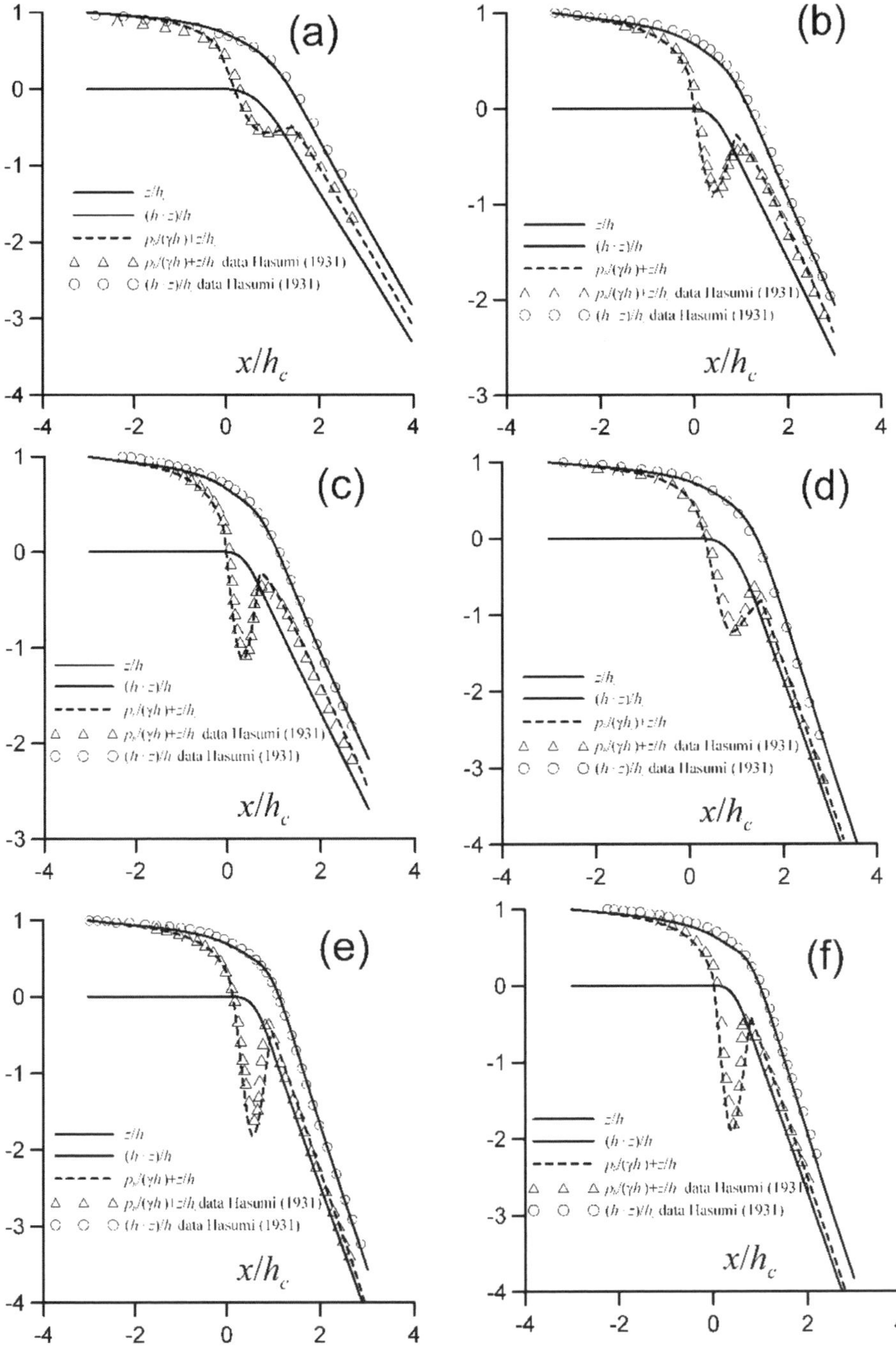

Figure 2 Comparison of computed $(h + z)/h_c$ and $(p_b/(\gamma) + z)/h_c$ distributions from Eqs. (3) and (5) with test data of Hasumi (1931) for $[R/h_c;S_o]$ = (a) [1.59;1], (b) [1;1], (c) [0.76;1], (d) [1.59;1.732], (e) [1;1.732], (f) [0.76;1.732].

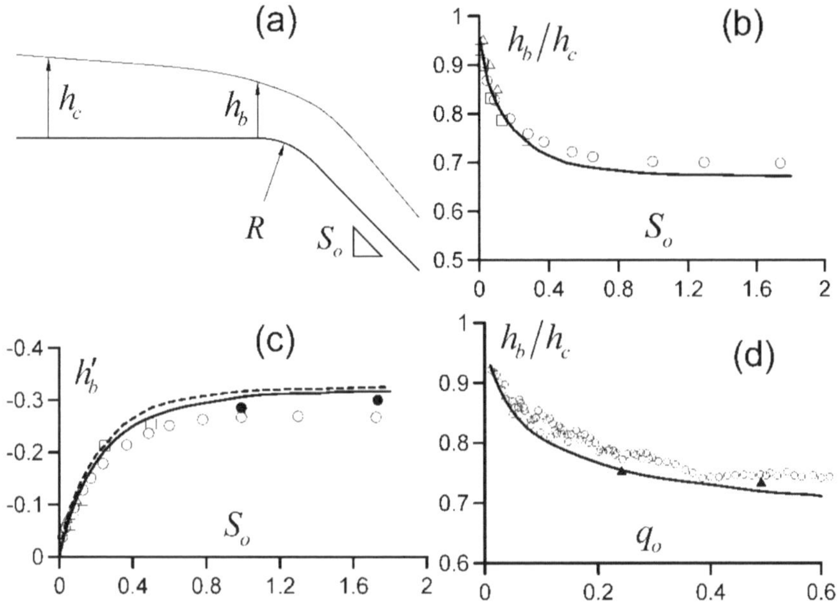

Figure 3 Brink depth results (a) definition sketch, (b) $h_b/h_c(S_o)$, (c) $h'_b(S_o)$ both from (–) Eq. (3) for $R/h_c = 1$, (---) Eq. (7), (d) $h_b/h_c(q_o)$ from (–) Eq. (3) for $S_o = 1.5$. (O) Laplace equation (Montes 1994), test data of (□) Weyermuller and Mostafa (1976), (Δ) Mandrup Andersen (1975), (•) Hasumi (1931), Westernacher (1965) (▲) potential flow solution, (O) test data.

The model results for the relative brink depth versus dimensionless discharge $q_o = q/(gR^3)^{1/2}$ are compared in Figure 3d with test data and results based on potential flow nets of Westernacher (1965). The computed curve is slightly below the latter because of viscous effects. Based on Figure 3a transition from mild to steep slopes may be used as a simple discharge measuring device, similar to a free overfall. As shown in Figure 3b, the brink depth ratio h_b/h_c varies only with R/h_c, or $q_o = (R/h_c)^{-3/2}$ if $S_o > 1$. Figure 3d relating to $S_o = 1.5$ indicates a relationship $h_b/h_c(q_o)$ without any effect of bottom slope. For practical purposes $h_b/h_c = 0.70q_o^{-0.06}$ may be used for $0.01 < q_o < 0.60$, resulting in $q(h_b)$.

2.5 Free surface control points

The energy head of potential flows is constant, i.e., $dH/dx = 0$. Differentiation of Equation (3) for the horizontal slope reach results in

$$1 = \frac{q^2}{gh^3}\left(1 + \frac{2hh'' - h'^2}{3} - \frac{h^2h'''}{3h'}\right) \tag{8}$$

This is equivalent to the minimum specific energy condition (Hager 1985, Castro-Orgaz *et al.*, 2008), associated with critical flow. However, there is no extreme in the channel bottom profile, as for weirs where $z'=0$ at the weir crest, implying that Equation (8) describes the entire free surface profile along a horizontal channel. The upstream critical depth section with parallel streamlined flow is a particular case of Equation (8), where $h' = h'' = h'''$, i.e., $q^2/(gh^3) = 1$. A critical flow approach with curvilinear streamlines therefore does not provide a useful solution for the transition from mild to steep slopes, given the absence of a fixed channel section for critical point computations.

2.6 Gradually-varied singular point method

Based on Equation (6), the gradually-varied flow equation in a sloping channel is

$$\frac{dh}{dx} = \frac{-z'}{1-\left[\frac{q^2}{gh^3}\left(1+z'^2\right)\right]} \tag{9}$$

From Massé's (1938) singular point method, the free surface slope at the critical point is

$$\frac{dh}{dx} = -\left(-\frac{h_c}{3}z''\right)^{1/2} \tag{10}$$

Equation (9) was solved numerically, starting at the critical point based on Equation (10). Obviously, for the horizontal bottom reach ($z' = 0$), Equation (9) yields $dh/dx = 0$, i.e., $h = \text{const} = h_c$. The results are compared in Figure 4 with Equation (3), from where the 1D and 2D solutions practically agree if $x/h_c > 1$, i.e.,

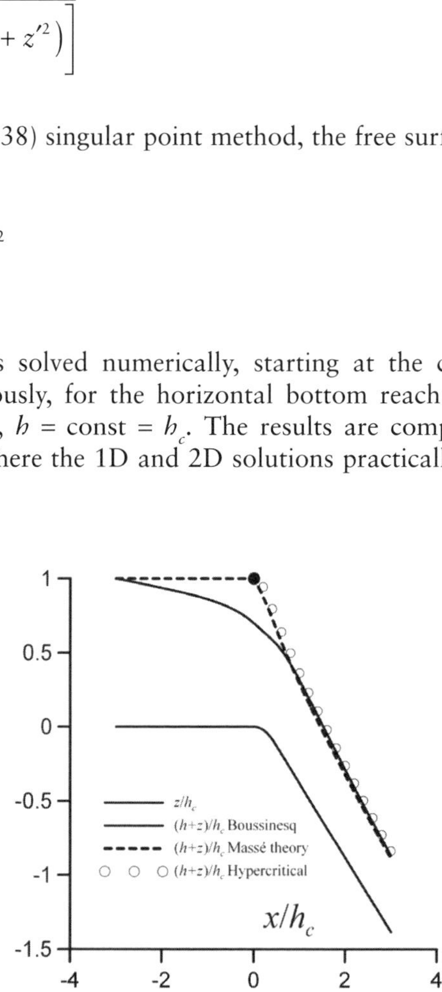

Figure 4 Comparison of computed profiles $(h + z)/h_c[x/h_c]$ from Eqs. (3) and (10) starting at the critical point (•) and Eq. (12) for $R/h_c = 1$ and $S_o = 0.5$.

streamline curvature effects are absent. For hypercritical flow with Froude numbers $F = q/(gh^3)^{1/2} > 3$ (Castro-Orgaz 2009) Equation (10) simplifies to

$$\frac{dh}{dx} = \frac{-z'}{-\left[\frac{q^2}{gh^3}\left(1 + z'^2\right)\right]} \tag{11}$$

whose general solution with $S_o = -z'$ is, using the boundary condition $h(x = 0) = h_c$,

$$\frac{h}{h_c} = \left(1 + \frac{2S_o}{1 + S_o^2}\frac{x}{h_c}\right)^{-1/2} \tag{12}$$

Equation (12) yields almost the same result as Equations (9) and (3) for $x/h_c > 1$, resulting in an accurate prediction of chute flows, to which the hypercritical approach applies.

3 TRANSITION FROM MILD TO STEEP SEWER REACHES

3.1 Computational assumptions

If the approach flow to a sewer manhole is subcritical and the outflow supercritical, critical flow occurs in the manhole itself. The slope of the approach (subscript o) channel S_{oo} is therefore smaller and the downstream (subscript u) channel slope S_{ou} larger than the critical slope. The transition from sub- to supercritical flow is examined here for a manhole which has to pass a large discharge Q. It is thereby not sufficient to determine only the critical depth h_c, but the *water surface profile* in the neighbourhood of the critical point must also be predicted. The downstream sewer has to be correctly designed so that a transition from free-surface to pressurized flow owing to a reduced diameter is avoided (Hager 1987).

Because the velocity in the downstream sewer is higher than of the approach flow, the downstream diameter D_u compared with the upstream diameter D_o may be reduced. Usually, a linear transition profile $D(x)$ between the two sewers is selected as

$$D = D_o - \theta x. \tag{13}$$

The coordinate origin $x = 0$ is set at the end of the approach flow sewer (Fig. 5) and the x-direction coincides with its direction. The contraction angle is $\theta = (D_o - D_u)/L_u$ with L_u as the length of the transition reach.

To prevent flow separation from the channel bottom, the transition curve from S_{oo} to S_{ou} is continuous. The simplest bottom transition profile $z(x)$ is a circular arc of radius R_u. For small differences in the up- and the downstream bottom slopes, the circular arc is approximated by the parabolic profile

$$z = -x^2/(2R_u). \tag{14}$$

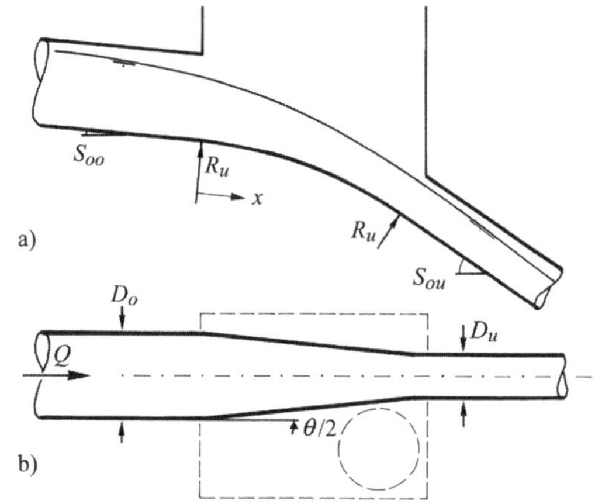

Figure 5 (a) Longitudinal section, (b) plan of control manhole with change from sub- to supercritical flow.

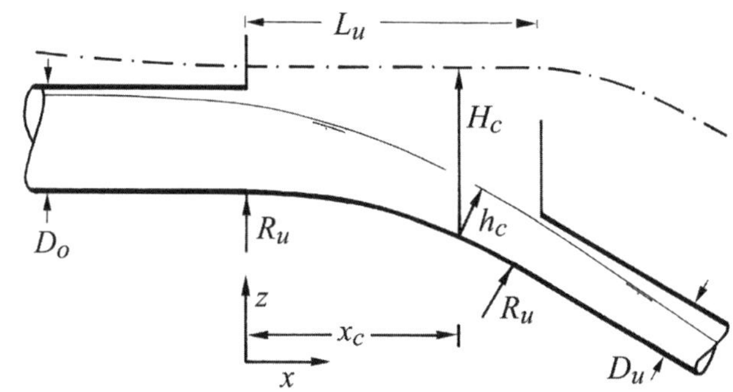

Figure 6 Hydraulically equivalent system with (—) flow surface, (- · -) energy head line.

The cross-section of the control manhole is usually U-shaped, formed by a rectangular top portion on a semi-circular bottom portion. Its cross-sectional area F can be approximated for filling ratios $y = h/D < 1.2$ by

$$F/D^2 = \frac{4}{3}y^{3/2}\left(1 - \frac{1}{3}y\right), \tag{15}$$

where the diameter D varies with x in accordance with Equation (13).

Since a control manhole is hydraulically a short structure, the change in the friction slope S_f along the manhole is small and its effect on the free surface profile even smaller. While $S_{oo} < S_f < S_{ou}$ holds for the slopes, the water depth relations are $h_o < h_{oN}$ at the upstream and $h_u > h_{uN}$ at its downstream end, with h_N as the uniform flow depth. To render simple computations, a constant energy head slope $S_E = S_{fm} = S_{oo}$ is assumed. If the manhole is also tilted by S_{oo}, the energy head line becomes horizontal so that $\mathrm{d}H/\mathrm{d}x = 0$. Figure 6 shows the hydraulic substitute system, the critical depth profile h_c, the corresponding critical energy head H_c as also the location of the critical point x_c.

3.2 Critical point

With these assumptions along $0 < x < L_u$, the energy head equation reads

$$H = -\frac{x^2}{2R_u} + h + \frac{Q^2}{2gF^2}, \tag{16}$$

$$\frac{\mathrm{d}H}{\mathrm{d}x} = 0. \tag{17}$$

The position of the critical point is determined by taking the derivative of Equation (16) assuming constant discharge and setting it according to Equation (17) equal to zero, thus

$$\frac{\mathrm{d}H}{\mathrm{d}x} = -\frac{x}{R_u} + \frac{\mathrm{d}h}{\mathrm{d}x} - \frac{Q^2}{gF^3}\left(\frac{\partial F}{\partial x} + \frac{\partial F}{\partial h}\frac{\mathrm{d}h}{\mathrm{d}x}\right) = 0. \tag{18}$$

Rearranging terms gives

$$-\left[\frac{x}{R_u} + \frac{Q^2}{gF^3}\frac{\partial F}{\partial x}\right] + \frac{\mathrm{d}h}{\mathrm{d}x}\left[1 - \frac{Q^2}{gF^3}\frac{\partial F}{\partial h}\right] = 0. \tag{19}$$

Since the expression in the second bracket is identical to $(1–\mathrm{F}^2)$, which equals zero at the critical point x_c,

$$\frac{x_c}{R_u} + \frac{Q^2}{gF^3}\frac{\partial F}{\partial x} = 0. \tag{20}$$

Setting $\mathrm{F}^2 = 1$ gives $Q^2/gF^3 = (\partial F/\partial h)^{-1}$. If this and the derivatives $\partial F/\partial x$ and $\partial F/\partial h$ obtained from Equation (15) are substituted into Equation (20), the result with $y_c = h_c/D_c$ is

$$\frac{x_c}{R_u} = -\frac{\partial F/\partial x}{\partial F/\partial h} = \frac{\theta}{3}y_c\left[1 + \frac{1}{3}y_c\right]\left[1 - \frac{5}{9}y_c\right]^{-1}. \tag{21}$$

The relative location of the critical point $x_c/(R_u\theta)$ therefore depends only on the filling ratio y_c at the critical point. Dimensionless parameters scaled with the approach flow diameter D_o are introduced as

$$X = \theta x/D_o,\quad y_o = h/D_o,\quad \rho_u = R_u\theta^2/D_o \tag{22}$$

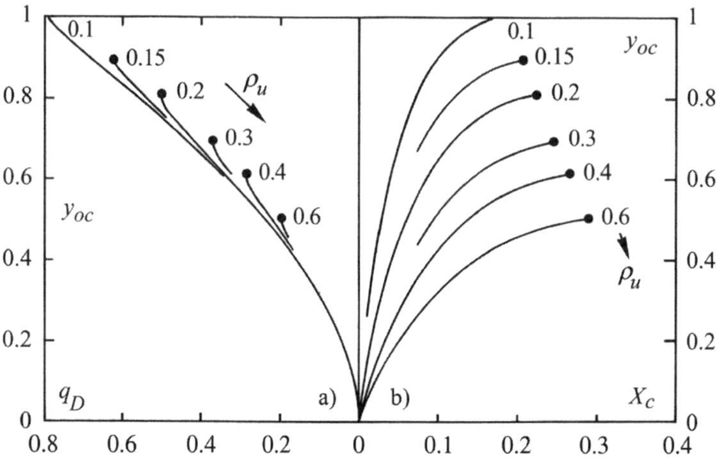

Figure 7 Relations at critical cross-section. Critical depth $y_{oc} = h_c/D_o$ as function of $\rho_u = \theta^2 R_u/D_o$ and (a) relative discharge $q_D = Q/(g_{D_o}^{\ 5})^{1/2}$, (b) critical distance $X_c = x_c\theta/D_o$. (•) Maximum.

Note that $D_c = D_o - \theta x_c = D_o(1 - X_c)$ and $y_c = h_c/D_c = h_c/[D_o(1 - X_c)] = y_{oc}(1 - X_c)$, where $y_{oc} = h_c/D_o$ varies only with x_c and ρ_u (Fig. 7b). With $y_{oc} = y_{oc}(X_c, \rho_u)$, the parameter h_c/D_c is eliminated, resulting in $y_{oc} = y_{oc}(q_D, \rho_u)$ with $q_D = Q/(gD_o^{\ 5})^{1/2}$ (Fig. 7a).

For a particular value of ρ_u the function $y_{oc}(q_D)$ breaks off as the maximum (subscript M) value q_{DM} is reached. A higher discharge across this manhole cannot be realized. The maximum relative discharge is approximated as (Hager 2010)

$$q_{DM} = 0.14\rho_u^{-0.8}. \tag{23}$$

or $Q_M = 0.14\ g^{1/2}D_o^{\ 3.3}/(\theta^2 R_u)^{0.8}$. The maximum discharge Q_M thus depends significantly on the approach flow diameter D_o and the contraction angle θ. Once the distance x_c and the critical depth h_c are known, the critical energy head results practically independent of ρ_u and related to relative discharge $q_D = Q/(gD_o^{\ 5})^{1/2}$ as

$$Y_{oc} = H_c/D_o = 1.28 q_D^{1/2}\left[1 + \frac{1}{4}q_D^{1/2}\right]. \tag{24}$$

3.3 Free surface profile

Considering Equation (24) in the system of Equations (16) and (17) implicitly yields for the free surface profile

$$1.28 q_D^{1/2}\left[1 + \frac{1}{4}q_D^{1/2}\right] = \frac{X^2}{2\rho_u} + y_o + \frac{(9/32)q_D^2}{(1-X)y_o^3\left[1 - \dfrac{y_o/3}{1-X}\right]^2}. \tag{25}$$

The free surface profile $y_o(x)$, with $y_o = h/D_o$, depends only on q_D and ρ_u. Figure 8 shows $y_o(x)$ for various relative discharges $q_D = 0.1$ to 0.4, with the critical points marked by bold circles.

For a given discharge Q and manhole geometry (D_o, R_u, θ) the parameters q_D and ρ_u are first determined to find the free surface. In particular, the flow depth at the downstream manhole end is determined from Figure 8, based on the smallest diameter that still guarantees free surface flow. For neighbouring values of ρ_u, the values of y_o from Figure 8 are interpolated to find an approximate value of y_o corresponding to the known value of ρ_u. An improved value of y_o is then determined by solving for Equation (25) and considering that there are two solutions, one for F < 1 and the other for F > 1.

For a given upstream diameter D_o, the maximum discharge can only be increased by reducing either θ or R_u of which both are coupled to the downstream bottom slope S_{ou}. Since $S_{ou} = -dz/dx$ at $x = L_u$, it follows $S_{ou} = L_u/R_u$, and the contraction angle is $\theta = (D_o - D_u)/(S_{ou}R_u)$. The required minimum (subscript m) downstream diameter D_{um} is obtained by setting the preceding expression for θ in Equation (23), thus

$$D_{um} = D_o - S_{ou}(R_u D_o)^{1/2}(0.14/q_D)^{5/8}. \qquad (26)$$

Therefore, the required minimum diameter is larger for smaller downstream bottom slope and larger relative discharge $q_D = Q/(gD_o^5)^{1/2}$.

Once the location of the critical point is determined, the flow depths h_o and h_u at the manhole entry and exit sections, respectively, can either be obtained from Figure 8 or determined from Equation (25). Surface profiles up- and downstream from the manhole are then obtained with a standard backwater and drawdown computation for reaches not including a change in the flow type. The computation of transitional flow thus involves simplifications thereby excluding the generation of shockwaves

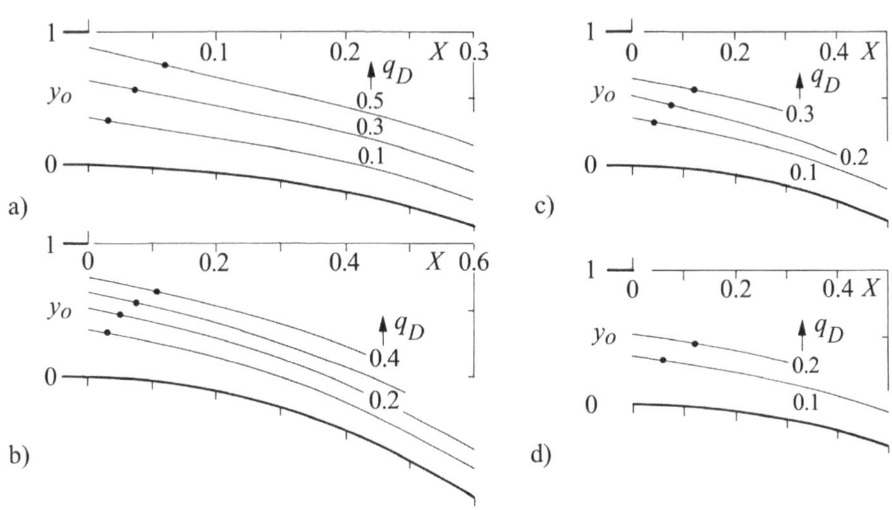

Figure 8. Dimensionless free surface profiles $y_o(X)$ versus relative discharge $q_D = Q/(gD_o^5)^{1/2}$ for $\rho_u = \theta R_u/D_o$ = (a) 0.1, (b) 0.2, (c) 0.3, (d) 0.4. (•) Critical point (Hager 1987).

originating from the converging transition profile to the downstream prismatic sewer, resulting in surface disturbances. Currently no experimental investigation on the height of these waves appears to exist. Further, the phenomenon of air entrainment in steep conduits should be considered (Hager 2010), yet Equation (26) allows for a preliminary design. The parameter R_u of this equation is a proper variable which, however, has a lower limit as to prevent flow separation from the channel bottom.

4 MANHOLE DESIGN FOR SUPERCRITICAL APPROACH FLOW

4.1 Introduction

Supercritical manhole flow is governed by either shockwaves generated at each flow discontinuity, or hydraulic jumps, if the discharge capacity is too small to convey a fully supercritical flow. Whereas shockwaves involve mainly a medium flow depth increase beyond a shock front, a hydraulic jump may result in both the collapse of the supercritical flow regime and a significant backwater effect. This is a serious problem for a sewer because of the abrupt change from free surface to pressurized two-phase flow. A choking phenomenon is accompanied with water hammer, a decrease of the discharge capacity finally resulting in so called *geysering* of wastewater from the manhole onto public space (Fig. 9), which must be avoided in any case (ATV 1996, 2000). The following presents the definite recommendation for the through-flow, the bend and the junction manholes, based on extensive hydraulic modelling at VAW, ETH Zurich.

4.2 Through-flow manhole

This is the simplest sewer manhole arrangement for control and maintenance purposes (Hager 2010). A manhole of U-shaped profile and length L is connected to equal up- and downstream sewers of diameter D. Figure 10 shows a sketch involving

Figure 9 Geysering of manhole in combined sewer (Hager & Gisonni 2005).

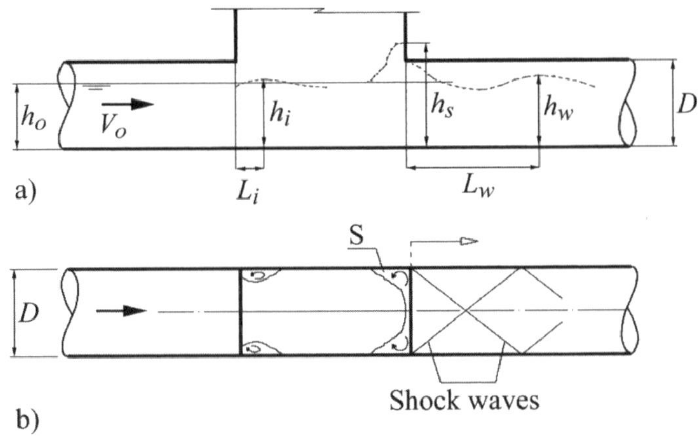

Figure 10 Through-flow manhole (a) section, (b) plan (Hager & Gisonni 2005).

the approach flow depth h_o and velocity V_o. For $y_o = h_o/D \leq 0.50$, the flow remains entirely in the circular-shaped pipe, whereas the flow abruptly expands at the manhole inlet for $y_o > 0.50$, forming a side depression, followed by a downstream shockwave of height h_i because of flow impact onto the side walls. Whereas this phenomenon is relatively small, a more dramatic feature occurs at the manhole outlet because of flow impact onto the upper portion of the circular profile, resulting in a shaft (subscript *s*) flow depth h_s. Depending on its height relative to D, the flow either continues as supercritical flow, or it breaks down due to the formation of an impact hydraulic jump. Choking then results at the manhole outlet because of jump formation and the breakdown of the air transport from the up- to the downstream sewer reaches (Fig. 11). If the discharge increases fast the choking phenomenon may initiate even geysering, as previously described.

Because U-shaped (subscript U) profiles correspond essentially to a rectangular channel, the governing Froude number is $F_U = Q/(gD^2h_o^3)^{1/2}$. The relative shaft outflow depth was experimentally determined to (Gargano & Hager 2002)

$$h_s/h_o = 1 + (1/3)(F_U y_o)^2. \tag{27}$$

Therefore, the relative wave amplitude $[(h_s-h_o)/h_o]$ increases quadratically with $F_U \cdot y_o$, or the ratio $[(h_s-h_o)/D]$ depends exclusively on F_U.

The manhole discharge capacity (subscript C) Q_C is of design interest. According to Equation (27) the approach flow filling y_o is relevant. The transition from free surface to pressurized manhole flow may be accounted for by the capacity Froude number $F_C = Q_C/(gD^5)^{1/2}$. Gargano and Hager (2002) proposed for $0.70 < y_o < 0.75$

$$F_C = 14.6 - 17.3y_o. \tag{28}$$

For all tests, no free surface flow resulted if $y_o > 0.75$, but choking never occurred for $y_o < 0.70$. In the average, the choking Froude number amounted to $F_C = 2$.

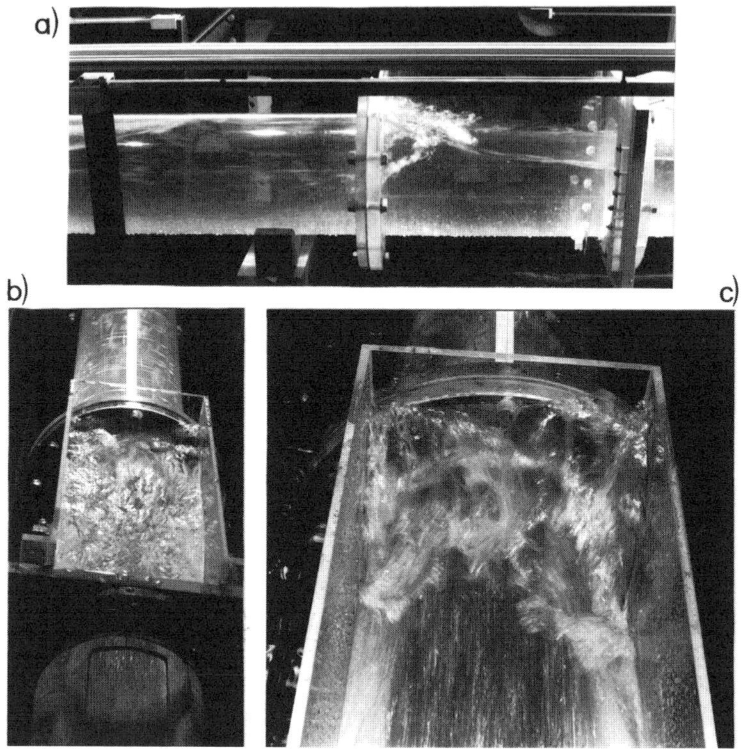

Figure 11 Choking at through-flow manhole outlet for $y_o = 0.75$ and $F_o = 1.30$ (a) section, (b) upstream view, (c) impact flow (Hager & Gisonni 2005).

The current sewer design practice accounts for the so-called full-flow approach, involving a relative sewer filling of some 85%, independent of the flow conditions. This condition was originally introduced for nearly uniform flows, which differ significantly from the above described supercritical flows. Observations indicate that the standard design procedure results always in the breakdown of manhole flow. Accordingly, supercritical flows in through-flow manholes must be limited both in terms of filling ratio and discharge capacity, to ensure the free surface flow regime.

4.3 Bend manhole

This manhole is often found in the urban infrastructure, given that roads are normally arranged in a rectangular grid. Of particular interest is the 90° bend manhole, but also the 45° deflection angle may be relevant. The average bend radius is usually $R_a = 3D$. One might think that the 90° bend manhole is more critical in terms of discharge capacity than the 45° manhole. Figure 12 shows a sketch involving the approach flow depth h_o and velocity V_o for a deflection angle of $\delta = 45°$. Two shockwaves form along the inner and the outer walls, yet in the following, only the wave along the outer wall of maximum height h_M is considered. Del Giudice *et al.* (2000) found with $F_U = Q/(gD^2h_o^3)^{1/2}$ that

$$h_M/h_o = [1 + 0.50(D/R_a)F_U^2]^2. \tag{29}$$

The angle Θ_M of maximum wave height is located between 35° and 55° measured from the manhole inlet (Hager 2010). The discharge capacity of the 45° bend manhole as compared with a 90° deflection is therefore dramatically reduced due to the presence of the maximum wave height at the manhole outlet.

To improve the manhole discharge capacity, a straight tailwater manhole *extension* of length $2D$ was added to the structure (Fig. 12). This extension resulted from detailed hydraulic tests, with a first wave maximum upstream of the manhole outlet and a second within the tailwater sewer which does not lead to flow choking, however. The manhole extension increases significantly the discharge capacity, which was determined for $y_o < 2/3$ from model tests to (Gisonni & Hager 2002a)

$$F_C = (3-2y_o)y_o^{3/2}. \tag{30}$$

This discharge capacity is thus significantly smaller than of the corresponding through-flow manhole, with a maximum of $F_{CM} = 0.90$ for $y_o = 0.67$, and only $F_C = 0.80$ for a typical sewer filling of $y_o = 0.60$. Note that in all tests, the flow across the bend manhole choked if the approach flow filling was in excess of 65%, as compared to 75% for the through-flow manhole. Figure 13 shows typical flow features in a bend manhole prior to flow choking. The discharge capacity may be increased if the tailwater sewer diameter D_d is increased thereby using a manhole extension length of $2D_d$ instead of $2D$.

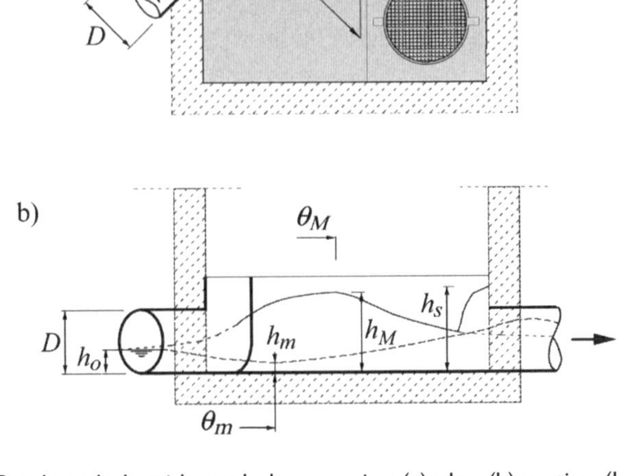

Figure 12 Bend manhole with manhole extension (a) plan, (b) section (Hager 2010).

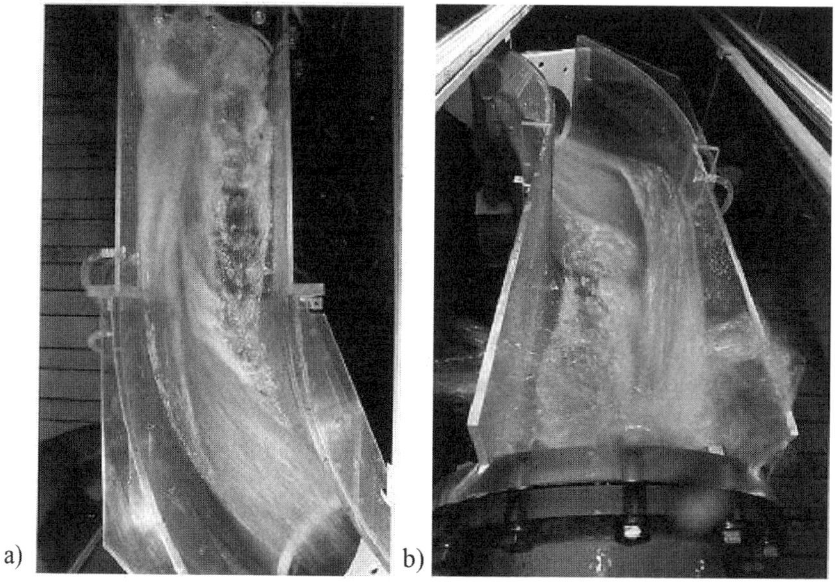

Figure 13 Flow features of bend manhole with manhole extension (a) plan, (b) downstream view (Hager & Gisonni 2005).

4.4 Junction manhole

This manhole may be considered hydraulically intermediate between the through-flow and bend manholes. The discharge capacity of the junction manhole is therefore also intermediate. However, its flow structure differs from the other two manhole types. Figure 14 shows an equal branch diameter manhole, including the upstream (subscript *o*) branch of approach flow depth h_o and velocity V_o, and the lateral (subscript *L*) branch with h_L and V_L, respectively. The junction angle between the two branches is δ, with a sharp-crested intersection at the junction point P. A data analysis of laboratory tests conducted at ETH Zurich indicated that the relevant Froude numbers are $F_L = Q_L/(gD_Lh_L^4)^{1/2}$ for the lateral (subscript *L*) branch, and $F_o = Q_o/(gD_oh_o^4)^{1/2}$ for the upstream branch.

Two distinctly different phenomena may occur in supercritical manhole flow, namely (1) Choking of manhole outlet due to swell generation associated with an abrupt breakdown of the supercritical manhole flow, and (2) Choking of one or even both branches due to flow blockage of the other branch or poor combining flow conditions. A complicated hydraulic jump then is generated submerging either one or even both branch pipes causing also the breakdown of the supercritical approach flow. In both cases, the breakdown may become so abrupt and strong that manhole geysering results. Figure 15 shows fully supercritical flow in a junction manhole, whereas Fig. 16 relates to choking type 1. The latter results from gate-type flow due to the swell impact onto the manhole outlet wall. Given the complexity of independent parameters, and the variety of flow regimes, the engineering design of these manholes must be simplified by retaining the main flow features.

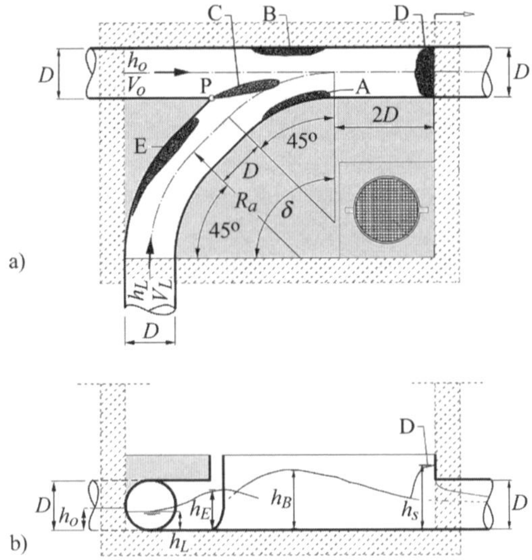

Figure 14 Definition sketch of junction manhole (a) plan, (b) section (Hager & Gisonni 2005).

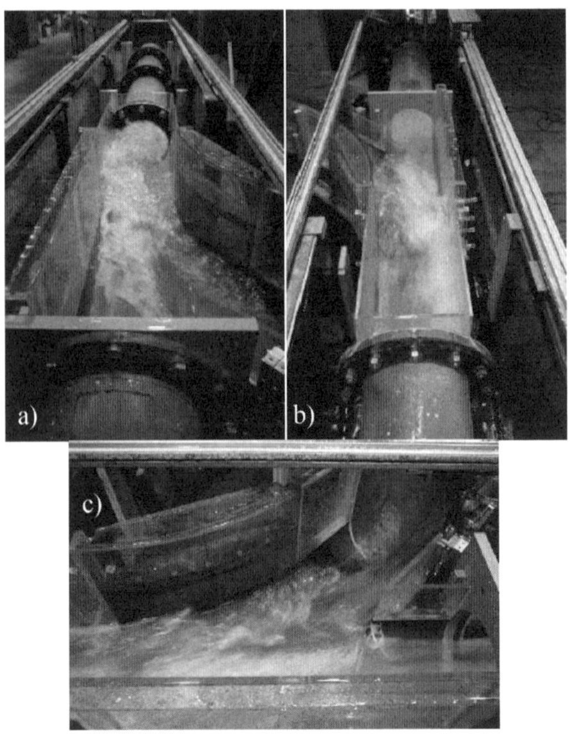

Figure 15 Typical supercritical flow in junction manhole for $y_o = y_L = 0.27$, $F_o = 5.95$ and $F_L = 2.84$ (a) upstream view, (b) downstream view, (c) plan (Hager & Gisonni 2005).

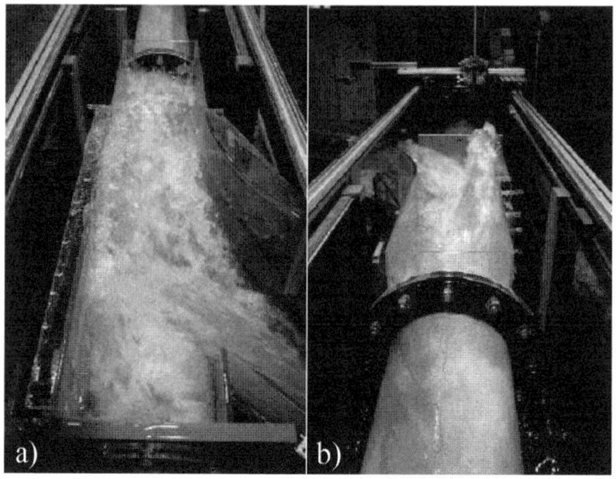

Figure 16 Choking of junction manhole outlet for $y_o = y_L = 0.34$, $F_o = 4.19$ and $F_L = 4.0$ with view from (a) upstream, (b) downstream (Hager & Gisonni 2005).

Table 1 Capacity froude numbers F_C and maximum filling ratios y_C for basic types.

Manhole type	*Through-flow*	*Junction*	*Bend*
Discharge F_C	2.0	1.4	0.8
Filling ratio y_C	0.75	0.70	0.65

4.5 Manhole discharge capacity

The previous results indicate that the 45° and the 90° bend and junction manholes are governed by similar flow mechanisms, provided that a manhole extension of length $2D$ is added beyond the lateral branch deflection (Figs. 12, 14). Further, both the 45° and the 90° bend and junction manholes behave hydraulically similar if an intermediate bend extension of $1D$ is added to the 90° deflection (Fig. 14). Then, the bend wave at roughly 45° from the manhole inlet is allowed to fall, and the flow becomes more uniform as compared with an abrupt flow deflection by 90°.

The maximum discharge capacity of junction manholes designed according to Figure 14 is $F_C = Q/(gD^5)^{1/2} = 1.4$, as compared to $F_C = 0.8$ for bend manholes. The through-flow branch thus lessens the effect of the lateral branch flow, and increases the discharge capacity of the junction manhole. It behaves intermediate to the through-flow ($Q_L = 0$) and bend manholes ($Q_o = 0$). The capacity Froude numbers and the maximum approach flow filling ratios y_C of the three basic manhole types are stated in Table 1. This contrasts strongly the current design basis, with the sewer filling of some 85%, independent of flow regime and manhole presence, and no limitation of discharge for manholes. Assuming therefore the traditional 'uniform flow concept' along with the full-flow sewer may result in undesirable and dangerous flow conditions for which the sewer was not designed.

4.6 Practical recommendations

Compared with the current manhole design involving circular sewers and an U-shaped manhole through-flow profile placed symmetrically within the structure, the designs shown in Figures 12 and 14 are asymmetric. The manhole wall opposite the bend branch is flush with the upstream branch with a vertical-sided through-flow profile, so that the bend wave may not flow onto the bench. The manhole entrance is therefore arranged opposite from this wall, with the manhole platform confined within the bend. Given that both the bend and junction manholes include a manhole extension of length $2D$, there is enough space for maintenance and sewer control works. The novel design performed excellently in the laboratory, and designs installed in prototypes have so far not resulted in problems. Long-time tests will indicate whether additions are necessary to improve their hydraulic performance. The present design may be considered a significant improvement of existing manholes subjected by supercritical flow.

5 CONCLUSIONS

This research deals with particular problems in environmental hydraulics, namely these of wastewater hydraulics. Two questions are addressed, one relating to the slope changes from mild to steep, and the other to junctions in sewer systems. Based on preliminary analysis, it was demonstrated that effects of streamline curvature in the first problem are insignificant if the focus is on the free surface profile. However, to detail the pressure line associate with a curved slope change, then the full Boussinesq approach needs to be applied. The result of the hydraulic approach allows to determine the minimum diameter of the downstream sewer to inhibit choking flow. The effects of various basic parameters on the reduction of the upstream to the downstream sewer are thereby also discussed.

The second problem is related to the hydraulics of special manholes under supercritical approach flow. These include the junction, bend and through-flow manholes. The flow structure was experimentally observed, yet it is so complex that a simplified design guideline was proposed based on the maximum approach flow filling ratio and the discharge capacity. It is found that their dimensionless values are significantly below those known for uniform flow, and that the junction manhole may be regarded as intermediate to the other two manhole types. Further, the manhole extension was identified as a simple structural means to improve the flow in these manholes. A generalized design procedure is proposed by which existing structures may be improved if required, and new manholes can be safely designed in terms mainly of the approach discharge. Based on these guidelines, the supercritical flow structure across these manholes may be conserved thereby resulting in free surface flow in which problems with flow choking are avoided.

REFERENCES

ATV (1996). *ATV-Handbuch Bau und Betrieb der Kanalisation* (Sewer construction and management), ed. 4. W. Ernst & Sohn: Berlin [in German].

ATV (2000). *Bau und Betrieb von Kanalisationen* (Construction and management of sewers). Ernst & Sohn: Berlin [in German].

Berger, R.C. and Carey, G.F. (1998). Free-surface flow over curved surfaces 1: Perturbation analysis. *Int. J. Numer. Meth. Fluids* 28 (2), 191–200.

Boussinesq, J. (1877). Essai sur la théorie des eaux courantes (Essay on the theory of water flow). *Mémoires* présentés par divers savants à l'Académie des Sciences, Paris 23, 1–680 [in French].

Castro-Orgaz, O. (2009). Hydraulics of developing chute flow. *J. Hydr. Res.* 47 (2), 185–194.

Castro-Orgaz, O. and Hager, W.H. (2009). Curved-streamline transitional flow from mild to steep slopes. *J. Hydr. Res.* 47 (5), 574–584.

Castro-Orgaz, O., Giraldez, J.V. and Ayuso, J.L. (2008). Higher order critical flow condition in curved streamline flow. *J. Hydr. Res.* 46 (6), 849–853.

Del Giudice, G., Gisonni, C. and Hager, W.H. (2000). Supercritical flow in bend manhole. *J. Irrigation and Drainage Eng.* 126 (1), 48–56.

Dewals, B.J., Erpicum, S., Archambeau, P., Detrembleur, S. and Pirotton, M. (2006). Depth-integrated flow modelling taking into account bottom curvature. *J. Hydr. Res.* 44 (6), 787–795.

Dressler, R.F. (1978). New nonlinear shallow flow equations with curvature. *J. Hydr. Res.* 16(3), 205–222.

Fawer, C. (1937). Etude de quelques écoulements permanents à filets courbes. *Thesis*, Université de Lausanne. La Concorde, Lausanne, Switzerland [in French].

Gargano, R. and Hager, W.H. (2002). Supercritical flow across combined sewer manhole. *J. Hydr. Eng.* 128 (11), 1014–1017.

Hager, W.H. (1985). Critical flow condition in open channel hydraulics. *Acta Mech.* 54 (3/4), 157–179.

Hager, W.H. (1987). Übergang von Flach- auf Steilstrecke in Kanalisationen (Transition from mild to steep sewers). *Gas-Wasser-Abwasser* 67 (7), 420–426 [in German].

Hager, W.H. (2010). *Wastewater hydraulics*: Theory and practice, 2nd ed. Springer, Berlin.

Hager, W.H. and Gisonni, C. (2005). Supercritical flow in sewer manholes. *J. Hydr. Res.* 43 (6), 659–666.

Hager, W.H. and Hutter, K. (1984). Approximate treatment of plane channel flow. *Acta Mech.* 51 (1), 31–48.

Hasumi, M. (1931). Untersuchungen über die Verteilung der hydrostatischen Drücke an Wehrkronen und -Rücken von Überfallwehren infolge des abstürzenden Wassers (Study on the distribution of hydrostatic pressures on weirs due to falling water). *Journal Dep. Agriculture*, Kyushu Imperial University, 3(4), 1–97 [in German].

Mandrup Andersen, V. (1975). Transition from subcritical to supercritical flow. *J. Hydr. Res.* 13 (3), 227–238.

Massé, P. (1938). Ressaut et ligne d´eau dans les cours à pente variable. *Rev. Gén. Hydr.* 4 (19), 7–11; (20), 61–64 [in French].

Matthew, G.D. (1963). On the influence of curvature, surface tension and viscosity on flow over round-crested weirs. *Proc. ICE* 25, 511–524. Discussion (1964) 28, 557–569.

Matthew, G.D. (1991). Higher order one-dimensional equations of potential flow in open channels. *Proc. ICE* 91 (3), 187–201.

Matthew, G.D. (1995). Discussion to A potential flow solution for the free overfall. *Proc. ICE* 112 (1), 81–85.

Montes, J.S. (1994). Potential flow solution to the 2D transition from mild to steep slope. *J. Hydr. Eng.* 120 (5), 601–621.

Montes, J.S. (1998). *Hydraulics of open channel flow*. ASCE Press, Reston Va.

Rouse, H. (1932). The distribution of hydraulic energy in weir flow in relation to spillway design. *MS Thesis*. MIT, Boston.

Westernacher, A. (1965). Abflussbestimmung an ausgerundeten Abstürzen mit Fliesswechsel (Discharge determination at rounded drops with a change of flow type). *Dissertation*, TU Karlsruhe [in German].

Weyermuller, R.G. and Mostafa, M.G. (1976). Flow at grade-break from mild to steep slope flow. *J. Hydr. Div.* ASCE 102(HY10), 1439–1448.

Chapter 14

Diffusive-type of double diffusion in lakes-a review

A. Wüest[1,2], T. Sommer[1,2], M. Schmid[1]
and J.R. Carpenter[1]
[1] Eawag, Swiss Federal Institute of Aquatic Science and Technology, Surface Waters—Research and Management, Kastanienbaum, Switzerland
[2] Institute of Biogeochemistry and Pollutant Dynamics, Environmental Sciences, ETH, Zürich, Switzerland

ABSTRACT: This chapter is a contribution in honour of Gerhard H Jirka, who has been fascinated by the amazing variety of small-scale structures that nature surprises us with, particularly in stratified natural waters. Here, we focus on the diffusive regime of double-diffusive convection that occasionally occurs in lakes. Preconditions are a permanent stratification by dissolved constituents - such as salinity and carbon-dioxide - and convective forcing by deep sources of heat. After setting the stage for double diffusion to occur, possible genesis processes of the diffusive regime are reviewed by explaining specific examples of this unusual stratification such as (i) the flushing of fossil seawater by freshwater, (ii) the solar-pond phenomenon in ice-covered lakes in Antarctica, and (iii) the deep release of salt and gases in volcanic regions. In particular, the two most prominent examples of natural waters in which double diffusion occurs, Lakes Nyos and Kivu, are reviewed in more detail. The generation and evolution of staircase layering are discussed in relation to experiences gained from laboratory experiments, DNS modelling, and analysis of data from natural waters.

1 INTRODUCTION

This chapter of the *Memorial Colloquium Book for Gerhard H Jirka* focuses on the diffusive-type of double diffusion (DD) in lakes. Besides being of particular interest for the specific systems considered - such as (i) the gas-containing "killer" lakes (Zhang, 1996), (ii) the weak mixing in Lake Kivu (Schmid *et al.*, 2005), or (iii) the ice-covered lakes on Antarctica (Green & Friedmann, 1993) - DD in lakes is also of fundamental interest to *Environmental Fluid Mechanics*. Although the finger regime has been observed in the upper thermocline of the Dead Sea (Anati & Stiller, 1991), it is primarily the diffusive regime which is relevant for DD staircase layering in lakes. Due to the typically strong stratification and subsequently low turbulence in the water column, DD in lakes can often develop in almost "undisturbed" form. Therefore, lakes provide ideal scale-up systems between laboratory (Brandt & Fernando, 1995) and large-scale geophysical systems such as oceanic flows (Kelley *et al.*, 2003) or geological flows relevant for ore-building processes (Turner, 1974). Besides these classical geophysical environments, DD occurs in various other systems. Evaporative island systems, where fresh groundwater can underlay superficial brines and cause unstable fingering (Bauer *et al.*, 2006), is another example of such a fascinating phenomenon. Additionally, the diffusive-type of DD has been thoroughly investigated in the 1970s when solar-ponds were the subject of intensive research focused on renewable energy production (Velmurugana & Srithar, 2008).

There are two conditions required for diffusive-type of DD to occur in natural waters: (i) the vertical gradients of temperature and dissolved substances (such as salinity, minerals, or carbon-dioxide) both have to increase with depth and (ii) these two constituents must have different molecular diffusivities ($D_T \approx 10^{-7}$ and $D_S \approx 10^{-9}$ m^2 s^{-1} for temperature and dissolved substances, respectively). Expressed in energetic terms, DD can develop relevant fluxes if the faster diffusing constituent (temperature) has a destabilizing profile, therefore producing "driving" buoyancy flux, while the slower constituent (dissolved substances) has a stabilizing profile and therefore consumes buoyancy flux. The physical setting is defined by the overall water-column stability N^2 (s^{-2}) and the density ratio R_ρ (–), which is the positive ratio of the slowly diffusing (positive) contribution to N^2 divided by the fast-diffusing (negative) contribution to N^2 (the latter is assumed to be due to temperature). As shown below, DD layering is usually found in natural waters if R_ρ ranges from $1.5 < R_\rho < 6$. Theoretically, DD-favourable conditions can exist for $1 < R_\rho < D_T/D_S$; however, due to the limited conversion of the "driving" buoyancy flux from heating to the "consuming" mixing of the stratifying component, the narrower range of R_ρ is more realistic and consistent with the mixing efficiency of plumes.

In the following, we will use several examples to show where the diffusive regime can form and become relevant in lakes. Thereafter, we summarize observations from the two most prominent examples of the diffusive-type of DD in lakes, Lakes Nyos and Kivu. Finally, we provide a short review of possible mechanisms for the generation of staircase layering and some preliminary ideas on how these mechanisms may be relevant in lakes.

2 GENESIS AND BOUNDARY CONDITIONS OF DOUBLE DIFFUSIVE REGIMES

There are various geophysical and bio-geochemical processes influencing the density stratification of natural waters. In the following we review – without intending to be comprehensive – those cases of the diffusive-type of stratification in lakes which have attracted the attention of environmental scientists.

Former fjord lakes – Permanent and diffusive-type of stratification is present in many coastal ex-fjord lakes on the west coasts of America and Europe, where seawater resides beneath freshwater (*Strom*, 1957; *Toth & Lerman*, 1975; Figure 1). A fine example is the ~340 m deep Powell Lake (British Columbia, Canada), which contains seawater at large depth, that has been trapped ~11,500 yr ago. As a result of the glacio-isostatic rebound, the fjord was lifted and separated from the ocean forming an inland-lake. The two downstream basins of the lake still show today deep-water salinities of ~17‰ and geothermal heating warmed the deep-water up to 9.4°C (Williams *et al.*, 1961; Sanderson *et al.*, 1986; Figure 1c). The freshwater, floating over the deep saline water, has maintained permanent stratification since the

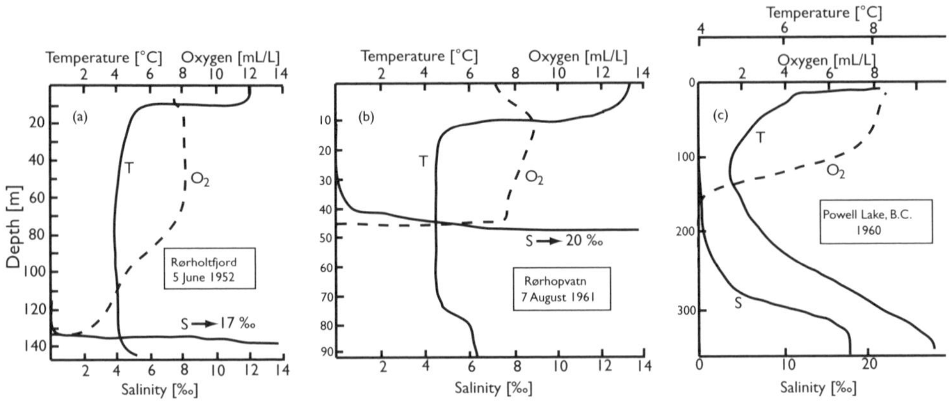

Figure 1 Three examples of diffusive-type of stratification in "former-fjord" lakes on the west coasts of the European and North-American continent: (a) Rørholtfjord in Southern Norway, (b) Rørhopvatn in Northern Norway, and (c) Powel Lake in Southern British Columbia, Canada. In all three lakes, at some depth (134, 43 and 150 m, respectively) the salinity drastically increases as a function of depth maintaining very strong and permanent stratification. Correspondingly below those depths, oxygen vanishes consistent with the lack of seasonal deep convection over thousands of years. Figures 1a and 1b are redrawn from Strøm (1962) and Figure 1c from Williams *et al.* (1961).

last ice-age, and therefore the gases CO_2 and CH_4 as well as other mineralization products have accumulated in the anoxic deep-water for more than 10 kyr of isolation. Today, R_ρ is still ~10 in the deepest layers, with some layering-like stratification but no clear evidence of DD layering. We expect, however, as salt is washed off and temperature and thermal expansivity increase, that R_ρ will decrease to levels of staircase layering in the future.

The well-known European analogy are ex-fjord lakes on the west coast of Norway (Strøm 1957; Høngve 2002). Profiles of temperature, salinity and oxygen, shown for Rørholtfjord (Southern Norway; Figure 1a) and Rørhopvatn (Northern Norway; Figure 1b) indicate the same characteristics of permanent diffusive-type of stratification with drastic salinity increases to former seawater at a critical depth, below which the trapped seawater is anoxic and warmer. The 147 m deep Rørholtfjord (Holtan 1965; Lande 1972; Barland 1991), isolated ~8000 yr ago, is by now elevated ~60 m above sea level and shows the transition to former ocean water at ~133 m depth (Figure 1a). In the much younger and 92 m deep Rørhopvatn, lifted only ~3 m above sea, the freshwater has eroded the oceanic salinity to a depth of ~44 m only (Figure 1b). Due to shorter exposure to freshwater and due to less intense warming and mixing, R_ρ is still far away (a few 1000 yr) from DD layering.

Ice-covered solar-pond lakes in Antarctica – There are numerous ice-covered lakes in Antarctica, which are density-stratified by salt brines, left behind by former evaporation in this dry region (Green & Friedmann, 1993). Overflowing freshwater sets an enormous buoyancy and below the wind-protecting ice the extremely weak mechanical energy input suppresses mixing to molecular level. As for Lake Vanda (Figure 2) with a depth of H = 75 m, the time-scale of diffusion of salt, H^2/D_T, becomes several

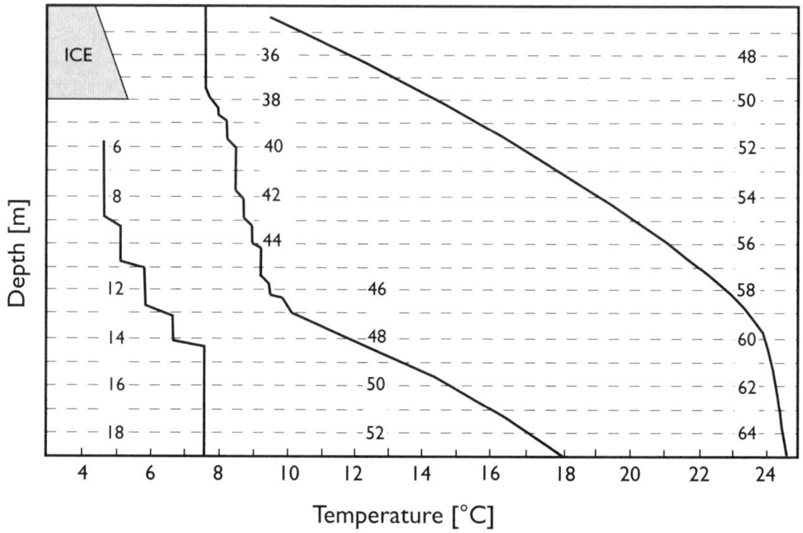

Figure 2 Lake Vanda is a prominent example of several ice-covered lakes in Antarctica which are permanently density-stratified of the diffusive-type due to "fossil" salt residing in the deepest layers (Hoare, 1966). As a result of the strong salinity stratification, the solar heat penetrating the thick ice-cover is accumulated in the deep-water of the lake. The temperature gradient results from a steady-state balance between heat deposited by solar radiation and the vertical divergence of the upward molecular and DD flux of heat. Between 6 and 47 m depth - where R_ρ is in a favourable range - the temperature profile shows a DD staircase with ~12 well-mixed convective layers. At greater depth, salinity increases up to 123 g L^{-1} (TDS) and stabilizes the water column so strongly that the heat fluxes are only of molecular nature. Redrawn from Huppert and Turner (1972).

10,000 yr. Therefore, independent of whether there is still today a salt source or not, the water column remains for a very long time in a diffusive-type of stratification.

Because the ice is transparent for sunlight (even under a thin snow-cover), solar radiation F_{SH} can penetrate into the underlying water and the absorption of the solar heat flux, $\partial T(z)/\partial t = -\partial F_{SH}(z)/\partial z$, warms the stratified water column at depth z. That the deep-water of Lake Vanda can reach a temperature of 25°C, despite the extremely low annually-averaged polar air temperature of −17°C is fascinating per se. In a steady-state, the rate of warming, $\partial T(z)/\partial t$, is equal to the vertical divergence of the upward heat flux, related to DD convection and molecular fluxes through steep gradients (Figure 2). Using Turner's laboratory results (Turner, 1968 and 1974) for quantifying DD convective fluxes, Huppert and Turner (1972) have been able to numerically simulate temperature profiles in excellent agreement with field observations Therefore, it was concluded, that these particular laboratory results can be applied to natural waters bodies of macroscopic scales. Besides Lake Vanda also the ice-covered Lake Miers in the Dry Valleys of the McMurdo Sound (Antarctica) shows distinct DD staircases, as excellently documented by Spigel & Priscu (1998).

Dissolved solids and gases from subaquatic springs – The two classical examples for this group of diffusive-type of stratification are Lake Nyos (Schmid et al., 2004; Wüest *et al.*, 2012) and Lake Kivu (Newman 1976; Schmid et al., 2010; Sommer *et al.*, 2012), where dissolved constituents have accumulated due to deep volcanic activities. In the following two sections we will discuss these examples in more detail, as they are at the same time the best studied lakes with DD staircases. This diffusive regime phenomenon occurs globally also in several marine environments (such as the Red Sea), where hot-saline water intrudes from the earth interior.

Groundwater and mining lakes – Different to the group of deep lakes indicated above, there are many lakes - both natural and manmade (such as mining lakes) – which are substantially supplied by groundwater originating from the surface and not, as above, from deep (geological) sources. Two intriguing, but very different, examples are show in Figures 3 and 4.

In Lake Banyoles (Catalonia), the groundwater inflow into the deepest reaches is warming the lake from below (Figure 3a) and importing fine (bentonitic) particles together with dissolved solids (Figures 3b, c). As discussed by Sánchez & Roget (2007), all three components (temperature, salinity, particles) substantially affect the density and form an example of triple-diffusive convection, as in Lake Nyos. The main difference to Nyos is that carbon-dioxide has a molecular diffusivity comparable to salt, whereas the particles in Lake Banyoles have a completely different diffusivity (if this term is at all appropriate for the motion of suspended particles). Three convectively-mixed layers have been observed in the deepest reaches, but it is expected that the phenomenon is varying with the groundwater inflow (Figure 3).

Stratification favourable to DD can be found in iron-meromictic lakes (small natural or mining lakes). If the groundwater is anoxic, it can - in addition to the example above - carry dissolved iron from the tailings into a lake. In the surface water, iron gets oxidized and hence precipitated back into the anoxic deep-water, where iron remains in solution. Precipitates of iron from above can be remobilized from the sediments by reduction with organic material (Boehrer *et al.*, 2009, Figure 4b). As a consequence, a gradient of dissolved substances is maintained, which stabilizes the density profile against the opposing effect of temperature during winter, when surface waters cool

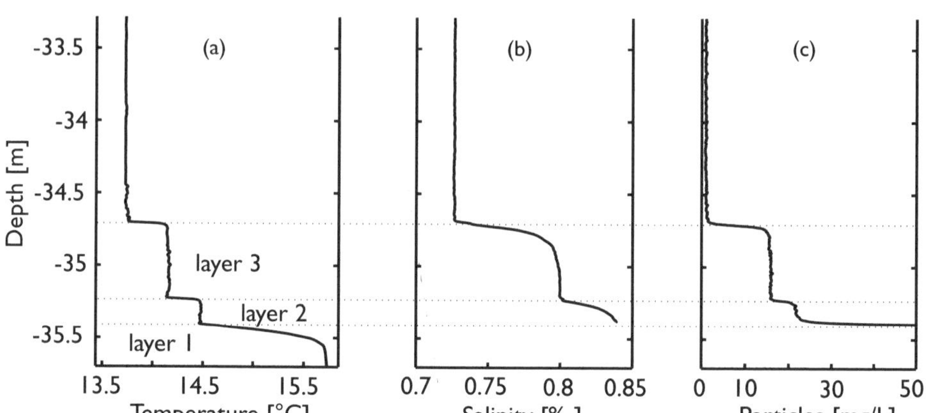

Figure 3 The deepest ~2 m thick zone of Lake Banyoles shows the characteristics of triple diffusive-type of stratification including three well–mixed layers as a result of heating from below (a) Whereas salinity (b) and particles (c) are stabilizing the water column, temperature (a) is destabilizing. Redrawn from Sánchez and Roget (2007).

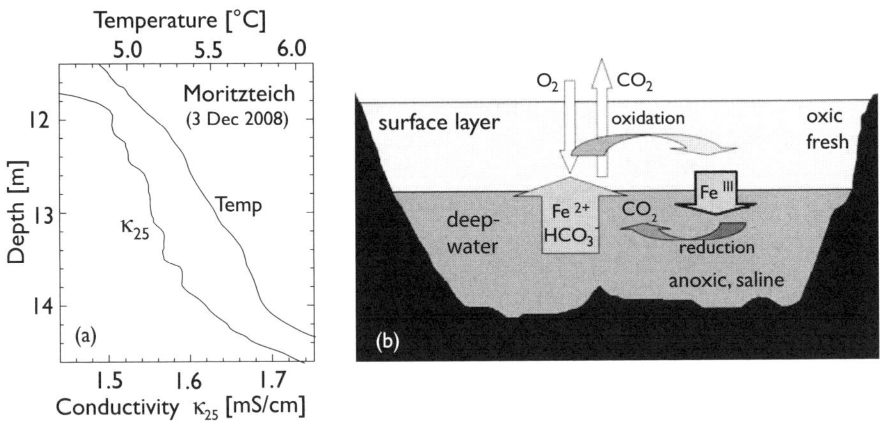

Figure 4 Example of an iron-meromictic lake: Moritzteich (Von Rohden *et al.*, 2009) shows five mixed (but not homogenous) layers during seasonal cooling supporting the temperature gradient for the diffusive-type of stratification (a) The iron-rich water contributes to the density stratification as upward diffusing dissolved iron (from the anoxic environment) form iron-oxides in the oxic surface layer, which is then precipitated back to the deep-water. Redrawn (a) from Von Rohden *et al.* (2009) and (b) from Boehrer *et al.* (2010).

below deep-water temperatures. This set-up can create a seasonal DD staircase structure as shown in Figure 4a (Von Rohden *et al.*, 2010).

3 LIMITED LIFETIME OF A DIFFUSIVE REGIME—THE EXAMPLE OF LAKE NYOS

Lake Nyos, a 208 m deep and 1.58 km^2 large crater lake in Cameroon, is permanently density stratified due to subaquatic springs supplying warm, salty and CO_2-enriched water, which is heavier than the receiving anoxic deep-water. The oxic surface layer is convectively mixed by evaporative cooling during the dry season down to a depth of ~50 m and separated by a strong permanent chemocline (rapid vertical transition in salinity, oxygen and CO_2) from the anoxic deep-water. In February 2002, surface cooling was especially pronounced causing a steep temperature gradient which increased the upward heat flux at the top of the chemocline. The density ratio R_ρ - which depends on the gradients of temperature, salt and CO_2 – dropped to low values immediately below the chemocline and triggered DD convection and the formation of staircase layering.

For the following ~30 months, the staircase zone expanded vertically to ~40 m thickness and its base reached finally down to ~92 m depth. After the initiation of the DD convection, the number of perfectly-homogeneous layers and interfaces of strong gradients grew within a few weeks / months to 27 ± 1 and remained constant for the next two years. Figure 5 shows a 15 m long subsection of a CTD profile (CTD = conductivity/temperature/depth) taken in December 2002 from within the DD zone: This particular CTD profile contained 26 well-identifiable homogeneous

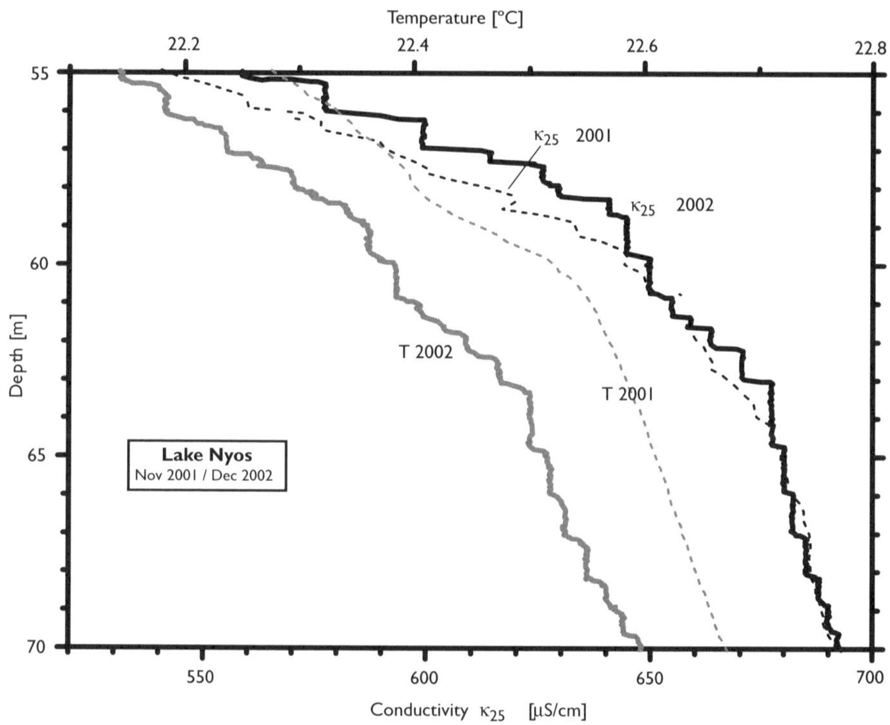

Figure 5 Conductivity and temperature observed in Lake Nyos in 55 to 70 m depth on 3 Nov 2001 (dashed lines) and 8 Dec 2002 (solid lines). Besides the formation of the DD staircase layering (19 well-mixed layers shown), the heat loss (increased vertical heat fluxes) is "well apparent", whereas salinity (conductivity) is hardly affected by the DD fluxes. This example also documents, that the salinity interfaces as well as the mixed layers are much better defined for salinity than for temperature. Redrawn from Wüest *et al.* (2012).

layers (*Schmid et al.*, 2004) with a variable thickness of 0.2 to 2.1 m which grew with time. This CTD subsection also shows that salinity steps were more distinctly structured compared to temperature interfaces (Figure 5). This phenomenon will be discussed in more detail below.

Also during the steady expansion - after the initial build-up of the DD zone had reached the number of 27 layers - new layers/interfaces were still continuously formed at the base of the downward growing staircase zone at a rate of ~1 month^{-1}. However, layers were also merging with the same rate within the DD zone, leaving the number of layers constant. Below we will argue that these two rates of forming and merging can well be explained by the upward heat fluxes, which drastically vary with depth over short distances and subsequently cause substantial local temperature and stratification changes.

The DD convection leads to an increased transport of the constituents out of the deep-water. These fluxes, calculated by the lake-internal heat budget and by using the DD flux laws (Kelley *et al.*, 1990; Schmid *et al.*, 2004), agreed well within the range of the uncertainties and increased from 0.1 W m^{-2} in the lowest layers to 0.5 W m^{-2}

in the uppermost layers of the staircase. This divergence of the heat flux caused a temperature decrease and the temperature profile in this DD zone was rapidly transformed (Figure 5). Due to the disparity of the fluxes (dissolved versus temperature) and the respective divergences of heat and dissolved substances, R_ρ increased rapidly and reached an average value over the DD zone of $R_\rho \approx 6$ after ~900 days of DD initiation. We can also understand the increase of R_ρ by considering the vertical (concave) form of the temperature profile. Expanding the staircase vertically reduces the average vertical temperature gradient over the DD zone (Figure 5) and thereby increases the average R_ρ. The heat fluxes continuously decreased and slowed the DD convective mixing until it collapsed after ~900 days at an $R_\rho \approx 6$ as the well-defined layers and interfaces (specifically their sharpness) eroded away. We do not know whether the DD staircase layering was ever active before or whether it was just not recognized. Earlier CTD profiles (Kusakabe *et al.*, 2008) indicate that it may have been active in the past.

4 THE STRUCTURE OF STAIRCASES—THE EXAMPLE OF LAKE KIVU

Lake Kivu (surface area of 2300 km^2, volume of 550 km^3, maximum depth 485 m) is the most voluminous and most extensive example of the diffusive-type of DD layering in terms of longevity, stability and distinctiveness. Four water constituents contribute to the permanent density stratification in the deep-water, which has a DD-favourable structure. Whereas salinity and CO_2 (volcanic origin) are the two stabilizing agents, temperature (geothermal) and dissolved methane (CH_4 from organic matter and CO_2 reduction) destabilize the water column. The increase of all four agents with depth is not linear but modified by several subaquatic springs. The major springs are located at the northern shore clustering at depths of ~250 m (~15 m^3 s^{-1} discharge) and ~180 m (~22 m^3 s^{-1}; Schmid *et al.*, 2005). These springs cause large variations of R_ρ over relatively short depth levels (Figure 6a). R_ρ covers the range of 2 to 5 (for > 80% of the profile) over which - according to the literature - DD staircase layering has been observed in natural waters.

The first observations of DD staircases in Lake Kivu were reported by Newman (1976), and the most recent measurements are from Schmid *et al.* (2010) and Sommer *et al.* (2012). Whereas Newman (1976) identified ~150 homogeneously mixed layers, Schmid *et al.* (2010) observed up to ~350, separated by sharp interfaces, in the northern basin below ~120 m depth, often reaching down to the sediment near maximum depth. Close to the shore, at ~2 km horizontal distance from the boundary, DD staircases fade away and vanish (Sommer *et al.*, 2012). Even though the staircase phenomenon per se persists most probably over very long time scales of at least the residence time of the deep-water (several hundred years), the detailed structure of the staircases change much more rapidly. Within the 32 years between the measurements in 1972 and 2004, the average mixed layer thickness decreased by more than a factor of 2 from ~1.3 m to ~0.4 m within the same depth interval (Newman, 1976; Schmid *et al.*, 2010).

Typical steps in temperature ΔT between two adjacent mixed layers, separated by a steep interface, are a few mK (Figure 6b). From 2004 (Schmid *et al.*, 2010) to 2011

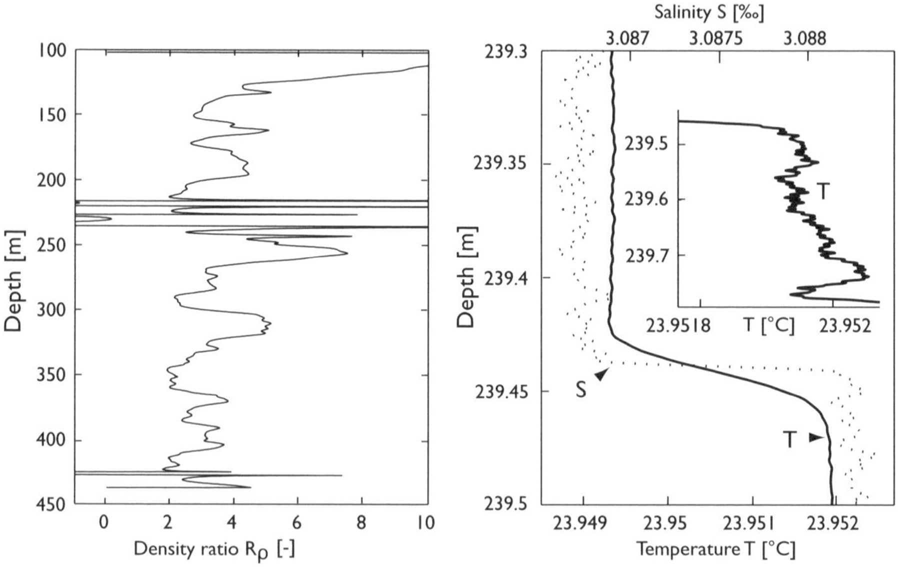

Figure 6 (a) R_ρ for the permanently stratified deep-water of Lake Kivu showing favouring conditions for DD layering in the range of $2 < R_\rho < 5$ (between 200 and 250 m there are zones, where R_ρ is jumping from plus to minus infinity because the temperature gradient changes sign). (b) An example of an interface (of about 300 over the entire depth) adjacent to two mixed layers. This 20-cm long profile section demonstrates that the temperature interface (here a few cm) is much thicker than the salinity interface (here a few mm). The inset shows the temperature fluctuations in the lower homogeneous layer indicating slight T differences between the sinking and rising thermals and the slight unstable background T gradient. Redrawn from Sommer *et al.* (2012).

(Sommer *et al.*, 2012) the changes in staircase properties are less pronounced, partly due to the short time span of only seven years. However, we could observe an increase in ΔT, which is most probably caused by increasing (local) heat fluxes. Although, we still have to proof this hypothesis, it is supported by the warming of the deep-water between 2004 and 2011 (~0.09 K in 420 m depth).

Considering lateral variations, individual mixed layers could be traced horizontally over distances of up to ~10 km in the northern basin within the depth range of 170 to 190 m (Sommer *et al.*, 2012). The temperature of those mixed layers increased along the tracks by ~0.05 K from north to south, whereas salinity was almost constant. We believe, that this horizontal gradient is caused by cold subaquatic spring inflows from the northern shore. We therefore plan to investigate the consequences of such lateral gradients for the horizontal and vertical fluxes of heat and salt. Similar coherent diffusive-type of staircase layering were also found in the Arctic with a horizontal extent of ~800 km (Timmermans *et al.*, 2008).

The main objective of Sommer *et al.* (2012) was to study the interface thicknesses h_T and h_S of temperature and salt, respectively. The idealized DNS modelling of Carpenter *et al.* (2012b) together with measurements of Padman and Dillon (1987) and Timmermans *et al.* (2008) suggest that the vertical fluxes of heat and salt through the core of the interface are of purely molecular nature (Carpenter *et al.*, 2012), and the vertical heat and salt fluxes can be accurately estimated by measuring the interfacial gradients ($\Delta T/h_T$, $\Delta S/h_S$). In agreement with the DNS modelling of Carpenter *et al.* (2012b) we found the temperature interface to be on average about twice as thick as the salinity interface (Figures 6b, 7b) for R_ρ between 3 and 6. Extremely thin interfaces of a few millimetres thickness were found for salinity especially for values of $R_\rho < 3$. Figure 6b shows such an example together with the much thicker temperature interface.

5 GENERATION AND EVOLUTION OF STAIRCASE LAYERING

We finally outline our current understanding of the generation and evolution mechanisms of staircase layering, based on observations in laboratory and natural waters, as well as DNS studies. A minimum of four staircase generation mechanisms can be identified for the diffusive-type of DD: (i) spontaneous generation from unstable linear gradients (Noguchi & Niino, 2010), (ii) formation by bottom-heating or top-cooling of a stable salinity gradient (Turner, 1968), (iii) intrusions of lateral T-S gradients (Merryfield, 2000, who describes in fact fingering), and (iv) the negative

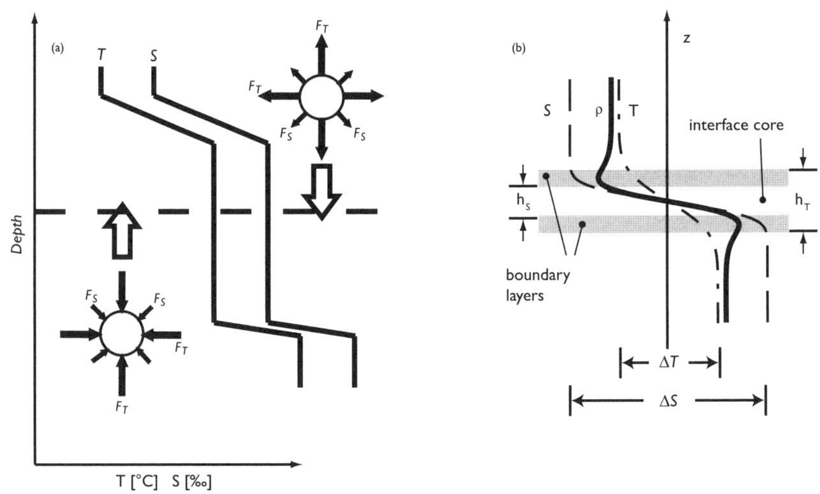

Figure 7 Two different interpretations of mixed-layer formation and/or maintenance: (a) if a water parcel would be dislocated macroscopically upward relative to its equilibrium position (horizontal dashed line) the parcel loses heat efficiently (bolt arrows) but loses only little salt (small open arrows) – the modified water parcel is then heavier and therefore pushed back towards the original equilibrium position. The same arguments cause a downward dislocated water parcel to become lighter and to return back upwards to its original location (dashed line). The net effect is a sharpening of the interface and a homogenising of the layers. (b) Due to the ~100-times faster diffusion of heat relative to salt, the temperature interfaces grow thicker than salinity interfaces and therefore two unstable boundary layers formed adjacent to the interface. These gravitational instabilities cause convection in both homogenous layers and maintain them well-mixed. Redrawn from Carpenter *et al.* (2012a).

turbulent diffusion mechanism, where due to the larger flux of heat relative to salt (Figure 7a), density gradients are sharpened, rather than smoothed (Linden, 2000). Although each of these represents a viable mechanism for staircase formation, it is very rare to observe the actual formation of a staircase in natural waters, and so it is difficult to identify one mechanism as more frequent, or likely, than another. One exception is Lake Nyos, as explained above, where we have observed the formation of new layers at the lower boundary of the expanding DD zone (Figure 5).

The first of these mechanisms follows from linear stability analysis of the linear profiles of temperature and salinity (Stern, 1960; Veronis, 1965; Nield, 1967; Baines & Gill, 1969). The instability of the water column to a linearly increasing temperature and salinity with depth can be understood as follows: If a water parcel is displaced by a small distance upwards it will easily release its heat to the cooler surroundings, but retain its salt (Figure 7a). This is due to the approximately 100-times larger molecular diffusion coefficient of temperature than of salinity. Once the heat of the parcel is released, it is denser than its surroundings, and denser than it was at its original (equilibrium) level, so it sinks deeper beyond this level. The parcel finds itself in a warmer environment, and can absorb heat, without gaining salt. The parcel will therefore rise, and a steadily growing oscillation is produced in the absence of friction. This is the basic linear instability of the diffusive convection regime. Recent DNS by Noguchi & Niino (2010) have shown that this growing oscillation breaks down into turbulence, and spontaneously generates a staircase structure.

Despite the clear identification of this spontaneous staircase generation mechanism from linear gradients, we believe that it is likely to be of limited relevance to natural waters and lakes in particular. This is due to the fact that the linear instability is only possible if $R_\rho < 1.16$ (Baines & Gill, 1969). This is a very strict requirement for the temperature and salinity profiles that is rarely, if ever, observed in natural waters, such as lakes. We find it much more likely that the staircases are formed from some source of heat in a salt-stratified water column, either from the vertical or lateral boundaries (see Lakes Nyos and Kivu above), or internally (see Lake Vanda for the example of solar radiation).

Once a staircase layering has formed, it is believed that a slightly different mechanism is responsible for maintaining its structure. The sharp steps of the staircase can be maintained against the smoothing action of diffusion by the following mechanism. Consider an interface across which sharp changes in temperature and salinity are present and these occur over a vertical length scale of h_T and h_S for temperature and salinity, respectively (Figure 7b). If we take $h_T = h_S$ initially, then the resulting density profile will be gravitationally stable everywhere (provided that $R_\rho > 1$). However, due to the larger molecular diffusion of temperature relative to salinity, over time h_T will grow larger than h_S. Since temperature is in a gravitationally unstable configuration, unstable boundary layers will develop on either side of the interface (Figure 7b). It is the convective breakdown of these boundary layers that drives turbulence in the mixed layers – from below and from above – and is responsible for maintaining sharp interfaces (Linden & Shirtcliffe, 1978). The recent studies of Carpenter *et al.* (2012a, b) support this idea, and in DNS modelling the interfaces were found to evolve to an average thickness ratio of $h_T/h_S \approx 2.6$ for R_ρ between 3 and 6 (Figure 7b). Therefore, a gravitationally unstable boundary layer is supported. They were found to have Rayleigh numbers characteristic on the order of 100. This observation of a pair of unstable boundary layers has also been made from microstructure measurements in Lake Kivu (Sommer *et al.*, 2012).

6 OUTLOOK

With the two studies on the Lakes Nyos and Kivu we have been in the comfortable situation to acquire a good sense of the macroscopic parameters involved in DD in natural waters and in lakes in particular. We could collect excellent statistics on the thickness of temperature (h_T) and salinity (h_S) interfaces (including their ratio h_T/h_S) and the mixed layers, as well as gradients, molecular interface fluxes and fluctuations (T', S') of temperature and salinity within mixed layers. In the next step we are investigating the relevant mechanisms by comparing these macroscopic parameters with results from DNS. First results are: (i) the unstable boundary layers have been identified both in DNS and field observations as the key for the quantification of the fluxes through the staircases (Linden and Shirtcliffe, 1979; Worster 2004); (ii) a good agreement exists between h_T/h_S in the field and DNS, (iii) fluxes in the interface cores seem to be purely molecular, and (iv) in-situ estimated heat fluxes agree with flux laws by Turner (1974) and Kelley (1990).

For understanding the temporal evolution of the DD process we have to consider two different scales: First, on a large-scale, the relevant boundary condition is the sustainable supply of buoyancy flux in order to drive the energetics and to maintain the stratification. In the case of Lake Nyos, the heat flux out of the staircase zone was larger than the supply by deep sources. Therefore the DD process existed only temporarily until it was energetically drained by DD-induced fluxes. In contrast, in Lake Kivu the scene is set by a permanent supply of heat and dissolved substances and therefore we can expect that DD staircase layering will exist as long as subaquatic fluxes are within a critical range.

Second, on a fine–scale the temporal evolution of the DD process has a different dynamics. A preliminary analysis provides evidence of mechanisms for (i) creating new mixed layers through interface splitting and for (ii) merging of existing layers. The vertically inhomogeneous upward heat flux within the DD staircase causes local convergence and divergence of heat, and subsequently temperature can evolve locally without affecting the overall structure of the staircase.

We therefore hypothesize that the DD staircase layering as a phenomenon can persist for very long time (hundreds of years in the case of Kivu), whereas the internal details within the interior of the staircase evolve comparably fast on time scales of days to months. Although we are still at the beginning, we hope to shed light on this picture within the next two years.

REFERENCES

Anati, D.A. and Stiller, M. (1991). The post-1979 thermohaline structure of the Dead-Sea and the role of double-diffusive mixing. *Limnology and Oceanography* 36 (2): 342–354.

Baines, P.G. and Gill, A. (1969). On thermohaline convection with linear gradients. *J. Fluid Mech.* 34, 289–306.

Barland, K. (1991). Trapped seawater in 2 Norwegian lakes – Kilevannet, a new lake with old trapped seawater, and Rørholtfjorden. *Aquatic Sciences*. [Online] 53 (1), 90–98. Available from: doi:10.1007/BF00877077.

Bauer, P., Supper, R., Zimmermann, S. and Kinzelbach, W. (2006). Geoelectrical imaging of groundwater salinization in the Okavango Delta, Botswana. *J Appl. Geophysics*, 60 (2): 126–141, DOI: 10.1016/j.jappgeo.2006.01.003.

Boehrer, B., Dietz, S., Von Rohden, C., Kiwel, U., Jöhnk, K.D., Naujoks, S., Ilmberger, J. and Lessmann, D. (2009). Double-diffusive deep water circulation in an iron-meromictic lake. *Geochemistry, Geophysics, Geosystems*. [Online], 10 (6), Q06006. Available from: doi:10.1029/2009GC002389.

Brandt, A. and Fernando, H.J.S. (1995). Double-Diffusive Convection, *Geophys. Monogr. Ser.*, 94, 334 p. doi:10.1029/GM094, American Geophysical Union, Washington, D.C.

Carpenter, J.R., Sommer, T. and Wüest, A. (2012a). Stability of a double-diffusive interface in the diffusive convection regime. *J. Phys. Oceanogr.*, in review.

Carpenter, J.R., Sommer, T. and Wüest, A. (2012b). Simulations of a double-diffusive interface in the diffusive convection regime. *J. Fluid Mech.*, submitted.

Gonzalez-Juez, E., Kerstein, A.R. and Lignell, D.O. (2011). Fluxes across double-diffusive interfaces: a one-dimensional-turbulence study. *Journal of Fluid Mechanics*, 677, 218–254.

Green, W.J. and Friedmann, E.I. (1993). Physical and biogeochemical processes in Antarctic lakes. *Antarctic Research Series 59*, American Geophysical Union, Washington, D.C.

Hoare, R.A. (1966). Problems of heat transfer in Lake Vanda, a density stratified Antarctic lake. *Nature*. [Online] 210 (5038), 787–789. Available from: doi:10.1038/210787a0.

Holtan, H. (1965). Salt water in the bottom layers of two Norwegian lakes. *Nature*, 207, 156–158.

Hongve, D. (2002). Seasonal mixing and genesis of endogenic meromixis in small lakes in southeast Norway, *Nord. Hydrol.*, 33 (2–3), 189–206.

Huppert, H.E. andTurner, J.S. (1972). Double-diffusive convection and its implications for the temperature and salinity structure of the ocean and Lake Vanda. *Journal of Physical Oceanography*, 2, 456–461.

Kelley, D.E. (1990). Fluxes through diffusive staircases: A new formulation. *Journal of Geophysical Research*, 95, 3365–3371.

Kelley, D.E., Fernando, H.J.S., Gargett, A.E., Tanny, J. and Özsoy, E. (2003). The diffusive regime of double-diffusive convection. *Progress in Oceanography*. [Online] 56 (3–4), 461–481. Available from: doi: 10.1016/S0079–6611(03)00026–0.

Kusakabe, M., Ohba, T., Issa, Yoshida, Y., Satake, H., Ohizumi, T., Evans, W.C., Tanyileke, G. and Kling, G.W. (2008). Evolution of CO_2 in Lakes Monoun and Nyos, Cameroon, before and during controlled degassing. *Geochemical Journal* 42 (1), 93–118.

Lande, A. (1972). "Trapped" sea-water in Rørholtfjorden. *Schweizerische Zeitschrift für Hydrologie*. 34 (1), 34–40.

Linden, P. and Shirtcliffe, T. (1978). The diffusive interface in double-diffusive convection. *J. Fluid Mech.*, 87:417–432.

Linden, P.F. (2000). Convection in the environment. In "Perspectives in Fluid Dynamics", Cambridge University Press, Ed. Bachelor, G.K., H.K. Moffatt & M.G. Worster, 289–345.

Merryfield, W.J. (2000). Origin of thermohaline staircases. *J. Phys. Oceanogr.* 30, 1046–1068.

Newman, F.C. (1976). Temperature steps in Lake Kivu: A bottom heated saline lake. *J. Phys. Oceanogr*, 6, 157–163.

Nield, D.A. (1967). The thermohaline Rayleigh-Jeffreys problem. *J. Fluid Mech.* 29, 545–558.

Noguchi, T. and Niino, H. (2010). Multi-layered diffusive convection. Part 1. Spontaneous layer formation. *J. Fluid Mech.* 651, 443–464.

Osborn, T.R. (1973). Temperature microstructure in Powel Lake. *J. Phys. Oceanogr*, 3, 302–307.

Padman, L. and Dillon, T. (1987). Vertical heat fluxes through the Beaufort Sea thermohaline staircase. *J. Geophys. Res.*, 92:10,799–10,806.

Sánchez, X. and Roget, E. (2007). Microstructure measurements and heat flux calculations of a triple-diffusive process in a lake within the diffusive layer convection regime. *Journal of Geophysical Research C: Oceans*. [Online] 112 (2), C02012. Available from: doi:10.1029/2006JC003750.

Sanderson, B., Perry, K. and Pedersen, T. (1986). Vertical diffusion in meromictic Powell Lake, British Columbia. *J. Geophys. Res.*, 91(C6), 7647–7655 [Online] Available from: doi:10.1029/JC091iC06p07647.

Schmid, M., Busbridge, M. and Wüest, A. (2010). Double-diffusive convection in Lake Kivu. *Limnology and Oceanography*, 55 (1), 225–238.

Schmid, M., Halbwachs, M., Wehrli, B. and Wüest, A. (2005). Weak mixing in Lake Kivu: New insights indicate increasing risk of uncontrolled gas eruption. *Geochemistry, Geophysics, Geosystems*. [Online] 6 (7), Q07009. Available from: doi:10.1029/2004GC000892.

Schmid, M., Lorke, A., Dinkel, C., Tanyileke, G. and Wüest, A. (2004). Double-diffusive convection in Lake Nyos, Cameroon. *Deep-Sea Research Part I. Oceanographic Research Papers*. [Online] 51 (8), 1097–1111. Available from: doi:10.1016/j.dsr.2004.02.010.

Sommer, T., Carpenter, J., Lueck R., Schmid, M. and Wüest, A. (2012). Temperature and salinity microstructure of a double-diffusive staircase. *J. Geophys. Res.*, in preparation.

Spigel, R.H. and Priscu, J.C. (1998). Physical limnology of the McMurdo Dry Valleys lakes. In: Priscu, J.C. (Ed.), Ecosystem Dynamics in a Polar Desert, Vol. 72. AGU, Washington, DC.

Stern, M. (1960). The salt-fountain and thermohaline convection. *Tellus*, 2, 172–175.

Strøm, K. (1957). A lake with trapped sea-water? *Nature*. [Online] 180 (4593), 982–983. Available from: doi:10.1038/180982a0.

Strøm, K. (1961). A second lake with old sea-water at its bottom. *Nature*, 189, 913.

Strøm, K. (1962). Trapped sea-water. *New Scient*. 274, 384–386.

Timmermans, M.-L., Toole, J., Krishfield, R. and Winsor, P. (2008). Ice-tethered profiler observations of the double-diffusive staircase in the Canada Basin thermocline. *J. Geophys. Res.*, 113(C00 A02):doi:10.1029/2008 JC004829.

Toth, D.J. and Lerman, A. (1975). Stratified lake and oceanic brines – salt movement and time limits of existence. *Limnology and Oceanography*, 20 (5), 715–728.

Turner, J.S. (1968) The behaviour of a stable salinity gradient heated from below. *J. Fluid Mech.* 33, 183–200.

Turner, J.S. (1974). Double-diffusive phenomena. *Annual Review of Fluid Mechanics*, 6: 37–56.

Velmurugana, V. and Srithar, K. (2008). Prospects and scopes of solar pond: A detailed review. *Renewable & Sustainable Energy Reviews* 12 (8), 2253–2263 DOI:10.1016/j.rser.2007.03.011.

Veronis, G. (1965). On finite amplitude instability in thermohaline convection. *J. Mar. Res.* 23, 1–17.

Von Rohden, C., Boehrer, B. and Ilmberger, J. (2010). Evidence for double diffusion in temperate meromictic lakes. *Hydrology and Earth System Sciences*, 14 (4), 667–674.

Williams, P.M., Mathews, W.H. and Pickard, G.L (1961). A lake in British Columbia containing old sea-water. *Nature*, 191 (4790), 830–832.

Wüest, *et al.* (2012). 900 days of double diffusion in Lake Nyos. *in preparation.*

Zhang, Y.X. (1996). Dynamics of CO_2-driven lake eruptions NATURE379 (6560) 57–59. DOI: 10.1038/379057a0.

Part 4

Jets and stratified flow

Chapter 15

Multiple jet interaction in stagnant shallow water

Adrian C.H. Lai[1] *and Joseph H.W. Lee*[2]
[1] *Singapore-MIT Alliance for Research and Technology Centre, Singapore*
[2] *Department of Civil and Environmental Engineering, Hong Kong University of Science and Technology, Clear Water Bay, Hong Kong SAR, P.R. China*

ABSTRACT: Multiple shallow water jets discharged from a unidirectional multiport diffuser can be found in a number of environmental hydraulics problems such as thermal discharges from once-through cooling water systems or pollutant discharges in a river. Previous studies have shown that dynamic interaction is significant for such jet groups which act like a line momentum source or two-dimensional actuator disk. Traditional integral jet models for a single free jet are not applicable for predicting the mixing of such jet groups.

In this paper, a recently proposed semi-analytical model to predict dynamic jet interaction is applied to the problem of multiple shallow water jet discharges in stagnant ambient. The flow field induced by jets is computed by a distribution of point sinks of strength equal to the entrainment per unit length along the jet trajectory and accounting for boundary effects. The jet flow is determined by an iterative solution of an integral jet model accounting for dynamic pressure and momentum flux changes. Model predictions are in good agreement with experimental data for several key jet group mixing characteristics.

1 INTRODUCTION

Closely spaced jets interact dynamically due to the pressure field induced by the jet group entrainment; they also interact kinematically when they merge with other jets. The resultant effect of the interaction is often a noticeable change in jet group mixing characteristics. For example, it was observed that the jets tend to attract each other in the near field of a multiple jet group; the trajectory of each individual jet can thus be different than that of a single free jet with the same discharge parameters (e.g., Liseth 1970; Baines and Keffer 1974; Kaye and Linden 2004; Lai and Lee 2008; Lai 2009). Merging of jets affect both the velocity and concentration field of a jet group; a typical example is the discharge from an array of equally spaced round jets, which eventually merge to form an equivalent slot jet with the same initial volume and momentum fluxes per unit diffuser length.

In general, both the dynamics and kinematics need to be considered for predicting the mixing characteristics of a jet group. The degree of dynamic interaction depends on both the jet discharge and ambient parameters, such as the total number of jets, jet spacing, momentum and buoyancy, degree of confinement, ambient current speed, orientation, and stratification. Some multiple jet discharges can be justified to have small dynamic interaction. For example, when there is a small number of jets and wide jet spacing without confinement, it is reasonable to simplify the problem by neglecting the dynamics and consider only kinematic interaction. Pani *et al.* (2009) successfully predicted the mixing of equally spaced unidirectional multiple jets (Wang 2000) with either large jet spacing or small number of jets in an unbounded co-flowing current by considering solely the jet merging using the superposition approach. For jets discharged into a moderate to strong crossflow, it is expected that the momentum imparted to a jet from the ambient crossflow governs the jet mixing, and the jet dynamics has a less important role. In a recent study, it is shown experimentally that the dynamic interaction of a rosette jet group with 6–8 horizontal nozzles mounted circumferentially on a riser is negligible for typical ocean outfall configurations and typical design crossflow velocities (Lai *et al.*, 2011).

Dynamic interaction can be significant in some problems, particularly when the number of jets is large, with small jet spacing, and under confinement. Lee and Jirka (1980) and Lee and Greenberg (1984) studied unidirectional multiple jets in shallow water in the context of environmental hydraulics. The number of jets considered was large (20–40) and the discharge is vertically confined with the diffuser length being much larger than the water depth. It was observed the outermost jet deflected strongly towards the jet group centre in the near field. Other mixing characteristics such as jet group-induced velocity and concentration were also observed to be affected due to the contraction of the jet group width – similar to a two-dimensional slipstream. The 2D momentum-induced flow can be predicted by a semi-analytical inviscid model which assumes a velocity discontinuity across the slip-streamlines. More recently, a semi-analytical model has successfully predicted the dynamic interaction and its effect on jet group mixing such as velocity and concentration field for various jet group configuration (Lai 2009; Lai and Lee 2011). The objective of this paper is to investigate the extent to which the model can be applied for predicting multiple jets in stagnant shallow water using the experimental data of Lee and Greenberg (1984) and Lee and Jirka (1980) as test cases.

2 PROBLEM DEFINITION

The problem considered is shown in Figure 1. The ambient water with infinite horizontal extent and depth H is bounded by a free surface and a solid bottom. An array of N jets with spacing s, each with port diameter D, discharging at a velocity u_o from a diffuser of length $L = s(N\text{-}1)$ at height h_o above the bottom into the otherwise stagnant ambient with total flow rate Q_o. A Cartesian coordinate system (x, y, z) is defined such that the diffuser is centred at (0, 0, 0) and aligned in the y direction, with jets discharging in the $+x$ direction.

In the near field close to the diffuser, the jet velocity induced by the i-th jet (u_{xi}, u_{yi}, u_{zi}) is three-dimensional. The jets entrain surrounding ambient fluid as they travel downstream; their jet width and volume flux increase along with decrease of jet velocity. The jets continue to grow in size and eventually merge with their neighbours and occupy the whole water depth; the jet group then becomes more like a

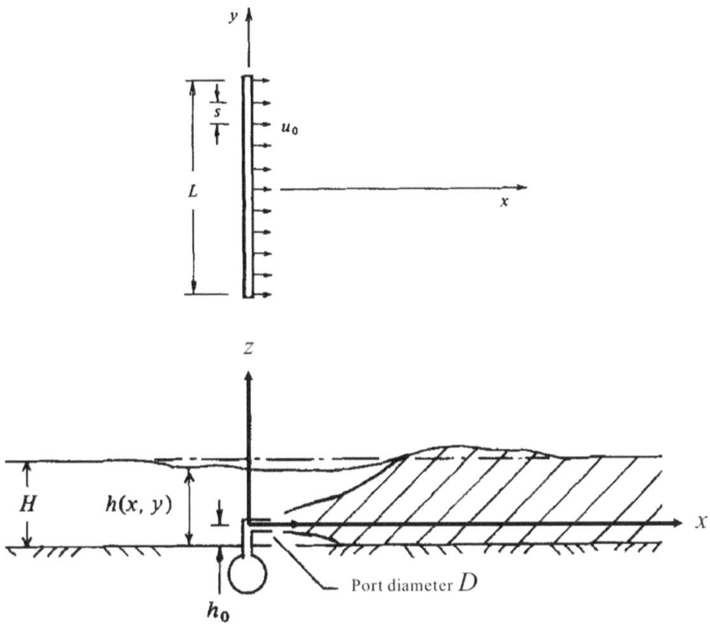

Figure 1 Multiple jets in shallow stagnant ambient.

two-dimensional jet with velocity (u_x, u_y). The decay of jet velocity slows down due to the restriction in entrainment; an ultimate velocity u_∞ would be attained at the end of the near field. The theoretically predicted distance for the jet group to attain u_∞ was shown to be $x \approx L$ (Lee and Greenberg 1984) in stagnant ambient. The jet group subsequently enters the "far field", where bottom friction effects become important. This paper considers only the near field. Key mixing characteristics of the jet group in the near field are the jet trajectory, velocity, and concentration.

3 THEORY FOR THE DYNAMIC INTERACTION OF MULTIPLE JETS IN SHALLOW WATER

The dynamic interaction of multiple jets is related to the velocity and pressure field external to the jet. We present a semi-analytical theory which predicts the external flow and pressure induced by a general jet group, and the subsequent use of an iterative solution to account for dynamic interaction of the multiple jets.

3.1 Jet group induced external flow field

Taylor (1958) showed that the external flow induced by a jet can be predicted by treating the jet as a line sink. To extend it to a general multiple jet group with arbitrary jet trajectory, we discretize the jet into a number of jet elements of length Δs, each being represented by a point sink with strength $m_i\,\Delta s$ equal to the entrainment.

The linearity of the irrotational external flow governing equation permits the use of the superposition principle; the jet induced external flow field can then be computed by summing up the flow field induced by the sink distribution. The theory can be extended to a jet group by simply summing up the flow field induced by every jet in the group.

The induced velocity by a distribution of sinks at (x_i, y_i, z_i) at point $P(x_p, y_p, z_p)$ is then given by:

$$u_{xj}\Big|_{(x_p,y_p,z_p)} = \sum_{i=1}^{N} \frac{m_i}{4\pi} \frac{(x_p - x_i)\Delta s}{[(x_p - x_i)^2 + (y_p - y_i)^2 + (z_p - z_i)^2]^{3/2}} \quad (1)$$

$$u_{yj}\Big|_{(x_p,y_p,z_p)} = \sum_{i=1}^{N} \frac{m_i}{4\pi} \frac{(y_p - y_i)\Delta s}{[(x_p - x_i)^2 + (y_p - y_i)^2 + (z_p - z_i)^2]^{3/2}} \quad (2)$$

$$u_{zj}\Big|_{(x_p,y_p,z_p)} = \sum_{i=1}^{N} \frac{m_i}{4\pi} \frac{(z_p - z_i)\Delta s}{[(x_p - x_i)^2 + (y_p - y_i)^2 + (z_p - z_i)^2]^{3/2}} \quad (3)$$

where $i = 1, 2, \ldots, N$ and N is the number of jet element.

And the jet group induced external flow field is:

$$u_{xmj} = \sum_{j=1}^{NJ} u_{xj}; \qquad u_{ymj} = \sum_{j=1}^{NJ} u_{yj}; \qquad u_{zmj} = \sum_{j=1}^{NJ} u_{zj}; \quad (4)$$

where NJ is the number of jets.

The pressure field is related to the velocity field by the Bernoulli's principle $P = -1/2\, \rho q^2$, where $q = \sqrt{u_{xj}^2 + u_{yj}^2 + u_{zj}^2}$.

The comparison of the present theory and Talyor's solution for the external flow field induced by a single jet is shown in Figure 2. A horizontal jet is discretized into a series of jet element each with length $\Delta s = 0.02H$, where H is the maximum horizontal extent of interest. With initial jet diameter of $D = 0.1$ m and velocity of 1.75 m/s, the flow field predicted by the present theory agrees with Taylor's solution for 3 significant digits. The theory has also been compared with the computation of a numerical model for a twin jet discharge; the model predicted flow field is comparable to the numerical results (Lai 2009), but it should be noted that the numerical model computation can only be used as reference, as the artificial zero pressure boundary conditions set at each boundary of the computational domain affects the resulting computation.

To account for the free surface and solid bottom boundary effects, the method of images is applied. Image pairs of the sink distribution are located at $(H, -H)$, $(2H, -2H)$, ..., $(kH, -kH)$ (assuming mid-depth discharges; see later discussion) such that no flux boundary conditions are satisfied at the water surface and bottom. This is particularly important in multiple jets in shallow water where the boundary effect is important. 16 pairs of images ($k = 16$) are needed for the present case, adding further images does not produce noticeable change in flow field. The large number of image pairs required when compared with cases in deep water (typically 3 pairs) reflects the degree of confinement.

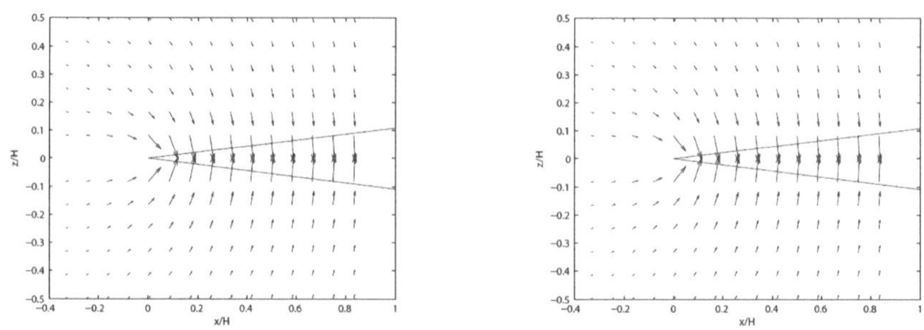

(a) Taylor's solution (b) Point sink distribution

Figure 2 Computed flow field induced by a jet.

3.2 A jet integral model for the dynamic interaction of jets

The change in jet momentum flux of an individual jet resulting from the jet group induced flow and pressure field can be computed by momentum balance over each of the jet element. The computed change in jet momentum flux is not sensitive to the choice of control volume (Lai 2009), and it can be conveniently be chosen as a cylinder with radius $B = \sqrt{2}b_g$ and length Δx = 1–2D, where b_g is the Gaussian jet-width. Figure 3 shows a control volume used in a jet at one particular computation step. The surface of the control volume is discretized into a number (20 in present model) of segments with length Δseg; Figure 3 shows 3 of the segments. The velocity vector $\mathbf{q} = u_{xmj}\mathbf{i} + u_{ymj}\mathbf{j} + u_{zmj}\mathbf{k}$ denotes the jet external flow computed by the distributed sinks, and the pressure p acting on each segment is obtained by the Bernoulli's equation. The segment normal vector $\mathbf{n} = n_x\mathbf{i} + n_y\mathbf{j}$ defines the orientation of the segment plane. The change in momentum in the y -direction along x can be found by (Milne-Thomson 1968):

$$\Delta M_y = -\left[\sum n_y \frac{p}{\rho}\Delta seg + \sum u_{ymj}(\mathbf{n}\cdot\mathbf{q})\Delta\, seg\right]\Delta x \tag{5}$$

The change in y-momentum is reflected in the deflection of a jet with a deflection angle of θ relative to the y-axis. The effect of the change in y-momentum can be reflected by incorporating the above term into a jet integral model cast in terms of jet volume, x- and y-momentum fluxes:

$$\frac{dQ}{ds} = \frac{d}{ds}\int udA = 2\sqrt{2\pi M}\alpha \tag{6}$$

$$\frac{dM_x}{ds} = \frac{d}{ds}\int (uu_x)dA = 0 \tag{7}$$

$$\frac{dM_y}{ds} = \frac{d}{ds}\int (uu_y)dA = -\sin\theta\left[\sum n_y \frac{p}{\rho}\Delta seg + \sum u_{ymj}(\mathbf{n}\cdot\mathbf{q})\,\Delta\, seg\right] \tag{8}$$

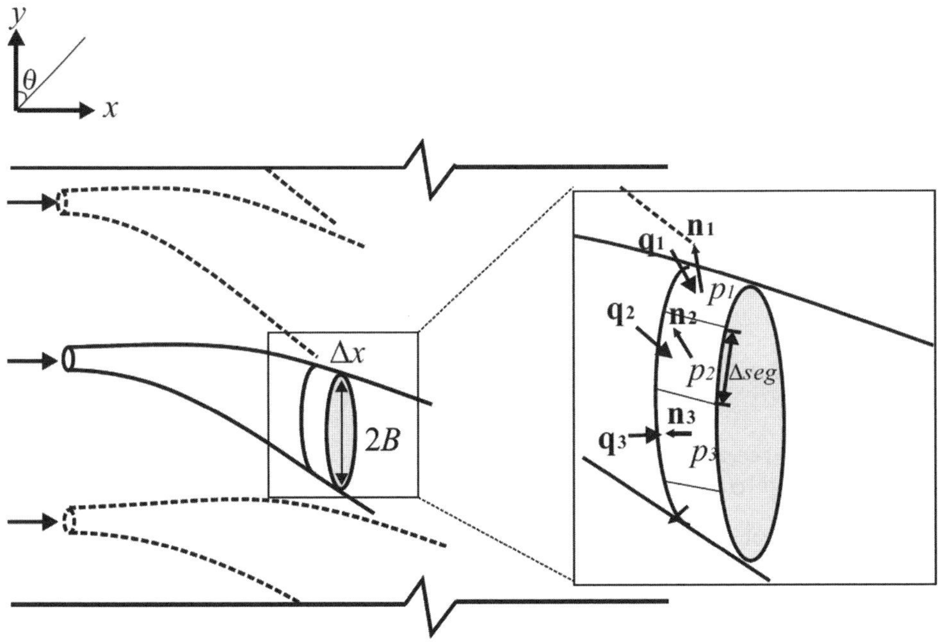

Figure 3 Control volume used in momentum balance.

$$\frac{dx}{ds} = \cos\theta; \qquad \frac{dy}{ds} = \sin\theta \tag{9}$$

where $\alpha = 0.057$ is the entrainment coefficient for a pure jet, u, u_x, u_y are the jet total, x- and y- velocity respectively.

Solving equations (6)–(9) give the interacting jet trajectory of the jet group. When the jet merges ($x_i + b_{g(i)} > x_{i+1} - b_{g(i+1}$ or $x_i - b_{g(i)} < x_{i-1} - b_{g(i-1)}$), the change in x-momentum is assumed zero. The x-momentum is also assumed to vanish at this point such that the resulting jet trajectory is straight after the jet merging.

If the jet deflection is significant, the change in distributed sinks is large and the jet group induced flow field in general needs to be recomputed. Equations (1)–(5) are used to re-compute the interacting jet group induced flow field and the results are incorporated into the jet integral model; the iterative solution converges when the maximum change in trajectory (x-coordinate) relative to the previous iteration is less than $0.01\Delta s$.

The distribution of sinks would not be extended to infinity in practice. In buoyant jet discharges when the jet eventually surfaced or trapped depending on the ambient stratification condition, the sink distribution can be assumed to stop at its maximum rise height, since afterwards the jet entrainment can no longer be modelled by the distributed sink flow. For the present problem, as the shallow water jets do not surface but spread over the whole water depth, alternative criteria for sink distribution termination is needed. It is expected that after the near field (with a length $x \approx L$), the bottom fiction gains importance and the jet momentum (which determines the entrainment) is gradually dissipated; the sink distribution is thus chosen to be stopped at $x = L$.

3.3 Superposition approach for jet group velocity

Solving the jet integral model equations results in the deflected trajectory of the jet group; other mixing characteristics such as jet centerline velocity u_c and jet-width b_g can also be obtained by relating them to the solved jet fluxes. The jet velocity profile along the perpendicular direction of s can be well approximated by $u(x,y,s) = u_c e^{-[(x-x_i(s))^2+(y-y_i(s))^2]/b_{gi}(s)^2}$, where x_i, y_i, and b_{gi} are the x-, y-, and jet width of the i-th jet at s respectively.

Using the Reichardt's hypothesis (e.g., Hinze 1975) on the momentum equation of a straight jet results in an equation linear in jet momentum u^2; it can be shown theoretically and experimentally that u^2 of each individual jet can be superimposed to obtained the jet group momentum (and thus velocity). This has been validated by a number of studies (e.g., Pani and Dugad 2002). In the case of shallow water jets with each jet being curved, we assume the x-momentum $(u_c(s)u_x(s) \approx u_x(s)^2)$ can be superimposed; this assumption should be justified ultimately by comparison with experimental data.

4 COMPARISON OF THEORY WITH EXPERIMENTS

The theoretical prediction is compared with experimental data of Lee and Greenberg (1984) and Lee and Jirka (1980). The experimental data of Lee and Greenberg (1984) are given in Table 1 for reference. For the experiments reported in Lee and Jirka (1980) the jet discharge parameters are: $N = 40$; $D = 0.002769$ m; $Q_o = 9.462 \times 10^{-5}$ m^3/s for each jet; $s = 0.0254$ m; $H = 0.0257$ m and 0.0381 m (for Experiment FF11 and FF13 respectively).

The predicted jet group trajectory of the final iteration is shown in Figure 4. The outermost jet is seen to deflect most significantly, and the degree of interaction decreases inwards, with the two jets at the centre showing only minor deflection. Same observation is made for other cases with different number of jets and diffuser length. Judging from the extent of outermost jet deflection, it can be seen that the degree of interaction for is greater for shallow water jets comparing to the discharges with weaker confinement (e.g., large water depth).

Table 1 Discharge parameters in shallow water jet experiments (Lee and Greenberg 1984).

Experiment	*N*	*Q_o (l/h)*	*s (cm)*	*H (cm)*	*u_o (cm/s)*	*u_∞ (cm/s)*
1	20	600	2.5	2.0	83.9	16.7
2	20	750	2.5	2.0	104.9	20.9
3	20	900	2.5	2.0	125.9	25.1
4	24	750	2.5	2.4	87.4	15.6
5	24	1050	2.5	2.4	122.4	22.3

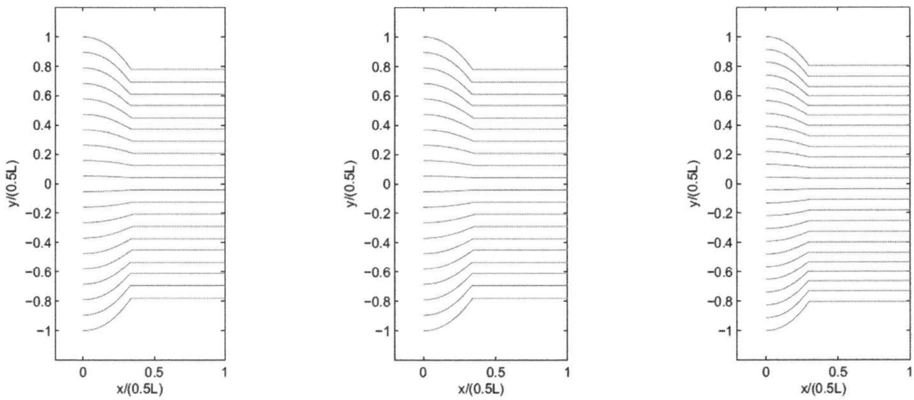

(a) Experiment 1 (b) Experiment 3 (c) Experiment 5

Figure 4 Predicted trajectories of multiple shallow water jets.

Figure 5 (a) shows the typical streamline patterns observed in an experiment of multiple shallow water jets (velocity magnitude indicated by length of the line segment). A strong flow is induced behind the diffuser and being drawn towards it; entrainment field at both sides of the jet group appear relatively weak. The model prediction (Figure 5(b)) shows the strong flow behind the diffuser, but the entrainment pattern from the side appears to be somewhat different from the observation - in particular the y-velocity being stronger and x-velocity being weaker. This may be attributed to the difference between modeling the diffuser as a line source of momentum (Lee and Greenberg 1984) or as a plane source of mass as in this model. Similar observation can be noted for a single jet. Batchelor (1967)'s solution represents a point source of momentum, which gives a weaker y-velocity but stronger x-velocity close to the jet when compared with Taylor's solution of a line sink. It may be interesting to note that Taylor's solution is a special case of Batchelor's solution when the jet spreading rate is zero.

More quantitative comparison can be made with the measured jet group velocity. Figure 6 shows the predicted and measured x-velocity within the jet group normalized by the ultimate velocity for all experiments in Lee and Greenberg (1984). The width and magnitude of the flat velocity region is seen to be well predicted by the model, and the predicted decay at the edge of the jet group is also in agreement with observation. A model without considering dynamic interaction is also shown; the prediction is done by using only superposition of momentum not accounting for the jet deflection. Such model noticeably over-predicts the jet group width and under predicts the jet group velocity. The prediction from Lee and Greenberg (1984) is also shown for reference; while the velocity at the flat region is reasonably predicted, the nature of the inviscid model does not predict the velocity decay at the jet group edge.

The predicted traverse y-velocity induced by the jet group near the diffuser is shown in Figure 7. In general the model prediction is about 5% higher than the

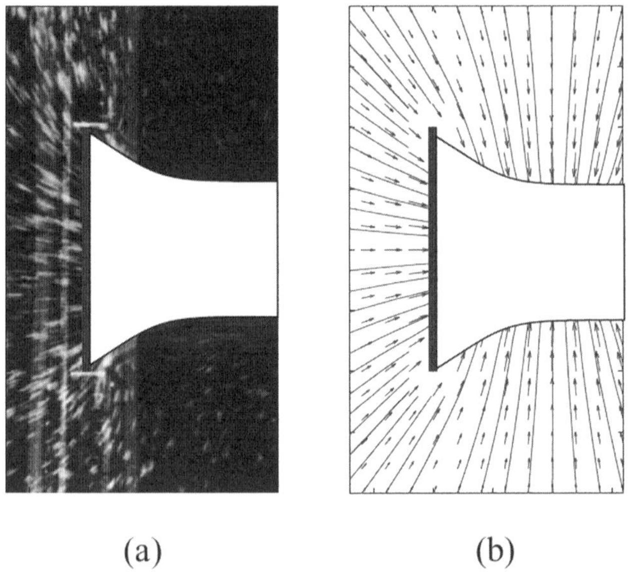

Figure 5 Streamline pattern in the vicinity of multiple shallow water jets (Experiment 1, Table 1); (a) observation; (b) computed by dynamic jet model.

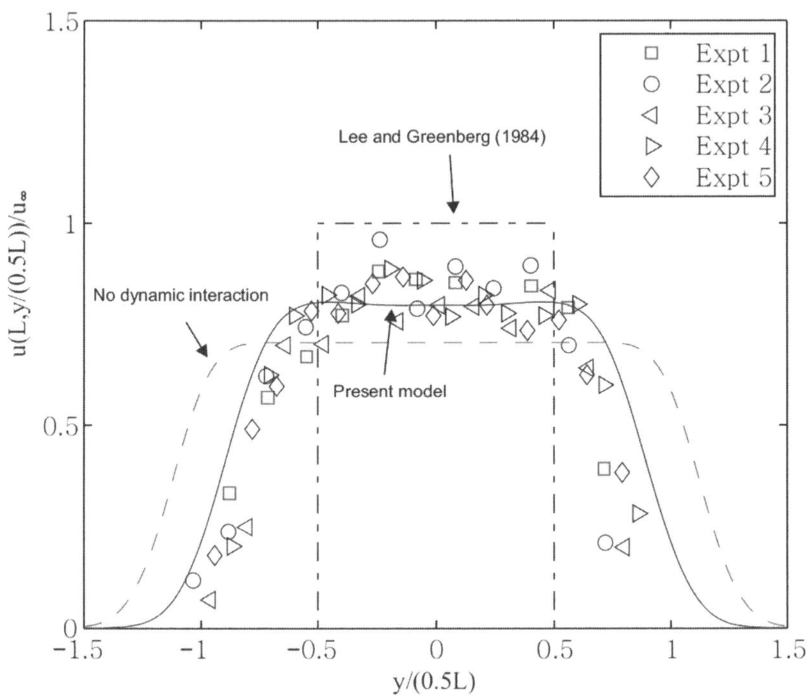

Figure 6 Predicted and observed normalized jet group *x*-velocity profile at $x = L$..

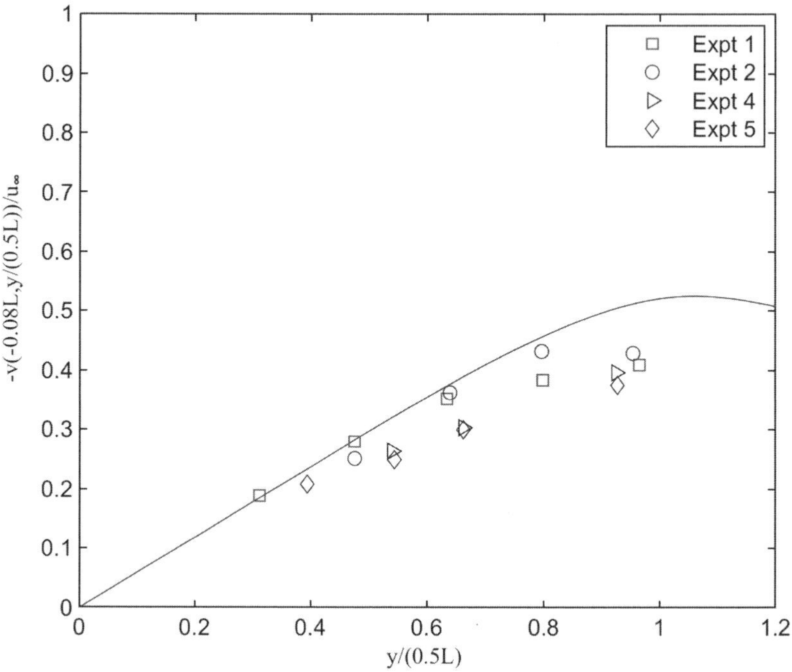

Figure 7 Predicted and observed jet group y-velocity profile at $x = -0.08\ L$.

observation; this is also evident in streamline pattern, in which the model appears to have predicted a slightly stronger flow in the y-direction.

Comparison between predicted and observed decay of the jet group maximum velocity are also made using the experimental data of Lee and Jirka (1980). 20 jets with equivalent volume and momentum flux per unit diffuser length are used to model the diffuser discharge (with 40 jets). Figure 8 shows that the jet group maximum velocity is well predicted in the near field; not accounting the dynamic interaction results in predictions about 10% lower than the observed velocity. In both experiments, the maximum velocity (located at the jet group centerline) decay level off at about 0.1 L. This indicates the jet group has completely merged, and also mixed completely vertically such that there is no further entrainment into the jet group central region. When comparing the two experiments, it is noted that experiment FF11 was conducted with identical discharge conditions but a smaller water depth, and hence the velocity decay level off sooner and also has a larger magnitude than that of experiment FF13. These are all captured in the model predictions.

Tracer concentration field c can be obtained by superimposing uc of each point, and divide by the jet group velocity u obtained by the superposition approach. The predicted concentration contour of $c/c_o = 0.05$ (where c_o is the source concentration) is compared with the measured isotherms of experiment FF13, $\Delta T/\Delta T_o = 0.05$ in Figure 9. The initial jet group contraction is well predicted, although the subsequently jet group spreading appears to be over-predicted. Without dynamic interaction,

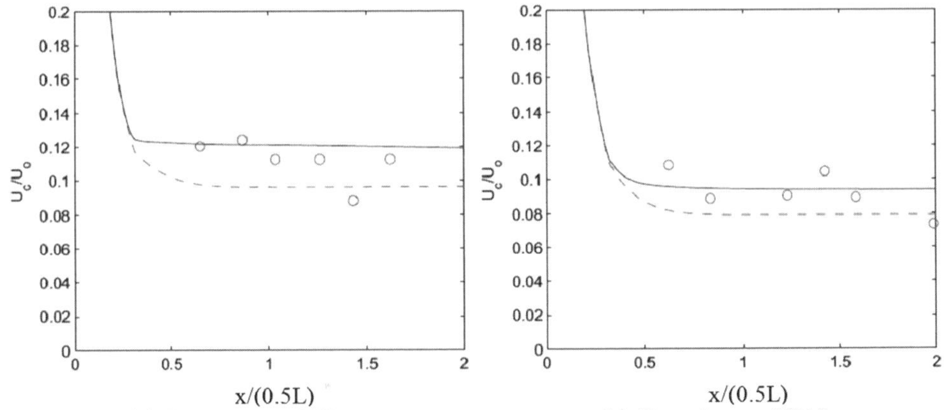

(a) Experiment FF11 (b) Experiment FF13

Figure 8 Decay of the jet group maximum velocity (solid and dashed line: with and without dynamic interaction respectively).

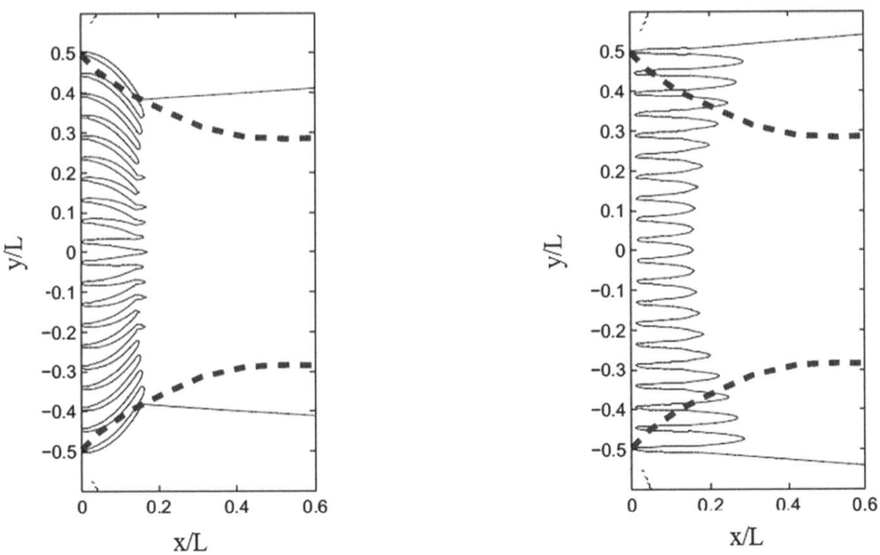

(a) With dynamic interaction (b) Without dynamic interaction

Figure 9 Predicted and observed isotherms for experiment FF13 (dashed line: observation).

neither the initial contraction nor the subsequent jet group spreading can be predicted. It should be noted that the measured data, which were intended for the far field study, are relatively sparse in the near field. The measurement resolution is about 0.5 L and 0.2 L in the x- and y-direction respectively; thus the comparison is more for reference and less conclusive on the model's capability in tracer concentration predictions.

5 SOME REMARKS OF THE MODEL

We present a few interesting model predictions for which no experimental data exist, as well as relevant details on the method of images accounting for the discharge depth.

5.1 Model predicted changes in jet mixing characteristics

The total momentum of a jet is predicted to increase when compared with a single free jet. Figure 10 shows the predicted volume flux Q and the jet centerline x-velocity u_c for the outermost jet at $y = (1/2)\ L$, the inner jet at $y = (1/4)\ L$, and the center jet at $y \approx 0$. The JETLAG (Lee and Cheung 1990; a validated Lagrangian jet model for a single free jet in arbitrary ambient conditions) predictions represent a single free jet discharge. The increase in Q and the slower decay of u_c (before merging) of the outermost jet comparing to the single free jet reflects the increase in M_y due to the interaction. The jets are being less affected by the interaction towards the diffuser center, with the center jet being almost identical to a single free jet.

Dynamic interaction computation between two jets should cease at some distance in the neighbourhood of the jet merging point; a momentum change to the impacting jets is also expected. The computation of dynamics of the present model ceases when the jet boundary (defined at b_g) of two neighbouring jets are in contact, afterwards both the momentum change due to the interaction ΔM_y and the momentum M_y are assumed zero. This assumption results in a sharp change in certain jet mixing characteristics at the point of merging. Figure 10 shows the changes in both Q and u_c at the point of merging ($x \approx 20D$). After merging Q can be seen to increase at the same rate for all jets as a single free jet.

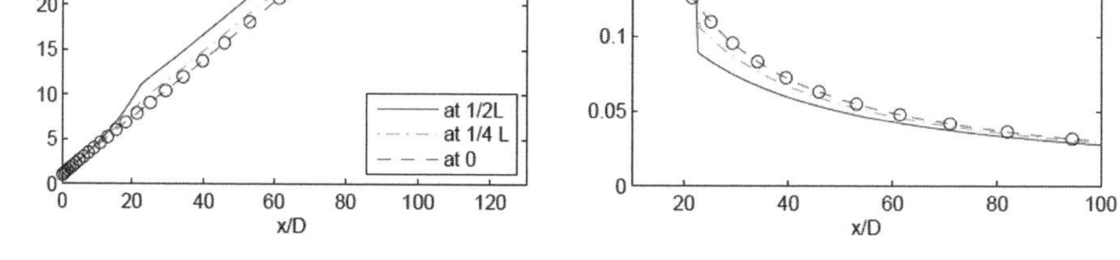

(a) Volume flux Q (b) centerline velocity U_c

Figure 10 Predicted mixing characteristics ((a): Volume flux; (b) centerline x-velocity) of selected individual jet (Experiment 1). Symbols are predictions of the Lagrangian model JETLAG.

5.2 Method of images accounting for the discharge depth

The model predicted results have assumed mid-depth discharges. In general the jets are discharged at a height h_o from the bottom; for example, Lee and Greenberg (1984)'s experiments were discharges at $h_o = 0.4$ cm. The mirror images of the distributed sinks are sinks at position that shifted vertically according to Table 2.

The resulting depth variation of jet group induced velocity at $(x, y) = (10D, 70D)$ (near the diffuser end) by placing the diffuser at mid-depth and at $h_o = 0.4$ cm is shown in Figure 11. The difference in the computed velocity in all directions is negligible, indicating that the 2D vertically mixed flow would be similar and not sensitive to the exact vertical location of the jets above the bottom. The main difference is the discharge close to the bottom will become attached to the bottom sooner than the mid-depth discharges.

Table 2 Position of the mirror images of distributed sinks.

	+z direction	−z direction
$k = 1, 3, 5 \ldots$	$2(H - h_o) + (k - 1)H$	$-2h_o + (k - 1)H$
$k = 2, 4, 6 \ldots$	kH	$-kH$

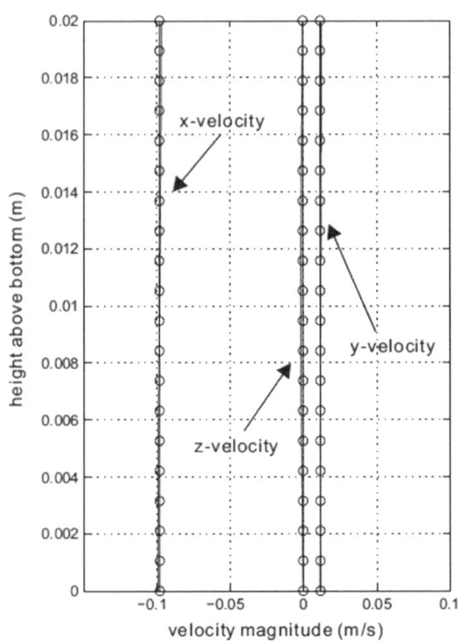

Figure 11 Depth variation of shallow water jet group induced velocities (Experiment 1) with diffuser at (i) mid-depth (solid line) and (ii) $h_o = 0.4$ cm (dashed line).

6 CONCLUSIONS

The dynamic interaction of multiple jets in stagnant shallow water is significant in the near field. The outermost jet of the jet group deflects strongly towards the center; deflection of inner jets weakens gradually towards the jet group center. The contraction of the interacting jet group leads to some changes to its mixing characteristics when compared to that without interaction.

A semi-analytical model is proposed to account for the interaction. The jet group induced external flow is obtained by treating each jet as a distribution of point sinks; the pressure and momentum flux changes due to the jet group induced flow of a control volume containing a jet element are tracked and incorporated into an iterative jet integral model. The confinement effect has been reflected in the model by the method of images. It is shown that the model is able to predict the strong jet deflection character and the jet group-induced velocities in the near field, although the observed strongly contracting flow cannot be reproduced.

REFERENCES

Baines, W.D. and Keffer, K.F. (1974). Entrainment by a multiple source turbulent jet. *Adv. Geophys.*, 18B, 289–298.

Batchelor, G.K. (1967). *An introduction to fluid dynamics*. Cambridge University Press.

Hinze, J.O. (1975). *Turbulence*. McGraw-Hill Book co.

Kaye, N.B. and Linden, P.F. (2004). Coalescing axisymmetric turbulent plumes. *Journal of Fluid Mechanic,* 502, 41–63.

Lai, A.C.H. (2009). *Mixing of a rosette buoyant jet group*. PhD thesis, The University of Hong Kong, Hong Kong.

Lai, A.C.H. and Lee, J.H.W. (2008). Dynamic interaction in a rosette buoyant jet group. *Proc., 2nd Int. Symp. on Shallow Water Flows* (CD-ROM).

Lai, A.C.H. and Lee, J.H.W. (2011). Dynamic interaction of multiple buoyant jets (submitted).

Lai, A.C.H., Yu, D. and Lee, J.H.W. (2011). Mixing of a rosette jet group in a crossflow. *Journal of Hydraulic Engineering, ASCE*, 137 (8), 787–803.

Lee, J.H.W. and Cheung, V. (1990). Generalized Lagrangian model for buoyant jets in current. *Journal Environmental Engineering, ASCE,* 116, 1085–1106.

Lee, J.H.W. and Greenberg, M.D. (1984). Line momentum source in shallow inviscid fluid. *Journal of Fluid Mechanic,* 145, 287–304.

Lee, J.H.W. and Jirka, G.H. (1980). Multiport diffuser as line source of momentum in shallow water. *Water Resources Research,* 16(4), 695–708.

Liseth, P. (1970). *Mixing of merging buoyant jets from a manifold in stagnant receiving water of uniform density* PhD Thesis, University of California, Berkeley, California.

Milne-Thomson, L.M. (1968). *Theoretical hydrodynamics*. Macmillan.

Pani, B.S. and Dugad, S.B. (2002). Turbulent jets: Application of point source concept. In: *Research persepctives in hydraulics and water resources engineering*. World Scientific, pp. 1–37.

Pani, B.S., Lee, J.H.W. and Lai, A.C.H. (2009). Application of Reichardt's hypothesis for multiple coflowing jets. *Journal of hydro-environment Research,* 3 (3), 121–128.

Taylor, G.I. (1958). Flow induced by jets. *J. Aero/Space Sci.,* 25, 464–465.

Wang, H.J. (2000). Jet interaction in a still or co-flowing environment. PhD thesis, The Hong Kong Univeristy of Science and Technology, Hong Kong.

Chapter 16

Evolution of turbulent jets in low-aspect ratio containers

S.I. Voropayev[1,2], C. Nath[1] and H.J.S. Fernando[1]
[1] Environmental Fluid Dynamics Laboratories, Department of Civil Engineering and Geological Sciences, University of Notre Dame, Notre Dame, Indiana, USA
[2] P.P. Shirshov institute of Oceanology, Russian Academy of Sciences, Moscow, Russia

ABSTRACT: The flow induced in a long cylinder by an axially discharging round turbulent jet was investigated experimentally, with applications to crude oil degas and refilling of the U.S. Strategic Petroleum Reserves (SPR). Homogeneous and stratified jets were used, and both symmetric and asymmetric (with jet radial offset) geometries were considered. It was found that the flow breakups up at a finite distance, does not reach a steady state and vacillates periodically. Digital video recordings, particle image velocimetry and conductivity micro-probes were used to delineate and quantify flow structures. Using the concepts of flow similarity, a model was developed and the results of measurements were parameterized via characteristic length and velocity scales based on the cylinder width and jet kinematic momentum flux. The model was extended to the case of stratified jets and verified experimentally. The scaling laws so developed could be used to extrapolate laboratory observations to SPR flows.

1 INTRODUCTION

The U.S. Strategic Petroleum Reserves (SPR) consists of a collection of approximately cylindrical underground caverns (roughly 700 m high and 70 m in diameter) artificially created by leaching sub-terrestrial salt deposits. Crude oil stored in these caverns is periodically removed from the bottom, degassed and then re-introduced near the top as a jet to maintain the oil quality (Lord & Rudeen, 2007). The depth of mixing, flow stability and the velocity distribution of such jets influence the efficiency of oil refilling, and the efficiency is maximum when there is no mixing between degassed and old oil. The cavern flow falls into the category of confined jet flows with limited depth jet penetration, periodic instability and oscillating behavior. Large coherent structures and recirculating cells appear even when the boundary conditions are held symmetric and steady (Villermaux & Hopfinger, 1994, Gebert et al., 1998, Denisikhina *et al.*, 2005). Such oscillations lead to strong mixing in the upper cavern, and are of importance for improving SPR performance and decreasing the cost of degassing operations. Although jet oscillations in long two-dimensional cavities have been documented for some 50 years (Molloy & Taylor, 1969) with more recent work delving into basic flow structures (Risso & Fabre, 1997, Lawson, 2001, Liu *et al.*, 1997, Mataouia & Schiestelb, 2009), no reliable parameterizations exist that can be used to extrapolate laboratory results to SPR caverns. In this communication, the evolution of a round turbulent jet released into a low-aspect-ratio cylindrical container filled with either homogeneous or stratified fluid is investigated experimentally. The goal is to model the degas flow in SPR caverns with the purposes of clarifying the general flow structure and to develop parameterizations for the main flow characteristics as a function of external parameters. Experiments were conducted first with homogeneous (distilled water) and then with stratified (by salt) water. Digital video recordings, particle image velocimetry and conductivity micro-probes were used to delineate and quantify flow structures and measure basic flow characteristics. Only the main results are reported below, and further details are given in Voropayev *et al.* (2011).

2 HOMOGENEOUS FLUID

2.1 Experimental set-up and method

The flow was generated in a circular glass cylinder of length L = 65 cm and inner diameter D = 10 cm. In some runs, a smaller cylinder (L = 45 cm, D = 4.5 cm) was used. To minimize optical distortions due to curvature, the cylinder was placed in a larger rectangular Plexiglas tank filled with distilled water. In most of the experiments, the bottom end of the cylinder was sealed by a glass disk, while the top was closed by a glass disk with a circular opening (diameter 1.5 cm) at the centre. The jet flow inside the cylinder was generated using a conical round nozzle (inner diameter $d = 0.165$ cm) placed at the centre of this opening. A precision pump takes water from the larger tank and feeds the nozzle with relatively small volume flux q, generating an intense turbulent jet with a substantial momentum flux J. The same amount of fluid, but with negligible momentum, leaves the cylinder through the hole in the top cover. Although most of the measurements were conducted with both ends closed, to better understand the mechanism of jet break up additional experiments were conducted with different (top/bottom) boundary conditions.

In the dimensional Cartesian coordinate system used (x, y, z) the x-axis is along the flow (with the origin at the nozzle exit) and y and z-axis are across the flow, with corresponding dimensional velocity components being (u, v, w). The instantaneous velocity component is presented as a sum of the mean and fluctuating parts, i.e., $u = \bar{u} + u'$. A similar notation is used for dimensionless velocity, i.e., $U = \bar{U} + U'$.

The water in the cylinder was seeded with highly reflective small Pliolite particles. A commercial PIV system (TSI Incorporated) was used, which includes: a Dual Nd: YAG Laser with optics to produce a thin light sheet that spreads along the cylinder axis, a Laser-Pulse Synchronizer and a CCD camera (PIVCAM 10–30, 1024 × 1024 pixels). The camera, laser and the synchronizer were connected to a control computer, and data processing was conducted using the TSI PIV software package Insight™. Pairs of images were captured at 15 Hz over the duration of the experiment (typically 120 s), and the velocity and vorticity fields were calculated for each pair of frames using Insight™. The data were obtained mostly in the $(x - y)$ plane along the cylinder axis. In selected runs, to clarify the mechanism of jet oscillations, additional data on cross-flow velocity components (v, w) were obtained in the $(y - z)$ plane.

In addition to PIV, digital video recordings were made by DVC-3400 camera, and a continuous laser was used for illuminating the flow field. The camera was free-running at 10 Hz and particle streak images were obtained from these recordings.

The experiments with PIV measurements and DVC recordings covered five jet Reynolds numbers (defined below): $Re = 10100$, 12700, 15300, 17900 and 23100. To obtain data on the frequency of flow oscillations, additional experiments at $Re = 2800$, 5900, 9100, 12200, 18400, 24600 were conducted where long DVC records were made and analysed.

2.2 General flow behaviour and scaling analysis

2.2.1 *General flow behaviour*

Upon initiation, the jet propagates relatively quickly along the cylinder with a characteristic spherical front. Nevertheless, in contrast to free jets, where the front propagates over large distances, the jet front in the cylinder stops at some critical distance x^* from the origin (see schematic in Fig. 1), loses its coherence and breaks down into smaller eddies forming, at $x \approx x^*$, the so-called weak *diffusive turbulence* (Risso & Fabre, 1997). Thereafter, fluid motions decay rapidly with distance, and at $x > x^*$ the fluid appeared still. In the upper part of the cylinder, $x < x^*$, however, the motion remains energetic and large 'coherent' eddies surrounding the central part of the jet are frequently visible. Visually the flow never reaches a steady state but changes periodically with a characteristic frequency f in rather complicated manner.

Typical streak images showing the central part of the jet in the $(x - y)$ plane at different times are shown in Fig. 2. At the first glance, the jet appears to be oscillating periodically in the $(x - y)$ plane, similar to flipping oscillations of two-dimensional jets. Additional observations in the $(y - z)$ plane, however, show that the flow is more complicated in that the jet is spiraling periodically around the cylinder axis

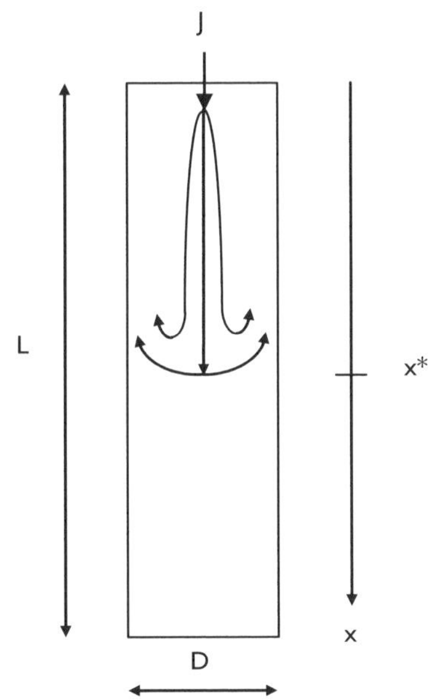

Figure 1 A schematic of mean flow in a long cylinder and external parameters.

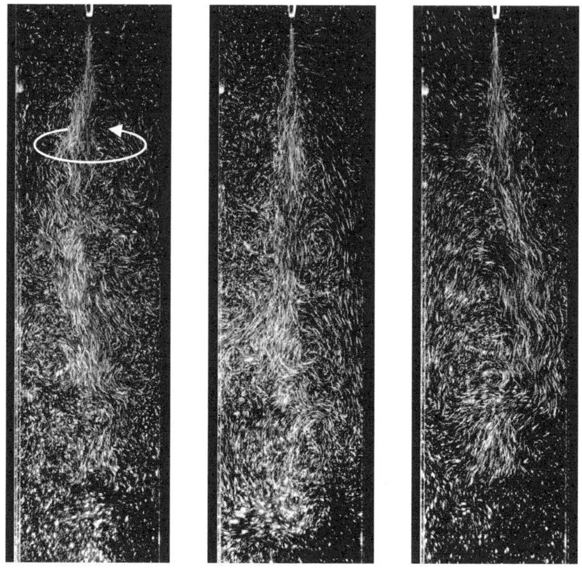

Figure 2 Particle streak images in the $(x - y)$ plane showing the central part of the jet at different times. In (a) – jet is tilted to the left and slowly rotates azimuthally (white arrow) around the cylinder axis, in (b) – jet continues its azimuthal rotation and becomes visible near the cylinder centreline, in (c) – it is tilted to the right. The direction of rotation changes with time. $Re = 10100$.

(precessing mode). Typical streak images and PIV data for instantaneous across flow velocity (v, w) and axial vorticity at $Re = 10100$ are shown in Fig. 3. Using PIV data the across flow $(y - z)$ velocity components (v, w) and the (kinematic) angular momentum M (per unit length of the cylinder) were calculated for different times t as $M(t) = \iint_S [v(t)z - w(t)y]\, dydz$ (S – cylinder cross section), and spectral analysis gave $f = 0.089$ Hz as the dominant frequency for precessing. This agrees well with the frequency of jet oscillations observed at $Re = 10100$ in the along flow $(x - y)$ plane (see Fig. 5 below).

Schematically, the flow consists of (i) the primary jet flow with relatively high velocity, and (ii) secondary recirculation cell(s) with smaller velocities. As shown schematically in Fig. 4a, the secondary flow is approximately axisymmetric with jet in the central part, accompanied by a toroidal recirculation cell that is visible in the $(x - y)$ plane as two symmetric elongated elliptical cells filled with smaller eddies. In (b) the flow loses its symmetry, central jet migrates to the left, toroidal cell is deformed and becomes visible in the $(x - y)$ plane as two asymmetric cells – intense to the right and weak to the left. In (c) the flow again passes through its approximately symmetric state, forming in (d) a large intense cell in the left and a small weaker cell to the right. Thereafter the process recurs. Symmetric state (a, c) is sometime unclear, and the flow vacillates between (b) and (d) by precessing around the cylinder axis as shown by the white arrow in Fig. 2. For clarity, dashed and solid lines in Fig. 4 show clockwise and anticlockwise motions, respectively.

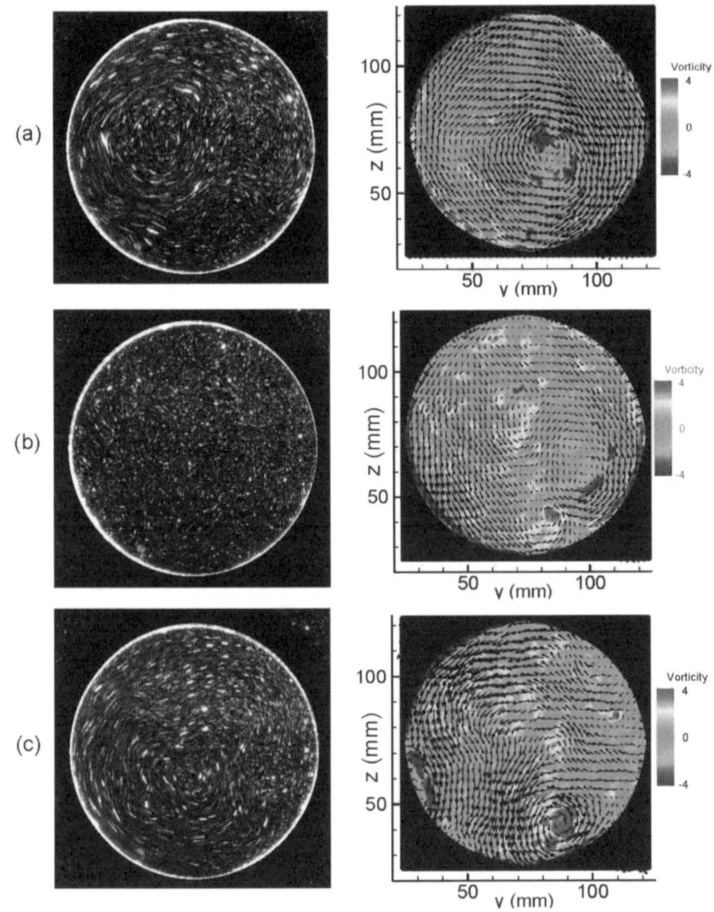

Figure 3 Left column – particle streak images in the $(y - z)$ plane. Right column – PIV data for vorticity (different colours) and velocity (black arrows) in the $(y - z)$ plane. (a) – fluid rotates mostly clockwise and the net angular momentum M is negative; (b) – direction of rotation changes and the net angular momentum is near zero; (c) – fluid rotates mostly counter clockwise and the net angular momentum is positive. $Re = 10100$. Thin near-wall boundary layers with sharp variation of azimuthal vorticity are seen in vorticity data.

The description above is based on observations in two planes. The real flow is three-dimensional and more complicated. The flow is unstable with a dominant rotational instability (precession) mode, and the global flow structure changes periodically with a characteristic frequency f.

2.2.2 Boundary conditions and jet dissolution

The driving force for this flow configuration is the jet (kinematic) momentum flux J. For a free jet, the momentum flux is conserved along the flow, which allows the jet to propagate a large distance from the origin. In contrast, the jet flow in a cylinder stops

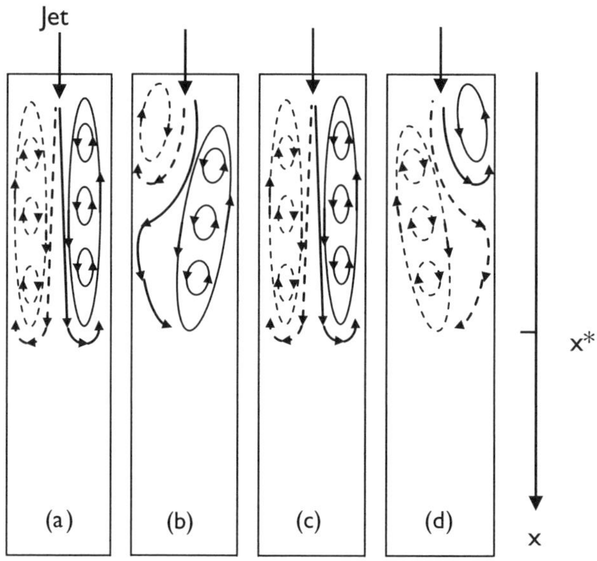

Figure 4 A schematic of the evolution of flow structures during an oscillation period (see explanation in the text).

at $x > x^*$. Prima facie this can be construed as due to side-wall influence, where the viscous drag reduces the axial velocity. Observations show, however, that velocity near the wall is in the opposite direction due to recirculation cells (see Fig. 4), and axial momentum decay cannot be accounted by the lateral viscous friction alone; hence pressure gradient induced by end walls is the only possible mechanism for jet stoppage. This contrasts the case of free jets, where outside pressure may be neglected.

To clarify the role of end walls and related pressure distribution on the flow, additional experiments were conducted with the same jet intensity J but with different boundary conditions: both ends of the cylinder were closed; the top of the cylinder open and the bottom closed; bottom open while the top closed. Observations showed that flows in all cases are visually identical, in that the stoppage occurs at $x = x^*$ followed by periodic jet oscillations with approximately the same frequency and global flow structure. Quantitative PIV data confirmed these conclusions.

The above results can be explained by considering the momentum flux transformation under different cylinder end conditions. When, e.g., only the bottom of the cylinder is closed, the momentum source J leads to an opposing pressure distribution in the cylinder (see Cantwell, 1986 for a related discussion). If viscous losses near lateral boundaries are neglected, then the conservation of momentum determines the mean pressure p_0 at the bottom of the cylinder as

$$p_0 \approx \rho J / S, \tag{1}$$

where ρ is the fluid density and $S = \pi D^2/4$, so that the net force at the bottom is ρJ. The momentum flux J transforms to a pressure distribution that opposes the motion, acts on the bottom/top of the cylinder and leads to the jet stoppage. Direct pressure measurements confirmed (1) and supported this analysis.

2.2.3 *Scaling analysis*

Consider jet-induced flow in a long cylinder (diameter D, length $L >> D$) with at least one end iclosed. The jet emanates from a nozzle of diameter d ($<< D$), with a volume flux q. The kinematic viscosity and density of the fluid are ν and ρ, respectively. The important governing dimensional parameters for the fully established flow are q, d, D, L, ν, ρ. Thus any flow characteristics A_i can be represented as $A_i = \Phi_i(q, d, D, L, \nu, \rho, x, y)$, where x is the axial distance, y the characteristic across flow distance and Φ_i are unknown functions. Experiments show that for $L >> D$ the length of the cylinder is not important, given the mean motion stops at a finite distance $x^* << L$. Also, for $d << D$, the volume flux q and nozzle diameter d are important only in combination $J = q^2/s$, $s = \pi d^2/4$, J being the jet momentum flux. In dimensionless form J is equal to the jet Reynolds number $Re = J^{1/2}/\nu$. When the flow is turbulent, $Re >> 1$, the molecular viscosity is assumed unimportant in the spirit of Reynolds number similarity, and hence only five dimensional governing parameters are possible: $A_i = \Phi_i(D, J, \rho, x, y)$, with only two independent parameters. Choosing the characteristic length l and time τ scales $l = D$, $\tau = D^2/J^{1/2}$, the velocity scale is $l/\tau = J^{1/2}/D$, and pressure $(l/\tau)^2\rho = J\rho/D^2$, which is consistent with (1). Using these scales, useful predictions are possible. For example, the mean dimensionless axial velocity $\overline{U} = \bar{u}D/J^{1/2}$, the dimensionless critical distance X^* and dimensionless flow oscillation frequency F become

$$\overline{U} = \overline{U}(X,Y), \quad X^* = x^*/D = C, \quad F = fD^2/J^{1/2} = C^* \tag{2}$$

where $\overline{U}(X,Y)$ is a function of the dimensionless coordinates $X = x/D$, $Y = y/D$, and C and C^* are constants. Similarly, for two-dimensional flows $l = D$, $\tau = D^{3/2}/I^{1/2}$, where I is the jet momentum flux per unit width, whence the predictions equivalent to (2) become $X^* = x^*/D = C_{2D}$, $F = fD^{3/2}/I^{1/2} = C^*_{2D}$. In general, the coefficients C_{2D}, C^*_{2D} may differ from C, C^* in (2).

2.3 Basic flow characteristics

2.3.1 *Frequency of flow oscillations*

Quantitative data on f were obtained using DVC recordings and PIV data. First, DVC recordings were made for all runs, replayed at a slow speed and the mean (over the recording period) f were obtained for each run. Thereafter, instantaneous velocity profiles were plotted as a function of time t and the mean f for each data set and the time correlation function for the axial velocity u at $y = \pm 0.25\ D$ [for $1 < x/D < 3$ and averaged over x] were obtained. Time autocorrelations $R_{LL}(y, \Delta t)$ at $y = -0.25\ D$ (near the left side) and $R_{RR}(y, \Delta t)$ at $y = 0.25\ D$ (near the right side) as well as time cross-correlations $R_{LR}(y, \Delta t)$ (between $y = -0.25\ D$ and $y = 0.25\ D$) were also calculated. The mean period $1/f$ of flow oscillations was obtained by calculating the time shift

Δt when the first maximum (for autocorrelation) or minimum (for cross-correlations) was observed.

In Fig. 5, f is plotted is as a function of the frequency scale $J^{1/2}/D^2$ for experiments with different jet intensities J. The agreement with (2) is clear, and the best fit gives $C^* = 0.1$. To our knowledge, no previous frequency data are available for 3D geometries, but some exist for 2D, which are included in Fig. 5 with appropriate modifications. To make direct comparisons with our 3D data, (2) can be written as $F = fD^{3/2}/(J/H)^{1/2} = f\beta^{1/2}/(J^{1/2}/D^2) = C^*_{2D}$, where $J = IH$ is the net momentum flux, $\beta = H/D$ the aspect ratio and H the width of the 2D jet. Using f and Re reported in literature, the modified frequencies $f\beta^{1/2}$ for 2D flow were calculated (Fig. 5), which fall on our best fit (solid) line, indicating that the coefficients are the same for both geometries. The agreement between results of 2D air experiments and our water jets supports the proposed similarity scaling, especially the Reynolds number similarity.

2.3.2 *Mean flow characteristics*

The mean dimensionless centreline axial velocity $U^*(X) = \overline{U}(X, Y = 0)$ as a function of dimensionless distance X is shown in Fig. 6 by the colored lines (1–5) for all runs conducted with larger cylinder. The mean of all is shown by black line (6), which illustrates independence on the Reynolds number. The data for smaller cylinder are shown by open circles (7). For comparison, the data from previous work are also shown by crosses (8). As can be seen, U^* smoothly decreases with X, crosses zero at $X \approx 3.2$ and shows (see black arrow) a small but measurable negative value until $X \approx 3.6$ before becoming vanishingly small. This gives an estimate for the critical distance $X^* \approx 3.6$, thus yielding $C \approx 3.6$ in (2). Empirically, the mean centreline axial velocity can be approximately fitted to the function

$$U^*(X) = A/X \exp(X), \qquad (3)$$

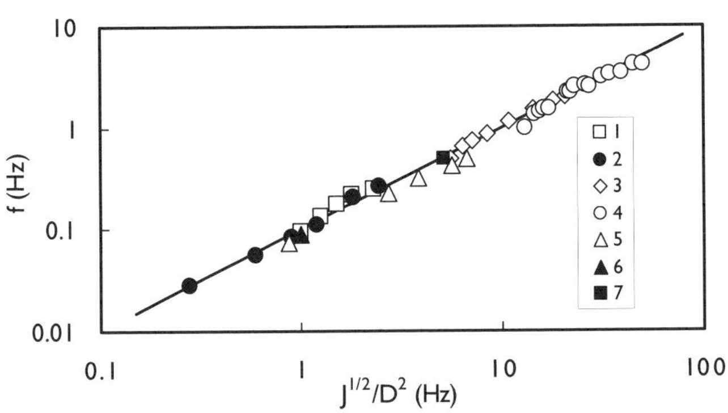

Figure 5 Frequency of oscillations f versus frequency scale $J^{1/2}/D^2$. Symbols – experimental data, solid line – equation (2) with $C^* = 0.1$. Data from: (1) – present experiments, $d/D = 0.016$ (PIV); (2) – present experiments, $d/D = 0.016$ (DVC); (3) – Villermaux & Hopfinger (1994) ($\beta = 0.2$), (4) – Villermaux & Hopfinger (1994) ($\beta = 0.3$), (5) – Mataouia & Schiestelb (2009) ($\beta = 1$), (6) – present experiments, $d/D = 0.016$ (angular momentum); (7) – present experiment, $d/D = 0.037$ (DVC).

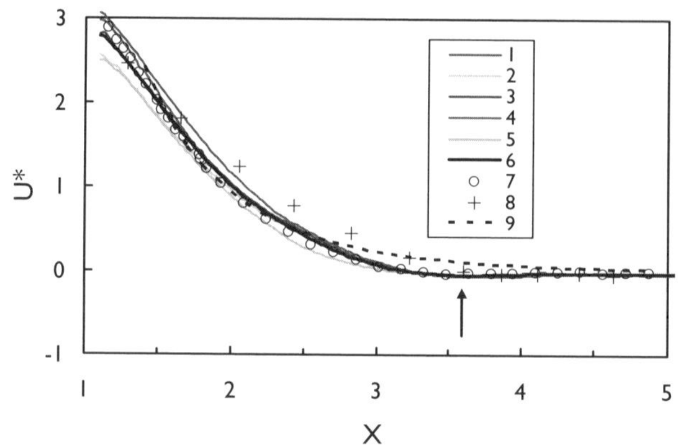

Figure 6 Decay of the mean dimensionless axial velocity U^* with dimensionless distance X are shown by colour lines (1–5) for the larger $d/D = 0.016$ at $Re = 10100$ (1), 12700 (2), 15300 (3), 17900 (4), 23100 (5), (6) – mean over all experiments with larger cylinder, (7) – smaller cylinder, $d/D = 0.037$, (8) – data from Risso & Fabre (1997), $Re = 150000$, $d/D = 0.195$, (9) – Equation (3).

where $A = 13.8$, which is shown in Fig. 6 by the dashed line (9). To parameterize $\overline{U}(X,Y)$ we may use the function,

$$\overline{U}(X,Y) = [A/X\exp(X)] \times [\sin(BY)/BY], \tag{4}$$

where $B = 10.2$. Although (4) does not capture the no-slip condition at the cylinder walls, it correctly describes the mean axial velocity profiles over a broad range of parameters, Re (10000–150000) and d/D (0.016–0.195) and employs a minimal number of *universal* coefficients.

2.3.3 *Turbulent flow characteristics*

The second order moments for U' and V' are shown in Fig. 7 as a function of X. The data for all Reynolds numbers satisfactorily collapsed (within ± 10%), and for clarity the average over all Reynolds numbers are shown.

For comparison, the mean values of U^{*2} are also shown. For U^* we used the averaged data given by black line (6) in Fig. 6. In Fig. 7 $< U'^2 >$ is significant only at smaller distances ($X < 2$) and comparable to U^{*2} while $< V'^2 >$ is smaller by a factor of 5. At larger distances both U^{*2} and $< U'^2 >$ decay in similar ways and remain comparable. Beyond $X \approx 3.6$ all flow characteristics become very small. The cross-correlations $< U'V' >$ are negligibly small at all distances.

2.4 Effect of jet asymmetry on the mixing length

In the above configuration the jet was positioned symmetrically, at the cylinder axis. In practice, however, the jet may be positioned not exactly symmetrically but with some radial offset $\Delta \neq 0$ relative to the cylinder axis. This may lead to a change of x^*,

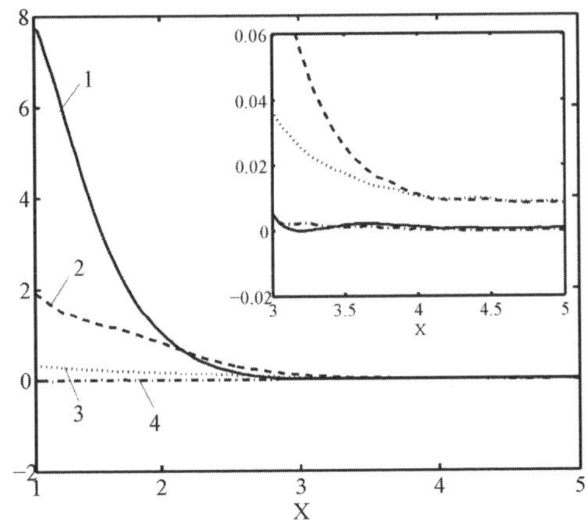

Figure 7 The maximum values of squared dimensionless mean axial velocity U^{*2} and turbulence statistics of U' and V' versus X: (1) – U^{*2}, (2) – $\langle U'^2 \rangle$, (3) – $\langle V'^2 \rangle$, (4) – $\langle U'V' \rangle$. Averaged values over all experiments conducted at different Reynolds numbers are shown. An enlarged view of flow characteristics at $X > 3$ is shown in the inset.

which in turn affects the mixing depth l during SPR oil degas and refilling. Given no work has been reported on this aspect, we conducted a series of experiments on this topic.

2.4.1 Experimental methods and results

Long glass cylinder of diameter D was filled with water solution containing the pH indicator thymol blue; it is dark blue when basic and yellow when acidic. The jet radial offset Δ relative to the cylinder centerline was changed from $\Delta = 0$ to $\Delta \approx D/2$. The water from the cylinder bottom was withdrawn via a precision pump and recirculated through the jet nozzle. After reaching the quasi-steady state, a few drops of NaOH solution was added to the cylinder top. Soon the upper mixed fluid becomes dark blue with a discernible boundary below that separates unmixed (yellow) fluid. Adding a few drops of HCl solution makes the fluid in the cylinder yellow again, and experiment could be repeated without changing the water. Photographs showing how the mixing depth l changes when the jet is shifted from the near wall position $\Delta \approx D/2$ (a, b) to the centerline position $\Delta = 0$ (c) are shown in Fig. 8A. There is a significant difference between l at $\Delta/D \approx 0.5$ and at $\Delta = 0$. A thin blue 'scar' from the previous mixing depth, which is well visible in Fig. 8A(c) indicates that there is no mixing below the upper blue region.

The PIV measurements were conducted thereafter (Fig. 8B), and the mixing depths l were determined therewith for different Δ/D; the results are shown in Fig. 9. For symmetric jets ($\Delta = 0$) the mixing depth l/D has a minimum, which practically coincides with the estimate for $X^* = x^*/D$ in (2). With the increase of jet asymmetry, the mixing depth increases and reaches maximum value at $\Delta/D \approx 0.4$. As will be discussed in Section 3 below, increased mixing will lead to increase of time needed to degas and refill SPR caverns.

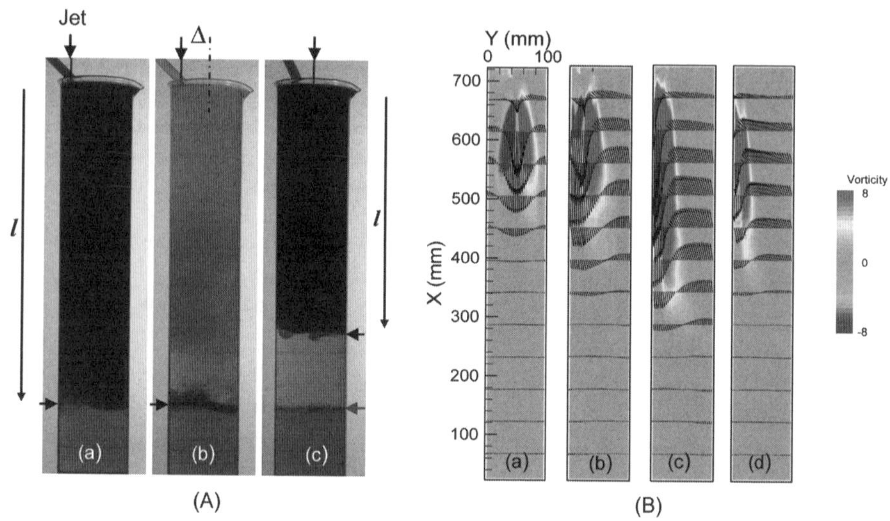

Figure 8 (A) – photographs showing how the mixing depth*l*changes when the jet is shifted from the near wall position, $\Delta \approx D/2$ (a, b) to the centerline position, $\Delta = 0$ (c) Black vertical arrows at the top show the nozzle; black horizontal arrows show the boundary between mixed and unmixed fluids, and blue horizontal arrow in (c) shows the 'scar' from previous mixing in (a, b). (B) – PIV data with velocity (small black arrows) and vorticity (color map) showing how the mixing depth *l* changes with the change of the jet offset: $\Delta/D = 0$ (a) 0.25 (b) 0.4 (c) 0.49 (d).

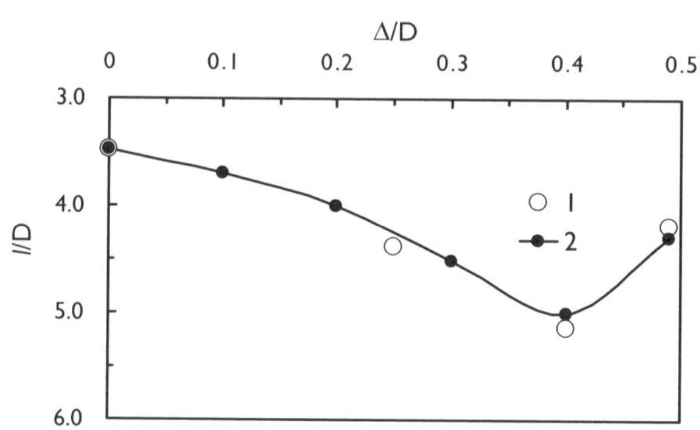

Figure 9 Dimensionless mixing depth *l/D* as a function of dimensionless jet offset Δ/D. (1) – PIV data, (2) – dye visualization. Each experiment was repeated five times and averaged values are shown.

3 STRATIFIED FLUID

In SPR caverns the degassed oil may have a different density from that in the cavern, leading to buoyancy effects. A series of experiments was conducted to investigate the buoyancy influence, wherein the cavern was initially filled with homogeneous fluid and a vertical jet with lighter/heavier fluid entered from the cavern top. At the same time the fluid from the cavern bottom was removed at the same flux rate. The flow evolution, especially the buoyancy influence on the mixing depth, was monitored using diagnostic techniques described before.

3.1 Experimental set-up and method

The working fluid was distilled water, with an equation of state $\rho = \rho^*(1+\beta S)$, where ρ is the density, S the salinity ($\%_0$), $\rho^* = 1$ gmcm^{-3} and $\beta \approx 0.0008(\%_0)^{-1}$. As shown in Fig. 10, the jet flow in the cylinder (1) was generated using a round nozzle (2) placed at the cylinder centreline. A precision pump (3) feeds the nozzle (2) with a volume flux q from a reservoir (4) filled with water of density ρ_0, generating a jet of momentum flux J. The fluid leaves the cylinder through a long vertical pipe (5), emptying into to another reservoir (6). A micro-scale conductivity probe is used to measure the salinity and hence the density. Initially the cylinder is filled with fluid of density ρ_1. At time $t = 0$, the jet of density ρ_0 is initiated. The jet mixes with the ambient fluid and mixed fluid of density $\rho = \rho(t)$ leaves the cylinder. Since degassed oil injected into SPR may be lighter or heavier than cavern oil, both the positive ($\rho_0 < \rho_1$) and negative ($\rho_0 > \rho_1$) cases of buoyancy were studied.

3.2 Experimental results and model

3.2.1 *General flow behaviour*

First, to clarify general flow structure, small amount of fluorescing dye was added to the jet. Sequence of images showing the formation of mixed layer of depth l in the

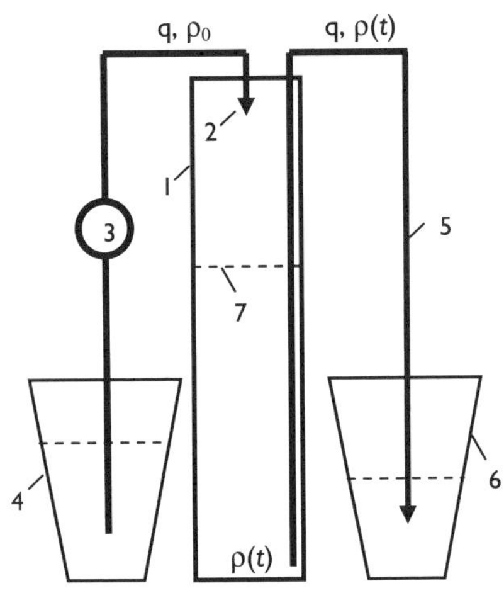

Figure 10 Schematic of the experimental set-up: 1 – closed from top and bottom long glass cylinder (length L = 118.5 cm, diameter D = 12 cm), which is initially filled with water of density ρ_1, 2 – vertical nozzle (diameter d = 0.165 cm) to generate jet, 3 – precision pump to control the volume flux q, 4 – reservoir with jet water of density ρ_0, 5 – long pipe to collect the water of density $\rho(t)$ taken near the cylinder bottom to reservoir 6, 7 – jet mixing boundary.

upper part of cavern for the case of lighter jet fluid is given in Fig. 11. In this case a mixed layer of depth l is built up rapidly at the top of the cylinder, with motions below remaining negligibly small at all times.

Similar observations for the case of heavier jet fluid, however, showed different flow behaviour. The fluid in the upper mixed layer is now heaver and deep convection develops into the cylinder. With time convective mixed layer reaches the bottom, and the entire cylinder of depth L is mixed. The upper part of the cylinder of depth l is rapidly mixed by jet action and the lower part of depth $(L\text{-}l)$ by convection.

3.2.2 *Model and comparison with measurements*

Let a long vertical cylinder is filled initially with fluid of density ρ_1. At time $t = 0$, the fluid of density ρ_0 is introduced from the top at $x = 0$ as a vertical jet with a volume flux rate q. At the same time fluid from near the cylinder bottom at $x = L$ is removed at a flux rate q. The interest is the density $\rho(x, t)$ of cavern fluid and its dependence on time t, depth x and other external parameters.

When the jet is positively buoyant, dye observations (Fig. 11) show immediate mixing to a depth l with fluid below remaining isolated. Thus we suppose that at $t > 0$ the fluid density $\rho(t)$ of the upper part of the cavern $x < l$ depends only on t and at $x \geq l$ the fluid column of density $\rho(x, t)$ moves down with the velocity $U = 4q/\pi D^2$. The solution to this problem can be found in the form of a propagating wave (e.g., Whitham, 1974) as

$$R(x,t) = \frac{\rho(x,t)-\rho_0}{\rho_1-\rho_0} = \begin{cases} \exp(-Ut/l), & 0 \leq x/l < 1 \\ \exp\{(x-l-Ut)/l\}, & 1 \leq x/l < 1+Ut/l, \\ 1, & 1+Ut/l \leq x/l \end{cases} \tag{5}$$

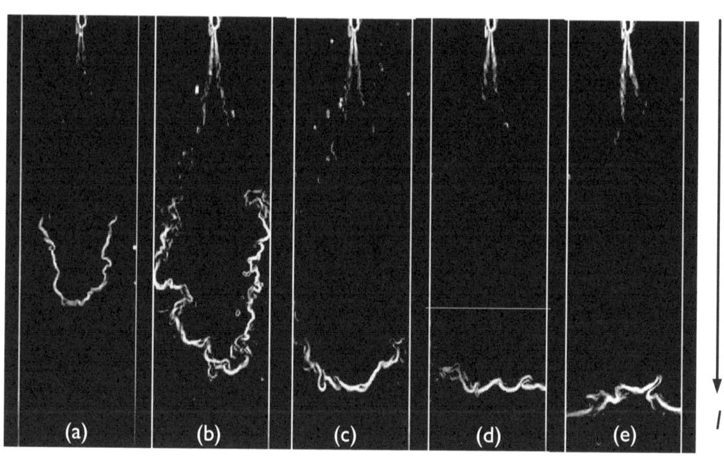

Figure 11 Sequence of images showing formation of mixed layer of the depth l in the upper part of the cylinder. Visualization – fluorescing dye, only the upper part of cavern is shown. $Re = 7000$, $S_0 = 0$, $S_1 = 105‰$; $t = 2.0$ (a) 4.3 (b) 12.1 (c) 46.1 (d) 74.6 s (e).

where $l = x^*$ is the mixing depth and x^* is given by (2), $R(x, t)$ is the dimensionless density and x/l the dimensionless depth. Note that the length L of the cavern does not enter the problem when $L \geq l$, which is physically obvious.

In practice, the time of oil processing during degas is measured in terms of time units $t_0 = V/q$, which is the time needed to recirculate one cavern volume V at a flux rate q. Thus (5) becomes

$$R(X, N) = \begin{cases} \exp(-NL/l), & 0 \leq X < 1 \\ \exp(X - NL/l - 1), & 1 \leq X < 1 + NL/l, \\ 1, & 1 + NL/l \leq X \end{cases} \tag{6}$$

where $N = t/t_0 = tq/V$ is the number of cavern volumes processed (dimensionless time) and $X = x/l$ is the dimensionless depth. Note that for $x = L$ (6) gives the density of fluid removed from the cavern bottom.

The results for the positive buoyant case are shown in Fig. 12. The dimensionless experimental data for all runs nicely collapsed, showing agreement with (6) that is shown by the solid line. The vertical density profiles were also measured and compared with calculations, as exemplified in Fig. 13. A time succession of profiles for run #2 is shown by thin lines in Fig. 13, which agree well with calculations based on (6). Some differences can be seen in the pycnocline area, $2 > X > 1$, which may be attributed to measured 'instantaneous' profiles being different from the calculated mean profile.

The solution (6) is rather general and can be used for the case of heaver jet fluid ($\rho_0 > \rho_1$) as well, where the stratification in the cavern is unstable and deep convection is the dominant factor. To the first approximation we may suppose that the mixing depth l is now equal to the cavern length L, for which (6) becomes

$$R(x, N) = \exp(-N), \quad N \geq 0, \tag{7}$$

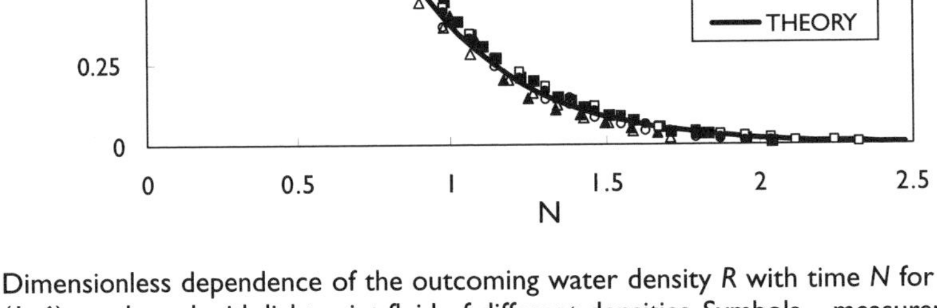

Figure 12 Dimensionless dependence of the outcoming water density R with time N for all six runs (1–6) conducted with lighter jet fluid of different densities. Symbols – measurements, solid line – solution (6).

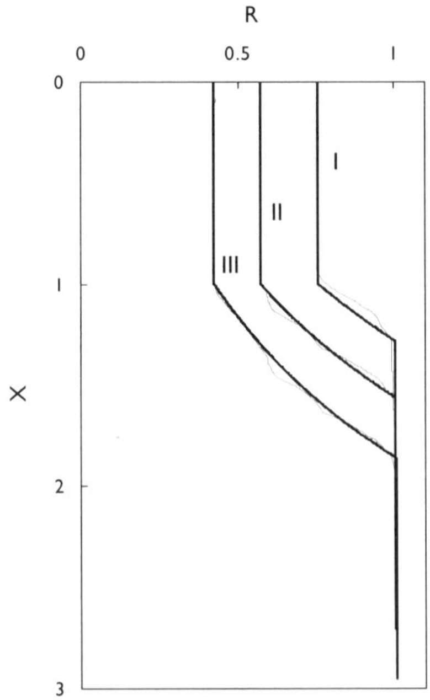

Figure 13 Comparison of measured (thin lines) and calculated (solid lines) vertical density profiles in cavern in run #2 for different times in dimensionless variables: $N = 0.1$ (I) 0.2 (II) 0.31 (III).

Comparisons showed a satisfactory agreement between measurements and the estimate (7) for all runs conducted with negatively buoyant jets.

4 CONCLUSIONS

The evolution of a turbulent jet released into a low aspect ratio cylinder was investigated experimentally using PIV and digital imaging methods. The study was focused on: (i) the general flow structure and instabilities that lead to periodic oscillations intrinsic to confined jets; (ii) the roles of (top/bottom) boundary conditions and resulting pressure adjustments that cause jet to disintegrate, (iii) parameterizations of flow velocities and flow oscillating frequency, (iv) the influence of the jet radial offset, and (v) the role of rotational instability on jet oscillations. A similarity model was developed, with characteristic length and velocity scales derived using jet kinematic momentum flux and cylinder width. The model proposed was extended to stratified jets and verified experimentally. The scaling laws so developed could be used to extrapolate laboratory observations to SPR flows.

This work was supported by the Sandia National Laboratories, which is operated by Lockheed Martin Corporation for the United States Department of Energy's National Nuclear Security Administration under contract DE-AC04-94 AL85000.

REFERENCES

Cantwell, B.J. (1986). Viscous starting jets. *J. Fluid Mech.*, 173, 159.

Denisikhina, D.M., Bassina, A., Nikulin D.A. and Strelets, M.K. (2005). Numerical simulation of self-excited oscillation of a turbulent jet flowing into a rectangular cavity. *High Temperature*, 43 (4), 568.

Gebert, B.M., Davidson, M.R. and Rudman, M.J. (1998). Computed oscillations of a confined submerged liquid jet. *Appl. Math. Modeling*, 22, 843.

Villermaux, E. and Hopfinger, E.J. (1994). Self-sustained oscillations of a confined jet: a case study for the non-linear delayed saturation model. *Physica D*, 72, 230.

Lawson, N.J. (2001). Self-sustained oscillation of a submerged jet in a thin rectangular cavity. *J. Fluid & Structures*, 15, 59.

Liu, H., Winoto, S.H., Dilip, A. and Shah, D.A. (1997). Velocity measurements within confined turbulent jets: application to cardiovalvular regurgitation. *Annals Biomedical Eng.*, 25, 939.

Lord, D.L. and Rudeen, D.K. (2007). Summary of Degas II Performance at the US Strategic Petroleum Reserve Big Hill Site. *Technical report, SAND2007-5564*, p. 53. Sandia National Laboratories. Albuquerque, NM.

Mataouia, A. and Schiestelb, R. (2009). Unsteady phenomena of an oscillating turbulent jet flow inside a cavity: effect of aspect ratio. *J. Fluids & Structures*, 25, 60.

Molloy, N.A. and Taylor, P.L. (1969). Oscillatory flow of a jet into a blind cavity. *Nature*, 224, 1192.

Voropayev, S.I., Sanchez, X., Nath, C., Webb, S. and Fernando, H.J.S. (2011). Evolution of a confined turbulent jet in a long cylindrical cavity: homogeneous fluids. *Phys. Fluids*, under revision.

Whitham, G.B. (1974). *Linear and Nonlinear Waves*. John Wiley and Sons, Inc.

Chapter 17

Modelling internal solitary waves in shallow stratified fluids

P.A. Davies[1] and M. Carr[1,2]
[1] Department of Civil Engineering, The University of Dundee, Dundee, UK
[2] School of Mathematics and Statistics, Mathematical Institute, University of St Andrews, St Andrews, UK

ABSTRACT: Recent laboratory modelling work by the authors on the behaviour of internal solitary waves (ISWs) in shallow, stably-stratified, multi-layer fluid systems is described, for cases in which the amplitude of the wave is comparable with the total depth of the system. The paper concentrates attention on particular aspects of shallow water ISW behaviour, namely, (i) the generation of transient boundary currents on the bottom solid boundary, (ii) the instability characteristics of such boundary currents, (iii) the differences in behaviour between ISWs of elevation and depression, (iv) the stability characteristics of ISWs in fluid systems of constant depth and (v) the consequences of wave breaking for vertical mixing. Experimental data are presented to illustrate the properties of the above flows and comparisons are made, where appropriate, with predictions from theory, numerical models and field observations.

1 INTRODUCTION

Internal solitary waves (ISWs) are nonlinear waves of permanent form that propagate on density interfaces in stably-stratified fluids. They are able retain their form over very large distances by maintaining a balance between nonlinearity and linear dispersion, processes that tend, in isolation, to respectively steepen and broaden the waves. In a geophysical context, they are observed in the Earth's atmosphere (Rottman & Grimshaw, 2001) and hydrosphere – for example in lakes (Imberger, 1998; Boegman *et al.*, 2003), the deep ocean (Apel *et al.*, 1995; Garrett & Kunze, 2007), shallow seas (Osborne & Burch, 1980; Apel *et al.*, 1985; Ostrovsky & Stepanyants, 1989; Stanton & Ostrovsky, 1998; Moum *et al.*, 2003; Klymack *et al.*, 2006; Lamb & Farmer, 2011), fjords (Farmer & Smith, 1978; Farmer & Armi, 1999), estuaries (Bourgault *et al.*, 2005; Groeskamp *et al.*, 2011) and in the vicinity of river plumes (Nash & Moum, 2005; Pan *et al.*, 2007).

In the ocean they are generated primarily (but not exclusively) by barotropic tidal flow over uneven bottom topography (Farmer & Armi, 1999; Apel, 2002, Vlasenko *et al.*, 2005) and they travel as rank-ordered (in amplitude) packets. Oceanic ISWs have attracted significant scientific interest since technological advances in synthetic aperture radar (SAR) imagery revealed the existence of such waves in most of the Earth's shelf seas and deep oceans (Apel, 2002). Ship-based observations have shown that ISWs may reach amplitudes comparable with the water column depth; see, for example, Duda *et al.* (2004) who report waves of depression of amplitude 150 m in a water depth of 340 m and Van Gastel *et al.* (2009) who measured a maximum vertical isopycnal displacement of 83 m in a total water depth of 124 m. ISWs may be waves of depression or elevation (*e.g.*, Moum *et al.*, 2007), depending upon whether the pycnocline is respectively above or below the mid-depth of the water column. Interest is focussed here on the former type. The comprehensive review by Helfrich and Melville (2006) provides a convenient entry point into the ISW literature.

Internal solitary waves are important in the marine environment because they have a controlling role in sediment suspension and vertical mixing of particulates, nutrients and contaminants (Bogucki & Garrett, 1993; Bogucki *et al.*, 1997; Bogucki & Redekopp, 1999; Stastna & Lamb, 2002; 2008; Hosegood & van Haren, 2004). For offshore installations such as oil and gas production and exploration platforms, exposure to ISW activity presents significant hazards. Direct hydrodynamic loading associated with the passage of ISWs is known to cause non-trivial tilting and displacement of rigs, increases in anchor tension, stresses on drill pipes, drill string breakages and vortex-induced vibrations leading to fatigue damage to moorings and risers. For sufficiently shallow waters, the suspension of bottom sedimentary material by the waves affects significantly the integrity of sea floor oil and gas infrastructures and the efficacy of sub-surface acoustic communications systems. In some locations, Soliton Early Warning Systems (SEWS) (Goff & Jeans, 2010) are in place to aid preparations for the arrival of ISWs (for example, so-called "weathervaning" of Floating Production, Storage and Offloading (FPSO) units to minimise ISW threats to the manoeuvrability of these vessels during shuttle tanker offloading operations).

Modelling investigations in which the characteristic amplitude, celerity and wavelength of the ISWs can be controlled and varied for different water column density configurations and relative water depths offer a means of understanding and predicting the behaviour and stability of these waves. The primary focus in this article is on laboratory modelling.

2 LABORATORY MODELLING

2.1 Experimental arrangement

Various laboratory methods have been developed to generate ISWs in stably-stratified fluid systems, including (i) moving ridge-like bottom topography through a quiescent fluid (Maxworthy, 1979), (ii) employing a mechanical displacement-type wavemaker (Koop & Butler, 1981), oscillating paddle (Gostiaux & Dauxois, 2007) or piston (Hütteman & Hutter, 2001) or (iii) inducing gravitational collapse of a locally-mixed region of fluid (Kao & Pao, 1979; Maxworthy, 1980).

The laboratory configuration and method of ISW generation to be adopted here is due initially to Kao *et al.* (1985) and is illustrated schematically on Figure 1. A long, rectangular-section, horizontal tank contains a homogeneous bottom layer of fluid of density ρ_3 and thickness h_3. Above this layer is a linearly-stratified layer of thickness h_2, within which the density varies between ρ_3 and ρ_1, the density of the upper fluid layer of thickness h_1 (Fructus *et al.*, 2009). A solid gate G placed close to one end of the tank isolates a region containing an excess volume V of fluid of density ρ_1 added behind the gate, as shown. The addition of the excess volume depresses

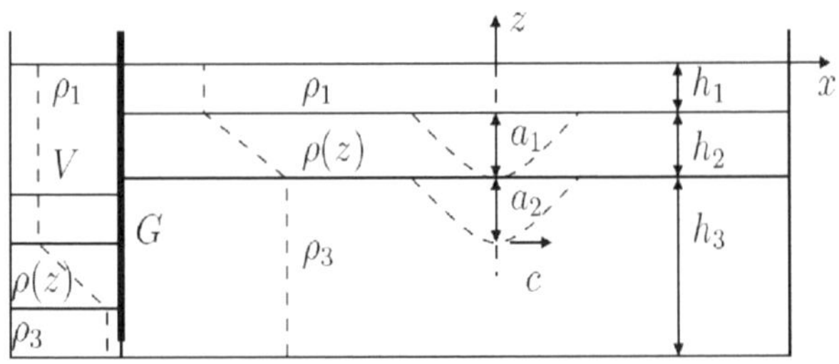

Figure 1 Schematic view of the laboratory model configuration – see text for definition of symbols.

the pycnocline behind the gate such that, when the gate is removed vertically, an internal solitary wave propagates with celerity c along the interface in the main part of the channel. The geometrical dimensions of the section of the tank behind the gate (and the magnitude of the excess volume V) are chosen (Kao *et al.*, 1985) to generate a single solitary wave. In all cases to be described here, the working fluids are immiscible (brine and fresh water) and a single ISW (rather than a packet of ISWs) is considered.

In this paper, attention is directed specifically at flow configurations in which the mean depth of the fluid system is constant. Investigations of wave distortion and breaking by ridge-like and corrugated bottom topography have been carried out by the authors (Sveen *et al.*, 2002; Guo *et al.*, 2004; Carr *et al.* but an important aspect of ISW behaviour that is excluded from consideration is shoaling resulting from the propagation of the waves on to a uniform slope. Many detailed modelling studies of ISW shoaling phenomena have been undertaken, both in the laboratory (*e.g.*, Helfrich, 1992; Michallet & Ivey, 1999) and with numerical simulation (*e.g.*, Aghsaee *et al.*, 2010), but these studies lie outside the scope of the present review.

2.2 Measurement systems

The behaviour of ISWs as they propagate along the pycnocline of the stratified fluid is typically characterised by (i) the distortions to the ambient density field and (ii) the velocity and vorticity fields induced by the passage of the wave. Laser- (or light-) induced fluorescence provides a suitable non-intrusive technique to determine quantitatively and synoptically the density disturbance field (see for example, De Silva *et al.*, 1996) but the costs of careful matching of the refractive indices of the working fluids (through the use of, for example, ethanol) is often prohibitive for the large working chambers required. Exploitation of refractive index variations between the working fluids offers an alternative approach to determining synoptic density fields and one that may be combined with particle image velocimetry (PIV) to obtain simultaneously velocity and density field data (see, for example, Dalziel *et al.*, 2007). PIV techniques to measure synoptically the velocity fields within the wave can be used successfully for non-breaking waves (Grue *et al.*, 1999), though, with breaking waves, optical distortions within the pycnocline caused by localised mixing preclude the use of PIV for this part of the flow domain (Fructus *et al.*, 2009).

Intrusive methods are used commonly to acquire sequential density profiles at prescribed locations within the wave. When the ambient stratification is achieved by fresh water and brine layer combinations (the usual arrangement), rapid-response micro-conductivity probes (Head, 1983) provide information on not only the form of the instantaneous density profile within the fluid at different locations but also the mixing properties (Guo *et al.*, 2004) of any disturbed region(s) generated by breaking waves (see below).

In the experiments carried out by the authors and reported below, channels constructed from transparent glass panels supported within a steel framework were employed. A light box illuminated a thin slice of the flow at the mid plane of the channel and the motions within this sheet of small, neutrally-buoyant tracer particles added to the fluid were recorded by an array of stationary digital video cameras mounted outside the tank, looking in through the side panels. Using *DigiFlow* software (see, for example, Sveen & Dalziel, 2005), particle image velocimetry (PIV) was applied to the digital camera records to derive synoptic velocity and vorticity fields within the illuminated sections; by overlapping the fields of view of the 3 cameras, it was possible to record and follow simultaneously the behaviour of the wave over at least one full wavelength.

3 BOTTOM BOUNDARY LAYERS

Laboratory studies (Grue *et al.*, 2000; Sveen *et al.*, 2002; Fructus *et al.*, 2009) have verified that vertical profiles of the velocity components $u(z)$, $w(z)$ of the disturbance velocity fields generated by ISWs in multi-layer, stratified systems may be predicted accurately from the computations of Grue *et al.*, (1997; 1999) and Fructus and Grue (2004), even for highly nonlinear ISWs. For cases in which the amplitude of the ISW is comparable with the total fluid depth, the bottom boundary plays an important frictional role in the spatial and temporal development of the wave. In particular, the boundary layer established to match the unsteady, spatially-varying interior ISW flow to the no-slip boundary condition generates a localised flow close to the bottom boundary, due to the existence of a spatially-varying pressure gradient induced by the wave. Such a boundary flow is generated *a priori* in the decelerating part of the wave, namely the region in which the horizontal disturbance pressure gradient is adverse.

The mechanism responsible for the generation of this localised, ISW-induced boundary flow was first identified by Bogucki *et al.* (1997) and discussed subsequently in a number of theoretical modelling investigations. In these investigations, separation of the boundary layer in the adverse pressure gradient region is shown to play an important role in the development of the flow. A number of numerical modelling studies of ISWs of elevation and depression (e.g., Bogucki & Redekopp, 1999; Wang & Redekopp, 2001; Stastna & Lamb, 2002, 2008; Diamessis & Redekopp, 2006), illustrate the susceptibility of the boundary flow to global instability for sufficiently high wave amplitudes and demonstrate that the manifestation of the global instability is the generation of a field of coherent vortices within the adverse pressure gradient region of the boundary layer. Such vortices are seen to be erupted from the boundary into the water column and to be advected subsequently by the interior flow. Recent numerical modelling studies (*e.g.*, Aghsaee *et al.*, 2012) have extended previous work to investigate ISWs of depression propagating on inclined as well as flat bottom bounding surfaces and have determined a unified threshold (based on the momentum thickness Reynolds number and the free stream pressure gradient at the point of separation) for vortex generation by global instability of the bottom jet.

The first laboratory demonstration of the boundary jet phenomenon (and the first measurements of the structure of the jet) was for large amplitude waves of depression in a shallow, two-layer fluid (Carr & Davies, 2006). The thin boundary layer was observed below the decelerating part of the wave (*i.e* the region in which the disturbance pressure gradient was adverse), flowing in a direction opposite to that of the interior of the bottom layer in which it was formed and in the same direction as that of the wave itself. Figure 2a illustrates the phenomenon with a time series plot of dimensionless horizontal velocity u/c_0 versus dimensionless time tc_0/h_2 at various distances z/h_2 from the flat horizontal boundary ($z/h_2 = 0$), at a given measurement station in the channel. Here, the linear long wave speed c_0 is defined by $c_0^2 = g'h_1h_2/(h_1 + h_2)$ and the modified gravitational acceleration g' by $g' = g(\rho_2-\rho_1)/\rho_1$, where $h_{1,2}$ and $\rho_{1,2}$ are the thickness and density respectively of the upper ("1") and lower ("2") layers and t is elapsed time. The plots show clearly (i) the asymmetry in the wave form between the fore and aft regions of the wave field and (ii) the dispersion in the flow profiles close to the boundary in the aft region caused by the generation of a boundary jet. Inspection of the different z/h_2 profiles shows that the jet is confined within about $(0.2–0.4)h_2$ of the bottom boundary. Measurements showed that the jet was stable throughout the forcing period.

Carr and Davies (2006) were able to interpret their results successfully in terms of classical boundary layer theory (Schlichting, 1979) and to confirm, in particular, the predictions of theory that, for the parameter ranges considered, (i) the boundary flow is associated with separation and (ii) the flow is stable. The experiments of Carr and Davies (2006) showed excellent qualitative agreement between the measured flow profiles close to the boundary and the numerical modelling predictions of Diamessis & Redekopp (2006), in spite of significant differences between the laboratory and numerical model systems. Specifically, the ISW amplitudes in the Diamessis & Redekopp (2006) study were significantly less than in the laboratory cases and the bottom layer in the numerical model was not homogeneous (as in the laboratory) but continuously stratified. In spite of this encouraging agreement with theory, questions remained about the relationship of the separation process to the form of the jet flow

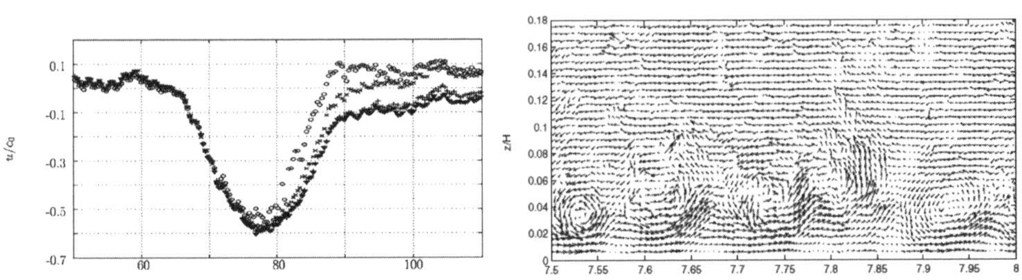

Figure 2 (a) Time series plots of u/c_0 versus tc_0/h_2 at z/h_2 = (o) 0.1, (x) 0.2, (+) 0.4 (from Carr & Davies, 2006) and (b) 2D velocity vector field of boundary vortices generated by global instability (from Carr *et al.*, 2008), ISW of depression.

observed. In particular as noted by Diamessis & Stefanakis (2011), the separated stable boundary layer in the experiments was observed to maintain a constant thickness and it did not reattach to form a separation bubble (in contrast to the predictions of Diamessis & Redekopp (2006)).

One of the most intriguing aspects of the boundary jet flow is its stability and the conditions under which, as predicted by the numerical models above, it becomes globally unstable and arrays of vortices are generated. Carr *et al.* (2008) were able to extend the earlier experiments of Carr & Davies (2006) on ISWs of depression to sufficiently high wave amplitudes a/H and wave Reynolds numbers $Re_w = c_0H/\nu$ (where H is the total fluid depth of the water column and ν is the mean kinematic viscosity of the stratified fluid) that global instability could be demonstrated (see Fig. 2b). The instability was manifested by a rapid roll up of the boundary layer vorticity into a row of coherent boundary vortices (see Fig. 2b) shed upwards and away from the boundary, as predicted by the numerical model simulations of Diamessis and Redekopp (2006) and others cited above. The laboratory data were obtained over a relatively small range of Re_w but they confirm that the critical amplitude required for global instability decreases as a function of increasing values of Re_w (as predicted by Diamessis & Redekopp (2006). The laboratory studies, however, indicate that the critical amplitude itself is significantly lower than the value predicted by Diamessis & Redekopp (2006) for the same Re_w. Such a quantitative discrepancy is not surprising since the numerical model employs weakly nonlinear forcing, whereas the internal waves generated in the laboratory study were fully nonlinear.

Thus far, the emphasis from the laboratory experiments has been upon ISWs of depression, where the principal applicability to oceanic ISWs lies. However, as indicated previously, a considerable body of relevant literature (Bogucki & Redekopp, 1999; Wang & Redekopp, 2001; Stastna & Lamb, 2001, 2008; Diamessis & Redekopp, 2006) also exists on boundary effects beneath waves of elevation. The nature of the boundary effects induced by this subset of ISWs requires clarification, since different numerical models give different predictions. For example, Stastna and Lamb (2005; 2008) have shown that global instability and boundary vortex formation occurs under the leading part of fully nonlinear ISWs of elevation but only in cases in which the magnitude of an oncoming sheared current exceeds a threshold value dependent upon the value of Re_w. For an unsheared opposing flow, the boundary layer is predicted to always remain attached. This finding contrasts with that of Diamessis and Redekopp (2006) who found that (initially) weakly-nonlinear ISWs of either depression or elevation propagating in an un-sheared flow induced separation of the boundary layer in the adverse pressure gradient region and the formation of vortices in both cases.

Laboratory experiments (Carr & Davies, 2010) carried out to resolve these questions have demonstrated that, even for cases where an ISW of elevation breaks at the pycnocline, the wave-induced boundary layer does not separate in the adverse disturbance pressure gradient region ahead of the wave peak. Furthermore, for values of Re_w comparable with the numerical results of Diamessis and Redekopp (2006), the experiments show no evidence of instability of the boundary jet and, in consequence, no formation of wall vortex structures. A ubiquitous feature of the ISW laboratory experiments, however, is the transient jet close to the bottom boundary, in the decelerating part of the wave (aft of the wave peak) as for the ISWs of depression.

Figure 3, in which the horizontal velocity data are plotted versus distance in a form that enables comparison with the computations of Liu *et al.* (2007), shows clearly the reversal of flow near the bed ($z = 0$) in the aft part of the wave ($ct > 0$, where c is the measured phase speed of the ISW).

The interpretation of the jet formation mechanism in terms of ISW-induced regions of adverse and favourable pressure gradients requires some care, since consideration of the non-hydrostatic contributions to these gradients leads to an expectation that the reverse jet should occur in the aft part of the wave of depression (as observed) but the fore part of the wave of elevation (as not observed), in the frame of reference of a stationary observer. The study by Liu *et al.* (2007) of the boundary layer flows generated under surface solitary waves provides useful insight into this matter. Liu *et al.* (2007) ascribe the boundary jet flow to a phase lag between the irrotational and rotational velocity components of the flow, a mechanism equivalent to that which operates when considering separation associated with the favourable and adverse *total* pressure gradients (including the hydrostatic gradients) induced by the wave. Considering the hydrostatic horizontal pressure gradient contributions alone, it is easily shown that the adverse pressure gradients exist in the aft part of the ISW for both depression and elevation cases – a property that is consistent with the results of the laboratory experiments.

Confirmation of the role of the total horizontal pressure gradients in determining (i) separation and (ii) the formation of the transient boundary jet flow has come recently from a set of numerical simulations of the Carr & Davies (2006) experiments with ISWs of depression, carried out by Thiem *et al.* (2011) using the Bergen Ocean Model (BOM). As Fig. 4a illustrates, the BOM simulation of the laboratory experimental run conditions associated with the velocity profile in Fig. 3a results in excellent qualitative agreement between the measured and computed horizontal velocity profiles. The reverse transient boundary jet is clearly shown between $z/h_2 = 0.2$–0.4. The simulations demonstrate that the hydrostatic pressure contribution to the total horizontal pressure gradient dominates the non-hydrostatic component, confirming that separation is related primarily to the total adverse pressure gradients within the flow; for *both* elevation and depression cases, these adverse pressure gradients occur aft of the propagating wave in the frame of reference of the stationary observer. Fig. 4b shows the horizontal total pressure gradients beneath the ISW of depression; in the figure the horizontal distance x is normalised by the tank length L.

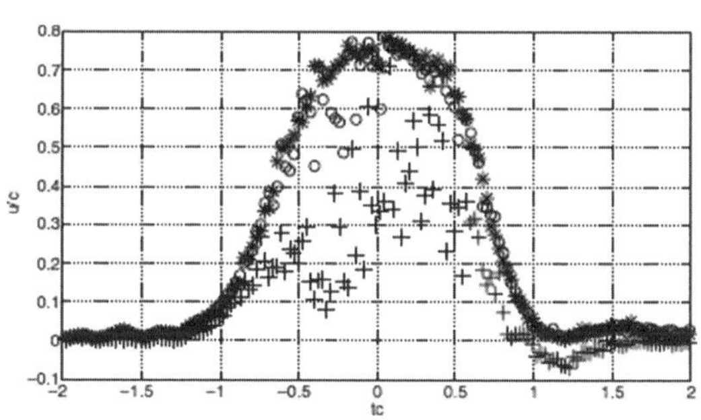

Figure 3 Time series plots of u/c versus tc at z/H = (+) 0.01, (O) 0.04, (*) 0.79 (from Carr & Davies, 2010), ISW of elevation. From Carr & Davies (2010).

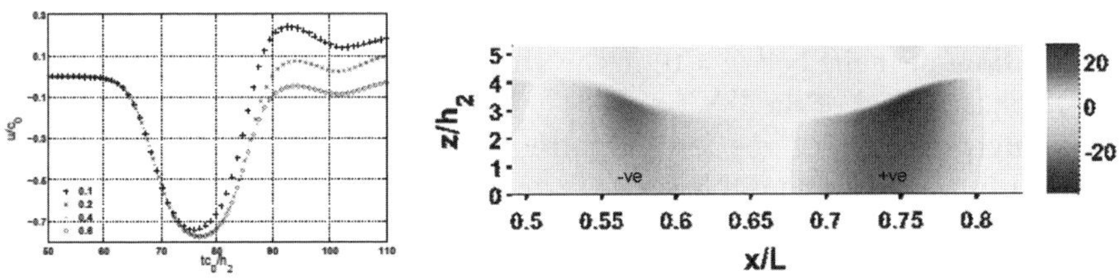

Figure 4 BOM (see text) simulations of Carr and Davies (2006) experiments for ISWs of depression (see Fig 3a), showing (a) u/c_0 versus tc_0/h_2 at various z/h_2 (see legend) and (b) horizontal (x) pressure gradients $\Delta p/\rho_2 c_0^2 L$ beneath the wave In (b) positive (negative) values of $\Delta p/\rho_2 c_0^2 L$ indicate increasing pressure to the right (left). From Thiem *et al.* (2011).

4 BREAKING WAVES

ISWs are known to break as a result of either convective or shear-driven instability (or a combination of both – see, for example, Carr *et al.*, 2008; Fructus *et al.*, 2009), but the condition for shear-induced breaking (the case of interest here) has been obtained only recently[1]. Grue *at al* (1999) reported the first laboratory experiment to document breaking in ISWs for $Ri_{min} = 0.07$, while noting that for $Ri_{min} = 0.23$ the wave was stable. (Here Ri is the local Richardson number, defined conveniently (Fructus & Grue, 2004) as $Ri = c(c-u)/\delta^2 N_\infty^2$, where c is the wave speed, u is the horizontal velocity, δ the vertical excursion of the streamline relative to rest in the steady frame of reference of the wave and N_∞ is the value of the buoyancy frequency in the far-field (obtained by tracing along a streamline in the steady-state)).

Laboratory and numerical investigations (Fructus *et al.*, 2009) in 3-layer stratified systems consisting of two homogeneous layers separated by a linearly stratified pycnocline have demonstrated subsequently that breaking occurs through shear instability when the horizontal width L_x of the pocket defined by $Ri < 1/4$ in the wave core exceeds 0.86 times the wave width λ at half amplitude – a condition verified subsequently by Barad and Fringer (2010) and Lamb and Farmer (2011). Most significantly, the theoretical analyses demonstrate that large amplitude ISWs may remain stable for values of the local Richardson number Ri much less than ¼, a finding in agreement with field observations (Duda *et al.*, 2004) and modelling studies of periodic internal waves (Troy & Koseff, 2005).

The breaking of ISWs due to shear instability is characterised by the formation of a sequence of rolled vortices (billows). Each of the billows grows with time until arrested by the stratification within the pycnocline, before breaking down to form a turbulent patch. In this regard the behaviour resembles qualitatively that associated with the classic Kelvin-Helmholtz instability of the interface between two fluids in parallel motion with different velocities and densities (see, for example, de Silva *et al.*, 1996; Strang and Fernando, 2001). Note, however, that the temporal and spatial development of the billows in the case of breaking ISWs is *a priori* affected significantly by the horizontal non-uniformity in velocity within each of the constituent layers.

1 Field observations showing breaking oceanic ISWs are reported by Moum *et al.* (2003).

The purpose of the present paper is to describe some recent measurements of the development of the density field within a breaking ISW billow. In particular, preliminary data are presented on the Thorpe scale L_T-a measure of the length scale of turbulent overturning events within the billow (Dillon, 1982; Thorpe, 2005). To this end, density profile data have been acquired for a breaking ISW of depression, using an array of micro-conductivity probes (see Section 2.2). Profiling is achieved by driving all probes vertically through the fluid at high speed when the aft portion of the wave comes within the measurement section. The composite image in Figure 5 illustrates the shear-induced breaking and billow formation processes and the false colour images in Figure 6 show a closer view of the sequence of isolated billows formed in the aft portion of the wave.

The development of each billow is characterised by distinct phases of growth, merging and decay. The initial perturbations grow spatially with time and develop a characteristic overturning form until a closed billow of fluid of lens-like shape is formed. In this development, the vertical growth of an individual billow is inhibited compared with the horizontal growth by the density stratification in the interface between the layers. During the growth and collapse processes, the velocity shear across the interface decreases with horizontal distance (x) from the maximum depression point of the wave (see, for example, Fructus & Grue, 2004) so the environment in which an individual billow develops will vary with elapsed time.

The distorted density profiles taken through the billows can be processed to provide values of the Thorpe scale L_T and the vertical extent (local billow height) L_B of the billow. The method of deriving the Thorpe scale L_T has been described extensively elsewhere (*e.g.*, Dillon, 1982; de Silva *et al.*, 1996). It consists of (i) sorting the measured profile data to generate the statically-stable state of minimum potential energy, (ii) calculating the vertical distances l_d (the Thorpe displacements) each fluid parcel

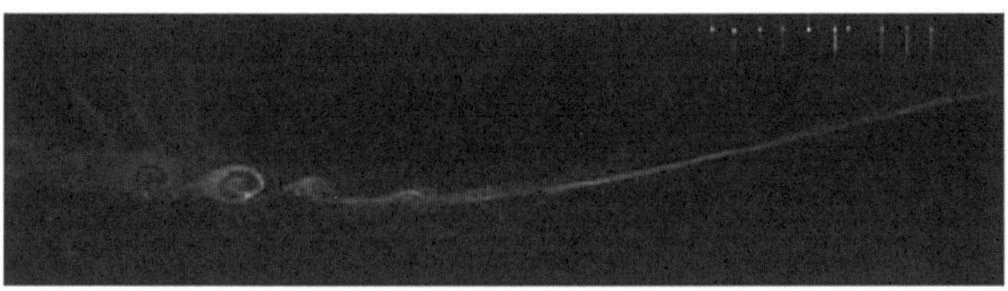

Figure 5 Composite image (side view) of breaking ISW of depression (L → R).

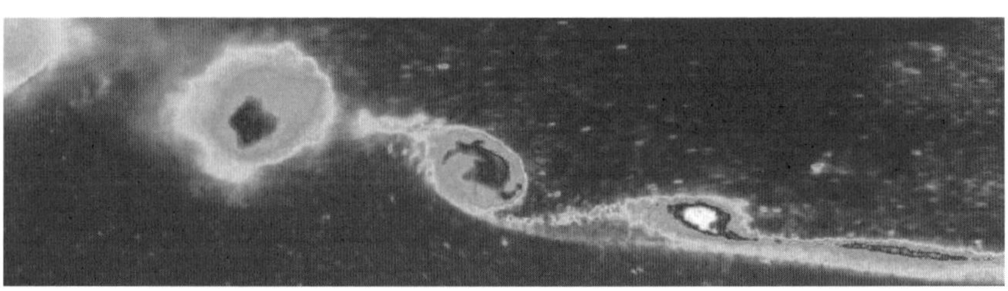

Figure 6 Billows in breaking portion of an ISW of depression. Direction of wave propagation: L → R.

has to be moved to achieve this sorted (monotonised) profile and (iii) determining L_T as the r.m.s value of all of the individual l_d values within the billow. The quantity L_B is measured directly from the original distorted profile. The composite plot in Figure 7 shows the relationship between L_B and L_T for all of the breaking ISW experiments conducted. The plot shows that, regardless of scatter and the amalgamation of data from different external conditions, the relationship between L_B and L_T is well represented by a linear relationship with a coefficient of 0.38 linking the two dimensions. This value may be compared with that (0.49 ± 0.03) obtained by de Silva *et al.* (1996) for the reference case of a stratified shear flow with uniform parallel layers.

5 NUMERICAL MODELLING

Recent studies by the authors (King *et al.*, 2010; Carr *et al.*, 2011) have developed numerical models employing a combination of contour advection and pseudo-spectral techniques (Dritschel & Ambaum, 1997; Dritschel & Fontane, 2010) to simulate shear-induced breaking of ISWs. Initial runs with the numerical model illustrates (see Figure 8) that it simulates well the formation of the billows near the trough of the wave and the downstream growth, decay and merging characteristics of the flow observed in the experiments.

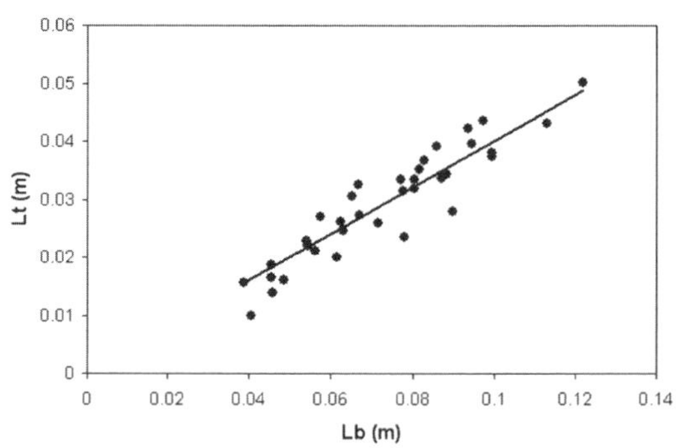

Figure 7 Dimensional composite plot (all experiments) of variation of L_T versus L_B, with best-fit linear regression.

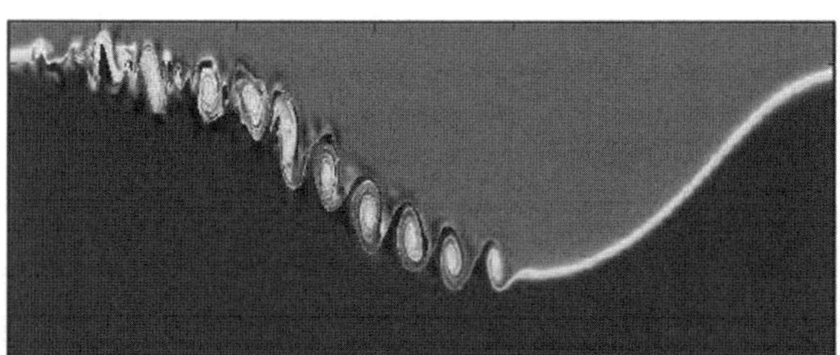

Figure 8 Numerical simulation of shear-induced instability and billow formation in an ISW propagating in a 3-layer stably-stratified fluid.

6 SUMMARY

In spite of the extensive observational and modelling efforts carried out in recent years, interest remains high in understanding the behaviour of ISWs in relatively simple geometrical systems and two or three-layer, stably-stratified density configurations. Important questions remain unanswered, particularly in the aspects of ISW behaviour considered in this contribution. Laboratory verification and parameterisation of the conditions for global instability of highly nonlinear ISWs in the absence of ambient flow and ambient shear is urgently needed (particularly for waves of elevation) in order to resolve discrepancies revealed by sophisticated numerical model investigations. Likewise, almost no laboratory experiments have been undertaken to determine the interaction(s) of ISWs with ambient flows – an aspect that is crucially important when comparing predictions of numerical models. Practically all laboratory experiments have been restricted to two-dimensional waves, even though many SAR-based observational studies (and many ship-based measurements) show that three-dimensional effects are likely to be very important in the oceans.

The consequences of wave breaking for mixing and energy loss are practically important for many ISW problems. In the laboratory, the available data for determining and quantifying these properties are inevitably qualitative, in the main. The preliminary experiments reported here offer opportunities to obtain more quantitative information but the practical difficulties in obtaining simultaneously the highly-resolved, synoptic density and velocity fields necessary for the computation of dissipation and mixing efficiency properties of the flow (particularly for the formation and development of turbulent patches that characterise stages of the wave-breaking processes) are formidable.

Finally, the studies outlined above (and others conducted with highly nonlinear ISWs in shallow two- and three layer fluids with ridge and corrugated bottom topography) offer intriguing opportunities for investigation of sediment bed responses to the passage of ISWs. This aspect of the consequences of ISW propagation has been considered widely in motivating ISW modelling studies but few laboratory data are available yet to inform this process.

ACKNOWLEGEMENTS

The authors are grateful to the Organizing Committee of the Gerhard Jirka Memorial Colloquium for their kind invitation to present this paper and for providing the opportunity to pay tribute to the memory of Gerhard Jirka, an outstanding, inspirational scientist and warm human being, whose positive approach to all aspects of life was an example to us all.

The authors acknowledge the support provided for their work on internal solitary waves by the UK Engineering & Physical Sciences Research Council (EPSRC) and the UK Northern Research Partnership. They are grateful to John Grue and his group at the University of Oslo, Marek Stastna, David Dritschel, Øyvind Thiem, Jarle Berntsen, Stuart Dalziel, Yakun Guo, Pete Diamessis, Larry Redekopp, Leon Boegman and Yong Sung Park for many fruitful discussions and productive collaborations on the topic. The important contributions of James Franklin and Stuart King to the work on shear-induced breaking of internal solitary waves are acknowledged with gratitude.

The technical assistance of John Anderson and Gary Conacher was invaluable in undertaking successfully the laboratory investigations reported here.

REFERENCES

Aghsaee, P., Boegman, L. and Lamb, K.G. (2010). Breaking of shoaling internal solitary waves. *Journal of Fluid Mechanics*, 659, 289–317.

Aghsaee, P., Boegman, L., Diamessis, P. and Lamb, K.G. (2012). Boundary layer separation and vortex shedding beneath internal solitary waves. *Journal of Fluid Mechanics* (in press).

Apel, J.R. (2002). Oceanic internal waves and solitons. In *An Atlas of Oceanic Internal Solitary Waves*, Global Ocean Associates, Washington DC, USA.

Apel, J.R., Holbrook, J.R., Tsai, J. and Liu, A.K. (1985). The Sulu Sea experiment. *Journal of Physical Oceanography*, 15, 1625–1651.

Apel, J.R., Ostrovsky, L.A. and Stepanyants, Y.A. (1995). Internal solitons in the ocean. *Journal of the Acoustical Society of America*, 98(5), 2863–2864.

Barad, M.F. and Fringer, O.B. (2010). Simulations of shear instabilities in interfacial gravity waves. *Journal of Fluid Mechanics*, 644, 61–95.

Boegman, L., Imberger, J., Ivey, G.N. and Antenucci, J.P. (2003). High-frequency internal waves in large stratified lakes. *Limnology & Oceanography*, 48 (2), 895–919.

Bogucki, D.J. and Garrett, C. (1993). A simple model for the shear-induced decay of an internal solitary wave. *Journal of Physical Oceanography*, 23, 1–10.

Bogucki, D.J. and Redekopp, L.G. (1999). A mechanism for sediment resuspension by internal solitary waves. *Geophysical Research Letters*, 26, 9, 1317–1320.

Bogucki, D.J., Dickey, T. and Redekopp, L.G. (1997). Sediment resuspension and mixing by resonantly generated internal solitary waves. *Journal of Physical Oceanography*, 27, 1181–1196.

Bourgault, D., Kelley, D.E. and Galbraith, P.S. (2005). Interfacial solitary wave run-up in the St Lawrence Estuary. *Journal of Marine Research*, 63, 1001–1015.

Carr, M. and Davies, P.A. (2006). The motion of an internal solitary wave of depression over a fixed bottom boundary in a shallow, two-layer fluid. *Physics of Fluids*, 18, 016601.

Carr, M., Fructus, D., Grue, J., Jensen, A. and Davies, P. A. (2008). Convectively induced shear instability in large-amplitude internal waves. *Physics of Fluids*. 20, 126601.

Carr, M., Stastna, M. and Davies, P.A. (2010). Internal solitary wave-induced flow over a corrugated bed. *Ocean Dynamics*. 60, 4, 1007–1025.

Carr, M., King, S.E. and Dritschel, D.G. (2011). Numerical simulation of shear-induced instabilities in internal solitary waves. *Journal of Fluid Mechanics*, 683, 263–288.

Dalziel, S.B., Carr, M., Sveen, J.K. and Davies, P.A. (2007). Simultaneous Synthetic Schlieren and PIV measurements for internal solitary waves. *Measurement Science & Technology*, 18, 533–547.

De Silva, I.P.D., Fernando, H.J.S., Eaton, F. and Hebert, D. (1996). Evolution of Kelvin-Helmholtz billows in nature and laboratory. *Earth & Planetary Science Letters*, 143, 217–231.

Diamessis, P.J. and Redekopp, L.G. (2006). Numerical investigation of solitary internal wave-induced global instability in shallow water benthic boundary layers. *Journal of Physical Oceanography*, 36, 784–812.

Diamessis, P.J and Stefanakis, T.S (2012). Near-bottom instabilities under strongly-nonlinear internal waves of depression. *Phys. Fluids (sub judice)*.

Dillon, T.M. (1982). Vertical overturns: a comparison of Thorpe and Ozmidov length scales. *Journal of Geophysical Research*, 87 (C12), 9601–9613.

Dritschel, D.G. and Ambaum, M.H.P. (1997). A contour-advective semi-Lagrangian numerical algorithm for simulating fine-scale conservative dynamical fields. *Quarterly Journal of The Royal Meteorological Society*, 123, 1097–1130.

Dritschel, D.G. and Fontane, J. (2010). The combined Lagrangian advection method. *Journal of Computational Physics*, 229, 5408–5417.
Duda, T.F., Lynch, J.F., Beardsley, R.C., Ramp, S.R., Chiu, C.S., Tang, T.Y. and Yang, Y.J. (2004). Internal tide and non-linear internal wave behavior at the continental slope in the northern South China Sea. *IEEE Journal of Oceanic Engineering*, 29, 1105–1130.
Farmer, D.M. and Armi, L. (1999). The generation and trapping of solitary waves over topography. *Science*, 283, 188–190.
Farmer, D.M. and Smith, J.D. (1978). Nonlinear internal waves in a fjord. In *Hydrodynamics of Estuaries and Fjords* (Ed. J.C.J. Nihoul), 465–493, Elsevier, NL.
Fructus, D. and Grue, J. (2004). Fully nonlinear solitary waves in a layered stratified fluid. *Journal of Fluid Mechanics*, 505, 323–347.
Fructus, D., Carr, M., Grue, J., Jensen, A. and Davies, P.A. (2009). Shear induced breaking of large internal solitary waves. *Journal of Fluid Mechanics*, 620, 1–29.
Garrett, C. and Kunze, E. (2007). Internal tide generation in the deep ocean. *Annual Reviews of Fluid Mechanics*, 39, 57–87.
Goff, M. and Jeans, G. (2010). Soliton Early Warning System (SEWS). *Marine Technology Reporter*, 53, 3, 26–33.
Groeskamp, S., Nauw, J.J. and Maas, L.R.M. (2011). Observations of estuarine circulation and solitary internal waves in a highly energetic tidal channel. *Ocean Dynamics*, 61 (11), 1767–1782.
Grue, J., Jensen, A., Rusås, P.-O. and Sveen, J.K. (1999). Breaking and broadening of internal solitary waves. *Journal of Fluid Mechanics*, 413, 181–217.
Grue, J., Jensen, A., Rusås, P.-O. and Sveen, J.K. (1999). Properties of large-amplitude internal waves. *Journal of Fluid Mechanics*, 380, 257–278.
Guo, Y., Sveen, J.K., Davies, P.A., Grue, J. and Dong, P. (2004). Modelling the motion of an internal solitary wave over a bottom ridge in a stratified fluid. *Environmental Fluid Mechanics*, 4, 415–441.
Gostiaux, L. and Dauxois, T. (2007). Laboratory experiments on the generation of internal tidal beams over steep slopes. *Physics of Fluids*, 19, 028102.
Head, M.J. (1983). The use of miniature four-electrode conductivity probes for high resolution measurement of turbulent density or temperature variations in salt-stratified water flows. Ph.D. Thesis, University of California, San Diego, USA.
Helfrich, K.R. (1992). Internal solitary wave breaking and run-up on a uniform slope. *Journal of Fluid Mechanics*, 243, 133–154.
Helfrich, K.R. and Melville, W.K. (2006). Long nonlinear internal waves. *Annual Reviews of Fluid Mechanics*, 38, 395–425.
Hosegood, P.J. and van Haren, H. (2004). Near-bed solibores over the continental slope in the Faeroe-Shetland Channel. *Deep-Sea Research* II, 51, 2943–297.
Hütteman, H. and Hutter, K. (2001). Baroclinic solitary waves in a two-layer fluid system with diffusive interface. *Experiments in Fluids*, 30, 317–326.
Imberger, J. (1998). Flux paths in a stratified lake: a review. In *Physical Processes in Lakes and Oceans* (Ed. J. Imberger), 1–17, American Geophysical Union, Washington, USA.
Kao, T.W. and Pao, H.P. (1979). Wake collapse in the thermocline and internal solitary waves. *Journal of Fluid Mechanics*, 97, 115–127.
Kao, T.W., Pan, F.-S. and Renouard, D. (1985). Internal solitons on the pycnocline: generation, propagation and shoaling and breaking over a slope. *Journal of Fluid Mechanics*, 159, 19–53.
King, S.E., Carr, M. and Dritschel, D.G. (2010). The steady state form of large amplitude internal solitary waves. *Journal of Fluid Mechanics*, 666, 477–505.
Koop, C.G. and Butler, G. (1981). An investigation of internal solitary waves in a two-fluid system. *Journal of Fluid Mechanics*, 112, 225–251.
Klymack, J.M., Pinkel, R., Liu, C.-T., Liu, A.K. and David, L. (2006). Prototypical solitons in the South China Sea. *Geophysical Research Letters*, 33, L11607.

Lamb, K.G. and Farmer, D.M. (2011). Instabilities in an internal solitary-type wave on the Oregon shelf. *Journal of Physical Oceanography*, 41, 67–87.

Liu, P.L.-F., Park, Y. S. and Cowen, E.A. (2007). Boundary layer flow and bed shear stress under a solitary wave. *Journal of Fluid Mechanics*, 574, 449–463.

Maxworthy, T. (1979). A note on the internal solitary waves produced by tidal flow over a three dimensional ridge. *Journal of Geophysical Research*, 84, 338–346.

Maxworthy, T. (1980). On the formation of nonlinear internal waves from the gravitational collapse of mixed regions in two and three dimensions. *Journal of Fluid Mechanics*, 96, 47–64.

Michallet, H. and Ivey, G.N. (1999). Experiments on mixing due to internal solitary waves breaking on uniform slopes. *Journal of Geophysical Research*, 104, 13467–13477.

Moum, J.N., Farmer, D.M., Smyth, W.D., Armi, L. and Vagle, S. (2003). Structure and generation of turbulence at interfaces strained by internal solitary waves propagating shore-ward over the continental shelf. *Journal of Physical Oceanography*, 33: 2093–2112.

Moum, J.N., Klymak, J.M., Nash, J.D., Perlin, A. and Smyth, W.D. (2007). Energy transport by non-linear internal waves. *Journal of Physical Oceanography*, 37:1968–1988.

Nash, J.D. and Moum, J.N. (2005). River plumes as a source of large amplitude internal waves in the coastal ocean. *Nature*, 437, 400–403.

Osborne, A.R. and Burch, T.L. (1980). Internal solitons in the Andaman Sea. *Science*, 208, 451–460.

Ostrovsky, L.A. and Stepanyants, Y.A. (1989). Do internal solitons exist in the ocean? *Reviews of Geophysics*, 27, 293–310.

Pan, J., Jay, D.A. and Orton, P.M. (2007). Analyses of internal solitary waves generated at the Columbia River plume front using SAR imagery. *Journal of Geophysical Research*, 112, C07014.

Rottman, J.W. and Grimshaw, R.H.J. (2002). Atmospheric internal solitary waves. Chapter 3 in *Environmental Stratified Flows* (ed R.H.J Grimshaw), Series: *Topics in Environmental Fluid Mechanics*, Vol. 3, Springer Verlag, Heidelberg, Germany.

Schlichting, H. (1979). *Boundary Layer Theory*, 7th ed., McGraw-Hill, New York.

Stanton, T.P. and Ostrovsky, L.A. (1998). Observations of highly nonlinear internal solitons over the continental shelf. *Geophysical Research Letters*, 25 (14), 2695–2698.

Stastna, M. and Lamb, K.G. (2002). Vortex shedding and sediment resuspension associated with the interaction of an internal solitary wave and the bottom boundary layer. *Geophysical Research Letters*, 29 (11), 1512.

Stastna, M. and Lamb, K.G. (2008). Sediment resuspension mechanisms associated with internal waves in coastal waters. *Journal of Geophysical Research*, 113, C10016.

Strang, E.J. and Fernando, H.J.S. (2001). Entrainment and mixing in stratified shear flows. *Journal of Fluid Mechanics*, 428, 349–386.

Sveen, J.K. and Dalziel, S.B. (2005). A dynamic masking technique for combined measurements of PIV and synthetic schlieren applied to internal gravity waves. *Measurement Science & Technology*, 16, 1954–1960.

Sveen, J.K, Guo, Y., Davies, P.A. and Grue, J. (2002). On the breaking of internal solitary waves at a ridge. *Journal of Fluid Mechanics*, 469, 161–188.

Thiem, Ø., Carr, M., Berntsen, J. and Davies, P.A. (2011). Numerical simulation of internal solitary wave-induced reverse flow and associated vortices in a shallow, two-layer fluid beneath benthic boundary layer. *Ocean Dynamics*, 61, 857–872.

Thorpe, S.A. (2005). *The Turbulent Ocean*. Cambridge University Press, Cambridge, UK.

Troy, C.D. and Koseff, J.R. (2005). The instability and breaking of long internal waves. *Journal of Fluid Mechanics*, 543, 107–136.

Van Gastel, P., Ivey, G.N., Meuleners, M.J., Antenucci, J.P. & Fringer, O. (2009). The variability of the large-amplitude internal wave field on the Australian North West Shelf. *Continental Shelf Research*, 29, 1373–1383.

Vlasenko, V., Stashchuk, N. & Hutter, K. (2005). *Baroclinic Tides: Theoretical Models and Observational Evidence*, Cambridge University Press, Cambridge, UK.

Part 5

Gravity currents

Chapter 18

Optical methods in the laboratory: An application to density currents over bedforms

J. Ezequiel Martin[1], *Tao Sun*[2] *and Marcelo H. García*[3]

[1] *Iowa Institute of Hydraulic Research-Hydroscience and Engineering, University of Iowa City, Iowa, USA*

[2] *ExxonMobil Upstream Research Co., Houston, Texas, USA*

[3] *Ven Te Chow Hydrosystems Laboratory, Department of Civil Engineering, University of Illinois at Urbana-Champaign, Illinois, USA*

ABSTRACT: Experimental results on density currents flowing over bedforms are presented, with the dual objectives of describing our findings for these flows as well as highlighting the strengths and challenges of the optical techniques used. Particle Image and Particle Tracking Velocimetry (PIV; PTV) provide the spatial and temporal resolution to examine the velocity field; Laser Induced Fluorescence (PLIF) does the same for the density field. Simultaneous measurements of these techniques – that resulted in the development of novel measuring and processing methodology – are extensively used to characterize the flow. Evidence regarding the effect of bed morphology on the mixing and entrainment capacity of the flow, is presented for an experiment designed with this objective, corroborating field results that indicate such effect.

1 INTRODUCTION AND MOTIVATION

There is an incomplete understanding of the effect of bedforms on the density and turbidity currents. While the general behavior of a gravity current can be described successfully using simplified one dimensional models, such description is dependent on the correct parameterization of the flow. These simplified models reduce the complexity of the phenomenon to few parameters, of which by far the most important are the friction coefficient between the current and the bed and the entrainment coefficient between the current and the ambient fluid. This type of parameterization produces a "hydraulic" or layer-averaged description of the current, with all the advantages that such models provide to the characterization of the flow. Typical examples of these models are those presented by Parker *et al.* (1986) that include a continuity equation, a mass conservation equation for the solute, a momentum conservation equation and an equation for the turbulent kinetic energy. The first three equations represent the simplest possible model, while the addition of a fourth equation allowed the authors to constraint the finding of their simplified model to realizable cases. This particular model was used to characterize the possibility of self-acceleration of the current; many authors (just to name two examples, see García and Parker, 1993; Kostic and Parker, 2006) have used similar models to investigate other characteristics of the flow with a tool that provides a straightforward parameterization in terms of a few quantities. It is easy to see then that the reliability of such parameters is paramount for the development and application of such models. The characterization of momentum exchange between the flow and the bed can be given, in terms of friction coefficient, bed shear stress, etc. Equivalently important is the characterization of the mass exchange between the current and the ambient, for which there is also a number of possible parameterizations, with an entrainment velocity or coefficient being the most commonly used. Many experimental works are available that deal with one or both of these quantities. The complexity of the problem however means that none of such works can be considered definitive for all flows and conditions, in particular due to differences in turbulence levels between laboratory and field scales, and to the limited amount of information that can be collected for field cases.

To tackle these difficulties, detailed measurements of density currents over different bed conditions were conducted, with particular emphasis on measurements of currents over bedforms.

2 FACILITY AND EXPERIMENTAL METHOD

Both the velocity and the concentration of the species causing the density difference are required to characterize a stratified flow. In the following sections a brief description of the measuring techniques used, as well as a brief description of the facility used is presented.

2.1 Description of the facility

The Gravity Current Front Flume in the Ven Te Chow Hydrosystems Laboratory is a recirculating flume, with a conveyor covering part of the bed and it has been used primarily to study arrested density current fronts (García and Parsons, 1996; Parsons and García, 1998; Martin and García, 2009): for the present study it has been modified to include a ramp with a 5% slope, a headbox at its upstream end to provide a steady dense water influx (fed from a constant head tank), and modified drainage of the main reservoir to keep the water level constant in the channel. The channel is 3 meters in length, 0.3 m wide and it has a maximum depth of 0.6 m. This facility is already fully instrumented to measure stratified flows, using a combination of Particle Image Velocimetry (PIV) and Plane Laser Induced Fluorescence (PLIF).

Measurements conducted in the flume include salinity density currents over smooth and rough bed, and over bedforms. The bedforms are constructed with individual triangular wedges and then covered with antiskid tape; this setup allows for an easy modification of height to length modification of the bedforms, as well as other characteristics such as actual length and asymmetry of the bedform. Results presented herein correspond to symmetric triangular bedforms, with a total length of 0.4 m and height of 0.01 m, which combined with the 5% slope the channel, resulted in approximately horizontal bed on the stoss side of the antidune, and twice the channel slope on the lee side of the bedform.

2.2 Experimental technique

Stratified flows present unique challenges to the experimental researcher. Flows with sharp to moderately sharp interfaces are easily disrupted by intrusive instrumentation, as they can be affected by blockage, and by additional mixing due to vortex shedding from the instrument. For gravity currents, another concern is the rather small extension of the un-mixed region of the current body, that in a typical laboratory setup is usually a small fraction of the total depth and does not exceed a few centimeters in height. For these reasons, non-intrusive measuring techniques present great advantages over intrusive ones. They also present unique challenges in the case of stratified flows, as the ideal conditions for optical methods are not usually met by these flows. For completely soluble species, such as salts, miscible liquids and also temperature, changes in the index of refraction affect optical measurements. Changes in refractivity can be alleviated to some degree by using two solutions of equivalent index of refractivity and different density as the current and ambient fluids. In the case of solid particles, in general the main concern is the attenuation of the signal as the microscopic characteristic of the fluid (density, index of refraction) are assumed unchanged for low to moderate sediment concentration. In this case, optical methods are typically severely affected and of limited applicability.

2.2.1 *Particle image velocimetry and particle tracking velocimetry*

Particle Image Velocimetry (PIV) and Particle Tracking Velocimetry (PTV) are two closely related velocimetry methods, and for the most part the acquisition process is indistinct, with differences only in the processing procedure. The basic idea of optical velocimetry methods is to capture a series of images of the flow and translate the information on those images to a velocity field. If the flow does not present any characteristics that can be traced, a seeding agent is included. The seeding particles need to be small enough and of similar density to the fluid to follow the flow; however too small particles present other challenges such as not scattering enough light or generating a particle image that is too small and produces "peak-locking" (i.e., pixel displacements for the particles are "locked" to integer values). It has been shown (Raffel *et al.*, 1998) that spherical particles producing a particle image with a diameter of at least 2–3 pixels reduce these effect. It is also necessary to illuminate the particle to captured their images; lasers are typically used for the high intensity illumination required by such small particles. The use of lasers also allows restricting the illuminated region of the flow to a small volume, usually a thin plane (although volumetric PIV works on a cube). The illumination is done as a burst, rather than continuously, providing a sharp image at a given time. Finally, this image is capture by a high sensitivity camera, that it is also capable of recording images at the required interval. PIV cameras are modified to capture two images in rapid succession (time intervals of microseconds), and velocity fields are estimated to first order, as the displacement of the particles in the shooting interval. In standard PIV the velocity field is calculated using a cross-correlation technique between the images, while in PTV individual particles are tracked.

For the particular case of stratified flows, the main challenge for PIV techniques is the effect of changes of the refraction index of the fluid. The refractivity index of

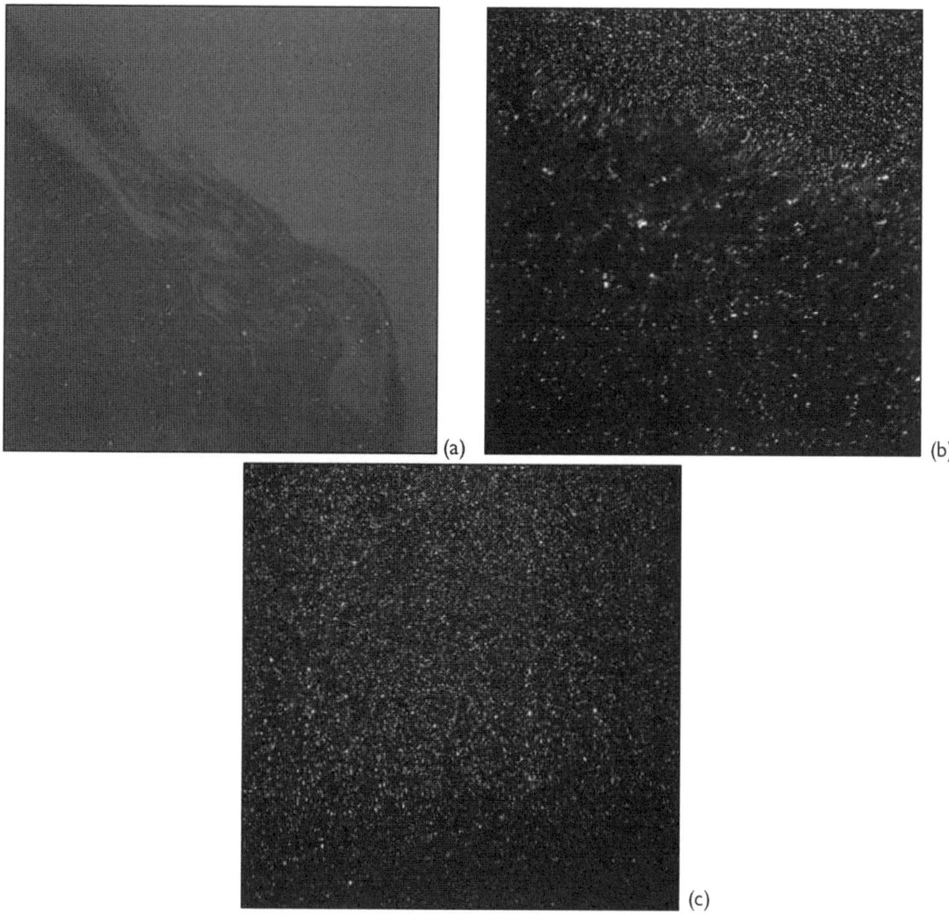

Figure 1 Examples of PIV images for stratified flows: (a) Saline density current front, with ethanol added to the ambient water to match refractivity index of the media. Background colors correspond to PLIF image captured simultaneously; (b) Density current front, without refractivity index matching; (c) body of the current shown in (b).

any solution is typically larger than that of pure water, and varies with concentration. Sharp changes of refraction index cause distortion of the particle images, resulting in larger particles diameters and typically smaller light intensity. The position of the particle also becomes uncertain, which results in erroneous velocity measurements as an apparent displacement due to the path of the scattered light is added to the actual particle displacement. Matching the index of refraction can alleviate this effect. This is particularly important in regions of the flow where strong interfaces occur and also when these interfaces are subject to shear resulting in very convoluted surfaces, as it increases the number of discontinuities in refractivity index that the scattered light must traverse. An example of this type of flow is the head of a density current: Figure 1 shows examples of images with and without matching conditions. The image in the case of matched conditions is virtually unaffected by the density changes and can be processed as any image from a case with uniform fluid properties. On the other hand, the deterioration of the particles' image for the unmatched case is so severe that measuring a velocity is not possible in the mixed region. It is worth mentioning that this "mixed" region is not mixed at all and is instead a region of strong discontinuities. As it can be observed in the last image of Figure 1, when the fluid properties change gradually, it is possible to obtain images with minimal distortion, and matching the index of refraction is not as critical. This is important as the cost and complexity of matching the index of refraction makes it prohibitively expensive for large facilities.

2.2.2 *Plane laser induced fluorescence*

While PLIF measures concentration and PIV velocity, the setup and equipment used is very similar, and in fact under certain circumstances it can be the same. For instance, Figure 1(a) in the previous section, shows a combined PLIF/PIV image. The implementation of the combined technique was developed previously for the study of density current fronts (Martin and García, 2009). The basic idea of PLIF is to introduce a florescent marker (usually a dye, such as Rhodamine) in known concentrations in the system and relate the measured intensity of light recorded by the camera to its concentration. Several challenges exist, as described in Martin and García (2009): attenuation of the signal; filtering of the PIV particles signal, if combined measurements are taken; variability of the source light intensity. Most of these effects can be easily corrected in the case of a density current front as two distinct areas of unmixed saline and ambient water occurred; for the case of the body of the current for which the saline concentration is unknown, a possible solution is to use a siphon to obtain a local average concentration to calibrate the PLIF measurements. Moreover, the usefulness of PLIF measurements is greatly reduced if the index of refraction of the solutions is not matched, and therefore PLIF measurements are only considered in this study for matched conditions only.

3 RESULTS

Velocity profiles are the most important piece of information to characterize the flow, as they are used to estimate friction and entrainment coefficients, as well as general parameters such as Reynolds and Richardson numbers. Different methods can be used to determine such quantities. In the following paragraphs a brief description of the profiles along with the methodology to determine the coefficients is given.

Figures 2 and 3 show examples of measurements with the Particle Image Velocimetry technique. In the case of flat bed, variations in the longitudinal direction can be neglected within the field of view of the camera, and the instantaneous velocity field can be averaged in time and in this direction, resulting on a unique profile for each condition. In the case of flow over bedforms, local variations near the bedform occur, but it is expected that at a certain distance from the bottom the flow is again quasi-uniform in the longitudinal direction and can also be averaged to produce a single profile. Examples of such profiles are presented in Figure 2. The local variation due to the presence of the bedform can also be observed on Figure 3: no separation of

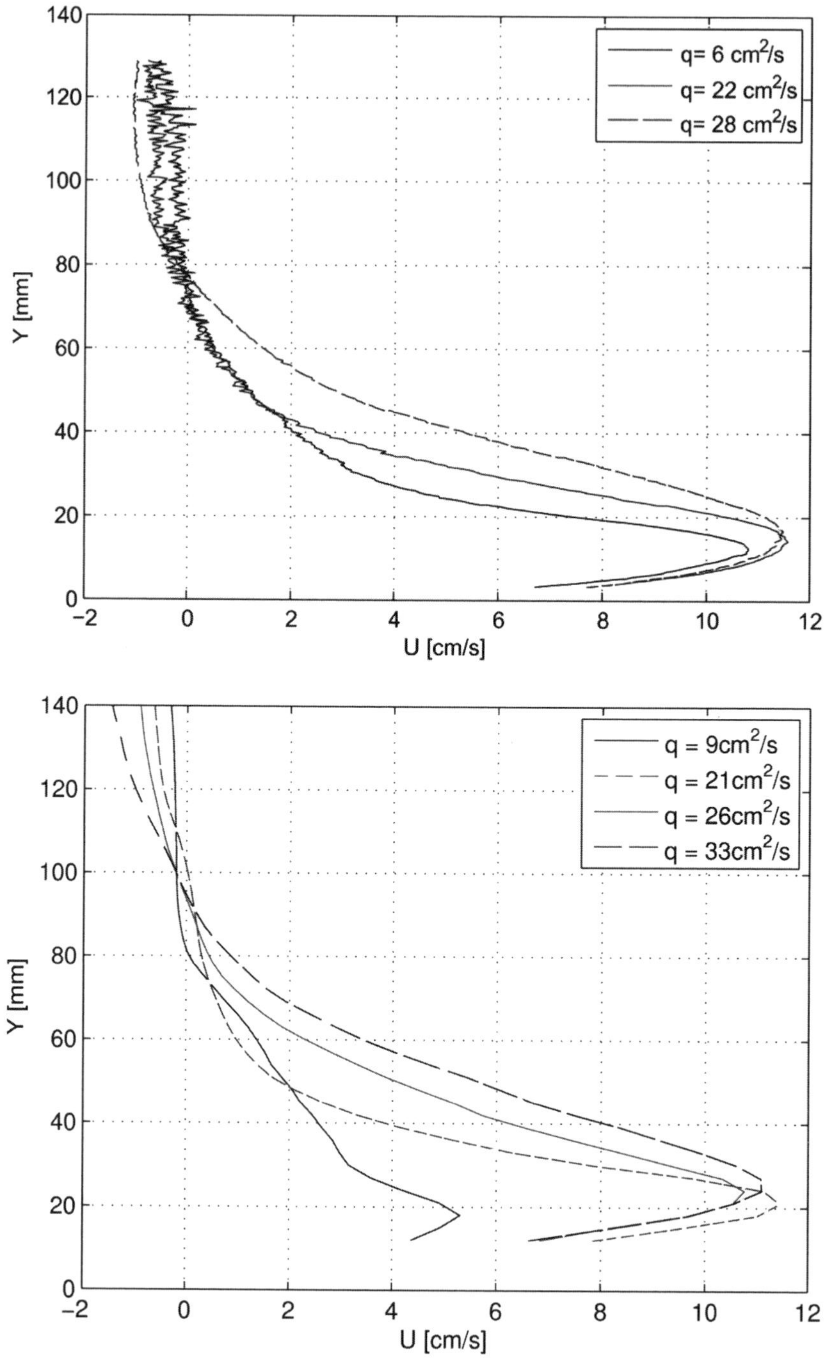

Figure 2 Examples of velocity profiles measured with PIV technique. Top: Smooth flat bed; Bottom: Flow over symmetric bedforms, with a height to length ratio of 0.05. The initial density difference in all cases is about 0.7%.

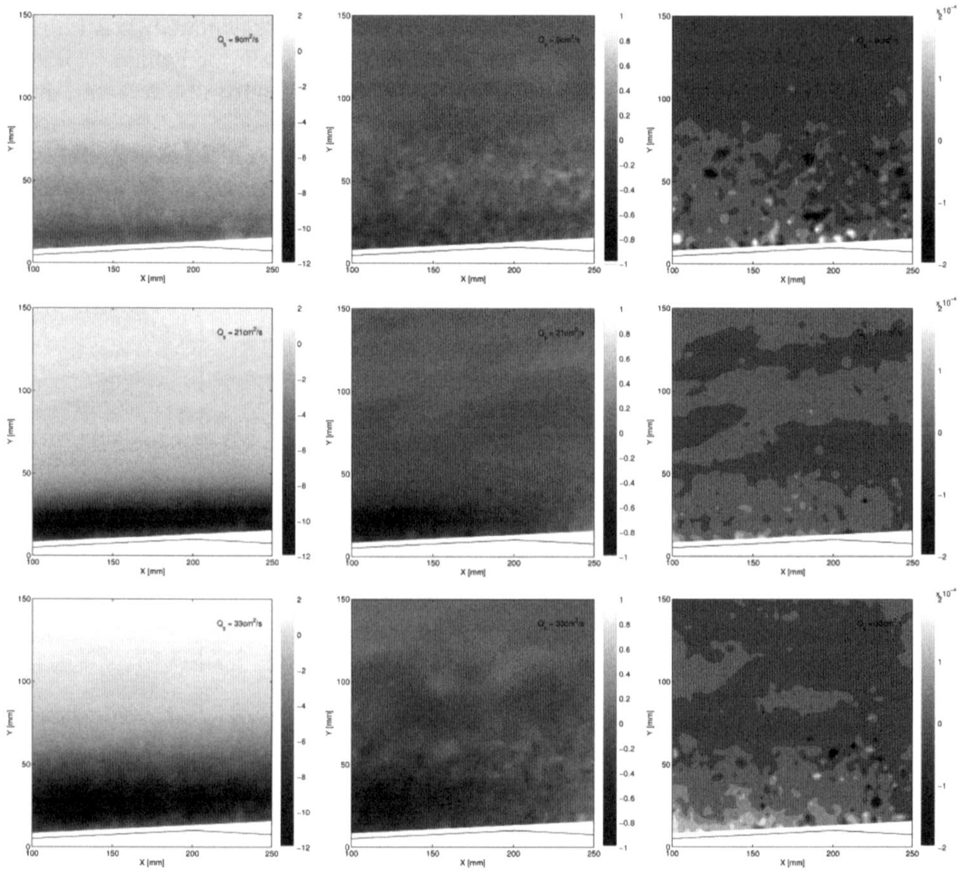

Figure 3 Examples of velocity profiles measured with PIV technique: from left to right, the time average horizontal velocity field, vertical mean velocity field and the Reynolds stress $\overline{u'v'}$, are shown. The flow rate for each case is indicated on the figures. The cases correspond to the mean profiles shown in Figure 2(b). The coordinates system is consistent with the overall slope of the flume, while the position of the bedform is schematically shown.

the flow is observed on the lee side of the bedform, as the height to length ratio of the bedform is only 5%, and the profile is not strongly affected away from the bed, as it can be observed in Figure 2. As the density difference for all the cases is nominally the same, changes in the strength of the stratification and the flow velocity are introduced by the initial discharge. For the lowest discharge ($q = 9$ cm^2/s) the flow follows the shape of the bedform, in the same way as a supercritical unstratified flow maintains an approximately constant depth when flow over anti-dunes. These flows present a large Richardson number and from the point of view of application to the field are of less interest than more energetic flows, as they are strongly affected by the scale of the model.

3.1 Friction coefficient

Figure 4 shows an example of how the friction coefficient is calculated. It is known that boundary layers flows present a region near the wall for which a linear approximation to the turbulent Reynolds stress $\overline{u'v'}$ is valid. It can be shown that in that case the friction velocity is related to $\overline{u'v'}$ by

$$\overline{u'v'} = \alpha y + u_*^2 \tag{1}$$

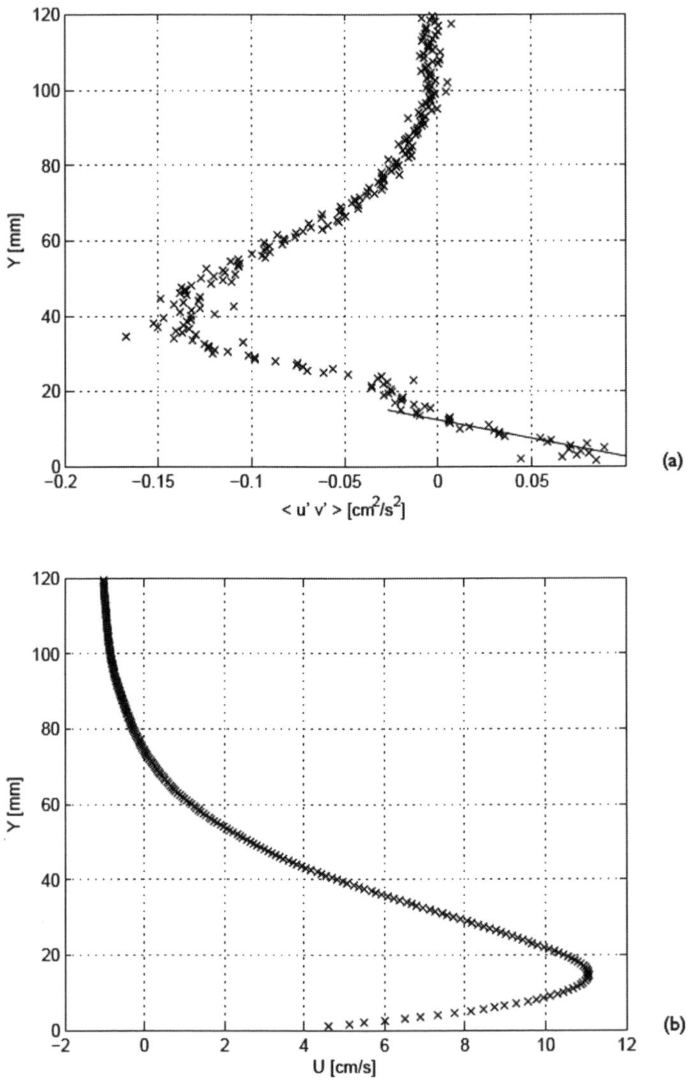

Figure 4 Estimation of the friction velocity u_* (a) Reynolds stress, $\overline{u'v'}$, showing the linear fitting near the bed used to estimate u_* as defined by Equation (1). The mean longitudinal velocity profile is also shown for reference.

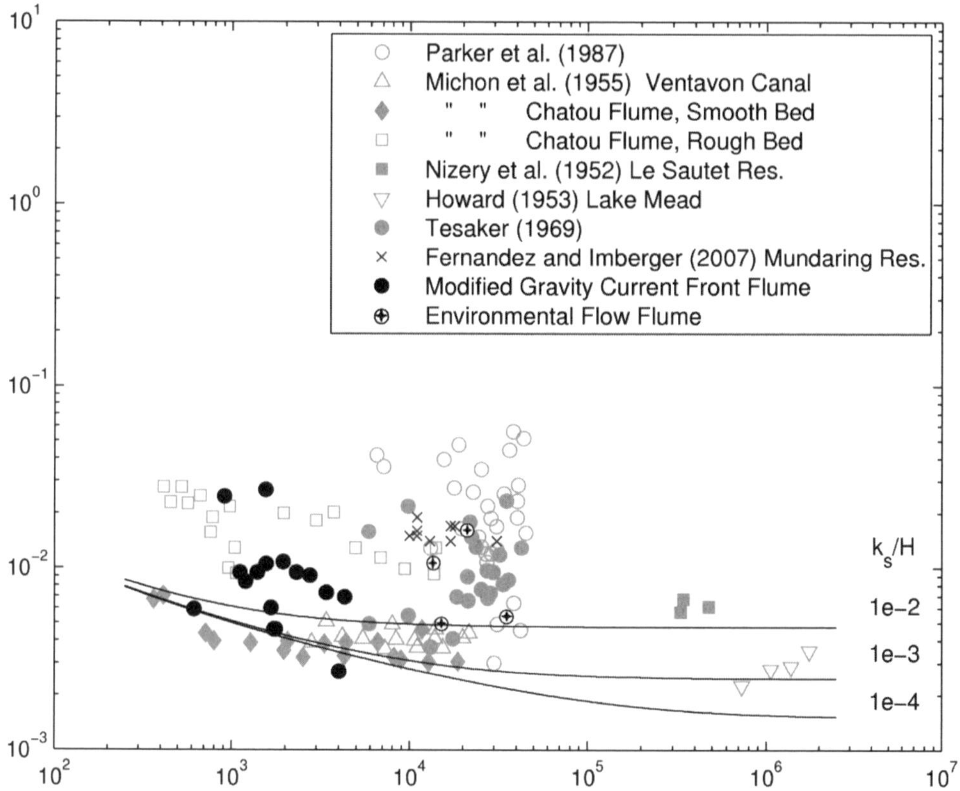

Figure 5 Friction coefficient as a function of the Reynolds number. Measurements from the present study are shown as solid blank circles, while previous results are shown in grey and data from another facility at the Ven Te Chow Laboratory are shown as circled plus signs. Datasets 1 through 7 are adapted from Parker *et al.* (1987), while newer field data from Fernandez and Imberger (2007) was obtained directly from that source. No distinction between roughness conditions or presence of bedforms is considered. As a reference, three predicted relationships (based on the Colebrook-White equation) are given for different roughness.

It is also found that assuming the validity of logarithmic law of the wall for the current gives good estimates of the friction velocity, and in those cases for which the length of the dataset is not long enough to produce accurate turbulent quantities, this method is also used to determine u_*.

Figure 5 summarizes some of the available information regarding the bottom friction coefficient for density and turbidity currents; a large dispersion of the data is noticeable. While part of this dispersion is due to the presence of bedforms, large values of friction coefficient appear for flat rough bed, in particular at transitional Reynolds numbers (for example, the data from Michon *et al.* (1955), and the highest *CD* measurements for the present set of data). The few data points at very large Reynolds number occur at ranges consistent with friction coefficients from unstratified open channel flow, but without further information on the bedforms that might develop in such flows it is difficult to determine if such parameterization is valid. From studies in unstratified flows, it is found that the added resistance due to the bedforms can exceed in up to an order of magnitude the skin friction; a similar variability is found in the case of gravity currents. It is important to notice that the parameterization presented in Figure 5 (adapted from Colebrook-White equation) does not include the effect of bedforms, just grain roughness: A shear stress partition of the skin and drag from is necessary to account the total friction coefficient.

3.2 Entrainment coefficient

The average entrainment coefficient $\overline{e}_w$ over a reach can be found by comparing the velocity profiles at the beginning and end of the channel reach, from a simple mass balance, as the change in layered averaged horizontal flux UH over the distance L of the reach (by dividing the result by the average velocity over the reach, a dimensionless entrainment coefficient rather than a entrainment velocity is obtained):

$$\overline{e}_w = \frac{\Delta(UH)}{L\bar{U}} \tag{2}$$

It is also possible, in theory, to determine the entrainment directly at a single location, from the vertical velocity, but in practice the typically small values of the entrainment velocity combined with the possibility of long period oscillations in the body interface, hindered that approach.

Equivalently to the information presented in Figure 5, Figure 6 summarizes the available results regarding the entrainment coefficient of ambient water by gravity currents. Data from previous research show that while most laboratories studies of gravity currents over flat bed follow the predictors based on the Richardson number, such as for instance,

$$e_w = \frac{0.075}{\left(1 + 718Ri^{2.4}\right)^{0.5}} \tag{3}$$

Field data however appears to indicate conditions for which a higher amount of mixing exists. It has been argued (Parsons and García, 1998) that mixing at the front of a density current can be hindered by scale effects; it is not unreasonable to expect a similar effect on the rest of the flow as observed for the case of salt wedges by Sargent and Jirka (1987). Data from Martin (2009) for the front of the current and from Strang and Fernando (2001) show that this relationship is by no means unique to all shear stratified flows. In particular, it should be noticed the existence of a critical Richardson number for which the entrainment is maximum. The effect of bedforms is again not considered in this figure, and all available data is included.

3.3 Shape factors

As it has been shown in the previous paragraph, the experimental determination of the entrainment coefficient for a current in a steady state does not require the

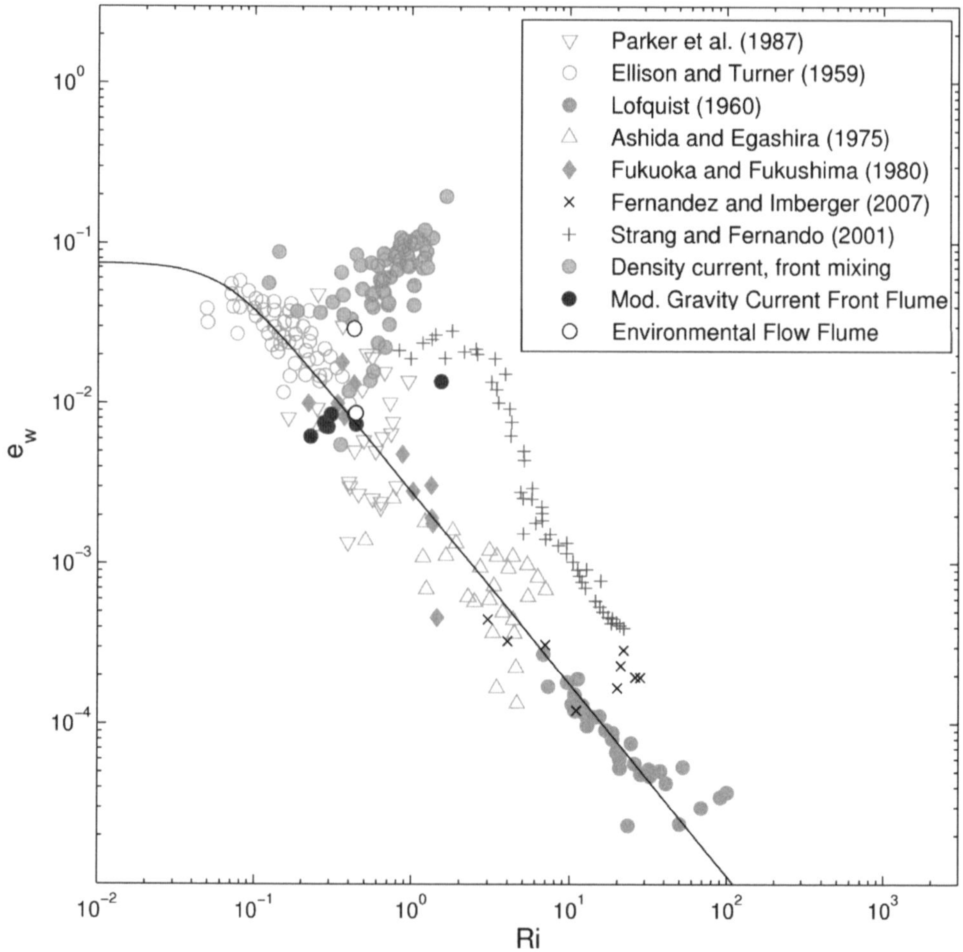

Figure 6 Entrainment coefficient as a function of the bulk Richardson number. Measurements from the present study are shown as solid black circles, while previous results are shown in grey. Results for other facilities at the Ven Te Chow Laboratory are also presented. Datasets 1 through 5 are adapted from Parker *et al.* (1987), while newer field data from Fernandez and Imberger (2007) was obtained directly from that source. Data from Strang and Fernando (2001) for a related experimental set up is also included. No distinction between roughness conditions or presence of bedforms is considered. The solid line corresponds to Equation (3), from Parker *et al.* (1987).

measurement of the density profile. The shape factors S_1 and S_2, as defined in the relation between the normal Richardson number and the entrainment coefficient (Parker *et al.*, 1986):

$$Ri_n = \frac{e_w + c_f}{S_1 S + S_2 / 2e_w} \tag{4}$$

do require the measurement of density profiles as they are defined by:

$$S_1 = \frac{U}{\Phi}\int_0^{\text{¥}} (\rho/\rho_0 - 1)dz \qquad (5)$$

and

$$S_2 = \frac{2U}{\Phi H}\int_0^{\infty}\int_z^{\infty} (\rho/\rho_0 - 1)dz'dz \qquad (6)$$

With U, H and Φ the layer averaged values of velocity, current depth and buoyancy flux. Experimental values for S_1 and S_2 have been reported respectively in the range [0.6–1.2] and [0.2–1.1] (Choi & García, 2002), justifying the approximation $S_1 = S_2 = 1$ usually found in layer-averaged modeling. Results for some of the conducted experiments are shown in the next section.

3.4 Effect of bedforms

In the previous paragraphs, the results for C_D and e_w were considered without distinguishing whether bedforms were present or not. As the effect of bedforms can be subtle in some of the calculated parameters, we restrict our analysis to experiments were all other variables except the bed condition are nearly identical. Another advantage of this set of data is that the concentration fields are also available with good resolution, which allows the characterization of the shape factors S_1 and S_2.

The velocity profiles over bedforms present significant variations near bed from those for flat bed, as shown in Figure 2. The profile is approximately linear instead of logarithmic, as non-homogenous regions are averaged. It is reasonable to expect from this condition that the form drag contributes significantly to the bed shear. Both the velocity and the concentration profiles are shifted upwards: this is consistent with the velocity reduction near the bed that generates a thicker unmixed region for the flow over bedforms. The limit between the unmixed core and the mixing region is given approximately by the position of maximum velocity. The density is assumed constant in the core region, and this assumption is corroborated by extraction of fluid samples, but cannot be verified from the optical measurements as bed reflections affect the PLIF measurements.

Table 1 summarizes the relevant parameters measured for those experiments conducted to determine the effect of bedforms on the entrainment coefficient. The drag coefficient is not included for these experiments as all the flows correspond to transitional conditions and the variability is high.

The measured entrainment coefficients for the case of flat bed are in good agreement with the prediction given by Equation (3); those for flow over bedforms are larger than the prediction. Even though the difference is small, they indicate an effect of the antidunes on the mixing, in particular as the Richardson number increases for those cases, for which Equation (3) predicts a reduction rather than an increase of e_w.

The values of the shape factors are consistent with those reported previously (Choi and García, 2002). No effect can be observed due to the presence of

Table 1 Calculated parameters for the two bed conditions considered. Ri_0 is the entry Richardson number, Ri is the calculated Richardson number in the measured region.

Bed	$q_0 = UH$ [cm²/s]	Ri_0 –	*Re* –	*Ri* –	$\overline{e_w}$ [Eq. 3]	e_w [Eq. 3]	S_1 [Eq. 5]	S_2 [Eq. 6]
Smooth, flat	16	0.42	2500	0.39	0.008	0.008	0.86	0.53
	21	0.36	3080	0.29	0.008	0.008	0.92	0.64
	28	0.3	3940	0.36	0.010	0.009	0.98	0.84
Antidunes	21	0.47	2920	0.45	0.009	0.007	1.04	0.79
	26	0.42	3490	0.62	0.010	0.008	0.98	0.67
	33	0.28	4110	0.58	0.010	0.007	1.00	0.74

bedforms. In particular S_1 is close to unity suggesting that the adimensional profiles of density and velocity are similar. This condition is consistent with comparable turbulent mixing for both fields; while this relation might not be true at large values of Ri, most cases of interest usually present moderate values of Richardson number.

4 SUMMARY

The results presented show that the effect of bedforms on gravity currents can be significant, not only because these features alter the drag, as is the case for unstratified flows, but also because the water entrainment capacity can be increased. A parameterization of the mixing parameters has not been attempted, first because the dataset is too small, but more importantly because such description is likely to be dependent on the characteristics of the bedform, which in term are affected by the current due to its transport capacity. A first attempt to characterize the interaction between flow and the resulting bed morphology was presented in Sequeiros (2007), but further work is needed. The importance of a correct parameterization of simplified layered average models for these flows should be stressed, as this type of models are realistically the only models available for the simulation of field cases with very large (i.e., geological) timescales.

REFERENCES

Choi, S. and García, M.H. 2002. k-ε Turbulence Modeling of Density Currents Developing Two Dimensionally on a Slope. *Journal of Hydraulic Engineering*, Vol. 128 (1), 55–63.

Fernandez, R.L. and Imberger, J. 2006. Bed roughness induced entrainment in a high Richardson number underflow. *Journal of Hydraulic Research*, 44 (6), 725–738.

García, M. and Parker, G. 1993. Experiments on the entrainment of sediment into suspension by a dense bottom current. *Journal of Geophysical Research*, 98(C3), 4793–4807.

García, M. and Parsons, J. 1996. Mixing at the front of gravity currents. *Dynamics of Atmospheres and Oceans*, 24 (1–4), 197–205.

Kostic, S. and Parker, G. 2006. The response of turbidity currents to a canyon-fan transition: internal hydraulic jumps and depositional signatures. *Journal of Hydraulic Research*, 44 (5), 631–653.

Martin, J.E. 2009. Particle image study of density current fronts. Ph.D. thesis, University of Illinois at Urbana-Champaign.

Martin, J.E. and García, M.H. 2009. Combined PIV/PLIF measurements of a steady density current front. *Experiments in Fluids*, 46 (2), 265–276.

Michon, X., Goddet, J. and Bonnefille, R. 1955. *Etude Théorique et Expérimentale des Courants de Densité*. Lab. Natl. Hydraulique, Chatou, France.

Parker, G., Fukushima, Y. and Pantin, H.M. 1986. Self-accelerating turbidity currents. *Journal of Fluid Mechanics*, 171, 145–181.

Parker, G., García, M. Fukushima, Y. and Yu, W. 1987. Experiments on turbidity currents over an erodible bed. *Journal Of Hydraulic Research*, 25 (1), 123–147.

Parsons, J.D. and García, M.H. 1998. Similarity of gravity current fronts. *Physics of Fluids*, 10 (12), 3209–3213.

Raffel, M., Willert, C.E. and Kompenhans, J. 1998. *Particle Image Velocimetry A Practical Guide*. Springer Verlag.

Sargent, F.E. and Jirka, G.H. 1987. Experiments on Salt Wedges, Journal of Hydraulic Engineering, 113 (10), 1307–1324.

Sequeiros, O.E. 2008. Bedload transport, self-acceleration, downstream sorting, and flow dynamics of turbidity currents. Ph.D. thesis, University of Illinois at Urbana-Champaign.

Strang, E.J. and Fernando, H.J.S. 2001. Vertical mixing and transports through a stratified shear layer. *Journal of physical oceanography*, 31 (8), 2026–2048.

Chapter 19

Extinction of near-bed turbulence due to self-stratification in turbidity currents: The dependence on shear Reynolds number

Mariano I. Cantero[1], S. Balachandar[2], A. Cantelli[3] and Gary Parker[4]

[1] *Bariloche Atomic Center, Institute Balseiro, San Carlos de Bariloche, Argentina*
[2] *Department of Mechanical and Aerospace Engineering, University of Florida, Gainesville, Florida, USA*
[3] *Shell International Exploration and Production, Houston, Texas, USA*
[4] *Department of Civil and Environment Engineering and Department of Geology, University Illinois Urbana-Champaign, Illinois, USA*

ABSTRACT: Turbidity currents are dense bottom flows driven by suspended sediment that occur in lakes and the ocean. Turbidity currents are self-stratifying in that the agent of the density difference, i.e., sediment, must be maintained in suspension if the flow is to be sustained. It has recently been shown using Direct Numerical Simulation that under appropriate conditions, the upward normal profile of suspended sediment may show a gradient sufficient to cause the extinction of turbulence near the bed. This extinction creates conditions favorable to the emplacement of massive turbidites, i.e., sediment deposits which show no evidence of reworking by the flow. The results were established using a shear Reynolds number of 180. Here the question of Reynolds invariance is studied by repeating the calculations with a shear Reynolds number of 400.

1 INTRODUCTION

Turbidity currents are dense bottom underflows driven by suspended sediment. These currents commonly occur in the oceans and lakes. They are stratified flows that fall into the same class as thermohaline bottom underflows, e.g., the massive flow emanating from the Bosporus onto the floor of the Black Sea (Flood *et al.*, 2008). Turbidity currents differ from thermohaline underflows, however, in an important way. In the latter case the agents of the density difference (between the flow and the ambient water) that drive the flow downslope are heat and salt, i.e., conserved quantities. In the former case, however, the agent is suspended sediment. The flow must maintain this sediment in suspension in order to sustain itself; were the sediment to settle out, the driving force for the flow itself would be lost.

Turbidity currents can also be considered to be subaqueous analogs of rivers laden with suspended sediment. In equilibrium river flow, suspended sediment concentration typically declines in the upward-normal direction. In so far as this sediment renders the water-sediment mixture heavier than sediment-free water, such a profile creates stable density stratification. It might be expected that a sufficiently strong upward normal gradient in suspended sediment concentration would damp turbulence, even in the case of dilute sediments. Indeed, this turns out to be the case for rivers, as noted by, among many others, Wright and Parker (2004).

Turbidity currents also tend to be stably stratified, with concentration decreasing upward from the bed. Turbidity currents and rivers are both examples of self-stratifying flows, i.e., they maintain their density variation in the vertical by suspending sediment that would otherwise reside on the bed. In the case of a river, stratification effects can damp near-bed turbulence, but cannot bring the flow itself to a halt. This is because nearly all of the driving force for the flow derives from gravity acting on the water phase itself.

Turbidity currents, however, are fundamentally much more fragile than rivers in regard to stratification. Since the driving force is obtained solely from gravity acting on the suspended sediment, sufficient damping of near-bed turbulence can prevent the re-entrainment of sediment as it settles onto the bed. Thus over time, the current can die as it loses the suspended sediment necessary to drive it. Turbidity currents are similarly more fragile than thermohaline underflows, for which turbulence suppression can relaminarize the flow, but not erase the driving force itself.

Recently Cantero *et al.* (2009, 2011) have performed numerical studies of turbidity currents using Direct Numerical Simulation (DNS) of the Navier-Stokes equations of fluid flow and associated equations describing mass conservation of a dilute suspension of sediment. They demonstrated that conditions do indeed exist for which near-bed turbulence is extinguished by self-stratification effects, leading to the sequestration of the suspended sediment at the bed. They showed that as turbidity currents flow over sufficiently low slopes, or in channels that are sufficiently wide, a broad range of conditions exist for which near-bed turbulence is extinguished. The transition from a flow with active turbulence to one with extinguished near-bed turbulence can be realized in terms of flow over a slope that decreases downstream, or flow within a channel that widens downstream.

Cantero *et al.* (2011) related near-bed turbulence collapse to the commonly-observed deep-sea deposit known as the "massive turbidite," i.e., a layer of sediment deposited by a turbidity current that shows little or no sign of being reworked by the flow as it was emplaced (Bouma, 1962). The lack of reworking structures such as ripples or laminations suggests that near-bed turbulence would have been very weak when the sediment was deposited. Cantero *et al.* (2011) applied their results to several field-scale cases to demonstrate the viability of the hypothesis of turbulence extinction as a cause of the emplacement of massive turbidites.

A major limitation of the analysis, and indeed any DNS analysis, is the fact that fully turbulent flows with only relatively small Reynolds numbers can be considered. Cantero *et al.* (2009) verified that in the absence of stratification effects, their calculations yielded fully turbulent flow obeying standard relations (such as the logarithmic velocity profile), so indicating behavior that was at or near the regime for near-invariance in the Reynolds number.

This notwithstanding, a repeat of the analysis at a substantially higher Reynolds number is an important step toward verifying the generality and field applicability of the results. Here this is performed using a Reynolds number that is over 2.2 times the value used in the original analysis.

2 TURBIDITY CURRENT WITH A ROOF

River flows laden with suspended sediment can be easily modeled in terms of a simple but insightful configuration; steady 2D flow that is uniform in the streamwise (x) direction, and shows variation in quantities averaged over turbulence in the upward normal (z) direction only. No such reference configuration exists for turbidity currents. This is because even in the case of 2D steady flow, entrainment of ambient water at the interface forces variation in the streamwise direction.

Cantero *et al.* (2009) introduced the artifice of the Turbidity Current with a Roof in order to overcome this limitation. As shown in Figure 1, the flow is driven entirely by the action of the downslope component of gravity on the suspended sediment, but a roof prevents the entrainment of ambient water. This configuration allows for flows that, when averaged over turbulence (AOT), are steady and uniform in the streamwise direction. The upward normal variation in concentration necessary to drive stratification effects is, on the other hand, captured by this configuration.

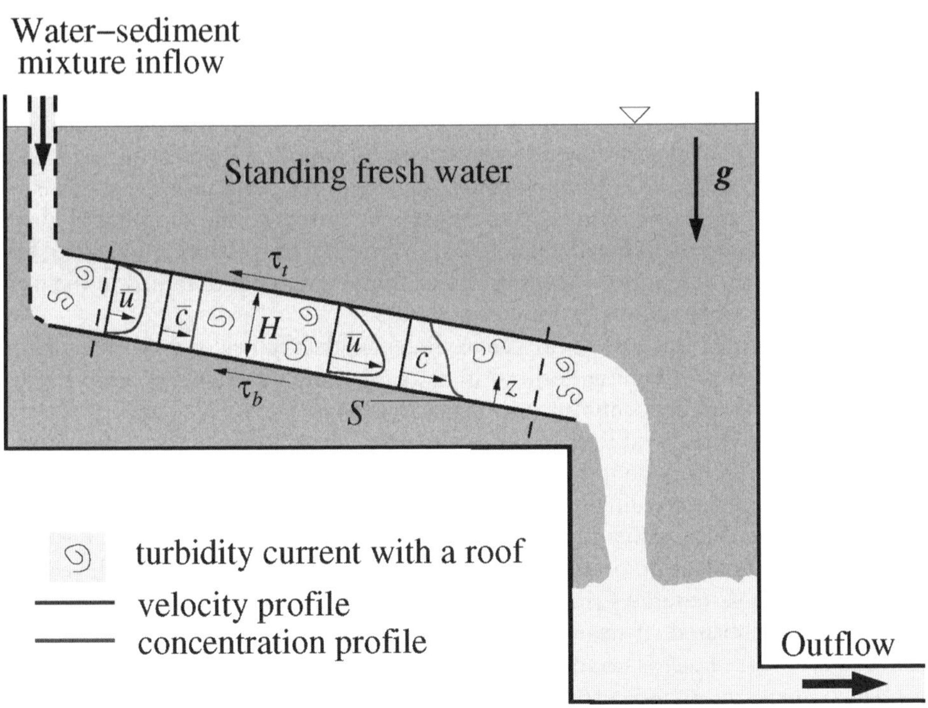

Figure 1 Definition diagram for turbidity current with a roof (TCR).

Relevant parameters are defined as follows. The gap height between the bed and roof is H, the bed and roof (top) shear stresses (AOT) are τ_b, and τ_t, respectively, and the channel slope is S. The suspended sediment has fall velocity V_s and density ρ_s; the associated submerged specific gravity R of the sediment (e.g., ~ 1.65 for quartz) is given as $(\rho_s-\rho)/\rho$, where ρ denotes the density of water in the absence of suspended sediment. The profiles of volume concentration of suspended sediment and streamwise flow velocity (both AOT) are denoted as $\bar{c}(z)$ and $\bar{u}(z)$ respectively. The suspension is assumed to be dilute, so that $\bar{c}(z) << 1$. The layer-averaged volume suspended sediment concentration is denoted as C. Bed, roof and nominal shear velocities u_{*b}, u_{*t}, and u_{*n}, are defined by the respective relations below;

$$u_{*b}^2 = \frac{\tau_b}{\rho}, \quad u_{*t}^2 = \frac{\tau_t}{\rho}, \quad u_{*n}^2 = \frac{1}{2}\left(u_{*b}^2 + u_{*t}^2\right) \qquad (1a, b, c)$$

The three dimensionless parameters governing flow behavior are the shear Richardson number Ri_τ, the shear Reynolds number Re_τ, and the dimensionless fall velocity of the sediment $\tilde{V}$, defined respectively below;

$$Ri_\tau = \frac{RgCH}{2u_{*n}^2}, \quad Re_\tau = \frac{u_{*n}H}{2\nu} \quad \tilde{V} = \frac{V_s}{u_{*n}} \qquad (2a, b, c)$$

where ν denotes the kinematic viscosity of the water and g denotes gravitational acceleration.

In order to illustrate the utility of this configuration, it is useful to consider the case for which $\tilde{V} \rightarrow 0$. In this limit, which corresponds to suspended sediment with no fall velocity (or analogously, dissolved salt), $\bar{c}$ is distributed uniformly in z and is thus everywhere equal to C. The downslope driving force per unit mass $RgCS$ is balanced equally by the bottom and top shear stresses τ_b and τ_t, and the velocity profile is symmetrical about the centerline $z = H/2$. No stratification effects are present, so in the case of fully turbulent flow, identical logarithmic layers are obtained in the upper and lower halves of the flow (Cantero *et al.*, 2009).

The symmetry of this flow is broken as dimensionless fall velocity $\tilde{V}$ exceeds zero. The concentration profile now declines in z, as illustrated in Figure 1, giving rise to self-stratification and concomitant turbulence damping.

It is important to be able to capture the case where turbulence is extinguished by stratification. This is done here by means of the artifice of a small "molecular" kinematic diffusivity of sediment κ. This defines a very thin boundary layer near the bed where suspended sediment can be sequestered even when the turbulence is extinguished. The sequestration of all or nearly all of the sediment within this layer is the way in which TOC describes the condition of a current that has died because all the sediment has deposited. If near-bed turbulence is absent, the sediment in this thin near-bed boundary layer is no longer available for re-entrainment into the flow.

The numerical methods used in this study are outlined in detail in Cantero *et al.* (2009), and are thus not repeated here.

3 RESULTS

The central case considered by Cantero *et al.* (2011) for the analysis of turbulence extinction was that for shear Richardson number $Ri_\tau = 11.4$ and shear Reynolds number $Re_\tau = 180$. As noted above, Cantero *et al.* (2009) verified that standard relations for fully turbulent flow were recovered as $\tilde{V} \to 0$. Cantero *et al.* (2009, 2011) studied the effect of gradually raising $\tilde{V}$ on the flow. More specifically, they considered the cases $\tilde{V} = 0.005$, 0.018, 0.022 and 0.023.

Their results are illustrated in Figure 2, in terms of the variation in z of the Reynolds stress τ_{Re} and Reynolds flux of suspended sediment F_{Re}, and the corresponding viscous stress τ_v and "molecular" flux F_v. These parameters are defined as

$$\tau_{Re} = -\rho\overline{u'w'}, \qquad F_{Re} = \overline{c'w'} \tag{3a, b}$$

$$\tau_v = \rho\nu\frac{d\bar{u}}{dz}, \qquad F_v = -\kappa\frac{d\bar{c}}{dz} \tag{4a, b}$$

where u', w' and c' denote the fluctuating components of streamwise velocity, upward normal velocity and volume concentration, respectively, and the overbars denotes AOT. The parameters plotted in Figure 2 are a) the ratio τ_{Re}/τ_v versus z/H (left panel) and the ratio F_{re}/F_v (right panel) versus z/H. In the case of fully turbulent flows, these ratios can be expected to be well above unity everywhere except near the two boundaries.

Consider the case $\tilde{V} = 0.005$ in Figure 2. Both ratios τ_{Re}/τ_v and F_{re}/F_v are indeed well above unity everywhere except near the bottom and top walls. In addition, both profiles are very nearly symmetric about mid-channel, indicating that stratification

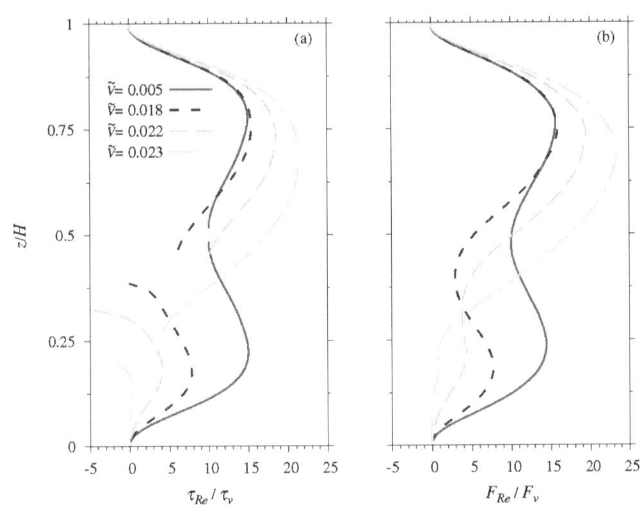

Figure 2 Results for the case $Re_\tau = 180$, with shear stress ratio in the left panel and suspended sediment flux ratio in the right panel.

effects are negligible. As $\tilde{V}$ increases to 0.018 and then 0.022, however, the symmetry is broken, and these ratios are progressively reduced in the lower half of the flow. The reduction is a direct measure of turbulence damping by self-stratification.

Figure 2 also includes plots of the ratios τ_{Re}/τ_v and F_{re}/F_v for the case $\tilde{V} = 0.023$. Evidently a threshold has been crossed between $\tilde{V} = 0.022$ and 0.023, at which the near bed turbulence is extinguished. This is evidenced for the case $\tilde{V} = 0.023$ by the substantial range of near-vanishing values of τ_{Re}/τ_v and F_{re}/F_v that extends upward from the bed, so as to encompass about one-fourth the thickness of the bottom half of the channel.

In Figure 3, these same calculations are extended for the same value of shear Richardson number $Ri_\tau = 11.4$ as in Figure 2, but with a shear Reynolds number Re_τ of 400. The values of $\tilde{V}$ used in the computations were 0.01, 0.02, 0.03 and 0.04. The overall patterns for the ratios τ_{Re}/τ_v and F_{re}/F_v observed in this case, and in particular the tendency for the damping of near-bed turbulence, are the same as those shown in Figure 2, where $Re_\tau = 180$. The same threshold effect appears; for sufficiently large values of $\tilde{V}$, the near-bed turbulence is again extinguished. The threshold value of $\tilde{V}$, however, is larger for the case with larger Re_τ. This value is somewhere between 0.03 and 0.04 for $Re_\tau = 400$; more precise delineation would require more numerical runs with DNS.

A comparison of Figures 2 and 3 allows the following conclusions: a) the same threshold phenomenon in regard to the extinction of near-bed turbulence is manifested for both Reynolds numbers; b) the threshold value of dimensionless fall velocity $\tilde{V}$ is larger for the larger shear Reynolds number; but c), the threshold values are of the same order of magnitude ($\tilde{V} = 0.023$ for $Re_\tau = 180$ and $\tilde{V} \sim 0.035$ for $Re_\tau = 400$).

In interpreting these results, it useful to recall that that in general, the scale of the fine turbulence is characterized not only by the Kolmogoroff scale, which delineates the finest scale of turbulence in general (including unstratified flow) but also the Ozmidov scale, which specifically delineates the smallest scale affected by density stratification effects (e.g., Cantero *et al.*, 2009).

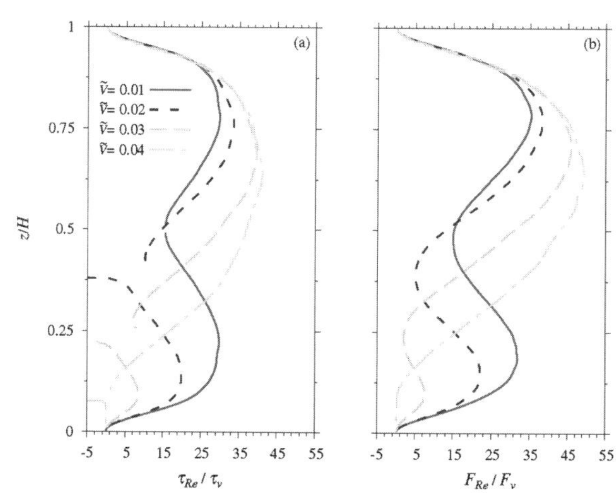

Figure 3 Results for the case $Re_\tau = 400$, with shear stress ratio in the left panel and suspended sediment flux ratio in the right panel.

4 CONCLUSIONS

Cantero *et al.* (2009) used Direct Numerical Simulation of water flow driven by a dilute suspension of sediment in the Turbidity Current with a Roof configuration to study conditions for the collapse of near-bed turbulence in turbidity currents. They delineated a threshold dimensionless fall velocity $\tilde{V}$ of about 0.022, beyond which density self-stratification extinguishes the near-bed turbulence. Under such conditions, sediment could be expected to rain out of the turbidity current and deposit on the bed, with little or no resuspension or reworking. Cantero *et al.* (2011) applied these results to field-scale turbidity currents, and in particular to the formation of massive turbidites. Direct Numerical Simulation is, however, limited in terms of the range of Reynolds numbers that can be considered. Here the work of Cantero *et al.* (2009), which was performed with a shear Reynolds number Re_τ of 180, is repeated for the larger value of 400. The same threshold effect was found, but the threshold value of $\tilde{V}$ was found to increase to about 0.035. Further studies will be required to determine whether or not this threshold value converges to a Reynolds-independent value for sufficiently large shear Reynolds numbers. In addition, further calculations are needed to delineate the dependence of the threshold value of $\tilde{V}$ on shear Richardson number Ri_τ.

ACKNOWLEDGEMENTS

This research was funded by Shell Innovation Research and Development. Additional support provided by the National Center for Earth-surface Dynamics, a Science and Technology Center funded by the US National Science Foundation (EAR-0120914). M.C. acknowledges research funding from CONICET, CNEA, ANPCyT and the University of Florida.

REFERENCES

Bouma, A. 1962. Sedimentology of some Flysch deposits: A graphic approach to facies interpretation. Elsevier, Amsterdam, p. 168.

Cantero, M.I., Balachandar, S., Cantelli, A., Pirmez, C. and Parker, G. 2009. Turbidity current with a roof: direct numerical simulation of self-stratified turbulent channel flow driven by suspended sediment. *Journal of Geophysical Research*, 114, C03008, doi:10.1029/2008 JC004978, p. 20.

Cantero, M.I., Cantelli, A., Pirmez, C., Balachandar, S., Mohrig, D., Hickson, T.A., T.-H. Yeh, H., Naruse, H. and Parker, G. 2011. Emplacement of massive turbidites linked to extinction of turbulence in turbidity currents. *Nature Geoscience*, in press.

Flood, R.D., Hiscott, R.N. and Aksu, A.E. 2009. Morphology and evolution of an anastomosed channel network where saline underflow enters the Black Sea. *Sedimentology*, 56 (3), 807–839.

Wright, S. and Parker, G. 2004. Density stratification effects in sand-bed rivers. *Journal of Hydraulic Engineering*, 130 (8), 783–795.

Chapter 20

Revisiting gravity currents and free shear flows

Johannes Bühler[1] *and Marko Princevac*[2]
[1] *Institue of Environmental Engineering, ETH Zurich, Zurich, Switzerland*
[2] *University of California at Riverside, Department of Mechanical Engineering, Riverside, California, USA*

ABSTRACT: The rate at which free shear flows widen along their path can be specified by invoking the concept of entrainment. A different approach was proposed by Prandtl, who considered the widening of jets as being due to a diffusion process. His diffusion relation in terms of the half-width has since been successfully modified for models of free shear flows averaged over the width. In the present contribution we apply it to gravity currents. A further topic we address here is that the structure of the conventional shallow water equations for gravity currents is consistent with the Bresse equations for open channel flows, but that the underlying depth and velocity scales of the two flows are different. For gravity currents the scales are derived from the velocity distribution in analogy to previous work on free shear flows, whereas they are based on the vertical extent, and flux, of the dense liquid phase in open channel flows. To avoid this disparity, we defined a set of scales for both flows which is based on the distribution and flux of excess mass. To compare entrainment and diffusion rates for gravity driven flows in terms of velocity- and mass-based scales, we reanalyze field data on katabatic winds, which are similar to gravity currents.

1 INTRODUCTION

Cooling water from thermal power plants, and treated wastewater, are suitably returned to coastal and inland waters via submerged diffusors, which eject the effluent through a series of riser pipes. Similar flows arise in the atmosphere due to flue gases released from smoke stacks. A model for the increase of the half-width of such flows along their path was proposed by Prandtl (1926), see also Abramovich (1963) and Schlichting (1979). His concept implies that free shear flows widen due to an outward diffusion of vorticity and momentum into the undisturbed ambient fluid. A quite different approach was chosen by Morton, Taylor and Turner (1956) [MTT] for jets and plumes in stratified surroundings. They distinguished the ambient stratification from the internal density distribution by introducing a flow boundary, and considered the distributions of velocity and density within this boundary as uniform. The mean width and velocity of the flow was derived from the velocity distribution, which was assumed to be Gaussian. In contrast to Prandtl's diffusion concept, the widening of these flows was attributed to an inflow of ambient fluid through the boundary. The corresponding entrainment velocity was assumed to be proportional to the mean streamwise velocity. The MTT approach has been widely used for studies of buoyant jets from diffusors, and of their interaction, Jirka (2004, 2006). Wright (1994) proposed a width-averaged flow model based on the diffusion approach of Prandtl. A similar concept was developed by Chu (1994), and modified by Lee and Chu (2003) to examine flows under more complex conditions.

Ellison and Turner (1959) [ET] extended the model of MTT to gravity currents on an incline, which can be due to an inflow of cold or saline water into a water body. Examples for such flows are gravity currents caused by the release of brine from desalination plants into coastal waters (Bleninger & Jirka, 2008). On a large scale, gravity currents arise from the overflow of salty water from shallow marine basins onto the abyssal plain, where they drive the oceanic circulation (Legg *et al.*, 2009). Turbidity currents are due to inflows rich in sediment, and to sediment erosion from the bottom (Garcia & Parker, 1993). The sediment load is deposited again when the slope becomes sufficiently small. In hydropower reservoirs part of the sedimentation occurs near the dam, and efforts are being made to control it (Oehy *et al.*, 2010). Katabatic winds in the atmosphere are similar gravity driven flows down a slope. They arise in the evening, after sunset, when the ground temperature decreases due to radiative cooling. An air layer adjacent to the slope then becomes colder than the air above it, and starts descending (Princevac *et al.*, 2005). In this contribution we extend the diffusion approach of Wright (1994) and Lee & Chu (2003) to gravity currents.

The entrainment by gravity currents on mild slopes is small, and they are similar to open channel flows. As a consequence, the shallow water equations derived by ET are consistent with the Bresse equations for open channel flows. A limitation of their model is, however, that their depth and velocity scales are based on the velocity distribution in analogy to previous work on free shear flows, whereas the depth of open channel flows is the vertical extent of the excess mass of the liquid phase, and the velocity is derived from its flux. To make the two models consistent with each other, ET's relations can be expressed in terms of mass-based flow scales (Bühler & Siegenthaler, 1986; Bühler *et al.*, 1991).

The diffusion concept for free shear flows is outlined in more detail in Section 2, and the entrainment relation of ET for gravity currents is discussed in Section 3. The diffusion approach is extended to gravity currents in Section 4. In Section 5 the diffusion relations are expressed in terms of mass-based flow scales. Previous results on shape factors for mass-based flow scales, and on mass- and velocity-based diffusion rates of free shear flows are summarized in Section 6. Entrainment and diffusion rates for katabatic winds are derived from available field data, and presented in Section 7. Conclusions are drawn in the final section.

2 PRANDTL'S JET DIFFUSION

The scales of jets in a co-flow of velocity u_a can be specified by their maximum excess velocity u_{em}, and the half-width b_u at which the excess velocity is $u_{em}/2$ (Figure 1). An early model for the widening rate of such jets was proposed by Prandtl (1926). He assumed that the transverse drift of turbulent structures which define the half-width b_u corresponds to the characteristic transverse velocity fluctuations v', which leads to

$$\frac{Db_u}{Dt} = v' \tag{1}$$

According to his mixing length concept the turbulent shear stress is $\rho\overline{u'v'} = \rho l^2 (\partial\bar{u}/\partial y)^2$, where l is the mixing length, ρ the fluid density, $\bar{u}$ the local flow velocity, y the transverse coordinate, and the overbars denote time-averaged quantities. As the transverse fluctuations are proportional to the streamwise ones, this leads to a closure relation

$$\frac{Db_u}{Dt} \propto l\frac{\partial\bar{u}}{\partial y} \propto \frac{lu_{em}}{b_u} \propto \alpha\, u_{em}, \tag{2}$$

where α is the constant ratio of the mixing length and the jet width. This implies that the mean lateral drift v_s of the structures near the outer boundary of the flow, and of their excess velocity u_{es}, are proportional to each other (Figure 1). This outward diffusion of momentum and vorticity also requires an inflow of undisturbed ambient fluid.

For plane jets and plumes in calm ambient fluids the drift velocity is proportional to the maximum velocity u_m, and

$$\frac{Db_u}{Dt} \propto \frac{db_u}{dx}u_m \propto u_m \tag{3}$$

which leads to

$$\frac{db_u}{dx} = const. \tag{4}$$

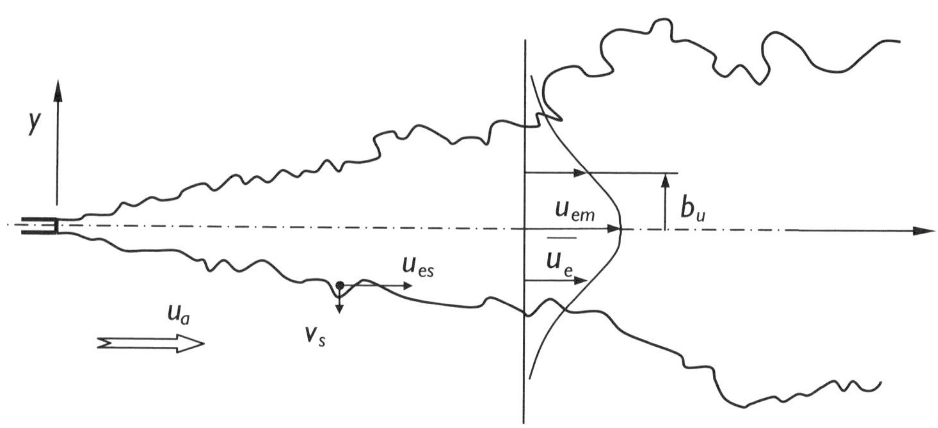

Figure 1 Jet in a co-flow. b_u, is the half-width, u_{em} the maximum excess velocity. Prandtl's model implies that the mean values of the transverse drift v_s of large structures, and of their excess velocity u_{es} are proportional to each other.

The velocity distributions were computed by means of the mixing length and similar assumptions (Abramovich, 1963; Schlichting, 1979).

The value of the constant in (4) is about 0.12 for plane jets. Rather surprisingly, this value agrees to within the experimental error with the one for plane plumes (Jirka, 2006). The same holds for the spreading rates of axisymmetric jets and plumes (Jirka, 2004). If one accepts the derivation of (2), the good agreement of α for jets and plumes can be expected because the velocity profiles of the two flows are very similar, and well approximated by a Gaussian distribution.

3 ET'S ENTRAINMENT MODEL FOR GRAVITY CURRENTS

As mentioned in the introduction, a quite different parameterization was chosen by MTT for axisymmetric jets and plumes in a stratified environment. To separate the density distribution within the flow from the one in the environment, they introduced a boundary between the two domains, and considered the density and velocity within the boundary as being uniform (top-hat profiles). Plumes and jets in unstratified environments were similarly analyzed by keeping the buoyancy flux constant or absent. The ratio of the entrainment and streamwise velocities is independent of the Reynolds number when the flow is fully turbulent. Conversely, the velocity through the highly contorted small-scale interface of these flows does depend on the Reynolds number, as it is dominated by a viscous process, and scales with the Kolmogorov velocity (Holzner & Lüthi, 2011).

Ellison and Turner [ET] extended the approach of MTT to gravity currents in deep and unstratified environments (Figure 2). For a source of discharge q_0 per unit width, and by allowing for a constant ambient velocity u_a, the following depth, velocity and buoyancy scales were used

$$
\begin{aligned}
UH &= \int (\bar{u} - u_a)dy \\
U^2H &= \int (\bar{u} - u_a)^2 dy \\
\Delta(U + u_a)H &= g\frac{(\rho_0 - \rho_a)q_0}{\rho_a}
\end{aligned}
\tag{5}
$$

by considering that the initial buoyancy flux was preserved in their experiments. The coordinates are defined in Figure 2, and the integration is carried from the bottom to a location where $d\bar{u}/dy$ vanishes.

MTTs entrainment relation was restated as

$$
\frac{d}{dx}[(U + u_a)H] = EU \tag{6}
$$

E is an entrainment function which depends on the Richardson number $Ri_e = \Delta H \cos\varphi / U^2$ based on the excess velocity and the slope φ. In the spirit of ET the conservation equations of momentum and buoyancy for a dilute gravity current in an ambient flow can be stated as

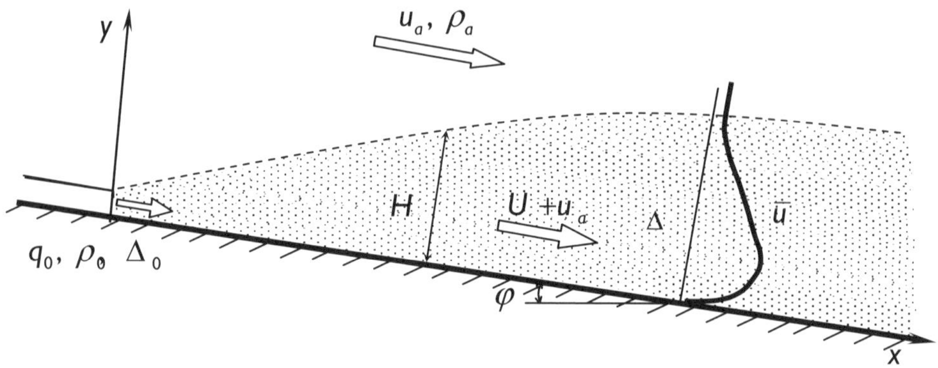

Figure 2 Definition sketch for a developing gravity current in unstratified and deep water. ET's top-hat scales for depth *H*, excess velocity *U*, and buoyancy Δ are based on the distribution of the excess velocity, $\bar{u} - u_a$ and on the buoyancy flux (from Princevac *et al.*, 2010).

$$\frac{d}{dx}\left[U(U + u_a)H + \frac{S_1}{2}\,\Delta H^2 \cos\varphi\right] = -\frac{du_a}{dx}(U + u_a)H + S_2\,\Delta H \sin\varphi - C_D(u_a + U)^2 \tag{7}$$

and

$$\frac{d}{dx}\left[\Delta(U + u_a)H\right] = 0 \tag{8}$$

where C_D is a Chezy-type drag coefficient. The shape factors S_1 and S_2 relate the excess pressure force and the excess bottom pressure to the remaining flow scales, *i.e.*

$$\begin{aligned} S_1 \Delta H^2 &= 2g \int \frac{(\bar{\rho} - \rho_a)}{\rho_a} y\,dy \\ S_2 \Delta H &= g \int \frac{(\bar{\rho} - \rho_a)}{\rho_a} dy \end{aligned} \tag{9}$$

By making use of (6) to (8), the variation of the depth and the Richardson number $Ri = \Delta H \cos\varphi / (U + u_a)^2$ can be computed. The Richardson number is approximately the ratio of the phase speed of long interfacial waves on a gravity current, and the velocity of that current. Flows with $Ri < 1$ are supercritical, those with $Ri > 1$ subcritical. Ri is related to the Froude number Fr of open channel flows by $Fr = Ri^{-1/2}$.

The focus of the present contribution is on the difference between entrainment and diffusion relations for gravity currents in an ambient flow. For illustrative purposes we also outline the resulting shallow water equations for the simpler case of calm environments, for which $u_a = 0$, and $Ri = Ri_e$. ET's relations are for this case, and can be stated as

$$\frac{dH}{dx} = \frac{E\left(2 - \frac{S_1}{2}Ri\right) - S_2 Ri \tan\varphi + C_D}{1 - S_1 Ri} \tag{10}$$

$$\frac{H}{3Ri}\frac{dRi}{dx} = \frac{E\left(1 + \frac{S_1}{2}Ri\right) - S_2 Ri \tan\varphi + C_D}{1 - S_1 Ri} \tag{11}$$

One feature of (10) and (11) is that the denominator vanishes when $Ri = 1/S_1$, which is close to one. Under these conditions, the flow is critical. An important result of ET is that on a constant slope *Ri* and d*H*/d*x* vary along the flow depending on the source conditions, and finally adjust to a constant value. In this final equilibrium state the flow is called uniform. Velocity and Richardson number of these inclined plumes are constant, as they are in free plumes.

4 APPLYING THE DIFFUSION APPROACH TO GRAVITY CURRENTS

4.1 Supercritical flows

Wall jets, and plumes on a vertical wall, are special cases of gravity currents, and the Richardson number $Ri = \Delta H \cos\varphi / U^2$ vanishes for both of these flows. As an estimate of the value of the entrainment function E_j for wall jets, ET adopted the one of 0.075 for free jets by Townsend (1956). They already noted that the values for free plumes, for which *Ri* also vanishes, appeared to be higher, but thought that insufficient data were available to make definite conclusions. More recent experiments show that $E_j = 0.0625$ for jets, and that the value of $E_p = 0.125$ for free plumes is indeed considerably larger (Jirka, 2006). Conversely, results by Patel (1971) show that the spreading rate d*H*/d*x* of wall jets is about 0.091 (with $C = 0.065$). Similar to what is found for free shear flows, this is again in excellent agreement with the value $dH/dx = 0.095 \pm 0.005$ of Grella and Faeth (1975) for wall plumes, which have a similar velocity distribution. These considerations suggest that we may apply the diffusion approach of Wright (1994) and Chu (1994) for free shear flows to gravity currents as well.

Provided that the excess velocity of the large structures determining the widening rate is *U*, their outward drift D*H*/D*t* can be expressed as $dH/dx\ (U + u_a)$. In the spirit of the approach of Wright (1994) for jets, (2) can then be modified for gravity currents as

$$\frac{dH}{dx}(U + u_a) = DU \tag{12}$$

where $D(Ri_e)$ is a diffusion function.

ET and later investigators determined the entrainment relation E(Ri) for equilibrium flows in calm water (see Fernandez & Imberger, 2006). For this case the two descriptions (6) and (12) are identical, and $Ri_e = Ri$. The relation dH/dx = D then describes the transition from a forced wall plume to a wall plume if the value of D at $Ri = 0$ is increased slightly from 0.075 to about 0.093. The relation can, however, also be applied to other slopes, as it becomes equivalent to (6) with $\mathrm{E}(Ri) = \mathrm{D}(Ri)$, as soon as the equilibrium state is reached.

By combining (12) with the momentum and buoyancy Equations (7) and (8), the variation of the Richardson number in calm environments can be expressed as

$$\frac{H}{3Ri}\frac{dRi}{dx} = \frac{\mathrm{D}\left(1 + \frac{S_1}{2}Ri\right) - S_2 Ri \tan\varphi + \mathrm{C_D}}{2 - \frac{S_1}{2}Ri} \tag{13}$$

whereas the nominator of this relation agrees with that in (11) for $\mathrm{E} = \mathrm{D}$, the denominators are quite different. In particular, the one in (13) no longer vanishes when gravity currents are critical, and the Richardson number is close to one, but when it is close to 4. The reason is that only (6), but not (12) is consistent with the continuity equation for open channel flows. Expression (13) is thus appropriate for supercritical flows only, which are jet-like and dominated by KH billowing. In subcritical flows on mild slopes the entrainment is small, and due to different processes, like Holmboe instabilities (Strang & Fernando, 2001), and turbulence generated at the solid boundary (Hebbert *et al.*, 1979). The transition between the two regimes also depends on the ratio of the thicknesses of the shear layers defined in terms of velocity and density (Negretti *et al.*, 2008).

For subcritical flows the diffusion model can be applied in a different way, as outlined in the following subsection.

4.2 Subcritical flows

The entrainment relation (6) by ET can be restated as

$$\frac{dH}{dx}(U + u_a) = \mathrm{E}U - H\frac{d(U + u_a)}{dx} \tag{14}$$

For uniform flows the last term vanishes, and this relation agrees with (12). The entrainment function E can then be replaced by the diffusion function D.

Entrainment and diffusion are small for subcritical flows, such that changes in slope or bottom roughness become relevant. The last term of (14) mainly accounts for the resulting acceleration or deceleration. For subcritical flows the entrainment relation (6) thus corresponds to the diffusion relation, and can be replaced by

$$\frac{d}{dx}[(U + u_a)H] = \mathrm{D}U \tag{15}$$

with D ≡ E. As a consequence, relations (10) and (11) remain valid for subcritical flows when restated in terms of D instead of E.

5 MASS-BASED FLOW SCALES

Shallow water equations similar to the ones proposed by ET are also used for open channel flows. A significant difference between the two descriptions is, however, that the flow scales U and H of ET are derived from the velocity distribution, whereas the depth h of open channel flows is the vertical extent of the dense liquid phase, and the velocity u is derived from its flux. Consistency of the two sets of flow scales can be achieved by deriving those for gravity currents from the distribution and flux of excess mass. A buoyancy scale g' and a depth scale h can be obtained from the excess bottom pressure and pressure force in (9). The buoyancy flux in (5) can then be used to determine the velocity u instead of the buoyancy Δ. This leads to

$$\begin{aligned} g'h^2 &= 2g\int \frac{(\bar{\rho}-\rho_a)}{\rho_0} y\,dy = S_1 \Delta H^2 \\ g'h &= g\int \frac{(\bar{\rho}-\rho_a)}{\rho_0} dy = S_2 \Delta H \\ g'uh &= g\frac{(\rho_0-\rho_a)q_0}{\rho_0} = \Delta(U+u_a)H \end{aligned} \tag{16}$$

Shape factors are now required to account for the excess volume and momentum fluxes in (5). A convenient choice is

$$\gamma h(u-u_a) = UH, \quad \gamma h\,u(\beta u - u_a) = (U+u_a)UH \tag{17}$$

where γ modifies the depth h. For open channel flows u_a can be set equal 0, $\gamma = 1$, and β corresponds to the momentum, or Boussinesq, coefficient (Chow, 1959). Princevac *et al.* (2009) also derived the shape factors β and γ for non-Boussinesq flows.

Altinakar (1993) conducted a series of experiments on supercritical gravity and turbidity currents on slopes from 0.73 to 1.7°, and also computed the mass-based depths h and γh. Ri was in the range from 0.23 to 0.79 for 13 experiments with saline gravity currents. Except for one of them, the values of γ ranged from 1.1 to 1.86, with a mean of 1.40. The flow can thus be considered as consisting of a dense layer of depth h, and a superimposed layer of ambient fluid of thickness $(\gamma-1)h$, which flows at the same velocity u (Figure 3). The approach is similar to the one used for sinking spheres, or thermals, in which the momentum of external fluid is accounted for by making use of the concept of added mass (Escudier & Maxworthy, 1973). Chu (1994) also invoked the concept of added mass for jet-like flows.

Noh & Fernando (1991) proposed a similar set of flow scales, except that they replaced the excess pressure force in (16) by the volume flux. As a consequence, the

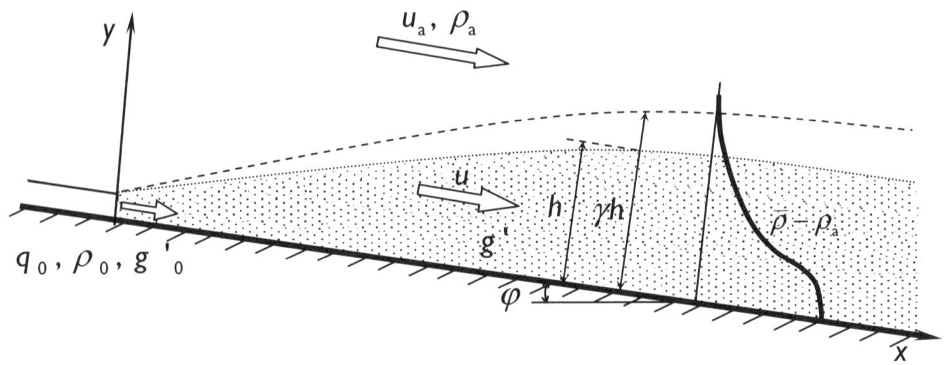

Figure 3 Definition sketch for mass-based flow scales of depth h, buoyancy g' and velocity u, which are derived from the excess density distribution $\rho - \rho_a$, and the excess mass flux. The excess volume flux is confined to the depth γh (from Princevac *et al.*, 2010).

buoyancy is distributed uniformly over the entire flow depth γh, and their Richardson number corresponds to the mass-based quantity $Ri^* = g'h\cos\varphi/u^2$.

An advantage of the mass-based set of scales in (16) is that the velocity u can be determined without carrying out velocity measurements when the buoyancy flux is known. Moreover, all three flow scales depend on the vertical buoyancy distribution, whereas the buoyancy Δ would be obtained by completely mixing the buoyancy flux with the volume flux at a given cross-section. A further difference between the two sets is that the entrainment can be zero or negative for special types of flows, whereas the mass flux infected by vorticity (or tracer) always increases in the flow direction (Head & Bradshaw, 1971).

The shallow water equations of ET can be written in terms of mass-based scales. The diffusion relation (15) for subcritical flows can be restated for the dense part of the flowing layer as

$$\frac{d}{dx}(uh) = \mathrm{D}^*(u - u_a) \tag{18}$$

where the star denotes definitions in terms of mass-based flow scales.

Relations (7) and (8) correspond to

$$\frac{d}{dx}\left[\gamma(\beta u - u_a)uh + \frac{1}{2}g'h^2\cos\varphi\right] = -\frac{du_a}{dx}\gamma uh + g'h\sin\varphi - C_D^* u^2 \tag{19}$$

$$\frac{d}{dx}(g'uh) = 0 \tag{20}$$

When the buoyancy flux is known, uncertainties related to velocity measurements affect the two shape factors in the momentum Equation (19) only, but not the flow scales.

Identities (5), (9), (16) and (17) can be used to derive the relation between the two sets of scales. For calm environments

$$h = HS_1/S_2,\ g' = \Delta S_2^{\,2}/S_1,\ u = U/S_2,\ \beta = S_2,\ \gamma = S_2^{\,2}/S_1, \\ D^* = DS_1/S_2,\ C_D^{\ *} = C_D S_2^{\,2},\ Ri^* = Ri/S_2^{\,3} \tag{21}$$

In the absence of ambient flows, relations (18) to (20) can be expressed as

$$\frac{dh}{dx} = \frac{D^*\left(2\beta\gamma - \frac{1}{2}Ri^*\right) - Ri^*\tan\varphi + C_D^{\ *}}{\beta\gamma - Ri^*} \tag{22}$$

$$\frac{h}{3Ri^*}\frac{dRi^*}{dx} = \frac{D^*\left(\beta\gamma + \frac{1}{2}Ri^*\right) - Ri^*\tan\varphi + C_D^*}{\beta\gamma - Ri^*} \tag{23}$$

As in (10) and (11), the denominators vanish when $Ri^* = g'h\cos\varphi/u^2 \approx 1$, and distinguish subcritical from supercritical flows. These relations now reduce to those for open channel flows for $D^* = 0$ and $\gamma = 1$. This also holds for equations related to hydraulic jumps (Bühler *et al.*, 2011). Princevac *et al.* (2009) derived the shallow water equations by considering ambient co-flows.

The mass-based equivalent of the diffusion relation (12) for supercritical flows is

$$\frac{dh}{dx} = D^*\frac{u - u_a}{u} \tag{24}$$

For calm ambient waters this can be combined with (19) and (20) to

$$\frac{dh}{dx} = D^* \tag{25}$$

$$\frac{h}{3Ri^*}\frac{dRi^*}{dx} = \frac{D^*\left(\beta\gamma + \frac{1}{2}Ri^*\right) - Ri^*\tan\varphi + C_D^{\ *}}{2\beta\gamma - \frac{1}{2}Ri^*} \tag{26}$$

The denominators in (23) and (26) again differ, as they do in (11) and (13).

6 PREVIOUS RESULTS

A first attempt to determine both velocity and buoyancy based flow scales in gravity and turbidity currents was undertaken by Altinakar (1993). His laboratory data were reviewed by Princevac *et al.* (2009), who also carried out a corresponding reanalysis of field data on katabatic winds by Doran *et al.* (2002). The values of γ for these latter flows range from 1.0 to 2.7, those of β from 1.01 to 1.23. In addition, these authors

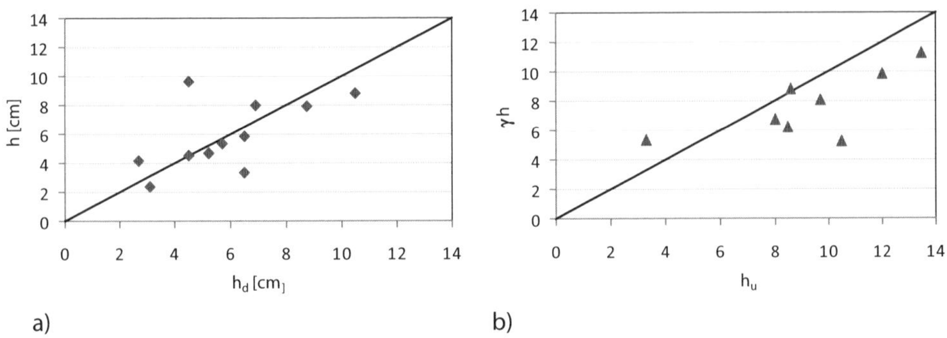

Figure 4 a) h vs. depth h_d of the lower part of the flowing layer, b) γh vs total layer depth h_u.

compared mass- based with other flow scales for arrested salt wedges (Sargent & Jirka, 1987), and a two-layer flow over an obstacle (Lawrence, 1993).

Altinakar (1993) distinguished two depth scales of gravity currents based on flow visualization. The observed height h_d of the bulk of the flow near the bottom, with a nearly uniform density distribution, was compared with the height h_u of its visual upper boundary, which includes the shear layer of gradually decreasing excess density. As shown in Fig. 4a, h_d agrees fairly well with h. Conversely, the height of the upper edge of the flow at h_u in Fig. 4b is somewhat larger than γh. Similar results were obtained for the turbidity currents. The depth $(\gamma\text{-}1)\ h$ can thus be used as a measure for the width of the shear layer.

Princevac *et al.* (2010) determined velocity and mass-based flow scales and widening rates for free shear flows by reanalyzing available data. Jets were considered as well by defining the mass-based flow scales in terms of the concentration of a nonbuoyant tracer added to the flow. The analysis indicates that the value of γ for both plane and axisymmetric jets and plumes is close to one, i.e. that the average outer boundaries of excess mass and velocity coincide. This suggests that mass-based scales, and the diffusion concept, could also be used for free shear flows in stratified environments as an alternative to the MTT approach. The flow scales could then be determined from the concentration distribution and flux of the tracer which is used for flow visualizations. Another result of their analysis is that the widening rates for plane and axisymmetric flows recommended by a number of investigators are more consistent with each other when expressed in terms of mass-based rather than velocity-based flow scales. This also holds for the mean concentrations, but not for all mean velocities.

To gain information on the widening rates of geophysical flows as well, previous data on katabatic winds were reanalyzed. The corresponding results on velocity- as well as mass-based diffusion rates are outlined in the following section.

7 DIFFUSION RATES OF KATABATIC WINDS

Doran *et al.* (2002) obtained field data for katabatic flows on the slope of a mountain in Utah by means of a tethersonde system. The data collected by the Pacific Northwest National Laboratory (PNNL) team at the Slope Site (Site 11) were considered here.

This slope site is characterized by a gentle ($\varphi = 1.58°$) and very smooth (aerodynamic roughness length ~ 0.1 m) slope. Typical measured profiles of wind speed and potential temperature (density) are shown in Princevac *et al.* (2005) and Princevac *et al.* (2009).

The focus of the present study is on the diffusion rates of these flows, i.e., on the values of D_l and D_s for large and small Richardson numbers according to (15) and (12), and on the corresponding mass-based quantities D_l^* and D_s^* in (18) and (24). These quantities were evaluated for the test section, which was 4 km in length. Measuring stations 1 to 4 were spaced 1 km apart, with station 1 at the lower end. The widening rates and velocity gradients were determined between stations 2 and 3 for stretch 2–3, as well as between other stations for stretches 3–4, 1–3, 2–4, and 1–4. The other required quantities were taken as the average of the values at the two end points. Data for stretch 1–2 were omitted as the widening rates were considerably larger than for the remaining stretches, and possibly influenced by winds along the valley.

All flows were supercritical, with values of Ri ranging from 0.19 to 0.56. The velocity u_a of the upper layer cannot be neglected, as its mean value of 1.7 m/s was about the same as the mean excess velocity of the flow (Princevac *et al.*, 2009), and the minimum value of Ri_e/Ri was 1.4. As a consequence, the widening rates are related to the Richardson number Ri_e based on the excess velocity. In contrast to Ri, many values of Ri_e exceed 1, as shown in Figure 5.

Figure 5 shows the diffusion function D_l for large Ri_e according to (15), which corresponds to the entrainment function E. To avoid separating the data for large and small Richardson numbers, values were computed for all flows. They are compared with the entrainment relation of Turner (1986), which was modified slightly to $E = (0.09–0.1\ Ri_e)/(1 + 5\ Ri_e)$ to account for the results of Patel (1971). For higher Ri_e values the relation $E = 1.9C_D^{3/2}/(Ri_e–0.3)$ by Hebbert *et al.* (1979) is shown for two values of C_D. It should be noted, however, that this relation was derived for a calm upper layer ($Ri_e = Ri$), and is included as a suggestion only. Based on Strang & Fernando (2001) one may expect that mixing is due to strong KH billowing when Ri_e is low, and that a transition to intermittent internal wave breaking or Holmboe

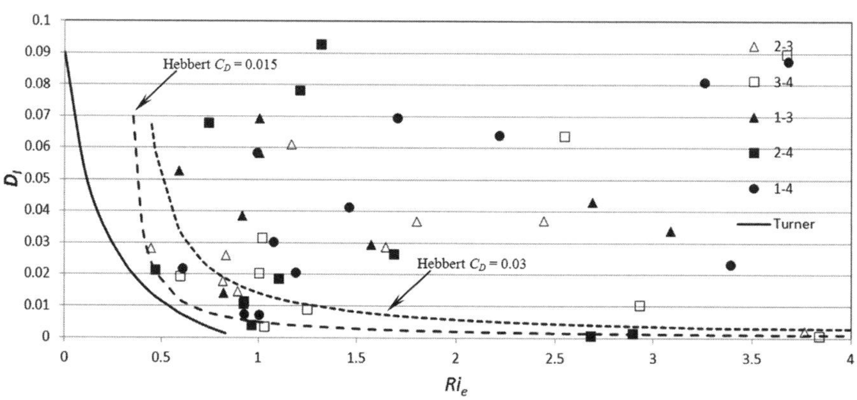

Figure 5 Velocity-based diffusion function $D_l(Ri_e)$ for large Ri_e, Eq. (15). Corresponds to $E(Ri_e)$.

instabilities occurs at Ri_e numbers above unity. Sequeiros *et al.* (2010) also observed that the shape of the excess density profiles of subcritical and supercritical flows is quite different. These results tend to support the use of different entrainment and diffusion relations for the two cases.

A distinctive feature of Figure 5 is that values of D_l are generally higher than expected according to the predictions. A similar discrepancy was found by Princevac *et al.* (2005) for these flows. Their results are not directly comparable with the present ones, however, as they used a different form of the entrainment relation than (6).

To explore the reasons for the difference between measurements and predictions we also reanalyzed the data according to (12) to obtain the diffusion function D_s for small Ri_e (Figure 6). The corresponding values should agree with those in Figure 5 for all uniform flows, but they differ considerably. In particular, the first figure includes a number of data points for $dH/dx < 0$, which are missing in the second one. The corresponding flows were thus not uniform, but accelerating. The data indeed show that $U+u_a$ and u, as well as u_a, tended to increase along the slope. A further feature

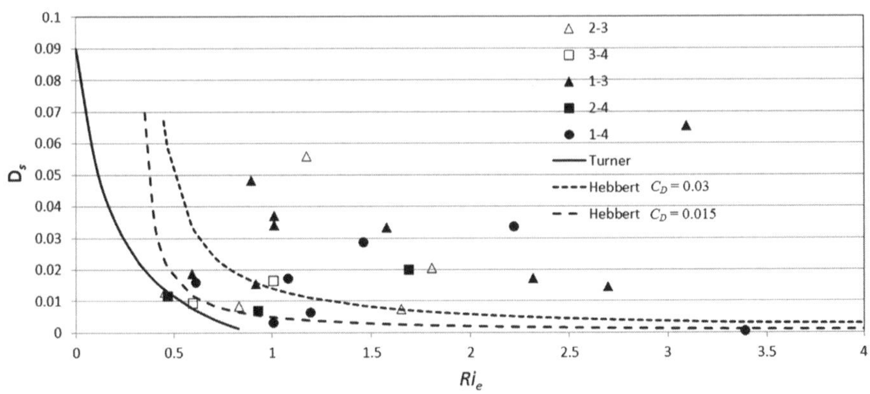

Figure 6 Velocity-based diffusion function $D_s\ (Ri_e)$ for small Ri_e, Eq. (12).

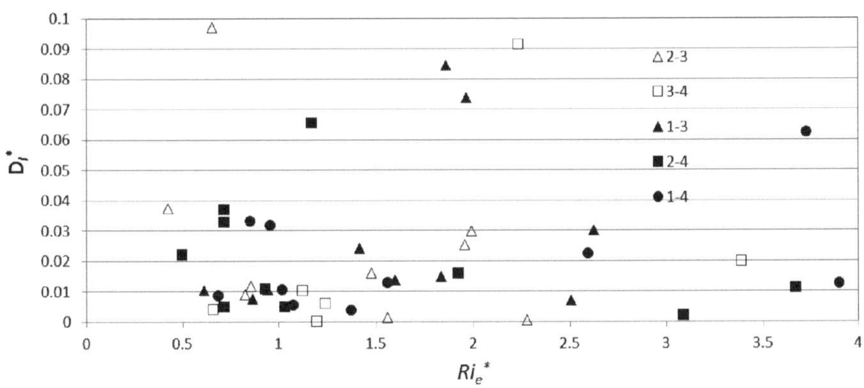

Figure 7 Mass-based diffusion function $D_l^*(Ri_e^*)$ for large Ri_e^*, Eq. (18). Corresponds to $E^*(Ri_e^*)$.

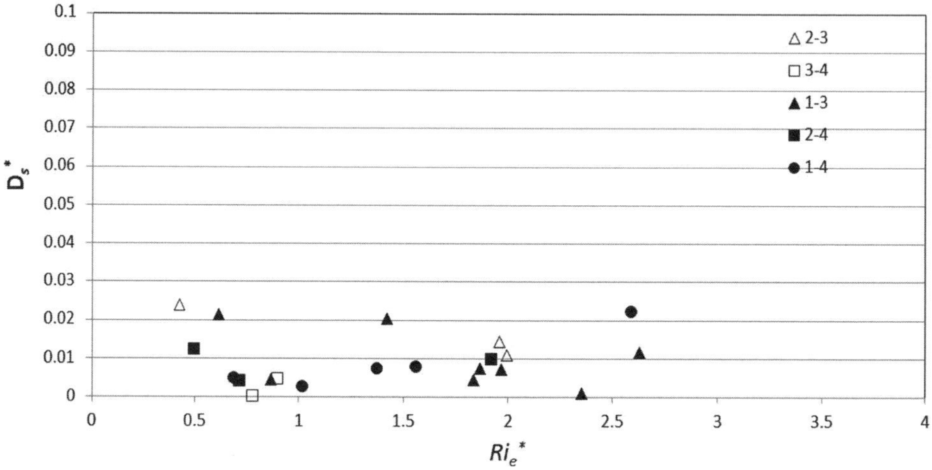

Figure 8 Mass-based spreading function $D_s^*(Ri_e^*)$ for small Ri_e^*, Eq. (24).

of the two figures is the large scatter, which is consistent with the fact that the flows were pulsating (Princevac *et al.*, 2008). A comparison of D_l and D_s for data on gravity currents thus provides a good idea on whether the flows were uniform and steady, or not.

The mass-based diffusion functions D_l^* and D_s^* according to (18) and (24) are shown in Figures 7 and 8, respectively. These values are smaller than the velocity-based ones in Figs. 5 and 6, as can be expected for values of γ in excess of unity. Again, there are more data in Figure 7 than 8 due to the acceleration, and the scatter in the first figure is larger due to the unsteadiness of the accelerating flows.

8 CONCLUSIONS

The traditional entrainment concept is compared with a concept of diffusion by Prandtl, which was later expressed as a model which averages over the width of jets and plumes. This model is applied to gravity currents because it agrees well with experimental data for wall jets and wall plumes. Field data on katabatic winds in a co-flow are re-evaluated to determine the diffusion function D. For flows with large Richardson numbers Ri_e based on the excess velocity, the diffusion function D_l corresponds to the entrainment function E. The diffusion rates for these flows are found to be larger than expected on the basis of traditional entrainment models. A reason for this discrepancy is that the flows were accelerating along the slope, and not uniform. This follows from a comparison of the data with those for the diffusion function D_s for small Richardson numbers. A comparison of the two diffusion rates is thus a convenient tool to determine the extent to which flows are uniform. The large scatter of the data also shows that the flows were pulsating, and unsteady. Another topic addressed here is that the flow scales for the depth and velocity of gravity currents are based on the velocity distribution, whereas the depth and velocity of open channel flows are derived from the distribution and flux of mass. To avoid this discrepancy, a set of flow scales for gravity currents is derived from the distribution and flux of excess mass as well. The field data are thus further evaluated to obtain the corresponding mass-based diffusion functions D_l^* and D_s^*. As expected, values of the velocity-based diffusion functions are larger, because the depth of flow γh of gravity currents is larger than the depth h occupied by their excess mass (Figure 3).

ACKNOWLEDGEMENTS

The first author thanks Prof. A. J. Schleiss in Lausanne for inviting him for a sabbatical, during which the idea for this contribution on gravity currents arose. Figures 2 and 3 were reproduced with kind permission from Springer Science+Business Media.

REFERENCES

Abramovich, G.N. (1963). *The theory of turbulent jets*. MIT Press.

Altinakar, M.S. (1993). *Weakly depositing turbidity currents on small slopes*. Dept. of Civil Engineering, Ecole Polytechnique Fédérale de Lausanne (EPFL), Lausanne, Switzerland. PhD thesis nr. 738.

Bleninger, T. and Jirka, G.H. (2008). Modelling and environmentally sound management of brine discharges from desalination plants. *Desalination*, **221** (1–3), 585–597.

Bühler, J. and Siegenthaler, C. (1986). Self-preserving solutions for turbidity currents, *Acta Mechanica*, **63** (11), 217–233.

Bühler, J. Oehy, Ch. and Schleiss, A.J. (2011). Jets opposing turbidity currents and open channel flows. Submitted to *J. Hydraul. Eng. ASCE*.

Bühler, J., Wright, S.J. and Kim, Y. (1991). Gravity currents advancing into a coflowing fluid. *Jour. Hydraulic Research*, **29** (2), 243–257.

Chow, V.-T. (1959). *Open channel hydraulics*. New York, McGraw Hill.

Chu, V.H. (1994). Lagrangian scaling of turbulent jets and plumes with dominant eddies. In: Davies, P.A. and Nieves V. (eds.) *Recent Research Advances in the Fluid Mechanics of Turbulent Jets and Plumes*, NATO ASI Series, Kluwer, pp. 45–72.

Doran, J.C., Fast, J.D. and Horel, J. (2002). The VTMX 2000 campaign. *Bulletin of the American Meteorological Society*, **83** (4), 537–554.

Ellison, T.H. and Turner, J.S. (1959). Turbulent entrainment in stratified flows. *J. Fluid Mech.* **6**, 423–448.

Escudier, M.P. and Maxworthy, T. (1973). On the motion of turbulent thermals. *J. Fluid Mech.*, **61**, 541–552.

Fernandez, R.L. and Imberger, J. (2006). Bed roughness induced entrainment in a high Richardson number underflow. *J. Hydraul. Res.*, **44** (6), 725–738.

Garcia, M. and Parker, G. (1993). Experiments on the entrainment of sediment into suspension by a dense bottom current. *J. of Geophys. Res*,. **98** (C3), 4793–4807.

Grella, J.J. and Faeth, G.M. (1975). Measurements in a two-dimensional thermal plume along a vertical adiabatic wall. *J. Fluid Mech.*, **71** (4), 701–710.

Head, M.R. and Bradshaw, P. (1971). Zero and negative entrainment in turbulent shear flow. *J. Fluid Mech.*, **46** (2), 385–394.

Hebbert, B., Imberger, J., Loh, I. and Patterson, J. (1979). Collie River underflow into the Wellington Reservoir. *J. Hydraul. Div. ASCE*, **105** (5), 533–545.

Holzner, M. and Lüthi, B. (2011). Laminar superlayer at the turbulence boundary. *Phys. Rev. Letters*, **106**, 134503.

Jirka, G.H. (2004). Integral model for turbulent buoyant jets in unbounded stratified flows. Part 1: The single round jet. *Env. Fluid Mech.*, **4**, 1–56.

Jirka, G.H. (2006). Integral model for turbulent buoyant jets in unbounded stratified flows. Part 2: Plane jet dynamics resulting from multiple port diffuser jets. Env. Fluid Mech., **6**, 43–100.

Lawrence, G.L. (1993). The hydraulics of two-layer flow over a fixed obstacle. *J. Fluid Mech.*, **254**, 605–663.

Lee, J.H.-W. and Chu, V.H. (2003). *Turbulent jets and plumes: a Lagrangian approach.* Dordrecht, Kluwer.

Legg, S., Brieglieb, B., Chang, Y. *et al.* (2009). Improving oceanic overflow representation in climate models. *Bulletin of the American Meteorological Society*, **90** (5), 657–670.

Morton, B.R., Taylor, G.I. and Turner, J.S. (1956). Turbulent gravitational convection from maintained and instantaneous sources. *Proc. Royal Society London*, **A234**, 1–23.

Negretti, E., Socolofsky, S.A. and Jirka, G.H. (2008). Linear stability analysis of inclined two-layer stratified flows. *Phys. Fluids*, **20**, 094104.

Noh, Y. and Fernando, H.J.S. (1991). Gravity current propagation along an incline in the presence of boundary mixing. *J. Geophys. Res.*, **96** (C7), 12586–12592.

Oehy, C.D., De Cesare, G. and Schleiss, A.J. (2010). Effect of inclined jet screen on turbidity current. *J. Hydraul. Res.*, **48** (1) 81–90.

Patel, R.P. (1971). Turbulent jets and wall jets in uniform streaming flow. *Aeronaut. Quart.* **XXII**, 311–326.

Prandtl, L. (1926). Ueber die ausgebildete Turbulenz. *Proc. 2nd Int. Congr. for Appl. Mech.*, Sept. 12–17 Zurich, 62–74.

Princevac, M., Bühler, J. and Schleiss, A.J. (2009). Mass-based depth and velocity scales for gravity currents and related flows. *Env. Fluid Mech.*, **9**, 369–387.

Princevac, M., Bühler, J. and Schleiss, A.J. (2010). Alternative depth-averaged models for gravity currents and free shear flows. *Env. Fluid Mech.*, **10**, 369–386.

Princevac, M., Fernando, H.J.S. and Whiteman, D. (2005). Turbulent entrainment into natural gravity-driven flows. *J. Fluid Mech.*, **533**, 259–268.

Princevac, M., Hunt, J.C.R. and Fernando, H.J.S. 2008: Quasi-Steady Katabatic Winds on Slopes in Wide Valleys: Hydraulic Theory and Observations. J. Atmos. Sci., **65**, 627–643.

Sargent, F.E. and Jirka, G.H. (1987). Experiments on saline wedge. *J. Hydraul. Eng. ASCE*, **113** (10), 1307–1324

Schlichting, H. (1979). *Boundary layer theory.* McGraw Hill, 7th ed.

Sequeiros, O.E., Spinewine, R.T., Beaubouef, T.S., Garcia, M.H. and Parker, G. (2010). Characteristics of velocity and excess density profiles of saline underflows and turbidity currents flowing over a mobile bed. *J. Hydraul. Engrg. ASCE*, **136** (7), 412–433.

Strang, E.J. and Fernando, H.J.S. (2001). Entrainment and mixing in stratified shear flows. *J. Fluid Mech.*, **428**, 349–386.

Townsend, A.A. (1956). *The structure of turbulent shear flow.* Cambridge University Press.

Turner, J.S. (1986). Turbulent entrainment: the development of the entrainment assumption, and its application to geophysical flows. *J. Fluid Mech.* **173**, 431–471.

Wright, S.J. (1994). The effect of ambient turbulence on jet mixing. In: Davies, P.A. and Nieves, V. (eds.) *Recent Research Advances in the Fluid Mechanics of Turbulent Jets and Plumes,* NATO ASI Series, Kluwer, pp. 13–27.

Chapter 21

On the effect of drag on the propagation of compositional gravity currents

G. Constantinescu
Civil and Environmental Engineering Department and IIHR Hydroscience and Industry, The University of Iowa, Iowa City, Iowa, USA

ABSTRACT: Highly resolved 3-D Large Eddy Simulation (LES) are used to study the effects of the additional drag induced by the presence of obstacles on the propagation of lock-exchange Boussinesq Gravity Currents (GCs) in a straight horizontal channel. Two types of configurations are considered. In the first case, a number of identical rectangular cylinders are uniformly distributed over the whole depth and length of the channel. This test case corresponds to a GC propagating in a porous medium of uniform porosity in which the additional drag induced by the cylinders acts over the whole height of the GC. In the second case, an array of identical obstacles in the form of square ribs or 2-D dunes is mounted on the bottom surface of the channel. The additional drag acts only over the lower part of the GC as the obstacle height is smaller than the height of the GC. Both cases are relevant for practical applications, as in most environmental applications GCs propagate over a rough bed (e.g., GCs at the bottom of rivers and oceans), interact with flow retarding devices (e.g., snow avalanches) or advance in a porous medium (e.g., a layer of vegetation). The study analyses the propagation of the GC during the slumping and drag dominated regimes.

1 INTRODUCTION

Gravity currents (GCs) forming on the bottom of rivers and lakes propagate in most cases over large-scale bedforms in the form of dunes. Arrays of obstacles are often used as protective measures on hilly terrains and on the skirts of mountains to stop or slow down gravity currents in the form of powder-snow avalanches. In other cases a gravity current advances through a layer of vegetation or through an array of flow retarding porous screens. As a result of the presence of these obstacles, the structure of the current, its front velocity, and its capacity to entrain sediment from the loose bed over which it propagates may change considerably with respect to the widely studied case of a gravity current propagating over a flat smooth horizontal bed. The latter case has received a significant amount of attention, especially for gravity currents in lock-exchange configurations (e.g., Shin *et al.*, 2004).

Experimental studies of lock-exchange gravity currents propagating in a porous medium in a horizontal channel were reported by Hatcher et al. (2000) and Tanino et al. (2005). Hatcher *et al.* (2000) developed a theoretical model for GCs propagating into a porous medium of uniform porosity in which the presence of the obstacles is modeled by a uniform drag force proportional to the square of the velocity and inversely proportional to the mean length scale of the obstacle. Analysis of the governing shallow flow equations showed that for currents with a finite volume of release, an intermediate buoyancy-turbulent drag flow regime may be present in between the buoyancy-inertia and viscous-buoyancy self-similar regimes. In this intermediate flow regime, the main forces that drive the evolution of the current are the turbulent drag and the buoyancy forces. For full-depth GCs with a large volume of release, the theoretical model predicts that the front velocity, U_f, during the drag dominated regime is proportional to $t^{-1/3}$ for high Reynolds number GCs (cylinder drag coefficient is assumed constant) and $t^{-1/2}$ for low Reynolds number GCs (cylinder drag coefficient is inversely proportional to cylinder Reynolds number).

The case of a GC propagating over a bed containing an array of large-scale bottom-mounted obstacles has received much less attention. The main difference with respect to the case of gravity currents propagating into a porous medium of uniform porosity is that the drag induced by the obstacles acts only over the lower part of the current. The additional drag force is dependent on the height, shape and spacing of the large scale roughness elements. The effects of the bottom obstacles on the flow within a turbulent bottom-propagating GC are, in many regards, similar to those observed for turbulent constant-density flows propagating over rough surfaces (Jimenez, 2004). In both cases, the presence of large-scale obstacles at the channel bottom provides an additional mechanism for energy dissipation. A category of relevant studies deals with the interaction between GCs and isolated surface-mounted obstacles (e.g., see Pawlak and Armi, 2000 for a review of these studies). Recently, this case was investigated numerically using high resolution simulations by Gonzalez-Juez *et al.* (2009, 2010). In particular, the numerical studies provided quantitative insight into the physical mechanisms generating the drag forces on the obstacle during the different stages of the interaction between the current and the obstacle.

The case of a GC propagating over an inclined bed containing large-scale sinusoidal bed deformations was considered by Ozgokmen *et al.* (2004) who investigated the entrainment rates of overflow GCs propagating in an unstratified environment. The numerical simulations were conducted at field scale conditions with constant but unequal viscosities/diffusivities in the horizontal and vertical directions. In a related study, Ozgomen and Fisher (2008) investigated the changes in the front speed, entrainment, drag forces and flow structure as a function of the bottom roughness for the case of a saline current propagating on an inclined rough surface into a temperature stratified ambient fluid after the nose starts lifting from the bottom and the current propagates at the level of neutral buoyancy.

In this work we use LES to investigate: 1) the effect of the presence of an array of uniformly distributed square cylinders on the propagation of a compositional lock-exchange GC. The effects of the Reynolds number and solid volume fraction on the propagation of the GC during the drag dominated regime is discussed; and 2) the effect of the presence of an array of identical bottom obstacles with the same height on the evolution of a lock-exchange compositional lock-exchange GC. The effects of the form (dunes vs. square ribs) and spacing of the bottom obstacles on the propagation of the GC during the slumping and drag-dominated regime are discussed.

2 NUMERICAL SOLVER AND SIMULATIONS SETUP

A finite-volume LES code is used to solve the governing equations on non-uniform Cartesian meshes. A semi-implicit iterative method that employs a staggered conservative space-time discretization is used to advance the equations in time while ensuring second order accuracy in both space and time. A Poisson equation is solved for the pressure using multigrid. A dynamic Smagorinsky model is used to estimate the subgrid-scale viscosity and diffusivity. All operators are discretized using central discretizations, except the convective term in the advection-diffusion equation solved for the concentration for which the QUICK scheme is used. Detailed validation of the code to predict the evolution of intrusive currents and bottom currents propagating over a flat surface at Reynolds numbers defined with the buoyancy velocity and the channel depth as high as Re = 10^6 are discussed by Ooi *et al.* (2007a, 2007b, 2009). Gonzalez-Juez *et al.* (2009) successfully reproduced the measured time-varying drag and lift coefficients of gravity current flows over isolated circular and square cylinders separated by a gap from the bottom wall.

The density difference between the lock fluid and the ambient fluid is small enough to use the Boussinesq approximation. The Navier-Stokes equations and the advection-diffusion equation for the concentration are made dimensionless using the lock-gate opening, h, and the buoyancy velocity, $u_b = \sqrt{g'h}$, where g' is the reduced gravity. The non-dimensional concentration is defined as $C = (\bar{C} - \bar{C}_{min})/(\bar{C}_{max} - \bar{C}_{min})$, where $\bar{C}_{max}$, $\bar{C}_{min}$ represent the maximum (lock fluid) and minimum (ambient fluid) concentrations in the domain and $\bar{C}$ is the dimensional concentration. As shown in the sketch of the lock exchange flow (Fig. 1), the lock gate is positioned in the middle of the computational domain (x/H = 0.0). In the full depth of release case investigated here, h = H, where H is the channel depth. The present paper discusses only the case of GCs with a high volume of release for which $x_0/H >> 1$. As the lock-exchange flow is close to anti-symmetric in all the simulations, only the evolution of the bottom propagating GC is discussed. The time scale used in the discussion of the results is $t_0 = H/u_b$.

The top and bottom surfaces are simulated as no-slip smooth (flat or deformed) solid surfaces. The flow is assumed to be periodic in the spanwise direction (z). A zero normal gradient boundary condition is assumed for the concentration at the top, bottom and at the two end boundaries. All simulations discussed in this paper were conducted with a value of the viscous Schmidt number of 600 corresponding to saline water. The flow field was initialized with the fluid at rest. The mesh contained between 40 and

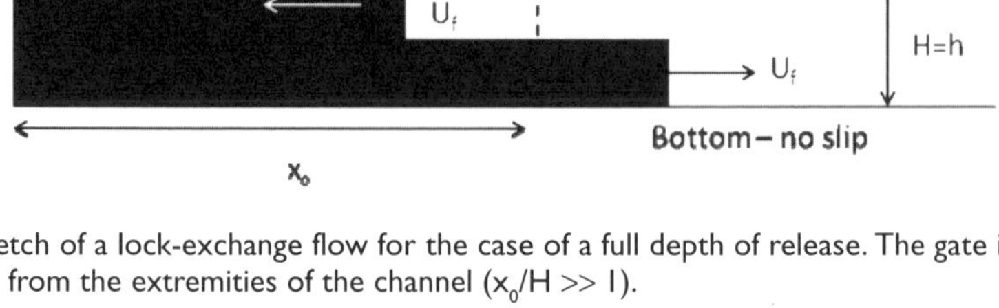

Figure 1 Sketch of a lock-exchange flow for the case of a full depth of release. The gate is positioned far from the extremities of the channel ($x_0/H >> 1$).

200 million cells and the mesh spacing in the wall normal direction was sufficiently small to resolve the viscous boundary layer (no wall functions were used).

3 RESULTS

3.1 Gravity currents propagating in a porous medium

The first series of simulations were conducted with a solid volume fraction (SVF) of 12%. The four simulations were performed for Reynolds numbers defined with u_b and H of 100, 375, 1,500 and 15,000. These simulations are denoted R100-P12, R375-P12, R1500-P12 and R15000-P12, respectively. A number of 597 spanwise-oriented cylinders of diameter d = 0.05 H were uniformly distributed in a staggered arrangement within the whole volume of the channel (Fig. 2). The channel length was 12 H. A second series of simulations were conducted with a Reynolds number of 15,000. The SVF in the R15000-P1, R150000-P5, R15000-P12 and R15000-P25 were 1.25%, 5%, 12% and 25%, respectively. The diameters of the cylinders were 0.035 H, 0.035 H, 0.05 H and 0.07 H, respectively.

Simulation results show that for Re = 100 (Fig. 2) and Re = 375 the interface height defined by C = 0.5 varies linearly with the streamwise distance from the lock gate (x = 0) up to the front position. This is in agreement with experiments conducted for low Reynolds number GCs propagating into a porous medium of uniform porosity (e.g., see Tanino *et al.*, 2005). For Re > 1,000 (e.g., see results in Fig. 3 for Re = 15,000), the linear variation of the interface height with the distance from the lock gate applies up to a certain distance from the front.

In simulations conducted with Re = 100 and Re = 375 (Fig. 4), the GC reaches a regime where $x_f \sim t^{1/2}$, corresponding to $U_f \sim t^{-1/2}$, where U_f is the front velocity. By the time this regime is reached ($t > 100t_0$), the cylinder Reynolds number (Re_d) defined with U_f and d is smaller than one for the cylinders situated within the body of the GC. The cylinder drag coefficient is within the linear drag regime in which $C_D \sim 1/Re_d$. Assuming the linear drag regime holds for all the cylinders, Tanino *et al.* (2005) found

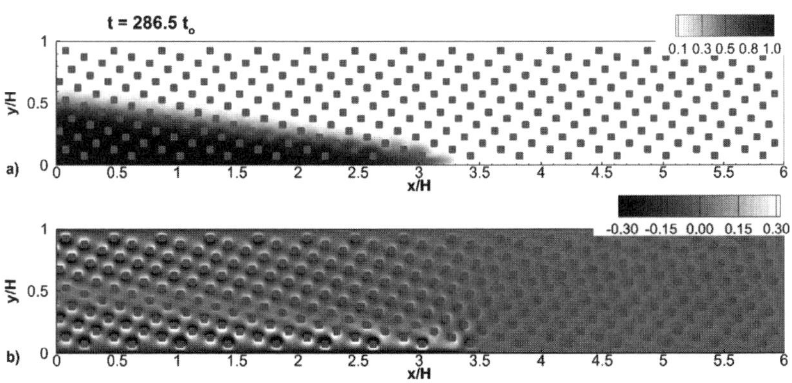

Figure 2 Visualization of the structure of the bottom propagating current in case LR1-P12 (Re = 100) at $t = 286.5t_0$. a) concentration, C; b) out-of-plane vorticity, $\omega_z/(H/u_b)$.

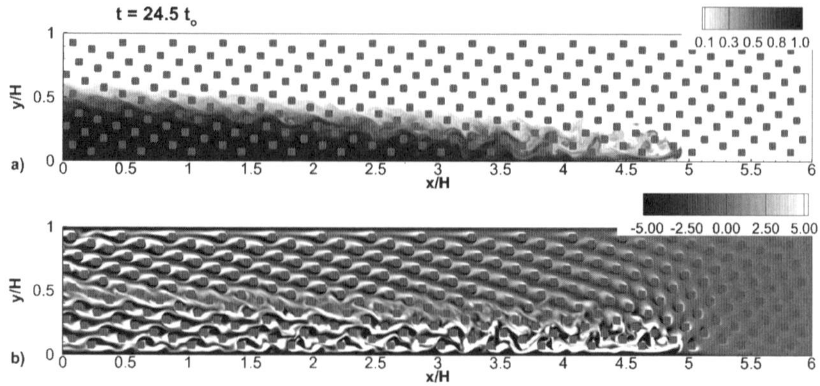

Figure 3 Visualization of the structure of the bottom propagating current in case R15000-P12 at $t = 24.5t_0$. a) concentration, C; b) out-of-plane vorticity, $\omega_z/(H/u_b)$.

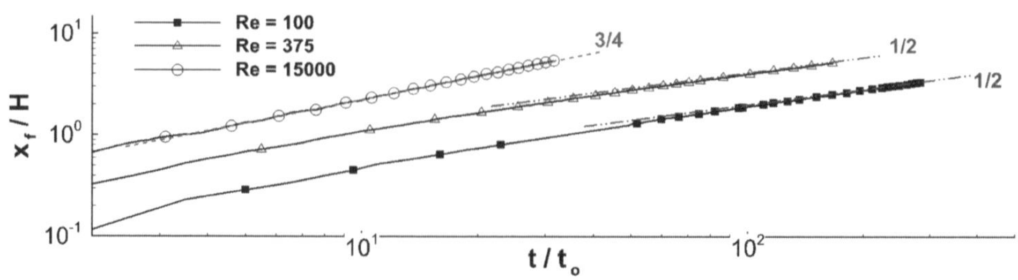

Figure 4 Front position, x_f/H, plotted versus nondimensional time, t/t_0, in the R100-P12, R375-P12, and R15000-P12 simulations conducted with SVF = 12%.

for low Reynolds number GCs propagating in a porous medium of uniform porosity an analytical solution that predicted $x_f \sim t^{1/2}$, consistent with the present numerical solutions.

A regime in which $x_f \sim t^{\alpha}$ is also reached in the Re = 15,000 simulations with SVF = 12%. However, the value of the coefficient α is close to 3/4 (Fig. 4). The main reason is that U_f remains sufficiently high during the whole duration of the simulation (until $x_f \sim 5.5$ H) such that $Re_d \gg 1$ for most of the cylinders situated in between $-x_f/H$ and x_f/H. Most of the empirical formulas for the drag coefficient of a cylinder assume a dependence with Re_d of the form $C_d = c_1 + c_2/Re^{\gamma}$ with $\gamma > 0$. Despite the fact that Re_d is not sufficiently high for a quadratic regime to be clearly established for the drag force on the cylinder in the Re = 15,000 simulations, assuming C_d = constant is an acceptable approximation.

Assuming C_d = constant for all the cylinders, shallow water theory predicts $\alpha = 0.67$ for the drag dominated regime of a high Reynolds number GC with full depth of release. In principle, this temporal decay applies at all streamwise locations within the layer of heavier fluid, including at the front of the GC. Though the difference between the two values of α is not very large, the result is somewhat surprising. One interesting result is that LES predicts that the discharge at the lock gate position (x/H = 0) decays

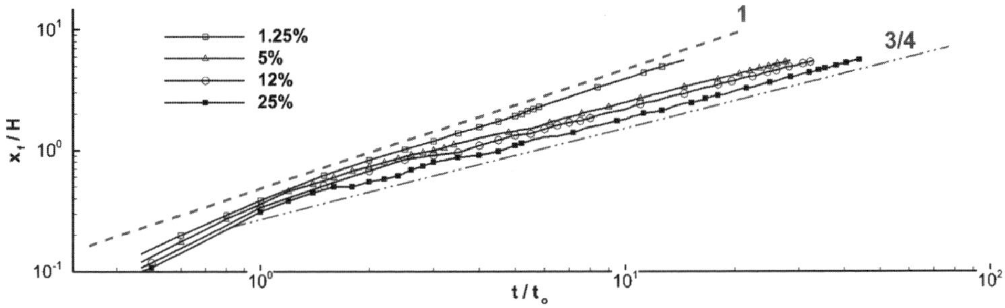

Figure 5 Effect of the SVF on the temporal evolution of the front position, x_f/H, in the R15000-P1, R15000-P5, R15000-P12 and R15000-P25 simulations.

with $t^{-0.33}$. The depth of both layers at x/H = 0 is equal to H/2 at all times. This means that the velocity within the two layers decays with $t^{-0.33}$, which is fully consistent with shallow water theory. This suggests that the GC in LES does not reach a self similar regime. Most probably this is because the large mixing occurring immediately behind the front in the high Reynolds number simulations of GCs advancing in a porous channel with SVF > 5% (e.g., see Fig. 3). Shallow water theory neglects the effect of mixing. This explanation is also supported by the fact that there is a change of the interface shape between regions close to the lock gate where the interface position varies linearly with x and regions close to the front where the variation is nonlinear.

As shown by the results in Fig. 5, a drag dominated regime characterized by a decay of $U_f \sim t^{-0.25}$ is reached in all the simulations conducted for Re = 15,000, provided that the SVF > 5%. The only exception is the simulations with SVF = 1.25% in which the GC remains in the slumping phase ($x_f \sim t$ or $U_f \sim$ constant) during the simulated time ($x_f < 5.5$ H). It is highly probable that in a longer channel the GC will eventually transition to the drag dominated regime.

3.2 Gravity currents propagating over surface-mounted obstacles

3.2.1 Gravity currents with a moderate drag force per streamwise unit length induced by the bottom obstacles

Two types of obstacles are considered. The 2-D dunes are representative of bedforms present at the bottom of rivers. The shape of the dunes is taken from the experiment of Mierlo and de Ruiter (1988) that focused on typical dunes observed in rivers. The ratio between the height, D, and the wavelength, λ, of the dune is 0.05, which is within the typical range observed for dunes in small and medium-size rivers. In the base case, denoted LR-F, the current propagates over a flat bed (Fig. 6a). In the LR-D15 (Fig. 6b) and LR-R15 (Fig. 6c) simulations, the dunes and, respectively, the square ribs are of equal height (D = 0.15 H) and equal wavelength (λ = 3 H, λ/D = 20). To understand the effect of increasing the total drag force per unit streamwise length on the evolution of the current, an additional simulation, denoted LR-R15-HD (Fig. 8), was performed. In the latter simulation the ribs were identical to the ones used in case LR-R15, but the

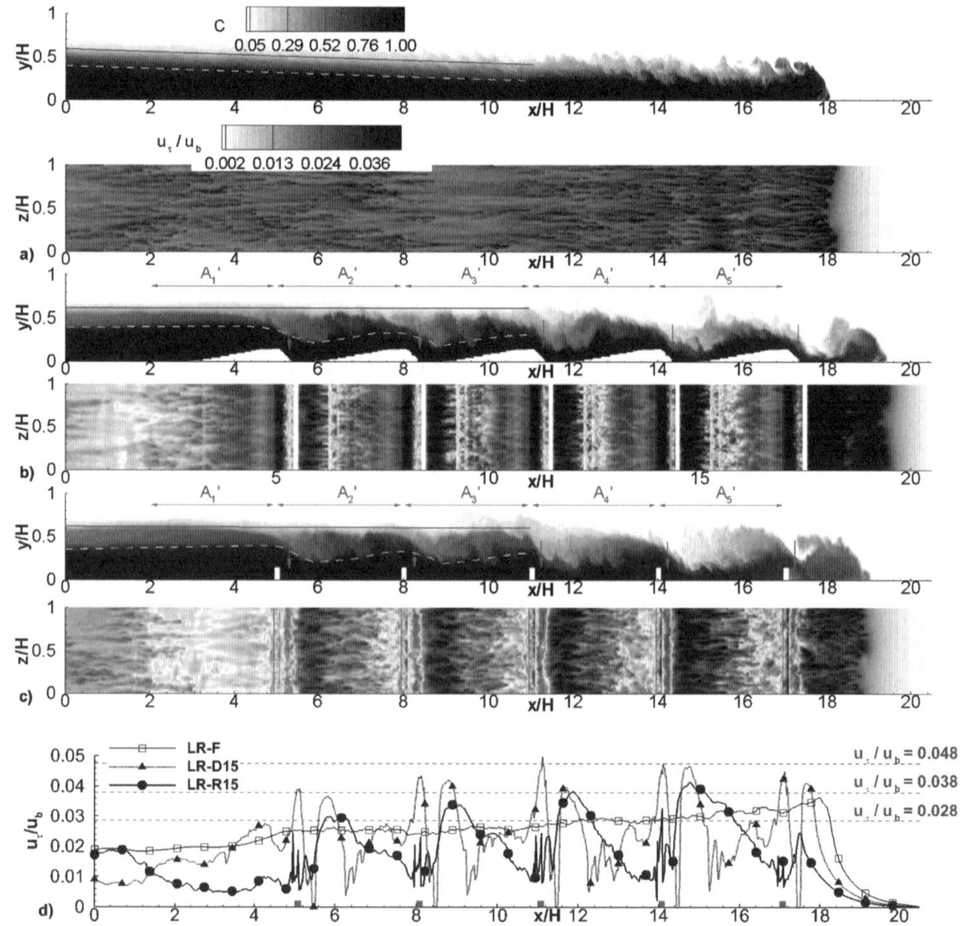

Figure 6 Distributions of concentration in a vertical (x – y) section and of bed friction velocity magnitude, u_τ/u_b, in the low Reynolds number simulations with a flat bed and with obstacles of height D = 0.15 H when $x_f \cong 19$ H. a) LR-F; b) LR-D15; c) LR-R15. The aspect ratio is 1:2 in the x – y and x – z plots. The solid line shows the interface between the ambient fluid and the layer of mixed fluid. The dashed line shows the interface between the layer of mixed fluid and the bottom layer of heavier fluid. The arrows in frames b and c point toward the jet-like flow forming downstream of the top/crest of the obstacles. Also shown in frame d are the distributions of the spanwise-averaged bed friction velocity magnitude in the three simulations at the time instants at which the currents are visualized in frames a, b and c.

spacing was reduced by a factor of three (λ = H). In all four simulations, the obstacles are placed on the bottom surface in the region with x/H > 0 and on the top surface in the region with x/H < 0. The Reynolds number defined with u_b and H is Re = 48,000.

The effect of the presence of the bottom obstacles on the structure of the GC can be inferred from the concentration distributions in Fig. 6 that shows the GC when $x_f/H \cong 18$. The obstacles and the propagation of the hydraulic jumps forming each time the front overtakes an obstacle induce the formation of a layer of mixed fluid of variable height. The mixing at the front and over the tail is larger for the ribs. This is because ribs have a higher degree of bluntness compared to dunes of equal height.

A short time after the removal of the lock gate, lock-exchange gravity currents propagating over a flat smooth horizontal surface reach a regime in which the front velocity, U_f, is close to constant (slumping phase). Figure 7 shows the temporal evolution of the front position, x_f/H, for the heavier current in the four simulations. In case LR-F, after the end of the initial acceleration phase ($t/t_0 = 3.5$), the front trajectory can be well approximated by a line of constant slope, until the end of the simulation (see Fig. 7). The non-dimensional front velocity, $\bar{U}_f = U_f/u_b$, during the slumping phase is 0.45, in very good agreement with experiments conducted for currents propagating over flat horizontal surfaces at Re ≅ 48,000. The presence of ribs or dunes slows down the advancement of the front compared to the flat bed case. The front trajectories are very close until the front approaches the first obstacle at x = 5 H. The trajectories start diverging for $t/t_0 > 10$. For $t/t_0 > 16$ the mean slope of the trajectory can be considered, to a good approximation, constant in the LR-R15 and LR-D15 simulations. This indicates that under certain conditions (e.g., depending on the values of D/H, λ/H, the shape of the obstacles) a slumping phase in which the front velocity is approximately constant is also present for GCs propagating over a bed containing obstacles.

In the simulations with obstacles, the slope of the front trajectory is subject to larger temporal variations as compared to the ones observed in the simulations with a smooth bed. This is because the front is decelerating as it approaches the crest of each dune or the upstream face of each rib, due to the adverse pressure gradients induced by the upstream surface of the obstacle. The front velocity increases above the mean value during the time the head is projected downwards, as it passes the crest of the dune or the top of the rib.

The mean value of $\bar{U}_f$, estimated based on the mean slope of the front trajectory ($t/t_0 > 20$) in Fig. 7, is 0.34 and 0.4 for cases LR-R15 and LR-D15, respectively. The mean value obtained from the front trajectory is in very good agreement with estimates of the mean front velocity based on the time it takes the front to advance

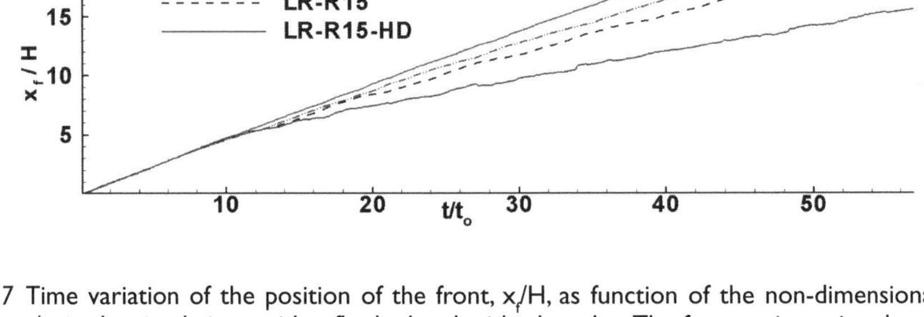

Figure 7 Time variation of the position of the front, x_f/H, as function of the non-dimensional time, t/t_0 in the simulations with a flat bed and with obstacles. The front trajectories show that a regime in which the mean front velocity is constant in time is reached in the LR-F, LRR15 and LR-D15 simulations. The solid gray line shows the front trajectory for case LR-R15-HD in which x_f increases proportionally to t^α ($\alpha = 0.72$) in the later stages of the evolution of the current.

between two successive obstacles. In the latter case, present results show that, past the second obstacle, the mean front velocity in between two successive obstacles is close to independent of the rank of the obstacle in the series. The mean front velocity in the LR-D15 and LR-R15 simulations is 12% and, respectively, 24% lower than the value ($\bar{U}_f = 0.45$) computed for the flat bed case (LR-F). The additional form drag induced by the bottom obstacles is the main cause for the reduction of the mean value of the front velocity during the slumping phase in cases LR-R15 and LR-D15 as compared to case LR-F. For obstacles of identical heights, the form drag is larger for the obstacles with a higher degree of bluntness.

Figure 6 compares the distributions of the nondimensional bed friction velocity, u_τ/u_b, during the later stages of the propagation of the GC in the LR-F, LR-D15 and LR-R15 simulations, for similar front locations, x_f/H. A strong jet-like flow forms downstream the top/crest of each obstacle, a short time after the front begins to move away from that obstacle. For obstacles of height D = 0.15 H, the flow becomes supercritical (local Froude number is larger than one) within the jet-like flow for those obstacles close to the front. The flow remains subcritical over the whole length of the GC in the absence of any obstacles.

In the tail sections far behind the front, significant differences are observed between flows with and without obstacles. In the absence of obstacles, a stably stratified tilted interface develops that is depleted of large-scale eddies (see also Ooi *et al.*, 2009). This is in contrast to GCs propagating over obstacles, for which a mixed layer of varying height develops in between the dense bottom fluid layer and the top layer of ambient fluid, as shown in Figs. 6b and 6c. The top of the mixed fluid layer is close to horizontal, and the jet-like flow becomes subcritical more than 2–3λ behind the front. The bottom of the mixed fluid layer undergoes quasi regular deformations with the period of the obstacles.

In case LR-F, streaks of high and low u_τ are present over most of the bottom wall surface in between $x/H = 0$ and the front of the gravity current (Fig. 6a). Consistent with the observations of Ooi *et al.* (2009), in the strongly turbulent sections of the GC the near-wall flow contains the usual coherent structures associated with a constant-density turbulent boundary layer. Streaks of high and low streamwise velocity exist in the vicinity of the bed. The streaks modulate the u_τ-distribution at the bed. In the present simulations conducted with a high volume of release, these streaks do not exist in the most upstream regions of the tail ($x/H << 0$, not shown), where the fluid is being accelerated from rest and the turbulence is still weak.

When obstacles are present, the contours of u_τ in Figs. 6b and 6c show that average values of u_τ change significantly in the streamwise direction. The streamwise velocity streaks are much stronger in those regions where the jet-like flow moves at a high speed parallel to the bottom surface, and they are weaker downstream of those regions. The vertical vorticity ω_y (e.g., see Fig. 8 for case LR-R15) is less sensitive to the streamwise variations of the spanwise-averaged values of the streamwise velocity close to the bed and thus to u_τ. Streamwise streaks of positive and negative ω_y are observed in Fig. 8 over most of the length of the bottom surface in the region $x/H > 0$. This confirms that, similar to the flat bed case, velocity streaks are present over the bottom surface in between $x/H = 0$ and the front. For the case of dunes, vorticity streaks cover the entire sediment bed, including the steep lee side, where the flow remains attached despite advancing in a strong adverse pressure gradient.

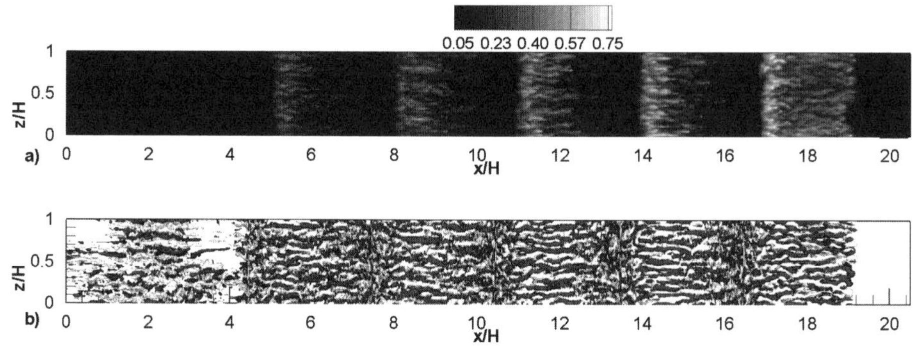

Figure 8 Visualization of the flow structure in the vicinity of the channel bottom in case LR-R15. a) streamwise velocity, u/u_b, contours showing the regions containing streaks of high and low streamwise velocity. The contours are shown in a surface situated at 0.006 H from the deformed bottom surface containing ribs; b) vertical vorticity contours, $\omega_y H/u_b$, on the bottom surface. The contours are shown on a surface situated at 0.006 H from the deformed bottom surface containing ribs. The view is from below the bottom propagating current. The light and dark vorticity contours correspond to $\omega_y = 2u_b/H$ and $\omega_y = -2u_b/H$, respectively. The aspect ratio x:z is 1:2.

The distributions of the instantaneous and spanwise averaged bed friction velocity magnitude, in Fig. 6 reveal strong qualitative and quantitative differences between the flat bed and cases with obstacles. In the flat bed case, after the stably stratified tilted layer has formed, the spanwise coherence of the billows is no longer strong enough to induce significant variations in the streamwise distributions of u_τ. Consequently, the rapid initial growth of u_τ at the current front is followed by a mild, nearly linear decay all the way to x/H = 0 (Fig. 6d). Away from the GC head, the distributions of u_τ in cases LR-R15 and LR-D15 are strongly modulated by the array of obstacles. Moreover, these distributions they depend on the obstacle shape (Fig. 6d). In contrast to the flat bed case, for obstacles the largest u_τ-values generally do not occur close to the front of the GC. Rather, u_τ peaks where the jet-like flow developing over the obstacles impinges on the bottom wall. With time, the distributions of u_τ between successive obstacles become nearly independent of the obstacle rank (Fig. 6d).

Figure 9 compares the total drag coefficient, C_D, induced by the passage of the bottom propagating current over the streamwise region associated with the first (2.5 < x/H < 5.5) obstacle in the series for the LR-F, LR-R15 and LR-D15 simulations. The total drag coefficient is calculated as

$$C_D = \frac{F_p + F_f}{0.5\rho u_b^2 L_3 \lambda} \tag{1}$$

where F_p and F_f are the streamwise components of the pressure force acting on the front and back surfaces of the obstacle and of the friction force, respectively. Both forces are calculated over a streamwise region of length λ. The width of the domain is L3 and the buoyancy velocity is u_b.

The pressure drag is equal to zero in the LR-F simulation. F_f increases until the front reaches the end of the streamwise region of length λ = 3 H over which the friction drag is calculated. It then remains approximately constant (C_D ~ 0.0015) for about $15t_0$ during which the head and dissipative wake regions propagate over that streamwise region. Then, F_f starts decreasing slowly.

The average levels of the friction drag in the LR-R15 and LR-D15 simulations are comparable to the one in the LR-F simulation. Results in Figure 9 show that in the simulations with obstacles of wavelength λ = 3 H, the pressure drag is about one order of magnitude higher than the friction drag. This confirms the important role played by the obstacles in the evolution of the current in cases LR-R15 and LR-D15.

The temporal variation of C_D over the streamwise region associated with the first rib in the LR-R15 simulation is, in many respects, qualitatively similar to that observed by Gonzalez-Juez *et al.* (2009) in a study of currents impinging on isolated obstacles. C_D increases exponentially during the impact stage, which lasts about $6t_0$ in case LR-R15. The increase in C_D is due to the interaction of the front with the bluff obstacle, which results in a strong pressure increase on the upstream face of the rib. The transient phase is characterized by large-scale temporal variations in the values of C_D. Then, C_D reaches a second maximum ($t \sim 12t_0$ in Fig. 9) during the transient phase. This second peak is induced by the decrease in the pressure on the downstream face of the rib, due to the increase in the strength of the recirculating vortex forming in between the heavier fluid from the splash that reaches the channel bottom and the downstream face of the rib. While in the case of an isolated obstacle C_D reaches fairly quickly a quasi-steady regime, in the present case the transient lasts for a much longer time. This is because of the passage of the backward propagating jumps that form each time when the front reaches a higher-rank obstacle. The strength of the transient induced by the passage of the backward jumps decreases with the rank of the obstacle at which the jump originated. In a good approximation, the quasi-steady regime around the first rib is reached by the time the backward jump originating at the third rib in the series passes the first rib.

The temporal history of C_D in the simulations with dunes and ribs of equal height is qualitatively similar during all the three phases. The largest quantitative difference

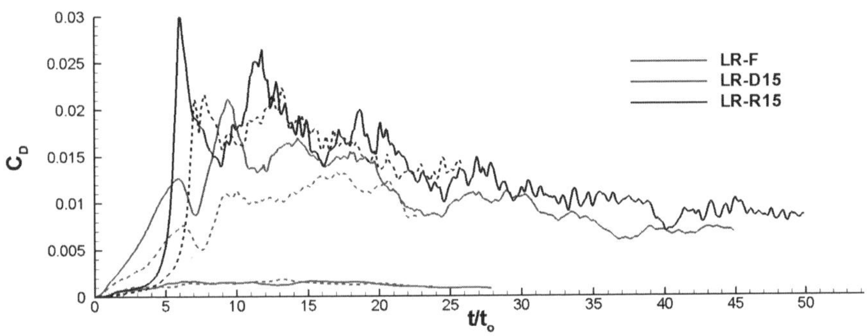

Figure 9 Temporal variation of the total drag coefficient over the region occupied by the first obstacle (2.5 < x/H < 5.5, solid line) in the LR-F, LR-D15 and LR-R15 simulations.

occurs during the impact stage. As a result of the lower degree of bluntness of the dunes, the front deceleration is smaller and the pressure forces on the upslope face of the dune are weaker than those induced on the upstream face of the corresponding rib in case LR-R15. This explains the observed decrease of the peak value of C_D at the end of the impact stage by about 70% with respect to case LR-R15. Moreover, the peak value of C_D in the simulation with dunes is not reached at the end of the impact stage, but rather during the transient stage. For a given obstacle rank, the maximum values of C_D during the transient phase in the simulation with dunes are only 30–50% lower than the overall maximum value of C_D in the simulation with ribs. The values reached by C_D during the quasi-steady regime in the two simulations are close.

3.2.2 *Gravity currents with a large drag force per streamwise unit length induced by the bottom obstacles*

The presence of a slumping phase is dependent on the total drag per unit streamwise length acting on the current. To prove that, an additional simulation (case LR-R15-HD) was performed in which the spacing between consecutive ribs was reduced by a factor of three. This simulation contained 22 ribs on the bottom wall compared to 5 in case LR-R15. The structure of the GC in case LR-R15-HD is visualized in Fig. 10 using concentration and out-of-plane vorticity contours. As opposed to the case of GCs propagating in a porous medium, the interface of the GC in case LR-R15-HD does not vary linearly with the distance from the lock gate. The temporal evolution of the front position in case LR-R15-HD is shown in Fig. 7 (linear-linear plot) and Fig. 11 (log-log plot). The current transitions to the slumping phase ($\bar{U}_f = 0.45$) before it starts interacting with the first rib in the series (t ~ $10t_0$). After the current overtakes the first few ribs, the front velocity is close to $0.28u_b$, which is significantly smaller than the front velocity ($U_f = 0.34u_b$) computed during the slumping phase in case LR-R15. However, while propagating over the ribs, the front velocity never reaches a regime where the front velocity is constant over time.

The log-log plot in Fig. 11 shows that, after the current overtakes the first couple of ribs in the series ($t > 20t_0$), $x_f(t) \sim t^{\alpha}$ with $\alpha = 0.72$. This value is close to the one ($\alpha = 0.67$) given by shallow water theory and the one ($\alpha = 3/4$) predicted for high Reynolds number GCs propagating in a porous medium during the drag-dominated regime. This is despite the fact that the structure of the GC in case LR-R15-HD (Fig. 10) is quite different than the one observed for GCs propagating in a porous medium in which the drag force acts over the whole height of the current (e.g., see

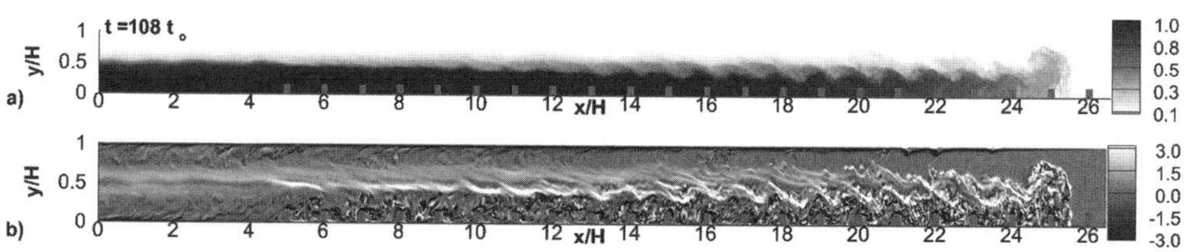

Figure 10 Visualization of the structure of the bottom propagating current in case LR-R15-HD at $t = 108t_0$. a) concentration; b) out-of-plane vorticity, $\omega_z/(H/u_b)$.

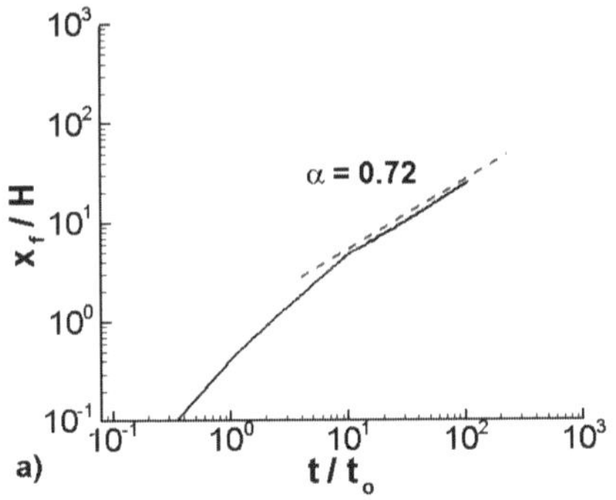

Figure 11 Time variation of the non-dimensional position of the front, x_f/H, as function of the nondimensional time, t/t_0, in the LR-R15-HD simulation plotted in log-log scale. The dashed line shows a best fit of the form $x_f = ct^{\alpha}$.

Fig. 3). The presence of a regime in which $x_f(t) \sim t^{\alpha}$ with $\alpha < 1$ means that a drag-dominated (non-linear) regime is present in case LR-R15-HD.

4 SUMMARY

This study used results of LES to investigate the physics of lock exchange Boussinesq compositional gravity currents with a large volume of release propagating: 1) in a porous channel containing uniformly distributed square cylinders; and 2) over an array of identical 2D obstacles (dunes and square ribs). In the simulations with obstacles, the form drag was much larger than the friction drag.

Simulation results conducted for GCs propagating in a porous channel showed that for a sufficiently high solid volume fraction, the interface elevation varies linearly with the streamwise position until some distance behind the front of the current. In the case of low Reynolds number currents for which the cylinder Reynolds number is of order one or lower, the linear variation is observed until the front position. For small solid volume fractions (e.g., SVF = 1.25%), the current propagated with constant front velocity (slumping phase) until it reached the end of the channel. This behavior is similar to the one observed for gravity currents with a high volume of release propagating over a flat horizontal smooth surface in channels with no cylinders. For larger solid volume fractions (SVF > 5%), the form drag induced by the cylinders was sufficiently large to induce the decay of the front velocity with time regardless of the Reynolds number.

Consistent with experiments and analytical models based on shallow water theory (e.g., Hatcher *et al.*, 2000, Tanino *et al.*, 2005), LES shows that for sufficiently high solid volume fractions low Reynolds number currents transition to a drag dominated regime in which $U_f \sim t^{\beta}$, where $\beta = -0.5$. By contrast, high Reynolds number GCs, for which the cylinder Reynolds number (Re_D) is high enough such that the drag coefficient on the cylinders can be considered constant, transition first to a drag dominated regime in which $\beta = -0.25$. Given the fact that the decay of the mean velocity in the forward and backward propagating GCs is consistent with shallow water theory at locations situated close to the lock gate, the difference between this value and the one predicted by shallow water theory ($\beta = -0.33$) was attributed to the large mixing occurring in the head region for GCs propagating in a porous channel at high Reynolds numbers. Shallow water theory does not account for mixing effects.

Simulation results showed that gravity currents propagating over an array of 2-D identical obstacles reach, under certain conditions, a slumping phase in which the front velocity is approximately constant. The front velocity is smaller than the one reached by the same gravity current propagating over a flat smooth surface (no obstacles) and is a function of the relative degree of bluntness of the obstacles in the series. The conditions for a slumping phase to be present for an extended time, during which the front propagates over a large number of obstacles, depend on the magnitude of the drag force per unit streamwise length induced by the obstacles. If the drag force per unit length increases due to an increase in the size of the obstacles or to a decrease in their spacing, then the slumping phase can be very short, or it may not be present at all. In contrast to the case of a current propagating over a flat surface for which the front velocity remains constant until the reflected disturbance reaches the front, the front velocity can start to decay with time due to the added drag force induced by the obstacles.

A simulation conducted with densely-spaced ribs in a long channel showed that after a short acceleration phase the current transitioned directly to a turbulent drag-dominated regime in which the front velocity decays proportionally to t^{β}, with $\beta \sim -0.28$. This value is close to the one predicted for high Reynolds number gravity currents with a large volume of release propagating into a porous medium of uniform porosity. This shows that the behavior of the gravity current is mainly a function of the total drag force per unit streamwise length induced by the obstacles. The way in which this force is applied on the gravity current (e.g., over its whole height or over its bottom part) plays a rather secondary role in determining the evolution of the front velocity during the drag dominated regime.

The next step will be to investigate the propagation of bottom propagating currents with a small volume a release in a porous medium and over bottom-mounted obstacles, a case that is very relevant for river and snow avalanches applications. Moreover, in most practical applications related to rivers, lakes and oceans, the GC propagates over a rough surface which is inclined with respect to the horizontal and/or in a stratified environment. In the case of rivers, GCs do not propagate in a still environment, but rather move within a turbulent open channel flow. The present model is being extended to study these cases that, while adding more complexity, will allow a better understanding of GCs occurring in the environment.

ACKNOWLEDGEMENTS

G. Constantinescu would like to thank Dr. T. Tokyay and Dr. A. Yuksel for their help in preparing this paper and for performing the simulations and data analysis. G. Constantinescu would like to acknowledge the TRACC facility at the Argonne National Laboratory and Taiwan's National Center for High Performance Computing (NCHC) for providing the computational resources needed to perform some of the simulations.

REFERENCES

Gonzalez-Juez, E., Meiburg, E. and Constantinescu, G. (2009). Gravity currents impinging on bottom mounted square cylinders: Flow fields and associated forces, *J. Fluid Mech.*, 631, 65–102.

Gonzalez-Juez, E., Meiburg, E., Tokyay, T. and Constantinescu, G. (2010). Gravity current flow past a circular cylinder: Forces and wall shear stresses and implications for scour, *J. Fluid Mech.*, 649, 69–102.

Hatcher, L., Hogg, A.J. and Woods, A.W. (2000). The effects of drag on turbulent gravity currents, *J. Fluid Mech.*, 416, 297–314.

Jimenez, J. (2004). Turbulent flows over rough walls, *Annual Review of Fluid Mech.*, 36, 173–196.

Mierlo, M.C. and de Ruiter, J.C. (1988). Turbulence measurements over artificial dunes, *Report Q789*, Delft Hydraulics Laboratory, Delft, Netherlands.

Ooi, S.K., Constantinescu, S.G. and Weber, L. (2007a). Two-dimensional large-eddy simulation of lock-exchange gravity current flows at high Grashof numbers, *J. Hydraulic Engineering*, 9, 1037–1047.

Ooi, S.K., Constantinescu, S.G. and Weber, L. (2007b). A numerical study of intrusive compositional gravity currents. *Physics of Fluids*, 19, 076602.

Ooi, S.K., Constantinescu, S.G. and Weber, L. (2009). Numerical simulations of lock exchange compositional gravity currents, *J. Fluid Mech.*, 635, 361–388.

Ozgokmen, T.M. and Fisher, P.F. 2008. On the role of bottom roughness in overflows, *Ocean Modeling* 20, 336–361.

Ozgokmen, T.M., Fisher, P.F., Duan, J. and Iliescu, T. (2004). Entrainment in bottom gravity currents over complex topography from three-dimensional nonhydrostatic simulations, *Geophysical Research Letters*, 31, L13212.

Pawlak, G. and Armi, L. (2000). Mixing and entrainment in developing stratified currents, *J. Fluid Mech.*, 424, 45–73.

Shin, J., Dalziel, S. and Linden, P.F. (2004). Gravity currents produced by lock exchange, *J. Fluid Mech.*, 521, 1–34.

Tanino, Y., Nepf, H.M. and Kulis, P.S. (2005). Gravity currents in aquatic canopies, *Water Resources Research*, 41, W12402.

Part 6

Mass transfer

Chapter 22

Gas transfer at water surfaces

B. Jähne
Heidelberg Collaboratory for Image Processing (HCI)
at Interdisciplinary Center for Scientific Computing (IWR)
and Institut für Umweltphysik, University of Heidelberg, Heidelberg, Germany

ABSTRACT: The exchange of inert and sparingly soluble gases including carbon dioxide, methane, and oxygen between the atmosphere and oceans is controlled by a thin 20–200 μm thick mass boundary sublayer at the top of the ocean. The hydrodynamics in this layer is significantly different from boundary layers at rigid walls since the orbital motion of the waves is of the same order as the velocities in the viscous boundary layer. Therefore there is no simple analogy between momentum and mass transfer. Starting with the knowledge available at the first International Symposium on "Gas Transfer at Water Surfaces" in 1983 at Cornell University, co-initiated by Gerhard Jirka, the parameters controlling air-sea gas transfer are discussed. Then it will be shown how in the wake of this symposium novel imaging techniques gradually evolved, which give direct insight into the mechanisms of the transfer processes at the air-water interface.

1 INTRODUCTION

It was at the first International Symposium on Gas Transfer at Water Surfaces in 1983 that I met Gerhard Jirka for the first time. Together with Wilfried Brutsaert, he has co-initiated a new symposium on a then emerging scientific topic, which was studied in different research fields without much contacts. In the foreword of the proceedings Wilfried and Gerhard wrote [5].

"*This interfacial mass transfer is, by its very nature, highly complex. The air and the water are usually in turbulent motion, and the interface between them is irregular, and disturbed by waves, sometimes accompanied by breaking, spray and bubble formation. Thus the transfer involves a wide variety of physical phenomena occurring over a wide range of scales. As a consequence, scientists and engineers from diverse disciplines and problem areas, have approached the problem, often with greatly differing analytical and experimental techniques and methodologies.*

It was against this background, that an International Symposium on Gas Transfer at Water Surfaces was held at Cornell University from June 13 to 15, 1983. The objectives of the Symposium were to summarize the state of the art and to promote scientific understanding of the gas transfer processes. In particular, it was intended as an open forum to stimulate dialogue and discussion among workers in different disciplines such as physical chemistry and chemical engineering, fluid mechanics and

Table 1 List of international symposia on "Gas Transfer at Water Surfaces", initiated by Gerhard Jirka.

	Date & year	*Place*	*Chairs [reference proceedings]*
1st	June 13–15, 1983	Cornell University, Ithaca	*W. Brutsaert & G.H. Jirka* [5]
2nd	Sept. 11–14, 1990	University of Minnesota, Minneapolis	*Wilhelms & Gulliver* [54]
3rd	July 24–27, 1995	Heidelberg University	*B. Jähne & E.C. Monahan* [27]
4th	June 5–8, 2000	University of Miami	*Donelan, Drennan, Saltzman & Wanninkhof* [9]
5th	May 2–6, 2005	Liege University	*A.V. Borges & R. Wanninkhof* [2]
6th	May 17–21, 2010	Kyoto University	*S. Komori & W. McGilles* [36]
7th	2015	University of Washington, Seattle	*A. Jessup*

hydrology, hydraulics and environmental engineering, geochemistry, oceanography, climatology and meteorology."

The conference was very successful and evolved into a series of symposia, which took place every five years (Table 1).

2 PARAMETRIZATION OF AIR-SEA GAS TRANSFER RATE

2.1 Thirty year hunt for wind speed relation

The amount of species exchanged between the air and water across the interface can be described by a quantity with the units of a velocity. It represents the velocity with which a tracer is pushed by an imaginary piston across the surface. This quantity is the *transfer velocity k* (also known as the *piston velocity*, *gas exchange rate* or *transfer coefficient*). It is defined as the flux density divided by the concentration difference between the surface and the bulk at some reference level:

$$k = \frac{j_c}{C_s - C_b}. \tag{1}$$

In the initial stages of the research on air-water gas transfer, the analogy between momentum and mass transfer was a key idea. Under most natural conditions, air water gas transfer is driven by the wind stress applied at the water surface. Therefore there should be a direct relation between the wind speed and the gas transfer.

Figure 1a shows one of the earliest collections of field data compiled already in 1982 [20]. This study resulted in a first semi-empirical relation between the gas transfer velocity and the friction velocity in air, u_{*a}, already four year before the well-known Liss-Merlivat relation [38]:

$$k = \begin{cases} 2.9 \cdot 10^{-3}\, Sc^{-2/3} u_{*a} & u_{*a} < u_{*c} \\ 2.9 \cdot 10^{-3}\, Sc^{-2/3} u_{*a} + 5.5 \cdot 10^{-3}\, Sc^{-1/2} (u_{*a} - u_{*c}) & u_{*a} \geq u_{*c} \end{cases} \tag{2}$$

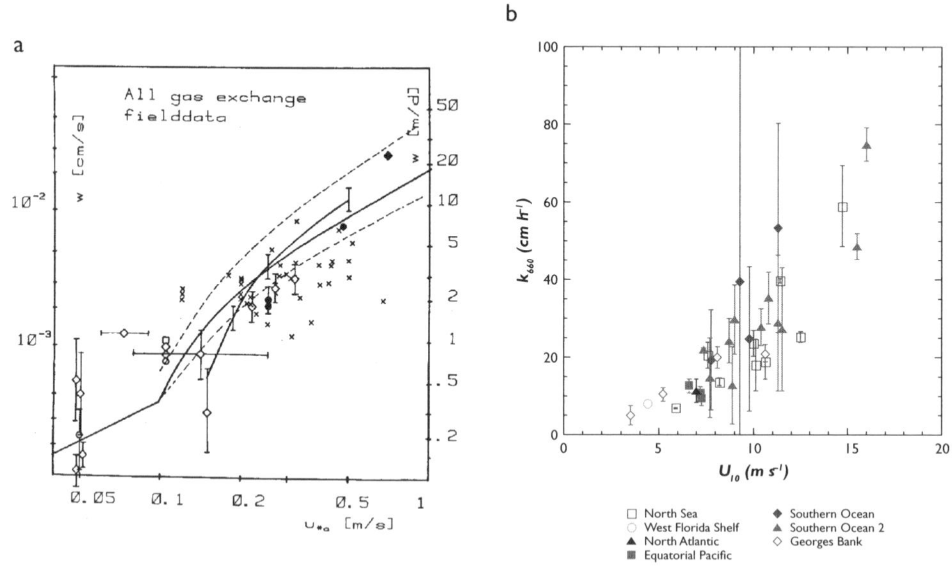

Figure 1 **a.** Early collection of gas exchange field measurements [20]. The solid line is the semi-empiric relation according to (2). The two dashed lines mark the lower and upper limits of the gas transfer velocities measured in wind/waves flumes with clean water surfaces. **b.** Latest collection of field measurements using dual-deliberate tracers ($^3He/SF_6$) in the review paper of Wanninkhof *et al.* [51].

with $u_{*_c} = 0.1$ m/s This relation never received much attention because it was only published in German in a report to the Battelle Institute in Frankfurt.

The data collection in Fig. 1a clearly shows that the gas transfer velocity increases with the friction velocity. But the significant scatter of the data with a variation by a factor of three at the same friction velocity is also evident. In the latest review paper by Wanninkhof *et al.* [51]—almost thirty years later—the picture is still the very same (Figure 1b). Therefore it is evident that there is no unique relation between the gas transfer velocity and the wind speed.

Gerhard Jirka's contribution in this field were experimental and modeling studies on the combined effect of stream flow generated turbulence and wind generated turbulence on gas exchange in riverine systems [6, 30].

2.2 Other parameters then wind speed

Already at the Cornell Symposium in 1983 it was evident that wind speed is not be the only parameter controlling air-water gas transfer and thus there were many contributions discussing the influence of other parameters then wind speed. This discussion is still going on today, because many details of mechanisms for air-sea gas transfer are still not resolved. Here the influence of the wave field, surface films, and wave breaking are discussed.

2.2.1 Wave influence

Wind waves cannot be regarded as static roughness elements for the liquid flow because their characteristic velocity is of the same order of magnitude as the velocity in the shear layer at the surface. This fact causes a basic asymmetry between the turbulent processes on the air and on the water sides of the interface [19]. Therefore the wave effect on the turbulent transfer in the water is much stronger and of quite different character than in the air. This was demonstrated by Jähne *et al.* [24], who showed that the gas transfer rate at a rough wavy surface is about 3 to 5 times faster than at a smooth surface at the same friction velocity. Field experiments presented at the Cornell Symposium in 1983 also clearly showed that the gas transfer does not only depend on the wind speed alone [23, 45].

The surface increase by the wavy surface cannot explain this enhancement by waves, because it is well below 20% [48]. Part of the enhancement can be explained, however, by a change in the Schmidt number dependency of the gas transfer rate, which is $\propto Sc^{-2/3}$ at a smooth surface, but $\propto Sc^{-1/2}$ at a wavy surface [24, 37].

When waves are generated by wind, energy is not only transferred via shear stress into the water but a second energy cycle is established. The energy put by the turbulent wind into the wave field is transferred to other wave numbers by nonlinear wave-wave interaction and finally dissipated by wave breaking, viscous dissipation, and turbulence [34]. It is not yet clear to what extent microscale wave breaking or micro Langmuir circulations [11] can account for the observed enhanced gas transfer rates at a rough, wind-driven wavy water surface. Experimental results suggest that the gas transfer rate is better correlated with the *mean square slope* of the waves as an integral measure for the nonlinearity of the wind wave field than with the wind speed [24, 26].

2.2.2 Monelcular surface films

A monomolecular film on the water surface creates forces that works against the contraction of surface elements. This is the point at which the physicochemical structure of the surface influences the structure of the near-surface turbulence as well as the generation of waves. As at a rigid wall, a strong film pressure at the surface maintains a two-dimensional continuity at the interface just as at a rigid wall. Therefore the transfer velocity remains $\propto Sc^{-2/3}$ in this case [24]. But still too few measurements at sea are available to established the influence of surfactants on gas transfer for oceanic conditions more quantitatively.

2.2.3 Wave breaking

At high wind speeds wave breaking with the entrainment of bubbles may enhance gas transfer further [4, 29, 39]. This phenomenon complicates the gas exchange between atmosphere and the oceans considerably [31, 57]. First, bubbles constitute an additional exchange surface. This surface is, however, only effective for gases with low solubility. For gases with high solubility, the gas bubbles quickly comes into equilibrium so that a bubble takes place in the exchange only for a fraction of its life time. Thus bubble-mediated gas exchange depends - in contrast to the exchange at the free surface - on the solubility of the gas tracer. Second, bubble-mediated gas transfer shifts the equilibrium value to slight supersaturation due to the enhanced pressure in the bubbles by surface tension and hydrostatic pressure. Third, breaking waves also enhance near-surface turbulence during the breaking event and the resurfacing of submerged bubbles.

Experimental data are still too sparse at high wind speeds for a reliable estimate of the overall influence of breaking waves on gas transfer. In the most recent review paper Wanninkhof *et al.* [51] therefore came to the conclusion "*The role of breaking waves at high wind speeds is recognized as important, but as yet there remains no reliable way to accurately quantify the effect of breaking waves on gas exchange.*"

2.3 Experimental challenges resulting from characteristic scales of gas transfer across the interface

At this point, the question arises why there had been not really much progress in our understanding of the mechanisms of air/water gas transfer in almost thirty year. The answer is strikingly simply. The characteristic scales of the transfer across the air/water interface are such that they make an experimental investigation of the mechanism a very difficult undertaking.

2.3.1 Mass balance time scale

Typical values of the transfer velocity across the water-side mass boundary layer are $10^{-6} - 10^{-5}$ m/s (1–10 m/day, see Fig. 1a). With respect to typical mixed layer depths in the ocean of about 100 m, gas transfer is a very slow process. It takes a time constant $\tau = h/k$ = 10–100 days for the concentration of dissolved gases in the mixed layer to come into equilibrium with the atmosphere. Thus a parametrization of the transfer velocity is only possible under steady state conditions over extended periods in the field. Therefore, mass balance methods in general are only poorly suited for the study of the mechanisms of air-water gas transfer.

2.3.2 Boundary layer thickness

The *boundary layer thickness* $\tilde{z}$ is defined as the thickness of a fictional layer in which the flux is maintained only by molecular transport: $j = D(C_s - C_b)/\tilde{z}$. Then with (1) the boundary layer thickness is given by

$$\tilde{z} = \frac{D}{k}. \tag{3}$$

Geometrically, $\tilde{z}$ is given as the intercept of the tangent to the concentration profile at the surface and the bulk (Fig. 2a). With thicknesses between 20–200 μm, the mass boundary layer is extremely thin. Because the diffusion constants D for dissolved gases are about three orders of magnitude smaller than the diffusion constant for momentum (kinematic viscosity ν), resulting in Schmidt numbers $Sc = \nu/D$ of

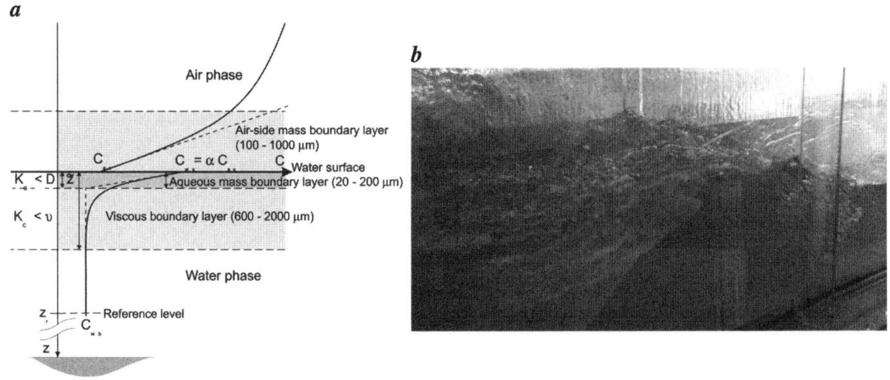

Figure 2 a. Viscous and mass sublayers at both sides of the air-water interface located at b. a wind-driven, rough and wavy water surface (Annular Heidelberg Air/Sea Interaction Facility, the "Aeolotron").

about 1000. Therefore the mass boundary layer is a a small sub layer within the viscous boundary layer at the water side (Fig. 2a).

In order to reveal the mechanisms of gas transfer it is therefore required to measure velocity fields and concentration fields with resolutions in order of 10 µm close to the water surface. How difficult such measurements are at a rough and wavy water surface becomes evident from Fig. 2b.

2.3.3 *Boundary layer time constant*

The *time constant* $\tilde{t}$ for the transport across the mass boundary layer is given by

$$\tilde{t} = \frac{\tilde{z}}{k} = \frac{D}{k^2}. \tag{4}$$

Typical values for $\tilde{t}$ are 0.04–4s. In order to reveal the processes taking place at these short time scales also very fast measuring techniques are required with time resolutions well below 10 ms.

The definitions of the three parameters k, $\tilde{z}$, $\tilde{t}$ are generally valid and do not depend on any models of the boundary layer turbulence. According to (3) and (4) they are coupled via the molecular diffusion coefficient. Therefore only one of them needs to be measured to get knowledge of all three parameters provided the molecular diffusion coefficient of the species is known.

3 IMAGING MEASURING TECHNIQUES

The Cornell Symposium was the first large international symposium I visited as a young post doctoral researcher and thus was very stimulating for me. At the conference no imaging methods were presented, but collaborations established in the wake of the symposium triggered the development of such techniques, given the urgent need to get a direct insight into the mechanisms of air-water gas transfer and to reveal them in this way. Cooperation of my research group with the Institute for Hydromechanics in Karlsruhe (IfH) was essential in this endeavor.

3.1 Active thermography

The ground to fast non-contact measurement techniques of the processes at the air/water interface was prepared by active thermography. The basic idea of this technique is to determine the concentration difference across the mass boundary layer when the flux density j of the tracer across the interface is known. The local transfer velocity can be determined by simply measuring the concentration difference Δc across the aqueous boundary layer (cold surface skin temperature) according to (1) with a time constant $\tilde{t}$ for the transport across the boundary layer (4). Heat proves to be an ideal tracer for this technique. The temperature at the water surface can then be measured with high spatial and temporal resolution using IR thermography. A known and controllable flux density can be applied by using infrared radiation. Infrared radiation is absorbed in the first few ten μm at the water surface. Thus a heat source is put right at top of the aqueous viscous boundary layer. The first active thermography measurements were performed in the wind/wave facility of the IfH using a chopped infrared radiator and a point measuring infrared radiation thermometer (Figs. 3 and 4).

3.2 Imaging active thermography

The real power of active thermography became only evident once it was possible to set it up as an imaging techniques. It was necessary to wait until IR cameras became

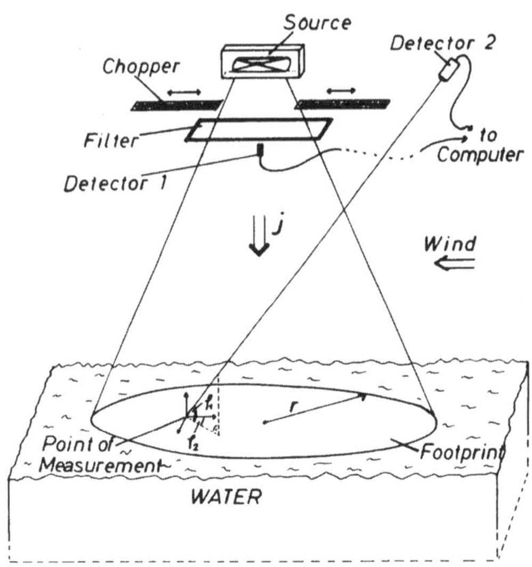

Figure 3 Experimental setup for active thermography, as it was first used at the wind/wave facility of the IfH, TU Karlsruhe in a cooperation experiment between E. Plate and the author with a chopped infrared radiator and an infrared radiation thermometer (from [25]).

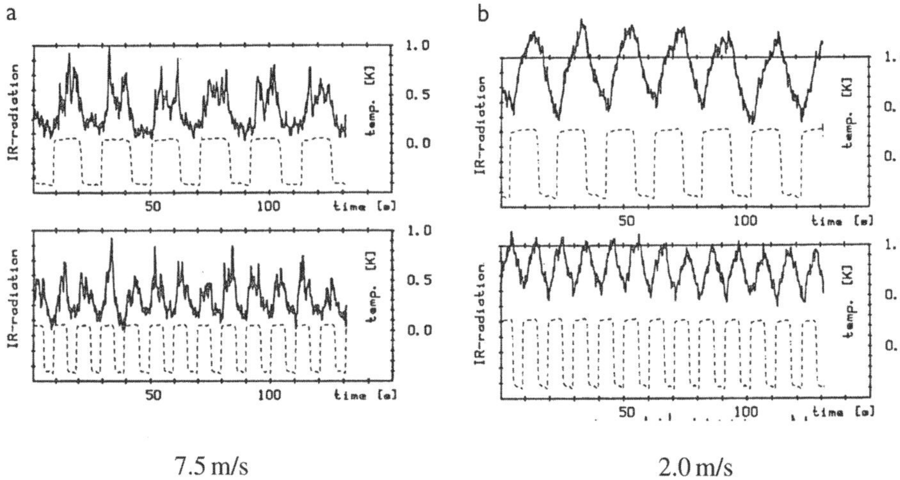

Figure 4 Response of the water surface temperature (solid line) to a heat flux variation (lower row) at the water surface (dashed line) with a period of 20 s (upper row) and 10 s. At 7.5 m/s the boundary layer time constant $\tilde{t}$ is short enough so that the surface temperature can follow the periodic heat flux variations, whereas this is not the case at the low wind speed. From talk of author at IfH, TU Karlsruhe, Nov. 15, 1985, unpublished.

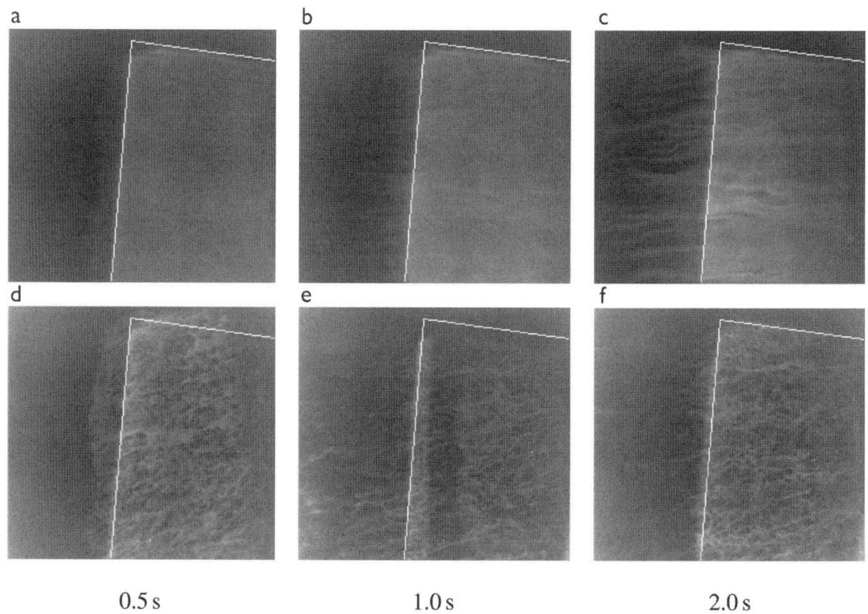

Figure 5 Imaging active thermography measurements in the Heidelberg Aeolotron [28, 42]. The area heated by a 100 W carbon dioxide laser is marked by a the two white lines. The time indicaed at the bottom marks the time after switching on the laser. The IR images in the upper low were taken at 2 m/s wind speed, the images at the lower row at 7 m/s. Wind is blowing from the left to the right.

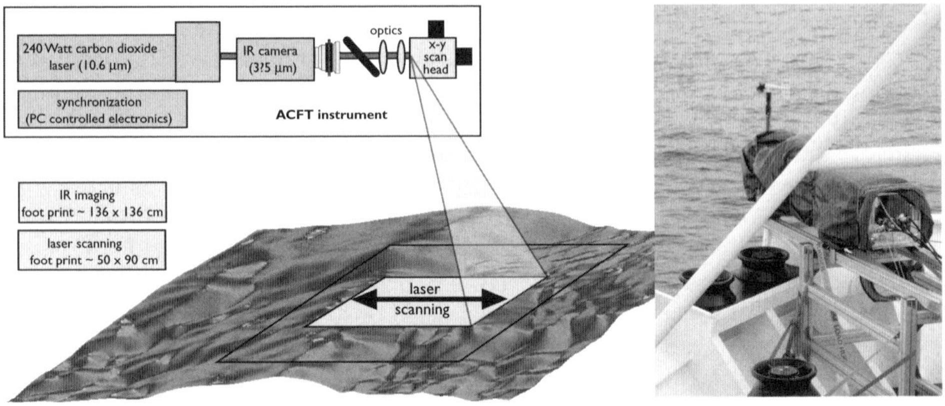

Figure 6 Experimental setup for active thermography as used in the Baltic Sea during the SOPRAN project, mounted at the bow of the FS Alkor. (from [46]).

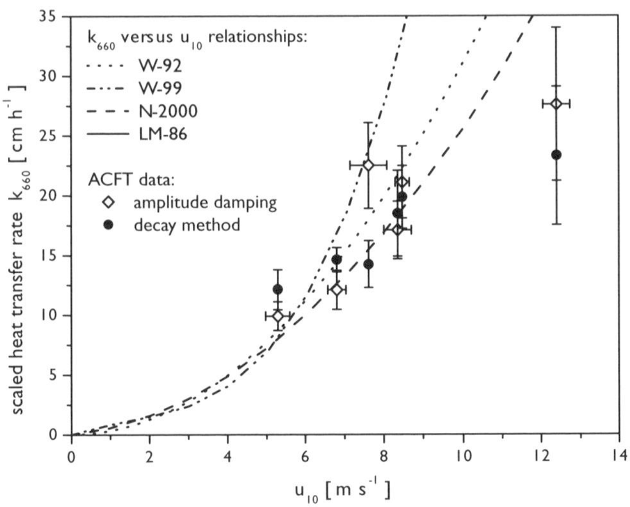

Figure 7 Transfer velocities extrapolated to a Schmidt number of 660, as measured by active thermography in the Baltic sea during the SOPRAN cruise in 2009. For comparison, some empirical wind speed relationships are shown: LM-86 [38], W-92 [50], W-99 [52], N-2000 [40] (from [46]).

available with a sufficiently good temperature resolution. The first successful measurements were reported only more than twenty years later from the Heidelberg Aeolotron [28, 42], (Fig. 5).

One of the biggest advantages of active thermography is that it can also be used in the field. Figure 6 shows the experimental setup used within the SOPRAN project in the Baltic Sea and Figure 7 a collection of measured transfer velocities corrected to a Schmidt number of 660.

3.3 Imaging of concentration fields

Measurements of concentration fields of dissolved gases are essential, because it is not clear whether heat transfer across the aqueous viscous boundary layer is governed by the same mechanisms as mass transfer. This doubt rises because the Prandtl number in water is about hundred times smaller than the Schmidt number of dissolved gases. Therefore the mass boundary layer is much thinner than the heat boundary layer and the gas transfer may be influenced by small-scale residual turbulence which has no influence for heat transfer because of the higher molecular diffusion of heat.

At the Cornell symposium a paper was presented by Pankow and Asher [41] using a fluorescent pH indicator to visualize gas exchange with carbon dioxide. This paper triggered research to image concentrations fields of dissolved gases in our research group. A literature research first revealed that such techniques have been used already back in the 60'ties in chemical engineering to investigate gas exchange with falling films [3, 16, 17, 18].

Oxygen quenching techniques to measure concentration fields of dissolved oxygen were first reported by Wolff and Hanratty [55], Wolff *et al.* [56] using pyrenebutyric acid (PBA). This technique was also used by Gerhard Jirka's group to investigate the gas transfer process with grid-stirred turbulence [13, 14, 15].

Recently, the oxygen quenching technique was improved considerably by [10]) using an organic ruthenium complex (Fig. 8). The fluorescence of this dye can be stimulated by blue light, it shows a high Stokes shift, is ten times more sensitive than PBA, has an excellent water solubility, and shows no surface activity. A time series of a high-resolution vertical profile is shown in Figure 9.

3.4 Imaging of slope of short wind waves

Short wind waves from millimeter capillary waves to short gravity waves can conveniently be measured by means of optical imaging of the gradient of the water surface elevation. The foundation for optical techniques to measure short waves was laid by the pioneering work of Cox. He not only used sun glitter images to infer the wave slope distribution and mean square slope of wind waves [7] (Fig. 10a),

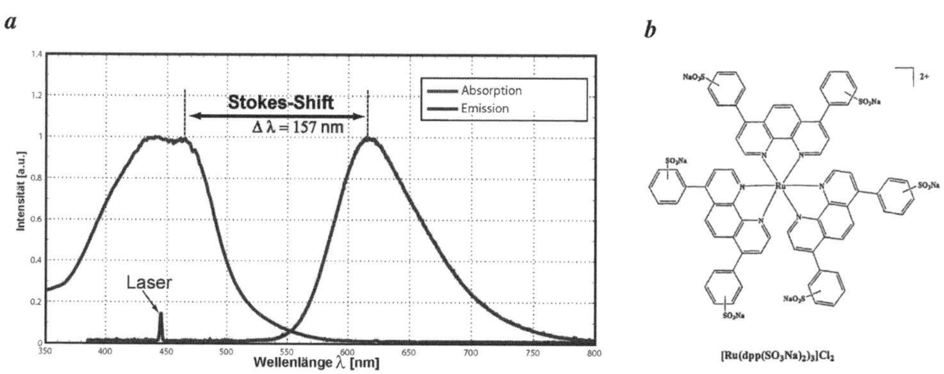

Figure 8 a. Absorption and fluorescence emission spectra of the ruthenium complex used for imaging of oxygen concentration fields [10]); b. Molecular structure of the ruthenium complex.

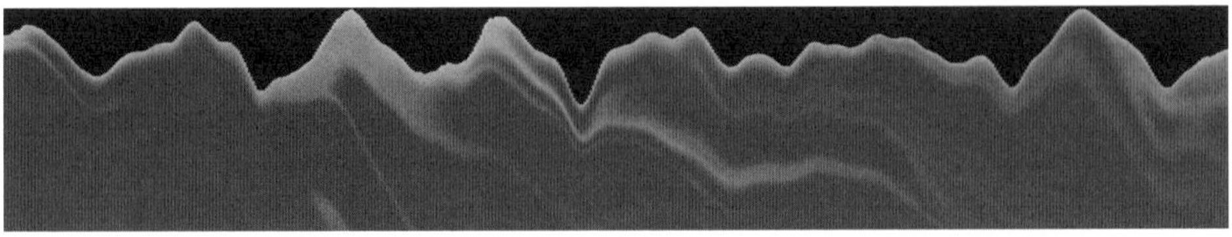

Figure 9 Time series of high-resolution vertical profiles of dissolved oxygen as measured in a small linear test wind/wave facility [53]). The shown image sector is 1 mm (vertical) × 860 ms (horizontal) with a resolution of 5.8 μm and 1 ms, respectively.

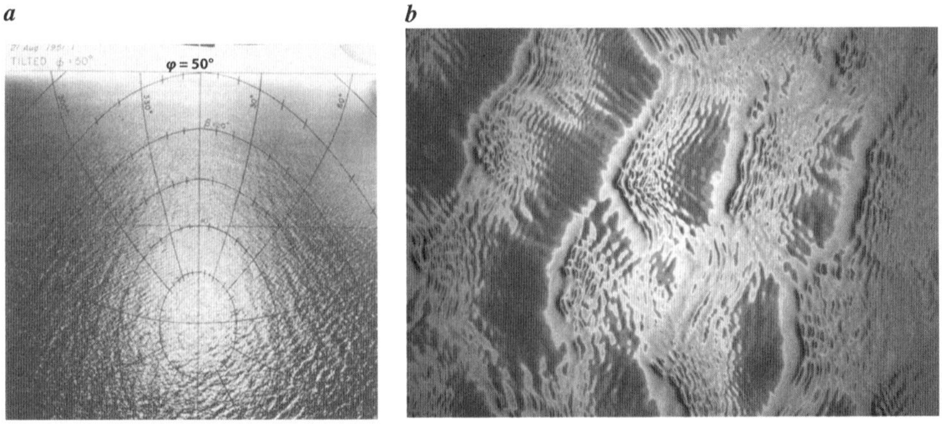

Figure 10 **a.** A sun glitter image from [7]; **b.** Wave slope image taken in the grande soufflerie d'interaction air-eau, IMST Marseille at 4.6 m fetch and 6.3 m/s wind speed [21]; wind is blowing from the left to the right; image size about 40 cm × 30 cm.

but also introduced the usage of optical techniques based on light refraction for the measurement of wave slope [8]. Keller and Gotwols [32] were the first to use refractive optical techniques to capture wave slope images.

Figure 10b show some of the first wave slope images taken by the author in the Marseille wind/wave flume in 1984. Jähne [22] conducted the first systematic study of 2-D wave number spectra of short wind waves taken in the large wind/wave flume of Delft Hydraulics, Holland. The first spatial wave slope measurements taken at sea were reported by Klinke and Jähne [35]. This technique, however, turned out to be complex to be used at sea on a routine basis. It continued to be used successfully in wind/wave facilities.

More recently, a three-color cameras and illumination were used to measure both components of the waves simultaneously [1, 59]. Rocholz [43] improved the technique further with respect to the frame rate and accuracy. The evolution of the waves can now be followed in time, even for waves in the capillary range. Thus, it is now possible to distinguish waves traveling in and against wind direction and to compute wave number frequency spectra to study the dispersion relation of wind waves.

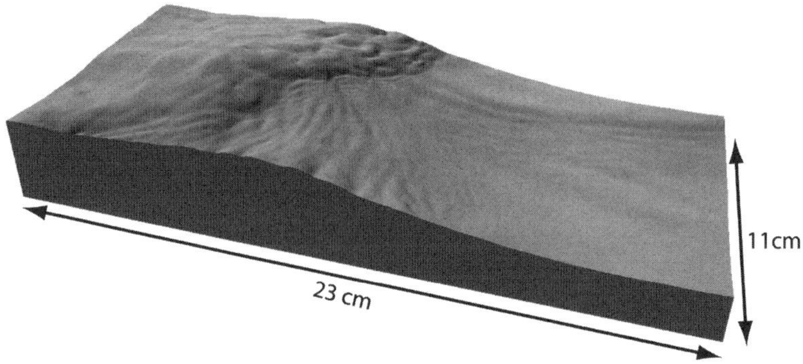

Figure 11 Combined wave slope and height imaging (from [44]).

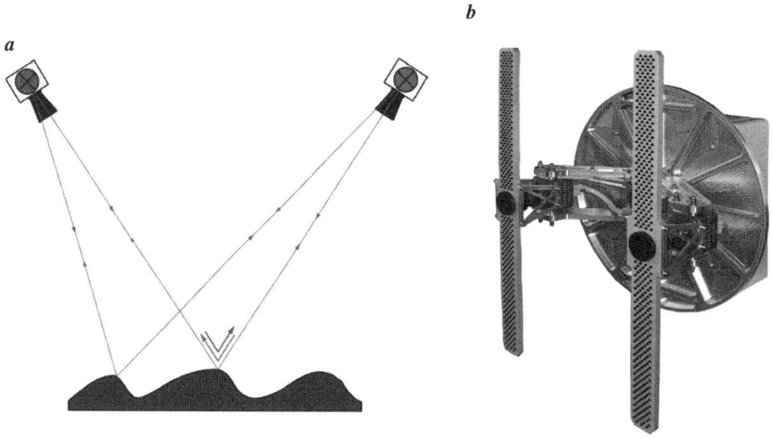

Figure 12 Modern variant of the sun glitter technique for combined wave slope/height statistics with a high-power IR LED light source for measurements at sea: **a.** Measuring principle, **b.** instrument with stereo camera and two light sources [33].

A light-reflection technique for wave slope imaging based on daylight illumination has been known in oceanography for a long time as Stilwell photography [47]. Gotwols and Irani [12] applied it to measure the phase speed of short gravity waves, but otherwise it found not much application. A promising extension of Stilwell photography is polarimetric imaging. Here both components of the wave slope can be determined by analysis the polarization state of light, given the known polarization state of the sky illuminating the ocean surface [58].

A modern extension of the sun glitter technique from Cox was first developed by Waas and Jähne [49] and later improved by Kiefhaber *et al.* [33]. Two cameras and two light sources are arranged in such a way that both cameras see the same specular reflections at the sea surface Fig. 12. The occurrence of speckles at a given image point is related to a certain slope probability. The instrument covers slopes up to 0.1. From this partial pdf, the mean square slope can be extrapolated. Furthermore, the local

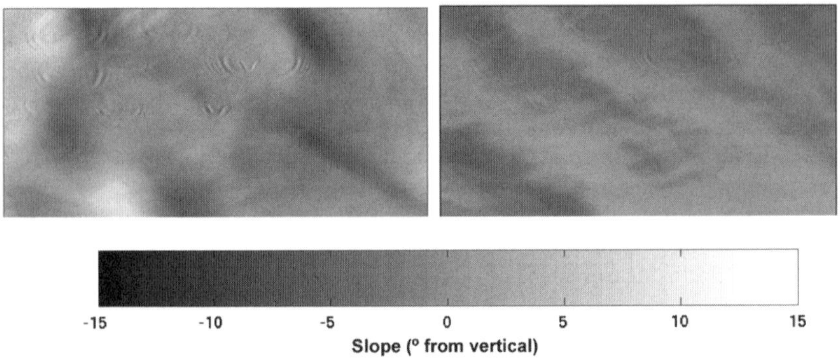

Figure 13 Horizontal and vertical slope of short waves inferred by polarimetric imaging (from [58]).

water height can be inferred from the parallax in the stereo images and the evaluation of speckle size and brightness gives information about surface curvature.

4 CONCLUSIONS AND OUTLOOK

In the past, progress towards a better understanding of the mechanisms of air-water gas exchange was hindered by inadequate measuring technology. However, new techniques have become available and will continue to become available that will give a direct insight into the mechanisms under both laboratory and field conditions. This progress will be achieved by interdisciplinary research integrating such different research areas as oceanography, micrometeorology, hydrodynamics, physical chemistry, applied optics, and image processing. Thus further development in this area critically depends on our ability to perform interdisciplinary research crossing the boundaries between classical research areas.

Optical and image processing techniques will play a key role because only imaging techniques give direct insight to the processes in the viscous, heat and mass boundary layers on both sides of the air/water interface. Eventually all key parameters including flow fields, concentration fields, and waves will be captured by imaging techniques with sufficient spatial and temporal resolution. The experimental data gained with such techniques will stimulate new theoretical and modeling approaches.

REFERENCES

[1] Balschbach, G., Klinke, J. and Jähne, B. (1998). Multichannel shape from shading techniques for moving specular surfaces. In *Computer Vision - ECCV'98*. Springer, Berlin.

[2] Borges, A.V. and Wanninkhof, R. (Eds.) (2007). *5th International Symposium on Gas Transfer at Water Surfaces*, Volume 66, issues 1–4 of *J Marine Systems*.

[3] Braun, D. and Hiby, J.W. (1970). Der gasseitige Stoffübergangskoeffizient am Rieselfilm. *Chemie Ingenieur Technik* 42 (6), 345–349.

[4] Broecker, H.C. and Siems, W. (1984). The role of bubbles for gas transfer from water to air at higher windspeeds. experiments in the wind-wave facility in hamburg. In W. Brutsaert and G.H. Jirka (Eds.), *Gas transfer at water surfaces*, Hingham, MA, pp. 229–236. Reidel.

[5] Brutsaert, W. and Jirka, G.H. (Eds.) (1984). *Gas Transfer at Water Surfaces*. Dordrecht: Reidel.

[6] Chu, C.R. and Jirka, G.H. (2003). Wind and stream flow induced reaeration. *Journal of Environmental Engineering-Asce 129* (12), 1129–1136.

[7] Cox, C. and Munk, W. (1954). Statistics of the sea surface derived from sun glitter. *Journal of Marine Research 13* (2), 198–227.

[8] Cox, C.S. (1958). Measurement of slopes of high-frequency wind waves. *J. Marine Res. 16* (3), 199–225.

[9] Donelan, M.A., Drennan, W.M., Saltzman, E.S. and Wanninkhof, R. (Eds.) (2002). *Gas Transfer at Water Surfaces*. American Geophysical Union.

[10] Falkenroth, A., Degreif, K. and Jähne, B. (2007). Visualisation of oxygen concentration fields in the mass boundary layer by fluorescence quenching. In C.S Garbe, R.A. Handler, and B. Jähne (Eds.), *Transport at the Air Sea Interface—Measurements, Models and Parameterizations*. Springer Verlag.

[11] Faller, A.J. and Perini, C. (1984). The roles of Langmuir circulation in gas transfer across natural water surfaces. In W. Brutsaert and G.H. Jirka (Eds.), *Gas transfer at water surfaces*, Hingham, MA, pp. 191–199. Reidel.

[12] Gotwols, B.L. and Irani, G.B. (1980). Optical determination of the phase velocity of short gravity waves. *Journal of Geophysical Research 85*, 3964–3970.

[13] Herlina. (2005). *Gas Transfer at the Air–Water Interface in a Turbulent Flow Environment*. Ph. D. thesis, University of Karlsruhe.

[14] Herlina and Jirka, G.H. (2004). Application of LIF to inverstigate gas transfer near the air–water interface in a grid-stirred tank. *Experiments in Fluids 37*, 341–348.

[15] Herlina, I. and Jirka, G.H. (2008). Experiments on gas transfer at the air–water interface induced by oscillating grid turbulence. *J. Fluid. Mech. 594*, 183–208.

[16] Hiby, J.W. (1968). Eine Fluoreszenzmethode zur Untersuchung des Transportmechanismus bei der Gasabsorption im Rieselfilm. *Wärme- und Stoffübertr. 1*, 105–116.

[17] Hiby, J.W. (1983). The chemical indicator: a tool for the investigation of concentration fields in liquid. *Ann.N.Y.Acad.Sci. 404*, 348–349.

[18] Hiby, J.W., Braun, D. and Eickel, K.H. (1967). Eine Fluoreszenzmethode zur Untersuchung des Stoffübergangs bei der Gasabsorption im Rieselfilm. *Chemie-Ing.-Techn. 39*, 297–301.

[19] Jähne, B. (1980). *Zur Parametrisierung des Gasaustauschs mit Hilfe von Laborexperimenten*. Dissertation, Institut für Umweltphysik, Fakultiät für Physik und Astronomie, Univ. Heidelberg. IUP D-145.

[20] Jähne, B. (1982). Dry deposition of gases over water surfaces (gas exchange). In D. Flothmann (Ed.), *Exchange of Air Pollutants at the Air-Earth Interface (Dry Deposition)*. Frankfurt: Battelle Institute.

[21] Jähne, B. (1985). *On the transfer processes at a free air-water interface*. Habilitation thesis, Institut für Umweltphysik, Fakultät für Physik und Astronomie, Univ. Heidelberg. IUP D-200.

[22] Jähne, B. (1990). Motion determination in space-time images. In O. Faugeras (Ed.), *Proc. Computer Vision – ECCV 90, Lecture Notes in Computer Science 427*, New York, pp. 161–173.

[23] Jähne, B., Heinz, G. and Dietrich, W. (1987). Measurement of the diffusion coefficients of sparingly soluble gases in water. *J. Geophys. Res. 92* (C10), 10,767–10,776.

[24] Jähne, B., Huber, W. Dutzi, A. Wais, T. and Ilmberger, J. (1984). Wind/wave-tunnel experiments on the schmidt number and wave field dependence of air-water gas exchange. In W. Brutsaert and G.H. Jirka (Eds.), *Gas transfer at water surfaces*, Hingham, MA, pp. 303–309. Reidel.

[25] Jähne, B., Libner, P. Fischer, R. Billen, T. and Plate, E.J. (1989). Investigating the transfer process across the free aqueous boundary layer by the controlled flux method. *Tellus 41B* (2), 177–195.
[26] Jähne, B., Münnich, K.O., Bösinger, R., Dutzi, A., Huber, W. and Libner, P. (1987, February). On the parameters influencing air-water gas exchange. *J. Geophys. Res. 92*, 1937–1950.
[27] Jähne, B. and Monahan, E.C. (Eds.) (1995). *Air-Water Gas Transfer—Selected Papers from the Third International Symposium on Air-Water Gas Transfer*, Hanau. AEON Verlag & Studio.
[28] Jähne, B., Popp, C. Schimpf, U. and Garbe, C. (2007). The influence of intermittency on air/water gas transfer measurements. In C.S. Garbe, R.A. Handler, and B. Jähne (Eds.), *Transport at the Air Sea Interface—Measurements, Models and Parameterizations*. Springer-Verlag.
[29] Jähne, B., Wais, T. and Barabas, M. (1984). A new optical bubble measuring device; a simple model for bubble contribution to gas exchange. In W. Brutsaert and Jirka, G.H. (Eds.), *Gas transfer at water surfaces*, Hingham, MA, pp. 237–246. Reidel.
[30] Jirka, G.H. and Brutsaert, W. (1984). Measuremnets of wind effects on water-side controlled gas exchange in riverine systems. In W. Brutsaert and G.H. Jirka (Eds.), *Gas transfer at water surfaces*, Hingham, MA, pp. 437–446. Reidel.
[31] Keeling, R.F. (1993). On the role of large bubbles in air-sea gas exchange and supersaturation in the ocean. *Journal of Marine Research 51*, 237–271.
[32] Keller, W.C. and Gotwols, B.L. (1983, November). Two-dimensional optical measurement of wave slope. *Applied Optics 22* (22), 3476–3491.
[33] Kiefhaber, D., Rocholz, R. Balschbach, G. Schaper, J. and Jähne, B. (2011). Measurement of ocean wave statistics with the reflective stereo slope gauge. In *DPG Frühjahrstagung Dresden*.
[34] Kitaigoradskii, S.A. and Donelan, M.A. (1984). Wind-wave effects on gas transfer. In W. Brutsaert and G.H. Jirka (Eds.), *Gas transfer at water surfaces*, Hingham, MA, pp. 147–170. Reidel.
[35] Klinke, J. and Jähne, B. (1995). Measurements of short ocean waves during the mbl ari west coast experiment. In B. Jähne and E.C. Monahan (Eds.), *Air-water Gas Transfer, Selected Papers from the Third International Symposium on Air–Water Gas Transfer*, Hanau, pp. 165–173. Aeon.
[36] Komori, S., McGillis, W. and Kurose, R. (Eds.) (2011). *Gas Transfer at Water Surfaces 2010*, Kyoto. Kyoto Univ Press.
[37] Ledwell, J.J. (1984). The variation of the gas transfer coefficient with molecular diffusivity. In W. Brutsaert and Jirka, G.H. (Eds.), *Gas transfer at water surfaces*, Hingham, MA, pp. 293–302. Reidel.
[38] Liss, P.S. and Merlivat, L. (1986). Air-sea gas exchange rates: Introduction and synthesis. In P. Buat-Menard (Ed.), *The role of air-sea exchange in geochemical cycling*, pp. 113–129. Boston, MA: Reidel.
[39] Monahan, E.C. and Spillane, M.C. (1984). The role of whitecaps in air-sea gas exchange. In W. Brutsaert and G.H. Jirka (Eds.), *Gas transfer at water surfaces*, Hingham, MA, pp. 495–503. Reidel.
[40] Nightingale, P.D., Malin, G., Law, C.S., Watson, A.J., Liss, P.S., Liddicoat, M.I., Boutin, J. and Upstill-Goddard, R.C. (2000). In situ evaluation of air-sea gas exchange parameterization using novel conservation and volatile tracers. *Glob. Biogeochem. Cycles 14*, 373–387.
[41] Pankow, J.F. and Asher, W.E. (1984). Carbon dioxide transfer at the gas/water interface as a function of system turbulence. In W. Brutsaert and G.H. Jirka (Eds.), *Gas transfer at water surfaces*, Hingham, MA, pp. 101–111. Reidel.
[42] Popp, C.J. (2006). *Untersuchung von Austauschprozessen an der Wasseroberfläche aus Infrarot-Bildsequenzen mittels frequenzmodulierter Wärmeeinstrahlung*. Dissertation, Institut für Umweltphysik, Fakultät für Physik und Astronomie, Univ. Heidelberg.

[43] Rocholz, R. (2008). *Spatiotemporal Measurement of Short Wind-Driven Water Waves*. Dissertation, Institut für Umweltphysik, Fakultär für Physik und Astronomie, Univ. Heidelberg.

[44] Rocholz, R., Wanner, S., Schimpf, U. and Jähne, B. (2011). Combined visualization of wind waves and water surface temperature. In S. Komori, W. McGillis, and R. Kurose (Eds.), *Gas Transfer at Water Surfaces 2010*, pp. 496–506.

[45] Roether, W. and Kromer, B. (1984). Optimum application of the radon deficit method to obtain air–sea gas exchange rates. In W. Brutsaert and G.H. Jirka (Eds.), *Gas transfer at water surfaces*, Hingham, MA, pp. 447–457. Reidel.

[46] Schimpf, U., Nagel, L. and Jähne, B. (2011). First results of the 2009. sopran active thermography pilot experiment in the baltic sea. In S. Komori, W. McGillis, and R. Kurose (Eds.), *Gas Transfer at Water Surfaces 2010*, pp. 358–367.

[47] Stilwell, D.J. (1969). Directional energy spectra of the sea from photographs. *J.Geophys. Res. 74*, 1974–1986.

[48] Tschiersch, J. and Jähne, B. (1980). Gas exchange trough a rough water surface in a circular windtunnel; wave characteristics under limited and unlimited fetch. In H.C. Broecker and L. Hasse. (Eds.), *Berichte aus dem Sonderforschungsbereich 94 Meeresforschung - Symposium on Capillary Waves and Gas Exchange, Trier July 2–6, 1979*, Number 17, pp. 63–70. Univ. Hamburg.

[49] Waas, S. and Jähne, B. (1992). Combined slope-height measurements of short wind waves: first results from field and laboratory measurements. In L. Estep (Ed.), *Optics of the Air-Sea Interface: Theory and Measurements*, Volume 1749 of *SPIE Proc.*, pp. 295–306.

[50] Wanninkhof, R. (1992). Relationship between wind speed and gas exchange over the ocean. *J. Geophys. Res. 97*, 7373–7382.

[51] Wanninkhof, R., Asher, W.E., Ho, D.T., Sweeney, C. and McGillis, W.R. (2009). Advances in quantifying air-sea gas exchange and environmental forcing. *Annu. Rev. Mar. Sci. 1*, 213–244.

[52] Wanninkhof, R. and McGillis, W.R. (1999). A cubic relationship between gas transfer and wind speed. *Geophys. Res. Lett. 26*, 1889–1892.

[53] Warken, P. (2010). Hochauflösende lif-methode zur messung von sauerstoffkonzentrationsprofilen in der wasserseitigen grenzschicht. Diplomarbeit, Institut für Umweltphysik, Fakultät für Physik und Astronomie, Univ. Heidelberg.

[54] Wilhelms, S.C. and Gulliver, J.S. (Eds.) (1991). *Air-Water Mass Transfer — Selected Papers from the 2nd International Symposium on Gas Transfer at Water Surfaces, Minneapolis Minnesota, Sept. 11–14, 1990*, New York. American Society of Civil Engineers.

[55] Wolff, L.M. and Hanratty, T.J. (1994). Instantaneous concentration profiles of oxygen accompanying absorption in a stratified flow. *Experiments in Fluids 16* (6), 385–392.

[56] Wolff, L.M., Liu, Z.C. and Hanratty, T.J. (1991). A fluorescence technique to measure concentration gradients near an interface. In S.C. Wilhelms and Gulliver, J.S. (Eds.), *AirWater Mass Transfer, selected papers from the 2nd Int. Symposium on Gas Transfer at Water Surfaces*, Minneapolis, pp. 210–218. ASCE.

[57] Woolf, D., Leifer, I., Nightingale, P., Rhee, T., Bowyer, P., Caulliez, G., de Leeuw, G., Larsen, S., Liddicoat, M., Baker, J. and Andreae, M. (2007). Modelling of bubble-mediated gas transfer: Fundamental principles and a laboratory test. *Journal of Marine Systems 66*, 71–91.

[58] Zappa, C., Banner, M., Schultz, H., Corrada-Emmanuel, A., Wolff, L. and Yalcin, J. (2008). Retrieval of short ocean wave slope using polarimetric imaging. *Meas. Sci. Technol. 19*, 055503 (13 p).

[59] Zhang, X. and Cox, C.S. (1994, August). Measuring the two-dimensional structure of a wavy water surface optically: A surface gradient detector. *Experiments in Fluids 17*, 225–237.

Chapter 23

Mass transfer from bubble swarms

John S. Gulliver
St. Anthony Falls Laboratory and Department of Civil Engineering
University of Minnesota, Minneapolis, Minnesota, USA

ABSTRACT: A technique that quantifies the mass transfer from a bubble swarm, without looking at individual bubble transfer, is described. The technique is applied to bubble plumes that are not affected by ambient boundary conditions. The bubble and surface volumetric mass transfer coefficients for oxygen are separately determined for 179 aeration tests, with sparger depths ranging from 2.25 m to 32 m, using a multiple component mass transfer model. Two empirical characterization equations are developed for surface and bubble mass transfer coefficient, correlating the coefficients to air flow, sparger depth, cross-sectional area and volume. The characterization equations indicate that the bubble transfer coefficient increases with increasing gas flow rate and depth, and decreases with increasing water volume. The mass transfer coefficient for fine bubble spargers is approximately six times greater than for coarse bubble spargers. The surface transfer coefficient increases with increasing gas flow rate and sparger depth. The characterization equations make it possible to predict the gas transfer that will occur across bubble interfaces and across the free surface with a bubble plume at depths up to 32 m and with variable air discharge in deep tanks and reservoirs.

1 INTRODUCTION

Low dissolved oxygen (DO) levels in water results in anoxia and can contribute to fish kills, odor, and other aesthetic nuisances. Submerged aeration systems with air diffusers or spargers are used as water quality enhancement devices in lakes and reservoirs, as well as wastewater treatment facilities, to increase dissolved oxygen levels and promote water circulation. Sparger systems have been used for many years in hydropower reservoirs to increase the oxygen concentration in the hypolimnion, where water is frequently withdrawn through the hydropower intakes and released downstream. The environmental benefits of aeration and oxygenation systems include promoting the survival of aerobic bacteria, decreased carbon dioxide concentrations, odor prevention through the oxidation of hydrogen sulfide, and improved living conditions for fish and other aquatic life. Two spargers, generally classified as fine bubble and coarse bubble spargers, are given in Figure 1.

Information on the performance of various sparger systems is needed for design. The industry standard is to disturb the equilibrium of dissolved oxygen with the atmosphere by reducing its concentration, turn on the sparger and measure the return of the water body to equilibrium or at least a steady value of dissolved oxygen

Figure 1 Fine bubble (left) produce bubbles that are less than 2 mm in diameter and coarse bubble (right) spargers produce bubbles that are greater than 5 mm in equivalent diameter.

(ASCE, 1992). An exponential relation is assumed, and two parameters are fit to the data, the steady state concentration and a "gas transfer coefficient."

There is one problem with the industry standard method, however, as illustrated in Figure 2. The sparger must be tested at the depth of the application. Because the bubbles are transferring the gas at depth, they are under the hydrostatic pressure of that depth. The bubble volume decreases and the concentration in the bubble increases. The equilibrium concentration associated with the bubble also increases accordingly, and the steady state overall water body concentration is higher than equilibrium with the atmosphere. It is somewhere in between equilibrium with the atmosphere and equilibrium with the sum of the rising bubbles. Thus, if the depth of the application is changed, both the steady state concentration and the "gas transfer coefficient" are different. The sparger tests developed for the 3–5 meter depths of sewage treatment are of little use for the application of lake and reservoir spargers. In addition, many applications require close to pure oxygen to avoid nitrogen supersaturation and dissolved gas bubble disease in fish. A more direct computation of surface transfer and bubble transfer is required for both applications.

When designing aeration systems at variable depths it is necessary to be able to separately calculate the bubble and surface volumetric mass transfer coefficients k_La_b and k_La_s. It is also necessary to conserve mass, and compute the concentration of oxygen and nitrogen inside of the bubbles. The purpose of this paper is to develop correlations between k_La_b, k_La_s and aeration/reservoir characteristics such as water volume, V, cross-sectional area, A_{cs}, sparger depth, h_d and gas flow rate, Q_a, so that aeration systems can be properly designed at large depths. Correlations are needed because design guidelines for aeration systems at depths above 7 m do not exist and because tests at these depths are difficult and expensive.

There are some challenges with determining mass transfer with bubble swarms that have not, as of yet, been dealt with completely. The first challenge is that a single

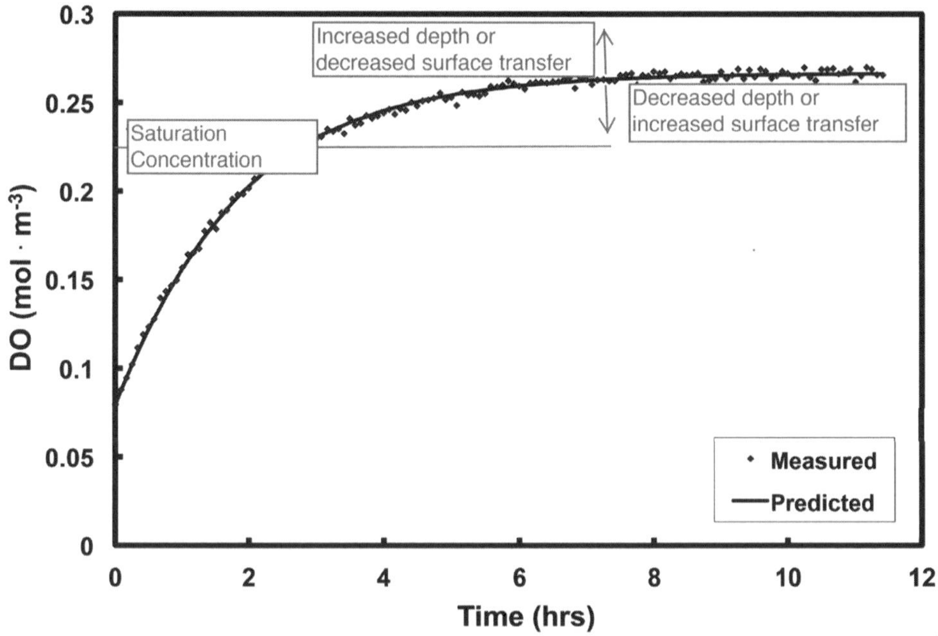

Figure 2 Measured dissolved oxygen data and the curve fit assuming an exponential relationship for a porous sparger operating at 8 meters depth. The steady state concentration is a combination of surface transfer versus bubble transfer and increases with depth.

bubble process times the number of bubbles does not equal the processes found in a bubble swarm. In a bubble swarm, the flow field is altered by the rise of bubbles, which typically move laterally as much as vertically. If the flow field is altered, the bubble size and shape, which is dependent upon the flow field, is also altered. Mass transfer depends upon the flow field, bubble size and bubble shape, so it is altered as well. The second challenge is the need to perform an overall mass balance on many compounds in both fluids. Air bubbles, for example, are primarily composed of oxygen, nitrogen, argon, and water vapor. There is a need to perform a mass balance that includes both phases on oxygen and nitrogen and possibly argon and water vapor plus any trace compounds that are of interest, and many of the concentrations are varying in both phases. There is no reason that we cannot perform this later challenge, except that it requires computational time and effort.

This paper will review some of the processes of mass transfer influenced by bubble swarms, and provide some simplifications that do not deny these physical processes. The approach will be to: 1) develop a mass balance in both gas and liquid phases on each important gas in the bubble swarm, i.e., oxygen, nitrogen and any other gas of interest, 2) incorporate a mass transfer boundary condition at the water surface, which separates bubble transfer from that across the water surface, 3) incorporate hydrostatic pressure into the equilibrium concentration of compounds in the bubble swarm, and 4) implement a bubble residence time relation that allows the entire bubble swarm to be modeled, rather that each bubble individually.

2 MODEL DEVELOPMENT FOR SPARGERS

Numerous disturbed equilibrium aeration tests can be analyzed by an improved mass transfer model (DeMoyer *et al.*, 2003; Shierholz, *et al.*, 2006) recognizing that there exist two distinct mass transfer processes in diffused aeration systems, the gas bubble mass transfer process and the free surface mass transfer process. Each of the processes must be separately analyzed and properly accounted for in the overall mass transfer model.

The two-process mass-transfer model includes the mass conservation of oxygen,

$$(1-\phi)\frac{dC_O}{dt} = \frac{k_L a_b}{h_d}\int_0^{h_d}\left(C_O^* - C_O\right)dz + k_L a_s\left(C_{sat(O_2)} - C_O\right). \tag{1}$$

where z is a variable distance from the sparger, t is time, Φ is the gas void ratio, C_O^* is the liquid-phase equilibrium oxygen concentration of the bubbles, C_O is the actual dissolved oxygen concentration, $C_{sat(O)}$ is the oxygen saturation concentration at atmospheric pressure, or equilibrium concentration with the atmosphere and k_La_b and k_La_s are the volumetric bulk transfer coefficients of oxygen transfer across the bubbles and the water surface, respectively.

The conservation of nitrogen should also be included:

$$(1-\phi)\frac{dC_N}{dt} = \frac{k_L a_b}{h_d}\left(\frac{D_N}{D_O}\right)^{1/2}\int_0^{h_d}\left(C_N^* - C_N\right)dz + k_L a_s\left(\frac{D_N}{D_O}\right)^{1/2}\left(C_{sat(N_2)} - C_N\right) \tag{2}$$

where D_N and D_O are the diffusion coefficients for nitrogen and oxygen, C_N is the dissolved nitrogen concentration, C_N^* is the liquid-phase equilibrium nitrogen concentration of the bubbles, and $C_{sat(N)}$ is the nitrogen saturation concentration. Argon, being approximately 1% of the atmosphere, is assumed to respond similar to nitrogen, since both are essentially inert gases. Since the volumetric bulk mass transfer coefficients for oxygen are used in Eq. (2), the ratio of diffusion coefficients to the ½ power is also required (Gulliver *et al.*, 1990). It is generally assumed that the bubbles are at 100% of the local relative humidity, so that the partial pressure of water inside the bubbles is the local water vapor pressure. For non-confined bubble plumes, Φ will be assumed small.

The oxygen and nitrogen equilibrium concentrations are given by:

$$C_O^* = C_{sat(O)}\frac{\left(P - P_{wv} + \rho g(h_d - z)\right)}{P - P_{wv}}\frac{y/(1+y)}{0.21} \tag{3}$$

$$C_N^* = C_{sat(N)}\frac{\left(P - P_{wv} + \rho g(h_d - z)\right)}{P - P_{wv}}\frac{1/(1+y)}{0.79} \tag{4}$$

where y is the gas-phase oxygen composition, which is the molar ratio of oxygen to other gases (primarily nitrogen) in the gas phase, P is atmospheric pressure, P_{wv} is water vapor pressure, ρ is the density of water and g is the acceleration of gravity.

The boundary condition applied to both Eqs. (3) and (4) is that the gas-phase oxygen molar ratio, y, is known when the bubbles first enter the water, or for air, $y = 0.266$ at $z = 0$.

In order to close the mass balance for oxygen and nitrogen, the vertical profile of y is computed.

The expansion of dy/dz is:

$$\frac{\partial y}{\partial z} = \left(\frac{1}{C_{Nb}} \frac{\partial C_{Ob}}{\partial z} - \frac{C_{Ob}}{C_{Nb}^2} \frac{\partial C_{Nb}}{\partial z} \right) \tag{5}$$

For most conditions, we can disaggregate temporal from spatial variations in molar ratio because temporal, relative to spatial, variations are small. Therefore $\partial C/\partial t \to 0$ and:

$$\frac{dC}{dt} = \frac{\partial C}{\partial z} \cdot U_r \tag{6}$$

Considering only bubble transfer from Eq. (1) and (2), Eq. (5) can then be rewritten using Eqs. (1), (2) and (6) as:

$$\frac{\partial y}{\partial z} = -\frac{k_L a_b h_d A_{cs}}{V_b U_r} \left(\frac{1}{C_{Nb}} \left(C_O^* - C_O \right) - \left(\frac{D_N}{D_O} \right)^{1/2} \frac{C_{Ob}}{C_{Nb}^2} \left(C_N^* - C_N \right) \right) \tag{7}$$

where A_{cs} is the cross-sectional area of the bubble plume or the reactor. The gas concentration in the bubble can be related to the liquid gas concentration using Henry's Law:

$$C_b = H \cdot C^* \tag{8}$$

where H is Henry's Law constant for the given gas and $V_b U_r$ can be expressed as measureable quantities using the definition of residence time (t_r):

$$t_r = \frac{h_d}{U} = \frac{V_b}{Q_g} \tag{9}$$

Cross-multiplying Eq. (9) allows the substitution of known quantities, $Q_g h$, for the unknown quantities, $V_b U_r$. This relatively simple substitution, first proposed by McWirter and Hutton (1989), enables us to perform a mass balance on the entire bubble swarm. Substituting Eqs. (8) and (9) into Eq. (7), a final expression for $\partial y/\partial z$ can be formed:

$$\frac{dy}{dz} = -k_L a_b \frac{A_{cs}}{Q_g} \left(\frac{1}{H_N C_N^*} \left(C_O^* - C_O \right) - \left(\frac{D_N}{D_O} \right)^{1/2} \frac{H_O C_O^*}{\left(H_N C_N^* \right)^2} \left(C_N^* - C_N \right) \right) \tag{10}$$

The variables C_O^* and C_N^* for each chamber are calculated using the following equations:[7]

With Eqs. (3), (4) and (10), the gas-phase oxygen or nitrogen composition, y, and the local equilibrium concentrations, C_O^* and C_N^*, can be calculated at all depths and times. The resulting values can then be used in either Eqs. (1) or (2), along with the experimental aeration data to obtain the best-fit values for the unknown bulk transfer coefficients, $k_L a_b$ and $k_L a_s$.

3 CHARACTERIZATION OF MASS TRANSFER RELATIONS

3.1 Bubble mass transfer

The bulk mass transfer coefficient for the bubble-water interface can be characterized by modifying a theoretical relationship developed by Azbel (1981) for bubble swarms:

$$k_L a_b = \beta_1 \frac{\varphi}{d_b} \frac{(1-\varphi)^{1/2}}{(1-\varphi^{5/3})^{1/4}} \frac{D^{1/2} U^{\eta}}{L^{1-\eta} \nu^{\eta-1/2}} \tag{11}$$

where Φ is the gas void ratio, d_b is bubble diameter, D is diffusivity, U is a characteristic turbulence velocity, L is a characteristic length of turbulence, and ν is kinematic viscosity. The original equation developed by Azbel was for bubble swarms where $\eta = 0.75$. Values for η have ranged from 0.6–1 (Akita and Yoshida, 1973; Deckwer *et al.*, 1982; El-Tamtamy *et al.*, 1984; Godbole *et al.*, 1984; Hughmark, 1967; Joseph *et al.*, 1984; Nakanoh and Yoshida, 1980; Kawase and Moo-Young, 1986; Thompson and Gulliver, 1997).

The following equation for bubble diameter is taken from Hinze (1955):

$$d_b \approx \left(\frac{\sigma}{\rho}\right)^{3/5} \left(\frac{L}{U^3}\right)^{2/5} \tag{12}$$

where σ is surface tension and ρ is liquid density. Equation (12) is determined by a balance of shear and surface tension forces as the bubbles try to increase in size due to reducing hydrostatic pressure as they rise.

The gas void ratio for bubble plumes in large water bodies is typically small. Thus the importance of the second and third Φ terms in Eq. (11) is considered negligible, and will hereafter be neglected. Substituting in Eq. (12) for bubble diameter, and using the sparger depth, h_d for the characteristic length, Eq. (11) becomes:

$$\frac{k_L a_b h_d^2}{D} = \beta_1 \varphi \left(\frac{U^2 h_d \rho}{\sigma}\right)^{3/5} \left(\frac{\nu}{D}\right)^{1/2} \left(\frac{U h_d}{\nu}\right)^{\eta} \tag{13}$$

The gas void ratio can be obtained from the following (Azbel 1981) correlation:

$$\varphi = \frac{2Fr}{1+2Fr} \approx Fr = \frac{U}{\sqrt{gh_d}} \tag{14}$$

The Froude number is small ($1 \times 10^{-3} \rightarrow 9 \times 10^{-7}$) and thus Eq. (14) essentially becomes the Froude number.

Equation 13 then becomes a relation that equates a Sherwood number, Sh to a Froude number, Weber number, We, Reynolds number, Re, and Schmidt number, Sc:

$$\mathrm{Sh} = \beta_1 \mathrm{Fr}^{\beta_2} \mathrm{We}^{3/5} \mathrm{Sc}^{1/2} \mathrm{Re}^{\eta} \tag{15}$$

For the velocity we will use, $U = Q_a/A_r$, the superficial gas velocity. Most of the turbulence generated in a sparger system is due to oscillations of the bubbles, and superficial gas velocity is a convenient means of representing rising bubbles. In addition, this Sherwood number is a bulk average for the water body, so it is necessary to have one term that is also a bulk average on the right-hand-side of Eq. (15). The coefficient β_2 was added to the Froude number in Eq. (15) because the gas void ratio correlation, Eq. (14), was developed for bubble columns and does not take into account the entrainment that occurs in bubble plumes in lake and reservoir aeration systems. This should have a significant effect upon the gas void ratio.

3.2 Surface mass transfer

Surface mass transfer depends upon similar parameters as bubble mass transfer, with the exception of the Froude and Weber numbers. Therefore, the results of the analysis of aeration data were fit to the following equation:

$$\mathrm{Sh}_s = \beta_3 \mathrm{Sc}^{1/2} \mathrm{Re}^{\beta_4} \left(\frac{A_s}{h_d^2} \right)^{\beta_5} \tag{16}$$

where Sh_s is the Sherwood number for surface transfer, $k_L A_s/(h_d D)$, and A_s is the surface area of the water body. Using Eqs. (15) and (16), both design parameters, $k_L a_b$ and $k_L A_s$, can be quickly determined from knowing V, A_{cs}, h_d and Q_a.

4 RESULTS AND DISCUSSION

4.1 Ratio of surface to bubble transfer

The ratio of $k_L a_s$ to $k_L a_b$ for the optimization of all experimental data is given in Fig. 3. It can be seen from Fig. 3 that the fine bubble spargers have a lower ratio of $k_L a_s$ to $k_L a_b$ than the coarse bubble spargers. As the Reynolds number gets up to 100,000, however, the value of $k_L a_s/k_L a_b$ seems to tend towards 1 for both course bubble and fine bubble spargers. The highest Reynolds number of the fine bubble spargers, however, was 36,000, where the ratio was 0.45.

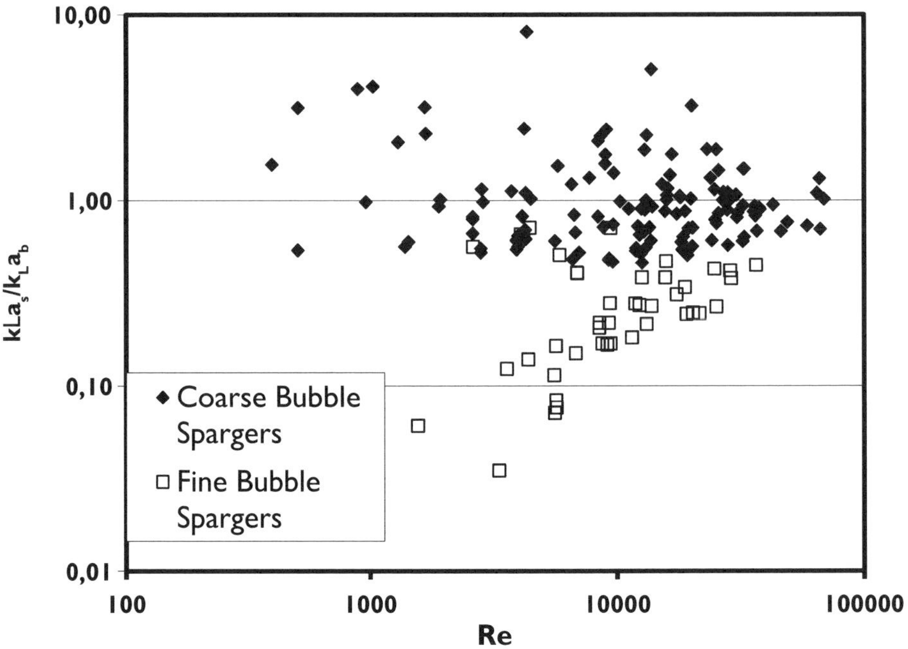

Figure 3 Ratio of reynolds number versus mass transfer coefficients for all tests, separated into coarse bubble and fine bubble spargers.

4.2 Bubble mass transfer

Performing a linear regression on Eq. (15), the adjustable coefficients β_1, β_2 and η were fit to the $k_L a_b$ results of the two-zone mass transfer model. Because there were only three tests at 32 m of depth, each of these was weighted by a factor of five in the regressions. The Reynolds number exponent, η, was determined to be 1.001, while the Froude number exponent, β_2, was determined to be –1.043. These exponents were then set to 1 and –1, respectively, and another regression was performed on the following equation:

$$Sh = \beta_1 \frac{We^{3/5} Sc^{1/2}\,\mathrm{Re}}{Fr} \tag{17}$$

β_1 was separately determined for the fine bubble and coarse bubble spargers. For the fine bubble spargers $\beta_1 = 0.165$ and for the coarse bubble spargers $\beta_1 = 0.027$. This indicates that for fine bubble spargers, $k_L a_b$ is approximately six times greater than $k_L a_b$ for coarse bubble spargers under otherwise similar conditions.

Figure 4 shows the correlation between Sherwood number (with $k_L a_b$) and the dimensionless parameters, Froude number, Fr, Weber number, We, Reynolds number, Re and Schmidt number, *Sc*. The figure separately identifies the coarse bubble and fine bubble spargers as well as the deep-water lock tests. The correlation plotted includes depths ranging from 2.25 m to 32 m. Note how well the large-depth lock

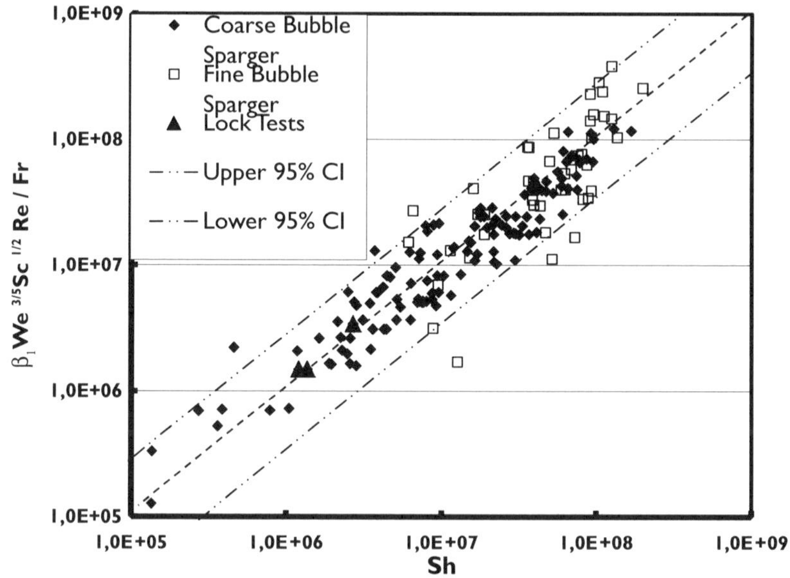

Figure 4 Correlation of the bubble mass transfer coefficients to dimensionless parameters for all tests, separated by coarse bubble and fine bubble spargers.

tests are predicted with this relationship. The 95% confidence interval shown in Fig. 4 corresponds to +/– 95% in the value of β_1.

Rearranging and separating each variable in Eq. (17) we get:

$$k_L a_b = \beta_1 D^{1/2} g^{1/2} \nu^{-1/2} \rho^{3/5} \sigma^{-3/5} Q_a^{6/5} A_{cs}^{-6/5} h_d^{1/10} \tag{18}$$

From Eq. (18) it can be seen that $k_L a_b$ increases with increasing airflow rate and depth; however, as the volume of water to be aerated increases, $k_L a_b$ decreases. This seems logical and is comparable with the results from the improved mass transfer model.

To more quickly determine $k_L a_b$ for design purposes, Eq. (18) can be simplified into the following form:

$$k_L a_b = \alpha \left(\frac{D}{\nu}\right)^{1/2} \left(\frac{Q_a}{A_{cs}}\right)^{6/5} h_d^{1/10} \tag{19}$$

where α includes the previously determined β_1 values, g, ρ at 20°C and σ at 20°C. The values of ρ and σ do not greatly change from 0°C to 30°C. The results only increase by 3.4% from 0°C to 30°C, so it is felt that the inclusion of ρ and σ into the coefficient α is acceptable. D and ν, however, do change more significantly with temperature, and they are therefore not included in α. α was determined from the β_1 values, g, $\rho_{.\ 20°C)}$ and $\sigma_{.\ 20°C)}$ to be 30.5 $m^{-13/10}$ $hr^{1/5}$ for the fine bubble system and 5.0 $m^{-13/10}$ $hr^{1/5}$ for the coarse bubble system.

Equation (19) would be difficult to use in reservoirs with a highly variable cross-section, A_{cs}. It can be converted in to an equation for the mass transfer coefficient, $k_L A_b$, where A_b is the surface area of the bubbles. This results in:

$$\frac{k_L A_b}{Q_a} = \alpha \left(\frac{D}{\nu}\right)^{1/2} \left(\frac{Q_a}{V}\right)^{1/5} h_d^{13/10} \qquad (20)$$

where $A_{cs} = V/h_d$ and V is the volume between the free surface and h_d.

4.3 Mass Transfer At The Free Surface

The best fit of the adjustable coefficients in Eq. (16) gave $\beta_4 = 0.996$. When that value was set to one, β_3 and β_5 in Eq. (16) resulted in values of 49.0 and 0.72, respectively. Then Eq. (16) becomes:

$$\mathrm{Sh}_s = 49\mathrm{Sc}^{1/2}\,\mathrm{Re}\left(\frac{A_s}{h_d^2}\right)^{0.72} \qquad (21)$$

or,

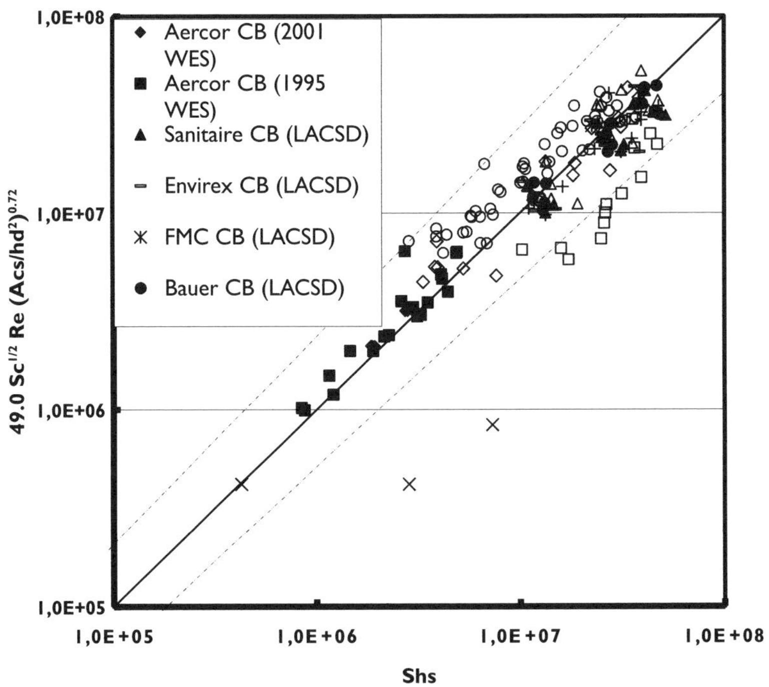

Figure 5 Correlation of the surface mass transfer coefficients to dimensionless parameters for all spargers. Included is the 95% confidence interval.

$$\frac{k_L A_s}{Q_a} = 49\left(\frac{D}{\nu}\right)^{1/2}\left(\frac{h_d^2}{A_s}\right)^{0.28} \tag{22}$$

This indicates that the surface transfer coefficient, $k_L A_s$, increases with increasing air flow rate and with sparger depth, and decreases with increasing cross-sectional area. Figure 5 illustrates the correlation of Eq. (21) for all experimental test results, and separately identifies each sparger tested. In general, each individual manufacturer's sparger tends to follow the perfect fit line. The exceptions are the Sanitare course bubble sparger (Sanitare tests), which tends to have a lower Sh_s, and the Norton fine bubble sparger, which tends towards a higher Sh_s, than that predicted. The LACSD tests of the Sanitare course bubble sparger, however, follows the perfect fit line more closely, likely because the methodology was similar to the other LACSD tests. The importance of Q_a on k_{Ls} is evident in Eq. (22). It does not include the influence of wind, which can be estimated through equations described, for example, by Wanninkhof *et al.* (1991).

5 CONCLUSIONS

Using the results from 179 experimental tests with spargers depths ranging from 2.25 m to 32 m, two equations characterizing the volumetric mass transfer coefficients for bubble transfer and free-surface transfer, $k_L a_b$ and $k_L a_s$, have been successfully developed to aid in the design of deep diffused aeration systems.

It was determined that the bubble mass transfer coefficient, $k_L a_b$, for fine bubble aeration systems is six times greater than those for coarse bubble aeration systems. It was also established that $k_L a_b$ increases with increasing depth and air flow rate, and decreases with increasing water volume. The surface mass transfer coefficient, $k_L A_s$, increases linearly with increasing gas flow rate and to the 0.28 power with increasing depth.

Applying the predictions to the DeMoyer *et al.* (2003) model, one can more effectively design diffused aeration systems for lakes, reservoirs, and wastewater treatment facilities at a variety of water depths through separate determination of the surface mass transfer coefficient and the bubble mass transfer coefficient.

The aeration tests did not obtain information that could be of more use in a computational fluid dynamics/mass transfer model because they were determining bulk quantities from disturbed equilibrium tests. However, a similar analysis of the tests could be used as calibration and/or verification of a computational fluid dynamics/mass transfer model.

REFERENCES

ASCE (1992). Standard for the measurement of oxygen transfer in clean water. America Society of Civil Engineers, New York, NY.

Akita, K. and Yoshida, F. (1973). Gas holdup and volumetric mass transfer coefficient in bubble columns. *Ind. Eng. Chem. Process Des. Develop.* 12 (1), 76–80.

Azbel, D. (1981). Two-phase flows in chemical engineering. 2nd Edition. Cambridge University Press.

Deckwer, W.D., Nguyen-Tien, K., Schumpe, A. and Serpemen, Y. (1982). Oxygen mass transfer into aerated CMC solutions in a bubble column. *Biotechnol. Bioengng*. 24, 461–481.

DeMoyer, C.D., Schierholz, E.L., Gulliver, J.S. and Wilhelms, S.C. (2003). Impact of bubble and free surface oxygen transfer on diffused aeration systems. *Wat. Res*. 37, 1890–1904.

El-Tamtamy, S.A., Khalil, S.A., Nour-El-Din, A.A. and Gaber, A. (1984). Oxygen mass transfer in a bubble column bioreactor containing lysed yeast suspensions. *Appl. Microbiol. Biotechnol*. 19, 376–381.

Godbole, S.P., Schumpe, A., Shah, Y.T. and Carr, N.L. (1984). Hydrodynamics and mass transfer in non-Newtonian solutions in a bubble column. *A.I.Ch.E. J*. 30, 213–220.

Hinze, J.O. (1955). Fundamentals of the hydrodynamic mechanism of splitting in dispersion processes. *A.I.Ch.E. J*. 1, 289–295.

Hughmark, G.A. (1967). Holdup and mass transfer in bubble columns. *Ind. Eng. Chem. Process Des. Dev*. 6 (2), 218–220.

Joseph, S., Shah, Y.T. and Carr, N.L. (1984). Two bubble class model for mass transfer in a bubble column with internals. *Inst. Chem. Engrs. Symp. Ser*. No. 87, 223–230.

Kawase, Y. and Moo-Young, M. (1986). Influence of non-Newtonian flow behavior on mass transfer in bubble columns with and without draft tubes. *Chem. Engng. Commun*. 40, 67–83.

McWhirter, J.R. and Hutter, J.C. (1989). Improved oxygen mass transfer modeling for diffused/subsurface aeration systems. *AIChE J*. 35 (9), 1527–1534.

Nakanoh, M. and Yoshida, F. (1980). Gas absorption by Newtonian and non-Newtonian liquids in a bubble column. *Ind. Engng. Chem. Process Des. Dev*. 19, 190–195.

Schierholz, E.L., Gulliver, J.S. Wilhelms, S.C. and Henneman, H.E. (2006). "Gas transfer from air diffusers," *Water Research*, 40 (5), 1018–1026.

Wanningkof, R., Ledwell, J. and Crusius, J. (1991). "Gas transfer velocities on lakes measured with sulfur hexafluoride," in Wilhelms, S.C. and Gulliver, J.S., Air-Water Mass Transfer, American Society of Civil Engineers, Washington, D.C., pp. 441–458.

Chapter 24

Modelling bacteria and trace metal fluxes in estuarine basins

R.A. Falconer[1], B. Lin[1,2], W.B. Rauen[1], C.M. Stapleton[3] and D. Kay[3]

[1] *Hydro-environmental Research Centre, Cardiff School of Engineering, Cardiff University, Cardiff, UK*

[2] *Department of Hydraulic Engineering, Tsinghua University, P.R. China*

[3] *Centre for Research in Environment and Health, Institute of Geography and Earth Sciences, Aberystwyth University, Aberystwyth, UK*

ABSTRACT: Details are given herein of the limitations of numerical hydro-environmental models in predicting faecal indicator and trace metal fluxes in near-shore coastal systems. Laboratory and empirical field investigations to improve on the predictive capabilities of such models for bacteria and trace metals are outlined, including developing relationships between: (i) bacterial decay, turbidity and suspended sediment levels, (ii) adsorption and desorption kinetics for bacteria, and (iii) partitioning coefficients and salinity effects on trace metal adsorption/desorption with sediments. The resulting laboratory and field study findings were applied to two U.K. case studies for bacteria and trace metals respectively, namely the Severn and Mersey estuaries. The new data provided by field and laboratory studies enabled refinements to numerical models, which led to improvements in model predictive performance.

1 INTRODUCTION

In recent years, there has been growing public concern about water quality in many coastal, estuarine and riverine systems worldwide, particularly in those parts of the world where aquatic systems have been increasingly used as receiving water bodies for the discharge of domestic effluents, industrial by-products, agricultural waste and urban drainage. Human health (WHO, 2003, Kay *et al.*, 2004) and aquatic life (CSTT, 1977) is often threatened by the transport of pollutants through watersheds to coastal waters, and it is therefore not surprising to find that water quality models have been used increasingly in recent years (Thomann and Mueller, 1987). This is primarily due to the fact that many people live close to, or interact with, coastal, estuarine and river waters. The flux of suspended sediments through watersheds can also provide an important transport pathway for both nutrients and adsorbed toxic substances, such as trace metals. Thus, a better understanding of sediment transport processes and their associated modelling are increasingly important for water quality modelling.

Some key general facts relating to water quality concerns that have attracted particular publicity in recent years have been summarised on the website water.org (1999), with some typical examples including: (i) that the ancient Romans had better water quality than half the people alive now; (ii) that waterborne diseases cause 1.4 million children's deaths every year; (iii) that women spend thousands of hours every year collecting and carrying water; (iv) that half of the world's hospital beds are occupied by patients with water related diseases; and (v) 70% of the world's fresh water supply is devoted to agriculture. In addition to these facts, the world's population is expected to increase from 6.5 to 9 billion by 2030, with the corresponding demand for water expected to increase by at least 30% during this period. This increase in water stress will occur in addition to that associated with the predicted effects of climate change.

In the developed nations, the spread of 'catchment-scale' regulation of water quality is seen world-wide and best exemplified by the US Clean Water Act (CWA) and the EU Water Framework Directive (WFD) (Anon, 2000, 2011; Kay *et al.*, 2007). Both require integrated management of drainage basins to achieve regulatory standards for the quality of environmental waters. The CWA presents approximately 20 years of implementation experience and it is interesting to note that the leading reason for water quality impairment, at the continental-scale of the USA, is microbial contamination of surface waters. This water quality impairment principally affects recreational and shellfish harvesting waters, where terrestrial catchment-derived flux of microbial pollutants into nearshore waters is the dominant process demanding predictive tools from the hydrodynamic modelling community.

Traditionally water engineers and researchers have focused their attention more on refining the fluid mechanics processes in hydro-environmental models, rather than the water quality processes, in addressing many of the challenges associated with the types of water quality problems occurring in river and coastal basin systems. For example, much emphasis has been focused on refining the turbulence and roughness processes in computational hydro-environmental models, whereas the treatment of bacterial decay is still generally treated as a constant, generally first order, decay rate, dependent only on day or night time conditions and expressed as two separate T_{90} values for day and night. This is the time for a 90% reduction in faecal indicator density in the hostile environment outside the human or animal gut (Kay *et al.*, 2005a). Furthermore, even with regard to the fluid mechanics associated with water quality models, the processes are often simplified where, for example, dispersion and diffusion coefficients are often treated as constants when we know from simple idealised flume studies that these parameters are at least a function of the shear velocity and depth (Fischer *et al.*, 1979). Despite these shortcomings in many computational water quality models, there is now a growing appreciation that hydro-environmental engineers and researchers need to focus more on improving the representation of the bio-/geo-chemical processes in future hydro-environmental impact assessment modelling studies, particularly in view of the growing global water security challenges and threats.

In this paper two such bio-/geo-chemical processes have been studied, refined and included in a hydro-environmental model. These process refinements include: (i) treatment of bacterial decay to include dynamic decay rates, where the decay, in the form of a T_{90} value, is related to the local turbidity, with the local bacteria concentration also being affected by the adsorption and desorption of bacteria from the sediments; and (ii) treatment of the adsorption and desorption of trace metals from sediments and the influence of salinity on the partitioning coefficient. Both of these studies involved field and laboratory studies and refinements to computational hydro-environmental models, with the model applications being to the Severn and Mersey estuaries respectively, both located in the U.K.

2 ENTERIC BACTERIA FLUXES IN THE SEVERN ESTUARY

In the first of these two studies the main focus of the research programme was to investigate the decay of bacteria more rigorously within a macro-tidal estuary and to include the effects of turbidity and irradiance in a dynamic form in the model; in this case as a varying T_{90} value. In many of the models widely used in practical hydro-environmental impact assessment studies the kinetic decay process is treated as a constant. However, bacterial decay is known to depend upon a wide range of parameters. Faecal indictor bacteria have evolved in the gut and environmental waters represent a low nutrient hostile environment. Lower temperature will depress microbial metabolism and prolong viability, but higher temperature reduces survival. Ultra-violet (UV) irradiance will enhance kill but parameters, such as turbidity, which attenuate UV penetration through the water column, which will produce longer microbial survival and hence T_{90} values. Other parameters such as pH and dissolved oxygen may also have some influence on T_{90} values but the dominant driver in waters of relatively stable temperature is likely to be the received dose of solar UV irradiance at any point in the water column (Hipsey *et al.*, 2008; Kay I, 2005a).

Bacterial transport processes are summarised in Figure 1, with the total disappearance of bacteria from the water column including: bacterial die-off based on a first order decay and bacterial disappearance due to sedimentation fluxes, calculated separately by employing two different sink terms. Moreover, an additional source term associated with the re-suspension of the bed sediments also has to be added to the bacterial inputs, while all of the other sources need to be considered in the usual manner. Bed sediment re-suspension is also potentially important in bathing water non-compliance.

Research was also focused on refining the treatment of the adsorption and desorption of bacteria into, or out of, the water column with the sediments, both when the latter were settled on the bed or suspended in the water column. Thus, the main project aims were to advance the current understanding, formulation and interaction of enteric bacterial fluxes with the sediments, for tidal and river flow variability, in a macro-tidal estuary. The project involved field sampling in the estuary, rivers and wastewater treatments works, hydrodynamic data acquisition, numerical modelling with dynamically linked one-dimensional (1-D) riverine and two-dimensional (2-D) estuarine models, with refinement and calibration of the models.

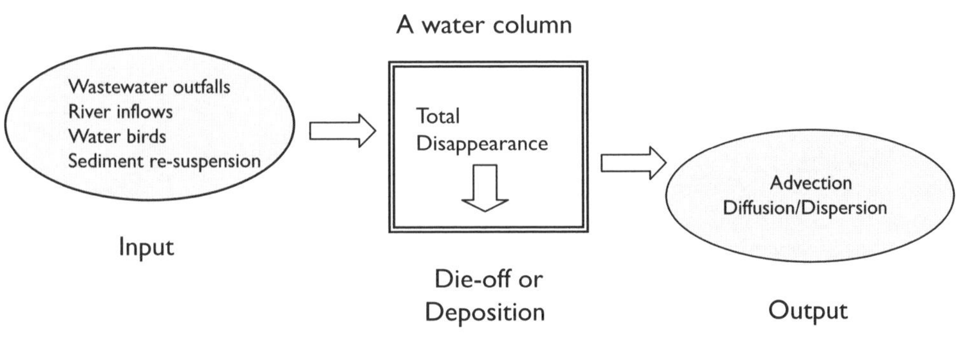

Figure 1 Conceptual model of enteric bacteria transport in natural waters.

The project was funded by the Environment Agency, U.K., following concerns about the relatively high bacteria levels along parts of the Severn Estuary—as with other U.K. estuaries—during intermediate river flow conditions. The study site is shown in Figure 2, highlighting the location of the main sampling sites. Along the estuary wastewater has been treated to a high level for at least the past 25 years, including secondary treatment plus ultra-violet disinfection. The tidal range in the estuary, particularly near Bristol, has the second highest spring tide in the world, with peak ranges in excess of 14 m, producing high velocities in near-shore tidal streams, which entrain high levels of suspended sediment with typical Suspended Particulate Matter (SPM) levels well in excess of 1,000 mg/l during spring tides.

2.1 Bacterial input estimation

In studying the bacterial fluxes and processes in the Severn Estuary and Bristol Channel, it was first necessary to estimate the bacterial inputs into the basin, both from the rivers and the wastewater treatment works, and for which there is a considerable number of inputs along both the Welsh and English coastlines. The main inputs are illustrated in the charts in Figure 3, for the rivers and wastewater treatment works, with further details being given in Stapleton *et al.* (2007a). In determining the data for the numerical model simulations, the data for the riverine inputs were obtained as follows: (i) discharges were obtained from Environment Agency discharge gauges, and (ii) empirically based models were used to estimate the enteric bacteria organism concentrations, based on the proportion of land-cover types and faecal indicator export coefficients (Kay *et al.*, 2005b, 2008a). Likewise, for the wastewater treatment works, the dry weather flows (DWFs) and maximum flows were obtained from the Environment Agency consent data and from the water companies' data and the bacteria concentrations were obtained from existing data archives describing faecal indicator concentrations in sewage-derived effluents (Kay *et al.*, 2008b).

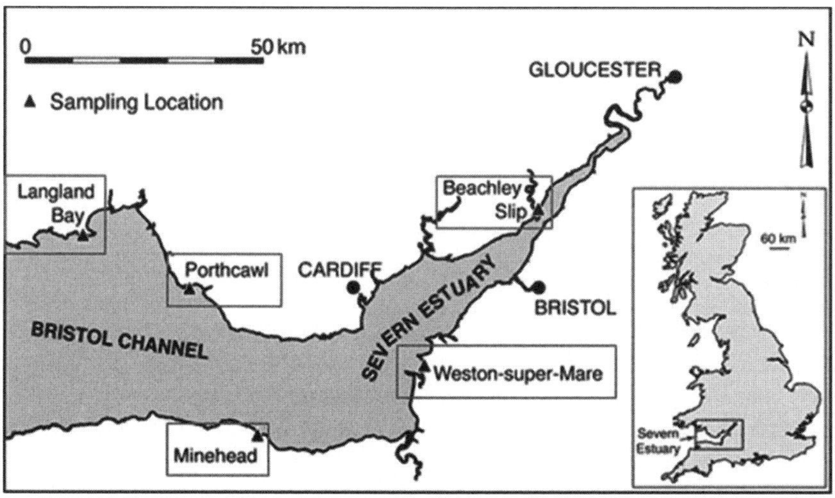

Figure 2 Map of the Bristol Channel and Severn Estuary.

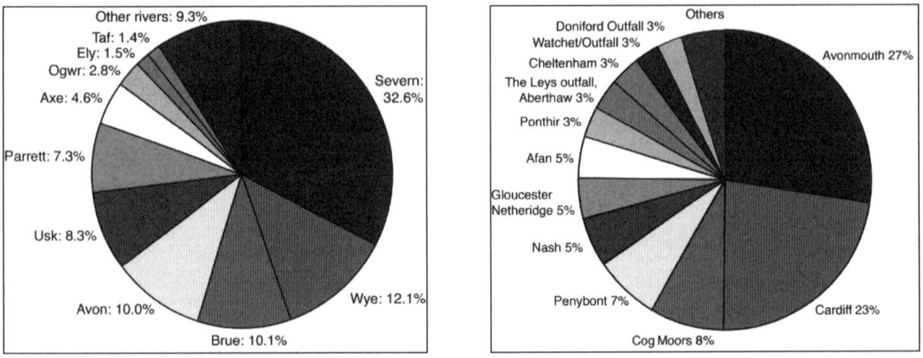

(a) Rivers (b) Wastewater Treatment Works

Figure 3 Riverine and wastewater treatment works source inputs into the Severn Estuary.

The main findings from these estimated inputs were that: (i) the majority of bacteria were delivered during high river flow conditions; (ii) relatively few sources contribute the majority of bacteria loads to the estuary; (iii) the wastewater treatment loads dominate during dry weather flow conditions, thereby providing a background bacterial flux in the estuary; and (iv) the riverine sources dominate for high (rainfall induced) flows, giving high coliform levels in turbid waters with low bacterial decay rates particularly with low irradiance conditions (i.e., high cloud cover or during night time).

2.2 Bacterial decay experiments

In studying the impact of bacterial decay processes in a highly turbid estuary, such as the Severn, where relatively high levels of diffuse source pollution were found (through measurements) to occur under high flow events, it was considered appropriate to study the effects of turbidity levels on decay rates and particularly as turbidity levels were known to vary considerably during the tidal cycle and particularly through the spring-neap cycle. Furthermore, the Severn Estuary has been considered, for many decades, as an ideal site for a barrage for renewable energy and such a barrage could considerably change the turbidity levels, both upstream and downstream of such a structure.

In assessing the impact of turbidity on the decay rate of enteric bacteria in the estuary, samples were taken from the five sites along the estuary, as shown in Figure 2. Samples were then irradiated in the laboratory at constant temperature under a simulated daylight wavelength spectrum characteristic of sunny conditions in the summer bathing season at this latitude. Dark and irradiated microcosms were tested four times for each site, with empirical relationships being developed for the estuary, relating T_{90} values to suspended sediment levels. The corresponding empirical relationships are shown in Figure 4, with the results showing considerable variation in the T_{90} value (in hours) with suspended solids levels in the water column (Kay *et al.*, 2005a). These empirical relationships were then discretised and subsequently included in the numerical model to enable the decay rate to be dynamically linked to the suspended sediment concentrations in the water column.

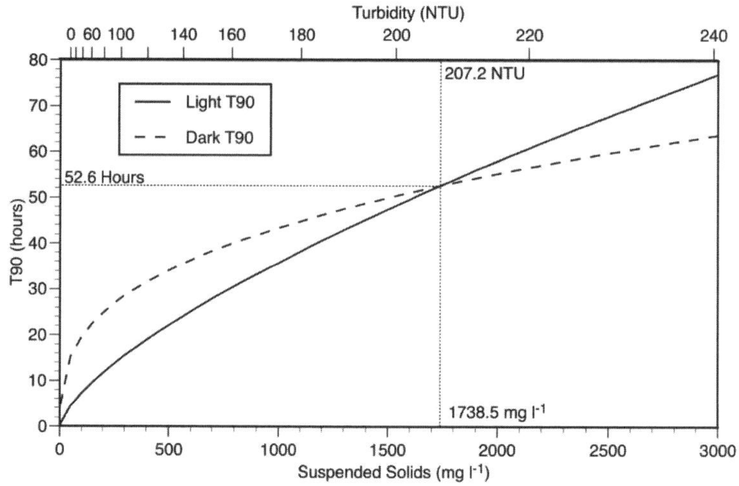

Figure 4 Measured variation of T_{90} value with suspended sediment in the Severn Estuary for light and dark conditions.

Fromures these experiments, a number of key findings were established, including: for light conditions: (a) for low turbidity saline waters, typical T_{90} values were around 7 hr, and (b) for high turbidity brackish waters, typical values were around 40 hr; likewise, for dark conditions: the corresponding values were: (a) 25 hr, and (b) 65 hr respectively. Other key findings were that the T_{90} values were lower in marine vis-à-vis estuarine waters and the *Enterococci* spp. decay rate in water where the turbidity levels were higher than 200 NTU were similar to those values obtained for dark conditions.

Following on from the decay rate experiments, a series of sediment associated experiments were then undertaken to establish the adsorption and desorption characteristics of the bacteria with the sediment. Sediment samples were investigated in two beakers in the laboratory, both incubated at 15°C and with one mixed and with the other allowed to settle. Samples were tested from two sites along the estuary. In the mixed beaker the samples remained fairly constant throughout the test, whereas in the settled beaker the concentrations for faecal indicator bacteria in the water column fell as the finer particles settled. The main findings from these results showed that: (i) the bacteria concentrations decreased as the suspended sediment particles settled, thereby confirming the process of bacteria adsorption and sediment deposition acting as a sink term; (ii) enteric bacteria in the water column appeared to associate mostly with particles having a mean d_{50} grain size of 5–14 μm (i.e., cohesive sediments); and (iii) the enteric bacterial concentrations in the intertidal bed sediments were in the range 3–1,000 cfu/g. Further details of the outcome of the field sampling and the experimental studies are given in Stapleton *et al.* (2007b).

2.3 Numerical model studies

In parallel with the field and experimental studies, numerical model refinements were undertaken, together with model calibration and verification, based on solving the three-dimensional (3-D) and 1-D Reynolds averaged Navier-Stokes (RANS) equations and the solute mass balance equation. Two 2-D/3-D models were used in the estuarine studies; the first being the Hydro-environmental Research Centre's (HRC) 3-D layer averaged RANS equations solver TRIVAST (Three-dimensional layer Integrated Velocities And Solute Transport), with this model being dynamically linked to the HRC's 1-D river model FASTER (Flow And Solute Transport in Estuaries and Rivers) for the rivers. Likewise, for the 2-D simulations the model DIVAST (Depth Integrated Velocities And Solute Transport) was used. These models are based on a regular finite difference grid solution in the horizontal plane and an irregular boundary fitting grid in the vertical plane. The models have been used extensively for coastal and estuarine studies, with further details of the 3-D model being given in Wu and Falconer (2000) and Lin and Falconer (2001).

More recently, the studies have been extended to use the HRC's unstructured grid model, a finite volume solution using a triangular mesh. The model uses a TVD (Total Variation Diminishing) algorithm, with a Roe-MUSCL scheme, with a predictor-corrector time stepping solution and domain decomposition for structures, such as tidal barrages. The model is ideally suited to predicting complex free-surface flows in estuarine and coastal basins, since it solves the 2-D RANS equations and both second- and third-order accurate and oscillation free explicit numerical schemes are included in the model to solve the shallow water equations, together with an algorithm to predict for flooding and drying of inter-tidal regions. Further details of the model are given in Xia *et al.* (2010a).

In modelling numerically the flux of bacterial indicator organisms, or trace metals etc. within river and estuarine basins, the conservation equation of a solute mass can first be derived in the general form for a 3-D flow field and then time averaged to give the following form of the equation for solution:

$$\frac{\partial \phi}{\partial t} + \frac{\partial \phi u}{\partial x} + \frac{\partial \phi v}{\partial y} + \frac{\partial \phi w}{\partial z} + \frac{\partial}{\partial x}\overline{u'\phi'} + \frac{\partial}{\partial y}\overline{v'\phi'} + \frac{\partial}{\partial z}\overline{w'\phi'} = \phi_s + \phi_d + \phi_k \tag{1}$$

where ϕ = time averaged solute (including suspended sediment and heavy metal) concentration; ϕ_S = source or sink solute input (e.g., an outfall); ϕ_d = solute decay or growth term; and ϕ_k = kinetic transformation rate for the solute.

The cross-produced terms $\overline{u\ \phi}$ etc. represent the mass fluxes of solute due to the turbulent fluctuations and are then layer (or depth or area) averaged to give a combined turbulent diffusion and longitudinal dispersion term. By analogy with Fick's law of diffusion, these terms are generally assumed to be proportional to the mean concentration gradient. For transport in the various directions a combined longitudinal dispersion and turbulent diffusion coefficient is included in the various forms of Equation (1), with typical values of the coefficients being found from field data or as given in Fischer (1973) and with dispersion and diffusion being expressed as a function of the shear velocity and depth.

In modelling bacteria (or similar) in the current study, the decay in Equation (1) was expressed as a first order function, in the form of the following formulation:

$$\phi_d = -k\phi \tag{2}$$

where k = coliform decay rate (s^{-1}), and with values of k being obtained dynamically from the relationships measured in the Severn Estuary, as given in Figure 4, and using the formulation below to convert the T_{90} value to k:

$$k = \frac{10}{e^{(T_{90} \times 3600)}} \tag{3}$$

Where T_{90} is the decay rate (hr).

In modelling enteric bacteria in river and estuarine waters, as for heavy metals, coliforms can also be associated with the sediments and, in the studies reported herein, the advective-diffusion Equation (1) was first used to predict the depth averaged cohesive and non-cohesive sediment fluxes and concentrations. For the cohesive sediment transport flux, the source term ϕ_s was equated to the net erosion-deposition, with resuspension and deposition rates being given by Sanford and Halka (1993). For the non-cohesive sediment transport flux the van Rijn (1984a,b) formulations were used to determine the bed load and suspended load concentrations. The corresponding source term in Equation (1) was expressed in terms of the product of the particle settling velocity and the difference between the sediment concentration at a reference level 'a' above the bed and the equilibrium sediment concentration at the reference level 'a'. The reference level 'a' was assumed to be equal to the equivalent roughness height k_s, with a minimum value being given by: a = 0.01H - where H is the depth of the water column.

A new conceptual model has been developed for the interaction between enteric bacteria and the suspended and bed sediments (see Yang *et al.*, 2008), with this model being further refined through current on-going studies. In developing new formulations of the link between enteric bacteria levels and suspended sediment concentrations in natural waters, the following assumptions were first adopted: (i) that the adsorption of bacterial organisms to suspended solids takes place instantaneously; (ii) that there are enough suspended solids surfaces in the water column to provide living places for the bacterial organisms; and (iii) that within the water column, the distribution of the suspended solids concentrations and bacterial populations are uniform along the water depth, therefore the bacterial populations absorbed onto the sediment surfaces are the same as that for each unit of sediment concentration. In applying these assumptions, and based on extensive field surveys, the formulation for the source term in the advective-diffusion equation, expressed in a form enabling the enteric bacterial levels in the water column to be calculated, can be written as:

$$\Sigma \phi_s = -kC - \frac{dC_d}{dt} + \frac{dC_r}{dt} + \sum_{n=1}^{N} \frac{Q_o C_o}{A_o H} \tag{4}$$

where $\Sigma\phi_s$ source or sink term, including bacterial decay, deposition disappearance, entrainment from the bed and wastewater treatment outfalls; C = depth averaged bacteria concentration (cfu/100 ml); dC_d = loss of bacteria population due to deposition of the suspended solids during a time interval dt; dC_r = bacterial population increase due to sediment resuspension in a time interval dt; Q_o = outfall discharge (m^3/s); C_o = outfall discharge concentration (cfu/100 ml); A_o = horizontal discharge area (m^2); H = water column depth (m); and N = total number of outfalls.

In setting up the numerical models of the region, the model domain was set up to cover the whole of the region illustrated in Figure 2, covering the area from an imaginary line drawn between Hartland Point (along the English coast) to Stackpole Head (along the Welsh coast), forming the seaward boundary, to the tidal limit at Gloucester at the head of the estuary. In the integrated 2-D/1-D regular grid model, the grid size was set at 200 m and the dynamic linking of the model occurred at Beachley slip. The 1-D model consisted of four reaches which were modelled using 351 cross-sections, with an average distance of 240 m between two consecutive cross-sections. The inflows for the four major rivers in the modelled area, namely the rivers Wye, Usk, Frome and Little Avon, were treated as lateral inflows.

The downstream boundary was specified as a tidal water elevation boundary, while the upstream boundaries were specified in the form of a velocity or flow boundaries. The models were then calibrated using Admiralty Chart data to find the optimum bed roughness height (k_s). The best agreement between the model predictions and the Admiralty Chart data were found when k_s was set equal to 35 mm. The model predictions were then validated against field data collected at four sites by Stapleton *et al.* (2007b). Typical comparisons between the measured and predicted tidal current components are shown for the speed at Minehead in Figure 5, where it can be seen that the agreement between both sets of results is encouraging. Comparisons were also taken of the measured and predicted coliform levels at the site, as shown in Figure 6, where encouraging predictions were obtained of the coliform levels particularly with the inclusion of the refinements to the processes. Full details of all the model comparisons are given in Stapleton *et al.* (2007c).

The key findings from this hydro-environmental study of bacterial levels in the Severn Estuary and Bristol Channel, based on comprehensive field, laboratory and numerical model studies, and as summarised in Stapelton *et al.* (2007a, b, c) can be summarised as follows: (i) the bacteria population levels in the Severn Estuary are closely linked to both the bacteria die-off and closely related sediment transport processes, with the die-off process enhanced by bacterial removal from the water

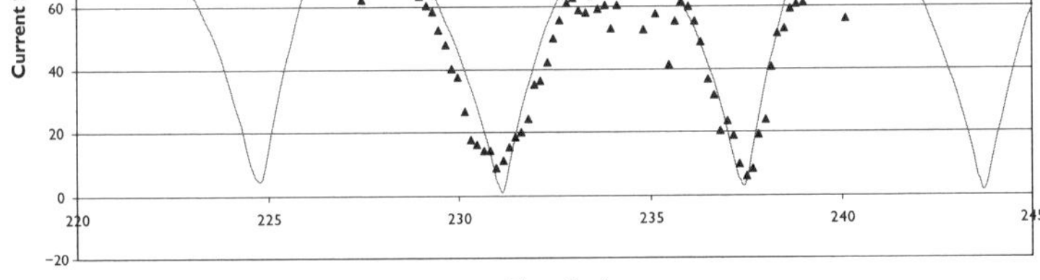

Figure 5 Comparison of measured and predicted tidal current speed at Minehead (see Figure 2).

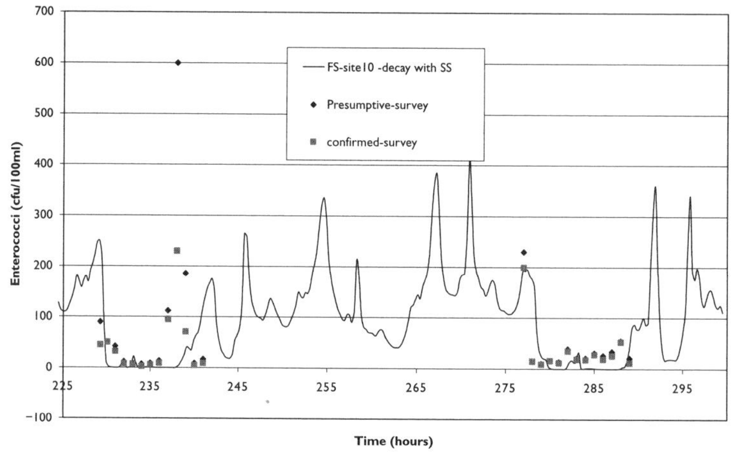

Figure 6 Calibration at Minehead Terminus (see Figure 2).

column by sedimentary deposition; (ii) the T_{90} value was a very important parameter and significantly affected the predicted bacterial concentration levels, with the T_{90} values being varied hourly with solar radiation and at each full timestep with turbidity in the numerical model; (iii) the bacteria population levels generally varied in a cyclical manner in phase with the tidal oscillations; and (iv) four days of site surveys were undertaken to obtain 8 groups of data pertaining to the enterococci levels, with these data being available for model calibration, and showing that the dynamic decay model results were in good agreement with the measured data at the four offshore survey sites.

In subsequent studies, the unstructured grid model has been extended to investigate the impact of a Severn Barrage on the tidal elevations, currents, suspended sediment, turbidity and bacteria levels in the estuary. Over the past 150 years, it has been proposed to build a barrage across this estuary for various reasons. However, since the early 1970s the case has been continuously presented for a tidal barrage to be constructed between Cardiff (along the Welsh coast) and Weston (along the English coast) to generate typically 5% of the UK's energy from tidal renewable resources (see Figure 7). It has high tidal currents and a large inter-tidal area with some 30 million tonnes of suspended sediments in suspension on spring tides and 4 million tonnes on neap tides. This severely limits bactericidal sunlight penetration through the water column and hence results in only slow decay of bacteria in the water column.

A wide range of studies have been undertaken over the past 30 years and, in the main, it has been proposed that the most effective way to operate the barrage would be to generate power on the ebb-tide only, thereby reducing the spring tidal range from 14 m to 7 m, with the effect being to create a significant loss of inter-tidal mudflats at low tide of approximately 14,000 hectares. The barrage would also reduce tidal currents both upstream and downstream of the structure and significantly reduce turbidity and suspended sediment levels. This would increase light penetration through the water column, through enhanced water clarity, producing reduced T_{90} and more rapid decay of bacterial concentrations (see Figure 4). Further details of

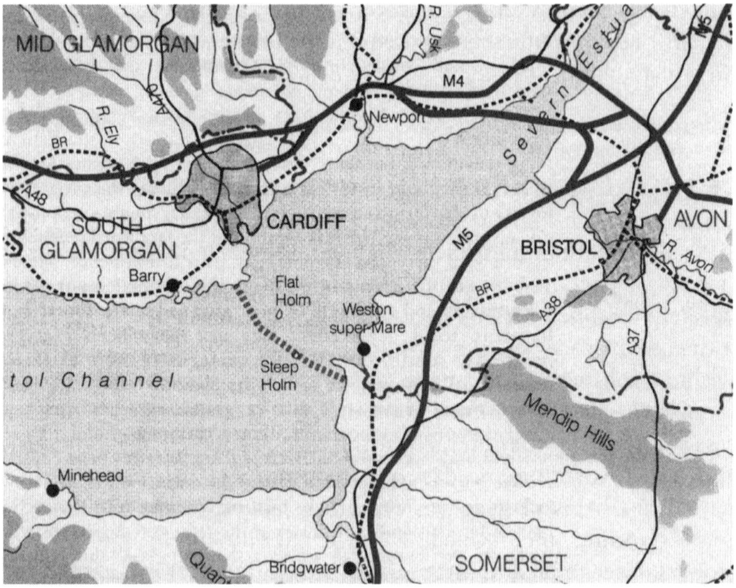

Figure 7 Proposed location of Severn Barrage (Courtesy: Severn Tidal Power Group).

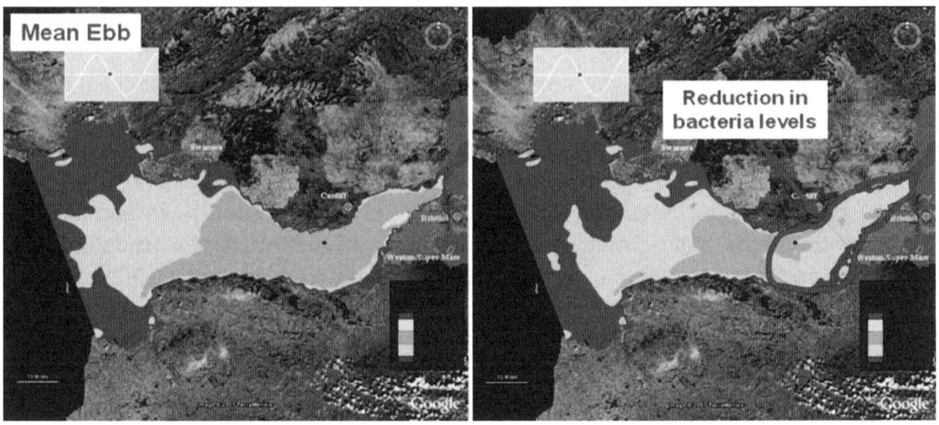

(a) Without Barrage (b) With Barrage

Figure 8 Predicted bacteria levels in Severn estuary: (a) without and (b) with a Barrage.

these modelling studies associated with the barrage proposal in the Severn estuary are reported in Ahmadian *et al.* (2010) and Xia *et al.* (2010b).

To investigate the effects of the barrage on the bacteria levels in the estuary, the model was run for a mean spring tide, both with and without a barrage, with the corresponding results being shown for an ebb-tide in Figure 8. The results showed that, for such a structure, the significant change in the turbidity levels would make a marked difference in the light penetration and this would also significantly affect the rate of bacterial decay within the water column. If this fundamental process was not included in the computational hydro-environmental model, the bacterial levels would erroneously show little change, between the simulations without and with a barrage.

3 TRACE METAL FLUXES IN THE MERSEY ESTUARY

In studying estuarine contamination in urban regions, the sediments within a basin often contain a legacy of particle-associated contaminants, discharged over many decades of industrial activity. Such sediments may undergo a series of complicated morpho- and hydro-dynamic processes, which ultimately determine their retention time and mobility within an estuary. The management of past and/or current industrialised estuaries requires an improved understanding and predictive modelling capability of these processes, including the interaction through the governing chemical processes, which include the adsorption and desorption of toxicants.

The main aim of this project was therefore related to the previous project, and involved understanding and modelling the capability of the fate of contaminated sediments for tidal, seasonal and inter-annual variability of trace metals in macro-tidal estuaries. The study involved laboratory flume tests, including both hydraulic and chemical analyses, field studies and computational model refinements and simulations of trace metal fluxes in estuaries, with particular application to the Mersey Estuary, in the U.K. The model was calibrated and refined using field and laboratory data, with field samples for the laboratory studies being taken from the Mersey.

The laboratory experiments were used to characterise the adsorption or desorption of trace metals to, or from, sediments in estuarine waters and the influence of salinity on the partitioning of contaminants between the adsorbed and the solute state. In solving the transport of trace metal fluxes in estuarine waters, Wu *et al.* (2005) developed a novel approach to modelling the flux of trace metals by solving the advective-diffusion equation for the total trace metal flux in the water column, including the components in the dissolved form and adsorbed onto the suspended sediments, and then decomposing the total flux into its two components through the partition coefficient, defined as:

$$K_D = \frac{P}{C} \tag{5}$$

where K_D = partition coefficient, P = contaminant concentration adsorbed on sediments and C = contaminant concentration in dissolved form. The value of the partition coefficient is known to be salinity dependent and in these studies the empirical relationship developed by Millward and Turner (1994) has been used, linking K_D with salinity as follows:

$$\ln(K_D) = b \times \ln(S+1) + \ln(K_D^0) \tag{6}$$

where b = empirical constant, S = salinity and K_D^0 = partition coefficient in freshwater.

3.1 Metal partitioning experiments

In the laboratory experiments undertaken to study metal partitioning in estuaries, an idealized estuary was set-up in a 17 m long and 1.2 m wide recirculating flume in the Hyder Hydraulics Laboratory at Cardiff University. A key geometric feature of the hydraulic model was the widening of the mean cross-sectional flow area along the downstream direction. As illustrated in Figure 9, the estuary model had three main regions, namely an upstream channel, a diverging channel and a downstream channel. A threefold increase in the flow width along the diverging channel allowed for a decrease of the mean flow velocity by a similar magnitude, which was key to characterization of the desired sediment erosive and depositional processes. A series of experiments was conducted to investigate the effect of varying the water depth and flow rate values on the net erosive and depositional patterns in the model. This procedure led to the selection of a water depth of 0.4 m and flow rate of 45l/s for further detailed experimentation, as this combination gave the desired simultaneous occurrence of net erosion in the upstream channel and net deposition elsewhere in the model. A hydraulic characterization of the experiments was made through measurements of the flow velocity, SPM concentrations and bed profiles (Rauen *et al.*, 2007; Couceiro *et al.*, 2009).

Centred at x = 4.5 m was a plug of sediments collected from the Mersey estuary and carefully prepared to: (i) mimic as closely as possible the range of particle sizes measured for the sediments placed in the rest of the model bed, which were obtained by sieving and had a median particle size of 0.13 mm; and (ii) contain a high pre-determined concentration of metals, which were obtained by chemical spiking under controlled laboratory conditions. The plug was situated in the region of peak net erosion in the model and spanned the width of the upstream channel, i.e., 0.3 m, while having a starting depth of 0.1 m and a length of 0.3 m. A typical experiment would last for 8 hours, during which time water sampling was conducted at periodical intervals at stations S1-S5, as shown in Figure 9, and at two depths, i.e., 1 cm above the sediment bed surface and at 40% of the local water depth, for posterior determination of the suspended sediment and metal concentrations, in both the diluted and adsorbed phases. At the end of an experiment, the flume water was slowly drained away and bed sediment cores were then collected at sampling stations along the flume, using a liquid nitrogen corer, cut into 0.01 m 'bed depth' slices and frozen, for posterior determination of the metal concentration profile in the bed sediments.

In these experiments, due to the unsteady morphodynamic processes – in particular relating to the erosion and deposition pattern of plug sediments – the results led to

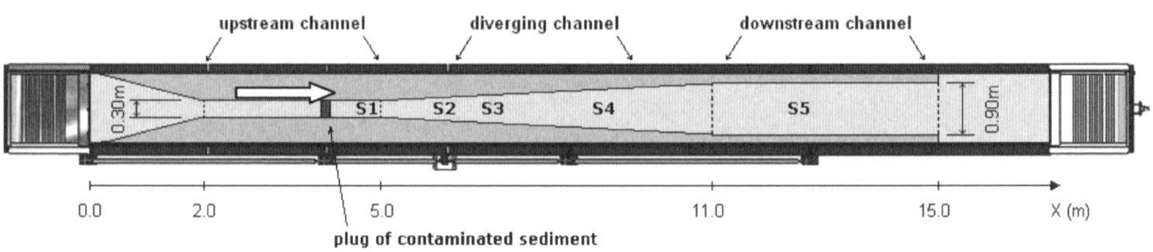

Figure 9 Illustration of estuary model, depicting main regions, sampling stations, streamwise flow direction and plugged sediment location.

the observation of a sub-surface peak in the contaminant concentration profile in the bed sediments. A similar pattern has been observed in estuaries with a history of contamination and has implications for the representation of bed layering processes in numerical models. Furthermore, the experimentation results obtained for the partition coefficient of nickel (Ni), which are exemplified in Figure 10, suggested that the K_D values did not depend upon the SPM concentration and were found to be typically 10^4 l/kg. This finding was in general agreement with previous observations and suggested that these flume experiments were an effective test-bed of metal partitioning processes at an environmentally relevant scale.

3.2 Numerical model experiments

In parallel with the field and experimental studies, numerical model refinements were again undertaken, together with model calibration and verification, based on solving the depth integrated 2-D and 1-D Reynolds averaged Navier-Stokes (RANS) equations and the solute mass balance equation. As before the HRC's linked 2-D and 1-D models DIVAST and FASTER were used to predict the hydrodynamic and solute transport processes. The depth and area integrated form of the solute transport Equation (1) was respectively included in these models to determine both the salinity and sediment transport concentrations (both for non-cohesive and cohesive sediments), and with the depth and area integrated forms of Equation (1) also being used to solve for the total trace metal concentration in the water column. Following predictions of the concentration field for the total trace metal concentration, the partitioning coefficient K_D - obtained from experimental results and using field samples - was then used to split the total concentration into the dissolved metal concentration in the water column and the metal concentration adsorbed on the suspended sediments. Further details of the model are given in Wu *et al.* (2005).

The model was then set up to model metal fluxes in the Mersey Estuary which is one of the most contaminated estuaries in the U.K., caused by past industrial activity in the region. The estuary is one of the largest in the U.K., with a catchment area of some 5,000 km^2, and includes the conurbations of Liverpool and Manchester, as well as other key towns along the north west of England. The estuary is a macro-tidal

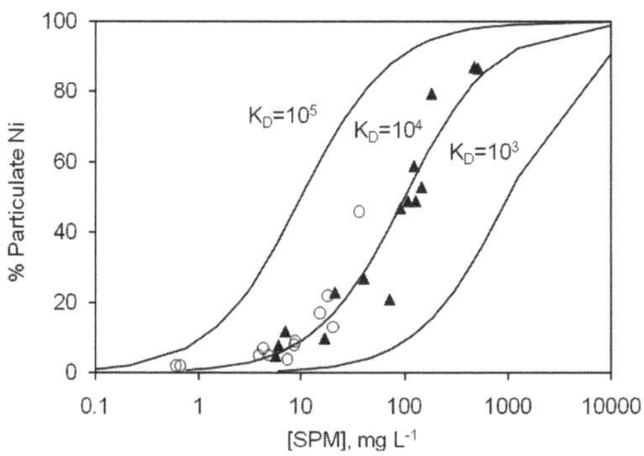

Figure 10 Partition coefficient results for particulate Ni as a function of the SPM concentration, as measured just above the sediment bed (circles) and at 40% of the depth (triangles).

basin, with tidal ranges at Liverpool varying from 10.5 m (extreme spring) to 3.5 m (extreme neap) over a typical spring-neap tidal cycle. Freshwater flows from the river Mersey into the estuary vary from between 10 m^3/s to 600 m^3/s at the extremes.

The model was set up and applied to simulate the tidal flow, salt, sediment and heavy metal distributions in the estuary from seawards of Gladstone to Howley Weir (landwards). The two-dimensional part of the model covered the region from the seaward boundary to the 2-D/1-D link location with a square mesh of 100 m. The one-dimensional part of the model covered the river from the 2-D/1-D link to Howley Weir, using 80 cross-sections and with extensive bathymetric data used at each section, collected from previous surveys undertaken by HR Wallingford Ltd and ABP. The Mersey Estuary and the monitoring sites are shown in Figure 11.

Six sets of hydrodynamic data were provided by the Environment Agency and were used to calibrate the flow model. Three datasets were measured in the spring, two were measured in the autumn and one was measured in the summer. Freshwater inputs from the river Mersey for these datasets covered both wet and dry season conditions and the model was calibrated against measured data at Waterloo, Eastham, Runcorn and Fiddlers Ferry, as shown in Figure 11. The optimum bed roughness, longitudinal dispersion coefficient and critical shear stresses for sediment deposition and erosion were calibrated by trial and error. For the sediments, SPM (extracted for total and available concentrations of metals) was sampled at several sites along the estuary and total dissolved metals were determined voltammetrically at similar sites. Bed sediment adsorbed metal concentrations were determined from liquid nitrogen cores. Predicted SPM concentration distributions are shown as an example in

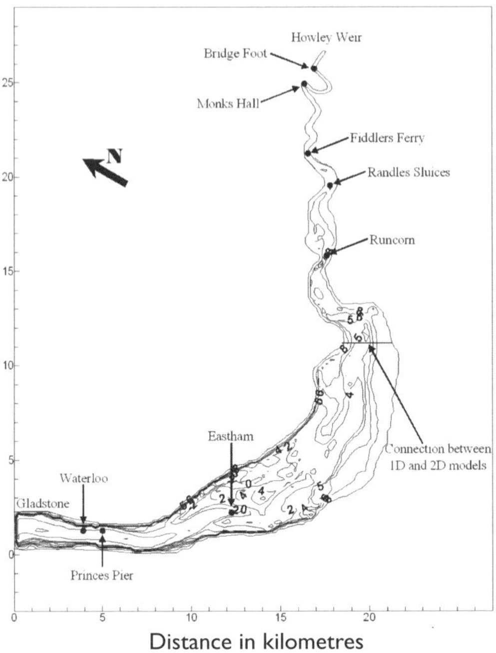

Figure 11 Map of the Mersey Estuary showing key monitoring sites.

Figure 12, with the comparisons generally showing good agreement for such a highly variable parameter.

Finally, using empirical coefficients from the flume laboratory studies and varying the partitioning coefficient K_D with salinity, as given in Equation (6), numerical model simulations were then undertaken for a range of trace metals in the estuary, including: zinc, cadmium and nickel, for both the dissolved and particulate phases. An example comparison between the measured and predicted results for dissolved cadmium is given in Figure 13, with similar results being obtained for the particulate concentration distributions and for the other master variables. On the whole, the model results compared well with the field data for the measured

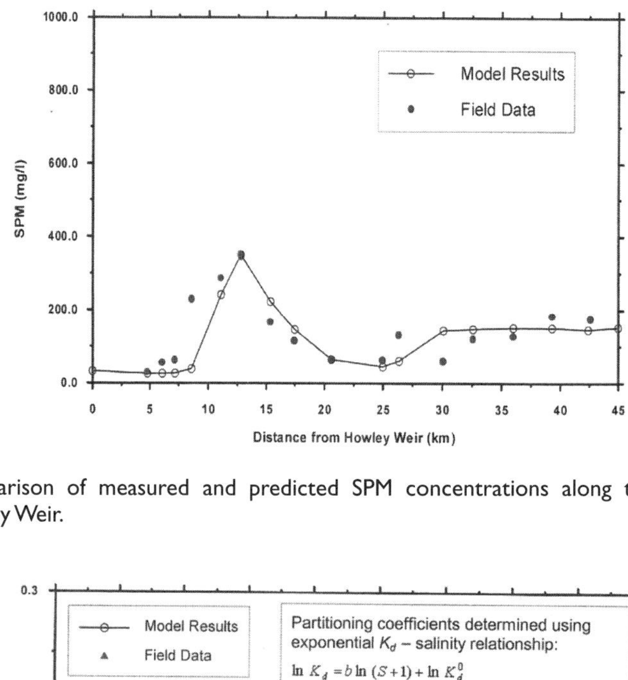

Figure 12 Comparison of measured and predicted SPM concentrations along the estuary from Howley Weir.

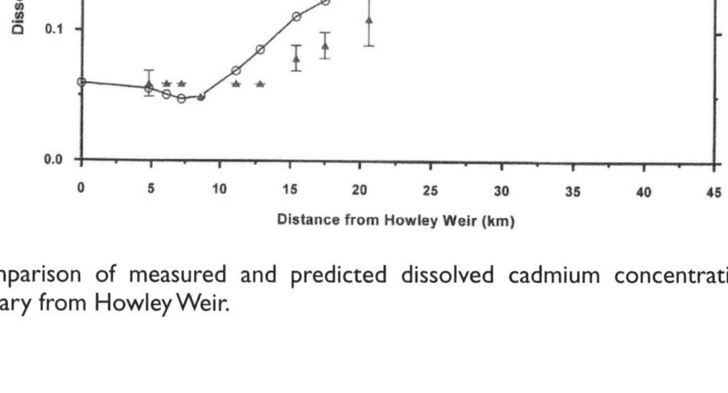

Figure 13 Comparison of measured and predicted dissolved cadmium concentrations along the estuary from Howley Weir.

metal concentrations and highlighted the importance of including salinity effects in the partitioning coefficient and the effectiveness of the novel approach of solving for the total concentration in the water column before splitting the concentration into the dissolved and particulate form. For further details and more results see Wu *et al.* (2005).

4 CONCLUSIONS

The paper highlights the growing application of hydro-environmental models for the improved management of water quality in estuarine and coastal basins and the concern that, whilst the hydrodynamic processes may be well understood and represented in such models, the complex bio- and geo-chemical processes are often oversimplified by engineers and environmental managers using such models for hydro-environmental impact assessment studies. Emphasis is focused in this study on the considerable scope for undertaking further experimental and field studies to develop better formulations, and improved representations, of the bio- and geo-chemical processes defining the fluxes of water quality constituents in estuarine and coastal waters, with particular emphasis in the paper on bacterial and trace metal fluxes.

Details are first given of experimental and field studies of bacterial decay and adsorption rates in the Severn Estuary, with the results showing that enteric bacterial occurrence and decay in the estuary were highly dependent upon turbidity and the adsorption/desorption rates with sediments. The decay rate (expressed in the form of T_{90} values) was measured and expressed in terms of the suspended sediment levels in the estuary and the corresponding relationships between T_{90} and suspended sediments were included in a numerical model along the estuary used to simulate bacterial levels and predict Bathing Water compliance. The corresponding enteric bacteria fluxes were therefore calculated using dynamic time dependent T_{90} values, which varied through the tidal cycle with turbidity and suspended sediments which in turn changed through the tidal cycle as a result of changing tidal currents and boundary shear stresses. The resulting model predictions showed that more accurate *Enterococci* spp. predictions were obtained within the estuary when the effects of turbidity and light penetration were included in the numerical model. These results were shown to be particularly important for the Severn Estuary, where proposals are currently in-hand for the siting of a large barrage, namely the Severn Barrage, for the purpose of generating large resources of marine renewable energy in the estuary - which has the second largest tidal range in the world.

In the second study, details are given of the impact of salinity on the partitioning of trace metals from the adsorbed to the dissolved phase (and vice versa), as tidal currents change the hydrodynamics and suspended sediment levels throughout the tidal cycle. Details are given of novel experiments to determine the partitioning coefficient for an idealised estuary with a contaminated plug and with sediment adsorbed and dissolved metal concentrations being measured downstream, at known locations and under controlled conditions. From the corresponding parameters measured in the laboratory, refinements were made to a numerical model set up to predict trace metal predictions in a contaminated estuary, namely the Mersey Estuary, with the results showing that the improved representation of the partitioning coefficient gave improved agreement with the field data for metal concentrations at a range of sites along the estuary.

The results for both of these studies have highlighted that there is currently a need, and considerable scope, for improving the process modelling of a wide range of water quality constituents, particularly macro-nutrients, between the bed sediments and water column, and within the water column, for improved hydro-environmental impact assessment studies.

ACKNOWLEDGEMENTS

The work relating to fate of bacteria in the Severn Estuary was funded by the EPSRC under Grant GR/M99774 and the Environment Agency of England and Wales. The authors are grateful to Drs Lei Yang, Mark Wyer and Guanghai Gao who also contributed to this project. The trace metal laboratory study was funded by the EPSRC Grant EP/C512316. The authors are grateful to Prof Geoff Millward and his team at the University of Plymouth and Dr Nigel Paul of Lancaster University for contributing to this project.

REFERENCES

Ahmadian, R., Falconer, R.A. and Lin, B. (2010). Hydro-environmental modelling of the proposed Severn barrage, UK. *Proc. of Inst. of Civil Engrs. Energy*. 163 (3), 107–117.

Anon (2000). Council of the European Communities. Directive 2000/60/EC of the European Parliament and of the Council of 23 October 2000 establishing a framework for Community action in the field of water policy. *Official Journal of the European Union* L327, 1–72.

Anon. (2011). US Clean Water Act Section 303d, Causes of Impairment: USEPA Watershed Assessment Tracking and Environmental Results. Accessed 24th August 2011. http://iaspub.epa.gov/waters10/attains_nation_cy.control?p_report_type = T.

Couceiro, F., Rauen, W.B., Millward, G.E., Lin, B., Turner, A. and Falconer, R.A. (2009). Transport and reactivity of nickel in estuarine sediments: results from a high capacity flume. *Marine Chemistry*. 117, 71–76.

CSTT. (1997). Comprehensive studies for the purposes of Article 6 & 8.5 of DIR 91/271 EEC, The Urban Waste Water Treatment Directive. 2nd Edition 1997.

Fischer, H.B. (1979). Longitudinal dispersion and turbulent mixing in open-channel flow. *Annual Review of Fluid Mechanics*, 5: 59–78.

Hipsey, M.R., Antenucci, J.P. and Brookes, J.D. (2008). A generic process-based model of microbial pollution in aquatic systems. *Water Resources Research*. 44, W07408, 1–26.

Kay, D, Bartram, J, Prüss, A, Ashbolt, N, Wyer, M D, Fleisher, JM, Fewtrell, L, Rogers, A. and Rees, G. (2004). Derivation of numerical values for the World Health Organization guidelines for recreational waters. *Water Research 38*, 1296–1304. (DOI:10.1016/j.watres.2003.11.032).

Kay, D., Crowther, J., Stapleton, C.M., Wyer, M.D., Anthony, S., Bradford, M., Edwards, A., Fewtrell, L., Francis, C., Hopkins, M., Kay, C., McDonald, A.T., Watkins, J. and Wilkinson, J. (2008a). Faecal indicator organism concentrations and catchment export coefficients in the UK. *Water Research*. 42, 2649–2661.

Kay, D., Edwards, A.C., Ferrier, B., Francis, C., Kay, C., Rushby, L., Watkins, J., McDonald, A.T., Wyer, M., Crowther, J. and Wilkinson, J. (2007). Catchment microbial dynamics: the emergence of a research agenda. *Progress in Physical Geography*. 31 (1), 59–76.

Kay, D., Stapleton, C.M., Crowther, J., Wyer, M.D., Fewtrell, L., Edwards, A., McDonald, A.T., Watkins, J., Francis, F. and Wilkinson, J. (2008b). Faecal indicator organism compliance parameter concentrations in sewage and treated effluents. *Water Research*. 42, 442–454.

Kay, D., Stapleton, C.M., Wyer, M.D., McDonald, A.T., Crowther, J., Paul, N., Jones, K., Francis, C., Watkins, J., Humphrey, N., Lin, B., Yang, L., Falconer, R.A. and Gardner, S. (2005a). Decay of intestinal enterococci concentrations in high energy estuarine and coastal waters: towards real-time T90 values for modelling faecal indicators in recreational waters. *Water Research*. 39, 655–667.

Kay, D., Wyer, M.D., Crowther, J., Stapleton, C., Bradford, M., McDonald, A.T., Greaves, J., Francis, C. and Watkins, J. (2005b). Predicting faecal indicator fluxes using digital land use data in the UK's sentinel Water Framework Directive catchment: the Ribble study. *Water Research*. 39, 3967–3981.

Lin, B. and Falconer, R.A. (2001). Numerical modelling of 3-D tidal currents and water quality indicators in the Bristol Channel. *Water and Maritime Engineering, Proc. of Inst. of Civil Engrs*. 48 (3), 155–166.

Rauen, W,B,, Lin, B. and Falconer, R.A. (2007). Contaminated sediment transport experiments in an idealised estuary model. *Proceedings of 5th International Symposium on Environmental Hydraulics, Tempe, AZ, USA*, Chapter Theme 4.2 1–6.

Sanford, L.P. and Halka, J.P. (1993). Assessing the paradigm of mutually exclusive erosion and deposition of mud, with examples from upper Chesapeake Bay. *Marine Geology*. 97, 37–57.

Stapleton, C.M., Wyer, M.D., Kay, D., Bradford, M., Humphrey, N., Wilkinson, J., Lin, B., Yang, Y., Falconer, R.A., Watkins, J., Francis, C.A., Crowther, J., Paul, N.D., Jones, K. and McDonald, A.T. (2007a). *Fate and Transport of Particles in Estuaries, Volume II: Estimation of Enterococci Inputs to the Severn Estuary from Point and Diffuse Sources*. Environment Agency Science Report SC000002/SR2, p. 50.

Stapleton, C.M., Wyer, M.D., Kay, D., Bradford, M., Humphrey, N., Wilkinson, J., Lin, B., Yang, Y., Falconer, R.A., Watkins, J., Francis, C.A., Crowther, J., Paul, N.D., Jones, K. and McDonald, A.T. (2007b). *Fate and Transport of Particles in Estuaries, Volume III: Laboratory Experiments, Enterococci Decay Rates and Association with Sediments*. Environment Agency Science Report SC000002/SR3, p.82.

Stapleton, C.M., Wyer, M.D., Kay, D., Bradford, M., Humphrey, N., Wilkinson, J., Lin, B., Yang, Y., Falconer, R.A., Watkins, J., Francis, C.A., Crowther, J., Paul, N.D., Jones, K. and McDonald, A.T. (2007c). *Fate and Transport of Particles in Estuaries, Volume IV: Numerical Modelling for Bathing Water Enterococci Estimation in the Severn Estuary* Environment Agency Science Report SC000002/SR4, p. 139.

Thomann, RV. and Mueller, J.A. (1987). *Principles of Surface Water Quality Modelling Control*. Harper Collins Publishers Inc., New York, p. 644.

Turner, A. and Millward, G.E. (1994). The partitioning of trace metals in a macro-tidal estuary: implications for contaminant transport models. *Estuarine, Coastal and Shelf Science*. 39 (1), 45–58.

van Rijn, L.C. (1984a). Sediment transport. Part 1: Bed load transport. *Journal of Hydraulic Engineering, ASCE*. 110, 1431–1456.

van Rijn, L.C. (1984b). Sediment transport. Part 2: Suspended load transport. *Journal of Hydraulic Engineering, ASCE*. 110, 1431–1456.

Water.Org. (1999). *Learn-about-the-water-crisis*. [Online] Available from: http://water.org/learn-about-the-water-crisis/facts/#water [Accessed 16th August 2011].

WHO. 2003. *Guidelines for safe recreational water environments Volume 1: Coastal and freshwaters* World Health Organisation, Geneva.

Wu, Y. and Falconer, R.A. (2000). A mass conservative 3-D model for predicting solute fluxes in estuarine waters. *Advances in Water Resources.* 23, 531–543.

Wu, Y., Falconer, R.A. and Lin, B. (2005). Modelling trace metal concentration distributions in estuarine waters. *Estuarine, Coastal and Shelf Science.* 64 (4), 699–709.

Xia, J., Falconer, R.A. and Lin, B. (2010a). Impact of different tidal renewable energy projects on the hydrodynamic processes in the Severn Estuary, UK. *Ocean Modelling.* 32 (1–2), 86–104.

Xia, J., Falconer, R.A. and Lin, B. (2010b). Hydrodynamic impact of a tidal barrage in the Severn Estuary, UK. *Renewable Energy*, 35 (7), 1455–1468.

Yang, L., Lin, B. and Falconer, R.A. (2008). Modelling enteric bacteria levels in coastal and estuarine waters. *Proc. of Inst. of Civil Engrs., Engineering and Computational Mechanics.* 161 (EM4), 179–186.